Department of the Environment

Acidic Emissions Abatement Processes

Manual of Acidic Emission Abatement Technologies

VOLUME 1: COAL-FIRED SYSTEMS

London: HMSO

First published 1991

ISBN 0 11 752213 9

Preface

This study of abatement technologies applicable at source to air pollution arising from the combustion of fossil fuels was commissioned by the Department of the Environment and undertaken by The Fellowship of Engineering. The intention has been to produce a comprehensive survey of abatement technologies giving details of processes, plant and equipment, input materials required and by-products arising and to present pollution emission data in the form of emission factors suitable for calculation of the effects of emission inventories.

The advantages and disadvantages, both technical and economic, of the many processes are stated wherever practicable, but these are necessarily dependent to a significant degree on the particular site, grade of fuel and size and type of emitter considered; in particular, costs should be regarded as indicative only. To facilitate comparative calculation, datum cases before the application of abatement processes are defined in detail and the format of each volume is standardised insofar as is practicable.

So far as is known, it is the first time that such an extensive survey has been undertaken in the UK. Data was collected over a period of more than a year. The work proceeded by questionnaire and inquiry of the many producers and licensors of processes as well as by the collection and analysis of published data. While the information in the four volumes is fairly comprehensive there are inevitably some gaps, and it should not be considered exhaustive. While the accuracy of all the data presented cannot be guaranteed, wherever possible references are given to enable specific items to be checked at source as necessary.

The work represented in this Volume was completed during the period 1986 to 1987 to provide data for the Department of the Environment and others in consideration and implementation of the European Community Directives on Air Pollution which were then under discussion. The emphasis is, therefore, primarily on the application of abatement technologies to existing large plants. Acidic emission abatement by energy saving means including combined heat and power and combined cycle technology was considered separately.

Preface

Contents

Acknowledgements

We are grateful to the following Companies, Institutions and Consultants who have provided information for incorporation in this report:

UNITED KINGDOM

Airoil-Flaregas Ltd, West Drayton, Middx.
Babcock Power Ltd, London
Beijer Institute, York
Berridge Environmental Laboratories Ltd, Chelmsford, Essex
British Gas, London
British Gypsum, Nottingham
Calor, Slough, Bucks.
Central Electricity Generating Board (GDCD), Barnwood, Glos.
Central Electricity Research Laboratories (CEGB), Leatherhead, Surrey
Chem Systems, London
Coal Research Establishment (British Coal), Stoke Orchard, Glos.
Department of the Environment, London
Dunphy Oil & Gas Burners Ltd, Rochdale, Lancs.
ETSU, Harwell
Fellowship of Engineering, London
Hamworthy Engineering Ltd, Poole, Dorset
International Energy Agency (Coal Research), London
John Zink Co. Ltd, St. Albans, Herts.
Johnson Matthey, Royston, Herts.
Lodge Cottrell, Birmingham
Lurgi (UK) Ltd, London
Metra Consulting, London
Nu-Way Ltd, Droitwich, Worcs.
Peabody Holmes Ltd, Maidstone, Kent
Pennwalt, Camberley, Surrey
Saacke Ltd, Portsmouth, Hants.
Stordy Combustion Engineering Ltd, Wolverhampton, West Midlands
Warren Spring Laboratory, Stevenage, Herts.

DENMARK

Niro Atomizer AS., Soeborg, Copenhagen.

FEDERAL REPUBLIC OF GERMANY

Deutsche Babcock Anlagen, Krefeld
G. Bischoff GmbH, Essen
Rheinisch-Westfalisches Elektrizitaätswerk, Essen
L & C Steinmuller GmbH, Gummersbach
Thyssen Engineering, Essen
Umweltbundesamt, Berlin
VGB, Essen

FINLAND

Lisop Oy, Kerava
Tampella, Tampere

FRANCE

CITEPA, Paris
Organisation for Economic Co-operation & Development, Paris
Societe Foster Wheeler Francaise, Paris

JAPAN

Chiyoda Chemical Engineering & Construction Co. Ltd, Yokohama
Ishikawajima-Harima Heavy Industries Co. Ltd, Tokyo
Mitsui Miike Engineering Corporation, Tokyo
Sumitomo Heavy Industries, Tokyo

THE NETHERLANDS

Concawe, The Hague
De Jong Coen BV, Schiedam
ESTS, Ijmuiden
Foster Wheeler International Corporation, The Hague

SWEDEN

AB Aroskraft, Vasteras
Flakt Industri AB, Vaxjo
National Swedish Environmental Protection Board, Solna

USA

Andersen 2000 Inc., Atlanta, GA.
Battelle, Columbus, Ohio
DB Gas Cleaning Corp., Orinda, CA.
Davy McKee, Lakeland, FA.
Electric Power Research Institute, Palo Alto, CA.
Foster Wheeler Development Corporation, Livingston, NJ.
General Electric Environmental Services Inc., Lebanon, PA.
Otto H. York, Parsippany, NJ.
Peabody Process Systems, Norwalk, CT.
R.E. Sommerlad (Consultant), Cranford, NJ.
Tennessee Valley Authority, Chattanooga, Tenn.

Study Group

The following assisted in the preparation of this manual:

FELLOWSHIP OF ENGINEERING

Steering Group :

Sir Frederick Page (Chairman)
Mr. M. Kneale (Project Manager and Nominated Officer for the Fellowship of Engineering)

Dr. F. Steele
Mr. J.R. Appleton
Mr. J.G. Dawson
Mr. G.A. Lee
Professor G.F.I. Roberts
Dr. J. Gibson
Dr. J.H. Chesters
Professor J.F. Davidson
Professor S. Eilon
Professor I. Fells
Mr. R.J. Kingsley
Mr. V.J. Osola
Mr. K.R. Vernon
Dr. D. Train

Dr. A.J. Apling (Nominated Officer of the Department of the Environment)
Mr. J. Murlis (Department of the Environment)

Coal Task Group :

Dr. D.R. Cope
Professor J.F. Davidson
Dr. J. Gibson (Chairman)

Oil Task Group :

Mr. P. Brackley
Professor I. Fells
Mr. G.A. Lee (Chairman)
Mr. J. Solbett

Gas Task Group :

Dr. C.G. James
Professor G.F.I. Roberts (Chairman)
Mr. P. Scott
Dr. F.E. Shephard
Dr. W.A. Simmonds
Professor A. Williams

Mobile Sources Group :

Professor G.P. Blair
Mr. J.G. Dawson (Chairman)
Mr. A. Silverleaf

FOSTER WHEELER POWER PRODUCTS :

Dr. R. Fletcher
Mr. K. Johnson
Mr. H. Luaw
Mr. D. McSherry
Mr. H.T. Wilson (Programme Manager)

HOY ASSOCIATES (UK)	:	Mr. D.W. Gill Mr. H.R. Hoy (Director) Mr. A.G. Roberts Mr. J.E. Stantan Mr. D.M. Wilkins
WARREN SPRING LABORATORIES	:	Dr. M. Williams Mr. J. Potter Dr. J.H. Weaving

Acronyms

AAF	American Air Filters
AFBC	atmospheric fluid-bed combustion
B&W	Babcock and Wilcox
BBF	biased burner firing
BF	Bergbau Forschung
BHK	Babcock-Hitachi K.K. (Japan)
BOOS	burners out of service
C-E	Combustion Engineering (USA)
CEA	Combustion Engineering Associates
CEC	Commission of the European Communities
CEGB	Central Electricity Generating Board
CFBC	circulating fluidised-bed combustion
COD	chemical oxygen demand
CONCAWE	Conservation of Clean Air and Water – Europe (Oil Companies' International Study Group)
DBA	dibasic acid
DMB	distributed mixing burner
DRB	dual register burners
DSAA	down-fired sequential air addition
ECE	Economic Commission for Europe (UN)
EDTA	ethylenediamine tetra-acetic acid
EGR	exhaust gas recirculation
EHE	external heat exchanger
EPA	Environmental Protection Agency (USA)
EPDC	Electric Power Development Co. Ltd (Japan)
EPRI	Electric Power Research Institute (USA)
ESP	electrostatic precipitators
FBC	fluidised-bed combustion
FBHE	fluidised bed heat exchanger
FGD	flue gas desulphurisation
FGR	flue gas recirculation
FRG	Federal Republic of Germany
FW	Foster Wheeler
GEESI	General Electric Environmental Services Inc. (USA)
ID	induced draft
IEA	International Energy Agency
IFNR	in-furnace NO_x reduction
IFP	Institut Francais du Petrole
IHI	Ishikawajima-Harima Heavy Industries (Japan)
IIP	Institut fur Industriebetriebsehre und Industrielle Produktion (University of Karlsruhe, FRG)
JBR	jet bubbling reactor (Chiyoda)
KHI	Kawasaki Heavy Industries (Japan)
KVC	Kawasaki volume combustion system
LEA	low excess air
LHV	lower heating value (net calorific value)
LIMB	limestone injection/multi-stage burner
LNB	low-NO_x burner
LNCFS	low-NO_x concentric firing system
LVHR	low volume heating rate

MACT	Mitsubishi Advanced Combustion Technology
MCR	maximum continuous rating
MHI	Mitsubishi Heavy Industries (Japan)
MIT	Massachusetts Institute of Technology (USA)
NATO-CCMS	North Atlantic Treaty Organisation–Committee on the Challenge of Modern Society
n.a.	not available
NCB	National Coal Board (British Coal)
NCR	non-selective catalytic reduction
OECD	Organisation for Economic Co-operation and Development
OFA	overfire air (injection)
OSC	off-stoichiometric combustion
PCF	primary combustion furnace
PENSYS	Pittsburgh Environmental Systems Inc. (USA)
PETC	Pittsburgh Energy Technology Centre (USA)
pf	pulverised fuel
PFBC	pressurised fluidised-bed combustion
PM	pollution minimum (burner)
PVC	polyvinyl chloride
RAP	reduced air preheat
SCR	selective catalytic reduction
SDA	spray dryer absorber
SGR	separate gas recirculation
SNR	selective non-catalytic reduction
TCA	turbulent contact absorber
TET	turbine entry temperature
TVA	Tennessee Valley Authority (USA)
UFI	upper fuel injection
UOP	Universal Oil Products
VDI	Verein Deutscher Ingenieure (Dusseldorf, FRG)

Units and Conversion Factors

Thermal		*Equivalent Units*
GJt	gigajoule	0.95×10^6 Btu
kWt	kilowatt	3,412 Btu/h
MWt	megawatt	3.4×10^6 Btu/h
TWt h	terawatt hour	3.4×10^{12} Btu
MJ/kg	megajoules per kilogramme	430 Btu/lb
therm	therm	1.0×10^5 Btu

Electrical		
GJe	gigajoule	278 kWh
kWe	kilowatt	1.34 hp
MWe	megawatt	1341 hp
MWh	megawatt hour	3.4×10^6 Btu
kW/MWt	kilowatt (consumption) per megawatt thermal (capacity)	–
kW/MWe	kilowatt (consumption) per megawatt electrical (capacity)	0.1%

Volumetric Flowrates		
Nm^3/h	normal cubic metre per hour, i.e. at 0°C and 1 atmosphere	0.59 ft^3/min (0°C, 1 atm)
m^3/h	actual cubic metre per hour	0.59 ft^3/min (at same temperature and pressure)
Nm^3/s	normal cubic metre per second	2119 ft^3/min (0°C, 1 atm)
l/Nm^3	litre of liquid per normal cubic metre of gas (liquid/gas ratio)	6.23×10^{-3} gallon/ft^3
m^3/MWh	cubic metre (water or effluent) per megawatt hour (electrical output)	–
l/GJt	litre (water or effluent) per gigajoule (thermal output)	0.23 gallon/10^6 Btu

Concentration		
mg/Nm^3*	milligramme (pollutant) per normal cubic metre (flue gas)	0.35 vppm (SO_2) 0.487 vppm (NO_2) 0.614 vppm (HCl)
mg/m^3	milligramme (pollutant) per actual cubic metre (flue gas)	–
g/GJ	gramme (pollutant) per gigajoule (thermal input)	2.33×10^{-3} lb/10^6 Btu
ng/J	nanogrammes per joule	1.0 g/GJ

lb/10^6 Btu	pound per million Btu	430 g/GJ
ppm*	part (pollutant) per million (parts of fluid). Typically by volume for gas (see also vppm); by weight for liquids	0.00001%
vppm	part per million by volume	2.86 mg(SO_2)/Nm^3 2.05 mg(NO_2)/Nm^3 1.63 mg(HCl)/Nm^3

*Although SO_2 and HCl concentrations are normally quoted in units of mg/Nm^3, it is common practice to express NO_x concentrations in ppm (vol) or vppm. This is because continuous NO_x monitors operate on a molar basis which relates directly to vppm units, and is independent of temperature, pressure and NO/NO_2 ratio. Separate detection of NO and NO_2 is therefore not required.

Mass

t	tonne (metric ton)	2205 lb
ton	U.S. (short) ton	2000 lb
g/l	gramme per litre	0.010 lb/gallon
kg/MWh	kilogramme (reagent) per megawatt hour (electrical output)	–

Efficiency Factor

GJe/tonne	gigajoule (electrical output) per tonne of fuel fired	–
GJt/tonne	gigajoule (thermal output) per tonne of fuel fired	–

Emission Factor

kg/tonne	kilogramme (elemental pollutant) per tonne of fuel fired	–

Pressure

mbar	millibar	0.0145 lbf/in^2 (psi)
mm H_2O	millimetre of water gauge	0.10 mbar
atm	atmospheric pressure	14.7 psig; 1.013 bar
Pa	pascal	0.01 mbar; 0.1 mm H_2O

Miscellaneous

mill	one thousandth of U.S. dollar	$ 0.001 (0.1 cent)
ha	hectare	2.47 acres 10,000 m^2
micron	micrometre (particle size)	0.000039 inch
m/s	metre per second	3.28 ft/s

Executive Summary

0.1 Background and Purpose of the Study

The Fellowship of Engineering has been commissioned by the Department of the Environment to prepare a factual Manual describing the essence of individual technologies or processes for abatement of acidic emissions from combustion of fossil fuel in mobile and stationary sources; the present study is concerned with emissions from stationary combustion systems. The objective is to consider the abatement technologies or processes currently available commercially, and those still undergoing development but showing promise of commercial adoption in the UK by the end of the century.

The aim has been to prepare the Manual in a form suitable for publication and for subsequent updating. The anticipated readership includes air quality professionals requiring information on current or likely future commercially developed technologies, and organisations needing a data base for assessing emission control economics. The Manual for stationary sources deals in separate volumes with combustion systems fired with coal, oil and gas; this Volume is concerned with coal combustion systems.

0.2 The Acidic Emissions

The emissions most important as precursors of acid deposition are sulphur dioxide, SO_2, and nitrogen oxides, NO_x. Estimates of the total UK inventory of man-made emissions for 1985 have been prepared by Warren Spring Laboratory (WSL). The estimates for SO_2 emissions are presented in Figure 0.1. It is

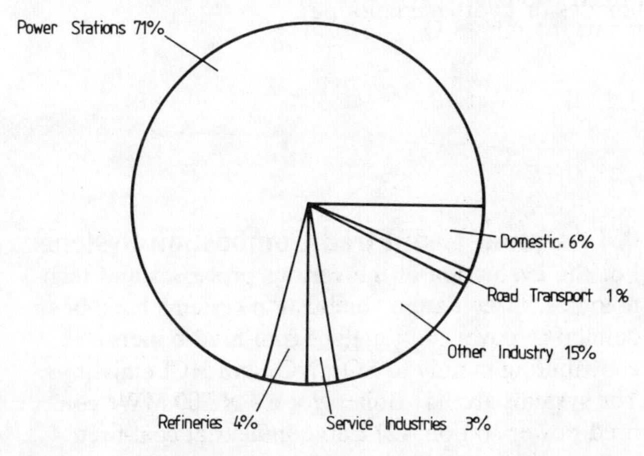

Figure 0.1 UK Man-made Sources of SO_2 Emission (1985). Total SO_2 Inventory 3.58 million tonnes

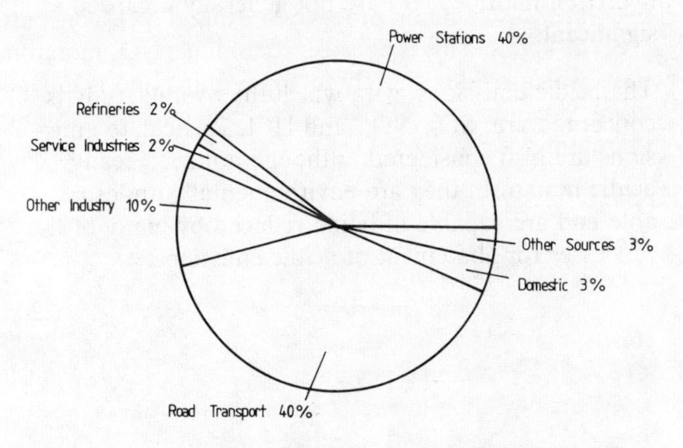

Figure 0.2 UK Man-made Sources of NO_x Emission (1985) Total NO_x Inventory (as NO_2) 1.84 million tonnes

seen that power stations are overwhelmingly the principal source, with industry (refineries, service and other industry) accounting for most of the rest.

The WSL estimates for the total NO_x emissions inventory are presented in Figure 0.2. When allowance is made for the high proportion of NO_x emissions originating from road transport, the remaining sources of emissions are found to be of similar relative significance for both SO_2 and NO_x, i.e. predominantly from power stations with a smaller but important contribution from industry. With regard to the latter, WSL estimates of NO_x emissions from combustion of various fuels in boilers generating up to 230 tonne/h steam are shown in Figure 0.3. The main contribution is from boilers fired with fuel oil, followed by coal,

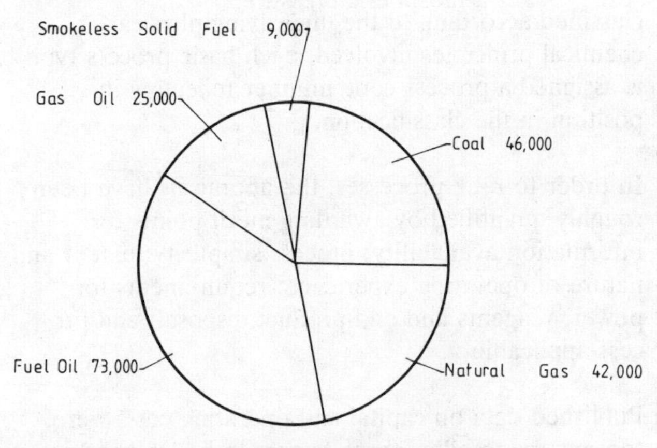

Figure 0.3 WSL Estimate of UK Commercial and Industrial NO_x Emissions (1982). Estimated Emissions, tonnes (as NO_2) from Various Fuels. Total Estimated (as NO_2) 195,000 tonnes

natural gas, gas oil and smokeless solid fuel. Together, however, these contribute a small proportion of the total, just over 10%.

UK coal, which is unusually rich in chlorine, releases HCl on combustion, providing most of the UK HCl emissions; only a minor contribution comes from the incineration of industrial waste containing PVC. The emissions are an order of magnitude smaller than those of SO_2 from power stations. Emissions of hydrogen fluoride, HF, are not generally regarded as significant.

The acidic emissions, with which the overall study is concerned are: SO_2, NO_x, and HCl. Particulate emissions are also considered; although not specifically acidic in nature, they are environmentally undesirable and are capable of being reduced by many of the processes for abatement of acidic emissions.

0.3 Content of the Manual

The Volume deals in turn with: the large number of processes for controlling emissions of SO_2, the gaseous pollutant that has received the greatest amount of attention worldwide; combustion technologies giving reduced NO_x emissions; flue gas treatment processes for NO_x removal; combined flue gas treatment processes for removing both SO_2 and NO_x; the increasingly important fluidised bed combustion technologies that can give reductions in SO_2 and NO_x emissions; and finally flue gas treatment processes for removal of acid halides. Only those abatement technologies applied during or after combustion are included, thereby excluding coal gasification (and combined cycle) systems.

The approach is to classify and describe the available processes or technologies, and to present appraisals based on published information. The processes are classified according to the underlying physical or chemical principles involved; each basic process type is assigned a process code number reflecting its position in the classification.

In order to rank processes, the appraisals have been roughly quantified by awarding merit points for: information availability; process simplicity; extent and nature of operating experience; requirements for power, reagents and end-product disposal; and process applicability.

Published data on capital and operating costs were too meagre to allow these factors to be included in the merit points system. The merit points system is presented in Table 0.1.

Table 0.1 Merit Points System

Scale of Merit Points	
0	Poorer than average
1	Average
2	Better than average
3	Outstandingly better than average
Categories for which points are awarded	
1	Information available
2	Process simplicity
3	Operating experience
4	Operating difficulty
5	Energy consumption
6	Reagent requirements
7	Ease of end-product disposal
8	Process applicability

It should be noted that because the average levels are different for different acidic emissions being abated, the merit rating for a flue gas desulphurisation process, for example, cannot be compared with those for NO_x abatement or for combined SO_2/NO_x abatement processes.

Those processes where the total number of points for the eight categories exceeded the arbitrarily defined limit of 10 were regarded as being potentially applicable commercially in the UK, and their application to either two or three typical coal-fired combustion systems has been evaluated in detail.

In each of the detailed evaluations, the processes are described and their status and operating experience indicated, together with process variations that have been adopted, and development potential. Process requirements for the specific applications are evaluated, the byproducts and effluents are identified and quantified, and then the emission and efficiency factors, and (where available data allow) capital and operating costs, are presented. Finally the process advantages and drawbacks are listed. The emissions (kg/h) and emission factors (kg/tonne fuel) are expressed throughout the Manual in terms of the mass of element (sulphur, nitrogen or chlorine), and not of the acidic compound (sulphur dioxide, nitrogen oxides or hydrogen chloride).

0.4 Datum Coal-Fired Combustion Systems

For the evaluation of the various processes and technologies, three datum combustion systems have been defined to cover the range of coal-fired boilers contributing mainly to SO_2, NO_x and HCl emissions. The systems are: (1) Boilers for a 4 × 500 MWe coal-fired power station; (2) Large industrial coal-fired boiler generating 450 tonne/h steam; and (3) Small factory coal-fired boiler generating 13 tonne/h steam.

The assumptions made for the datum systems, and the performance and emissions to be expected from them (without emissions abatement) are summarised in Table 0.2.

0.5 Flue Gas Desulphurisation

Flue gas desulphurisation is the most widely practised of acidic emission abatement processes and technologies, and of the large number of processes that have been developed, twenty-seven have been appraised. Six of the processes had more than 10 merit points and were selected for detailed evaluation; they are described briefly below, and their appraisals are summarised in Table 0.3.

Sea water scrubbing (Process Code No. S11.1)

Gas is scrubbed with sea or river water; the effluent, containing sulphites and sulphates, is pumped to sea.

Limestone or lime slurry scrubbing (Process Code No. S21.1)

Gas is scrubbed with limestone or lime slurry; forced atmospheric oxidation of the effluent slurry produces gypsum.

Lime spray dryer absorption (SDA–Process Code No. S22.1)

Reaction of hot gas with lime slurry in a spray dryer absorber produces solids containing absorbed SO_2 as sulphite and sulphate.

Table 0.2 Assumptions for Datum Coal-Fired Combustion Systems: Data for Systems Without Acidic Emissions Abatement

Case 1: 4 × 500 MWe coal-fired power station boilers
Case 2: Large industrial coal-fired boiler generating 450 tonne/h steam
Case 3: Small factory coal-fired boiler generating 13 tonne/h steam

Datum System No.	1	2	3
Output, MWe (net)	1860	–	–
Output, tonne/h steam	–	450	13
Steam pressure, bar	–	100	10
Steam temperature, °C	–	480	180
Useful Energy Output, GJe/h	6696	–	–
GJt/h	–	1468	35.3
Coal Composition (% by weight)			
Moisture	9.9	10.6	10.6
Ash	16.7	8.0	8.0
Carbon	60.3	67.7	67.7
Hydrogen	3.6	4.5	4.5
Nitrogen	1.3	1.4	1.4
Sulphur	1.6	1.5	1.5
Chlorine	0.25	0.26	0.26
Gross CV, MJ/kg	24.6	28.0	28.0
Boiler Type	Water Tube	Water Tube	Twin Fire Tube
Coal State	Pulverised	Pulverised	Smalls
Firing Method	Tangential	Opposed	Chain Grate
Coal fired, tonne/h	756	60.3	1.68
MWt (gross)	5168	469	13.1
Excess air, %	25	25	35
Flue gas flow, Thousand Nm^3/h			
Wet	6096	551	16.52
Dry	5704	513	15.46
Water vapour	392	38	1.06
Gas temperature, °C	150	150	230
Actual gas volume, Thousand m^3/h	9448	854	30.44
Emissions, mg/m^3 dry gas			
SO_2	4240	3520	3255
NO_x (as NO_2)	1025	1440	515
HCl	340	315	290
Particulates	115	115	660
Elements in acidic emissions, kg/h			
Sulphur	12096	904	25.2
Nitrogen	1784	225	2.42
Chlorine	1892	156	4.37
Particulates	656	59	10.2

Table 0.3 Appraisal of Flue Gas Desulphurisation Processes

1 Code No.	S22.1	S21.1	S11.1	S31.1	S41.1	S51.1
2 Reagent or Process	Lime Spray Dryer	Limestone or Lime Slurry	Sea Water	Wellman-Lord	Magnesia Slurry	Active Carbon Absorption
Appraisal						
3 Merit Points	13	12	11	11	11	11
4 Merits	1,2*,5,8	1,2,3*,8	2*,6	1,3,6,7,8	1,6,7,8	2,5,6,7,8
5 Drawbacks	4	4	–	4,5	3	3,4
Status						
6 No. of Units	94	478	14	35	8	10
7 MWe Range	3–1400	1–1150	10–230	35–1550	120–360	0.3–370
8 Total MWe	17,000	126,000	1310	7425	1485	502
9 % S. Capture	70–95	50–99	80–99	90–99	90–98	80–98
Detailed Evaluation						
10 System No.	1,2,3	1,2**	1,2	1,2	1,2	1,2

Notes

Line 1: Process Code No. denoting position of basic process in the classification.
Line 2: Principal reagent, or name of process.
Line 3: Total number of merit points (see Table 0.1 for scale of merit points).
Line 4: Categories with above average merit rating (see Table 0.1); asterisk indicates outstandingly better than average (i.e. 3 merit points).
Line 5: Categories with below average merit rating (see Table 0.1).
Line 6: Number of FGD units erected or on order worldwide.
Lines 7 & 8: Range of combustion plant sizes, and total capacity installed or on order, in equivalent MWe; 1 MWe equivalent = 1 MWe of power produced or 3000 Nm^3/h of gas treated or 2.5 tonne/h of steam produced.
Line 9: Range of sulphur capture performance reported.
Line 10: Datum combustion systems for which application of the process has been evaluated in detail.

**Three variants of this process evaluated in detail.

Wellman-Lord process (Process Code No. S31.1)

Gas is scrubbed with sodium sulphite solution, forming sodium bisulphite; the absorbent is regenerated thermally, with evolution of SO_2 for sale or conversion to sulphuric acid or elemental sulphur.

Magnesia scrubbing (Process Code No. S41.1)

Gas is scrubbed with magnesia slurry; the absorbent is regenerated by calcination with carbon, releasing SO_2 for conversion to sulphuric acid.

Active carbon adsorption (Process Code No. S51.1)

SO_2 is adsorbed by active carbon, which is regenerated thermally, releasing SO_2 and CO_2. The SO_2 can be sold or converted to sulphuric acid or elemental sulphur.

The processes evaluated in detail were those considered to have potential for application in the UK Only the lime slurry spray dryer absorption process (13 merit points) was evaluated for all three Systems, as it was considered to be the only process applicable to all three operating scales.

Two other basic types of process with non-regenerable reagents were selected for Combustion Systems 1 and 2: limestone or lime slurry scrubbing (12 merit points) and sea water scrubbing (11 merit points). Because of the overwhelming preponderance of FGD plants that are based on limestone or lime slurry scrubbing, and the wide range of designs that have been developed, three representative variants of the limestone scrubbing process were evaluated: the IHI limestone/gypsum, Chiyoda Thoroughbred 121 limestone-gypsum, and Saarberg-Holter-Lurgi (S-H-L) lime-gypsum processes.

Regenerable reagent processes, which because of their complexity are generally suitable only for the largest operating scale (System 1), could also be considered for a smaller scale (System 2) if adoption of FGD led to creation of an absorbent regeneration industry. Three such FGD processes (all with 11 merit points) were therefore evaluated for Systems 1 and 2: Wellman-Lord, magnesia slurry scrubbing and active carbon adsorption.

There are therefore eight individual processes evaluated in detail. In each instance, a sulphur capture was assumed that was within the range of capability of the process, with the most rigorous performance for System 1 and the least for System 3. The resulting emission factors for sulphur, together with those for nitrogen, chlorine and particulates, are summarised in Table 0.4 for the application to the three Datum Combustion Systems; the processes are listed in the Table in descending order of merit.

Table 0.4 Emission Factors for FGD Processes

Code No.	FGD Process	Merit Points*	Emission Factors, kg/tonne coal fired			
			Sulphur	Nitrogen	Chlorine	Particulates
Datum System 1: 4 × 500 MWe coal-fired power station boilers						
–	None	–	16.0	2.36	2.50	0.87
S22.1	Lime SDA	13	1.51	2.36	0.50	0.11
S21.1a	IHI	12	1.51	2.17	0.50	0.11
S21.1b	Chiyoda	12	1.51	2.35	0.50	0.11
S21.1c	S-H-L	12	1.51	2.35	0.50	0.11
S11.1	Sea water	11	1.51	2.35	0.50	0.11
S31.1	Wellman-Lord	11	1.51	2.35	0.50	0.11
S41.1	Magnesia	11	1.38	2.16	0.46	0.10
S51.1	Active carbon	11	1.48	2.14	0.49	0.11
Datum System 2: Large industrial boiler–450 tonne/h steam						
–	None	–	15.0	3.72	2.60	0.98
S22.1	Lime SDA	13	4.19	3.66	0.50	0.13
S21.1a	IHI	12	2.71	3.36	0.50	0.13
S21.1b	Chiyoda	12	2.73	3.68	0.51	0.13
S21.1c	S-H-L	12	2.74	3.69	0.51	0.13
S11.1	Sea water	11	2.75	3.70	0.51	0.13
S31.1	Wellman-Lord	11	2.73	3.68	0.51	0.13
S41.1	Magnesia	11	2.56	3.45	0.47	0.12
S51.1	Active carbon	11	2.72	3.37	0.50	0.13
Datum System 3: Small factory boiler–13 tonne/h steam						
–	None	–	15.0	1.44	2.60	6.07
S22.1	Lime SDA	13	4.49	1.41	0.51	0.13

* See Table 0.1

Table 0.5 Efficiency Factors and Costs for FGD Processes

Code No.	FGD Process	Merit Points*	Efficiency Factor per tonne coal	Capital Cost, £million		Operating Cost** £/kWe-year
				New Built	Retrofit	
Datum System 1: 4 × 500 MWe coal-fired power station boilers			GJe			
–	None	–	8.89	–	–	–
S22.1	Lime SDA	13	8.65	99.1	129.3	24.90
S21.1a	IHI	12	8.62	118.6	189.0	20.75
S21.1b	Chiyoda	12	8.74	120.0	177.1	18.90
S21.1c	S-H-L	12	8.76	111.8	163.4	19.78
S11.1	Sea water	11	8.81	n.d.	n.d.	n.d.
S31.1	Wellman-Lord	11	8.65	157.8	206.6	27.88
S41.1	Magnesia	11	8.05	178.9	261.9	22.34
S51.1	Active carbon	11	8.63	n.d.	n.d.	n.d.
Datum System 2: Large industrial boiler–450 tonne/h steam			GJt			
–	None	–	24.4	–	–	–
S22.1	Lime SDA	13	23.9	20.1	27.0	45.47
S21.1a	IHI	12	23.8	21.5	34.2	32.25
S21.1b	Chiyoda	12	24.0	17.0	25.1	26.48
S21.1c	S-H-L	12	24.1	23.4	34.1	33.60
S11.1	Sea water	11	24.1	n.d.	n.d.	n.d.
S31.1	Wellman-Lord	11	22.9	35.5	56.5	52.15
S41.1	Magnesia	11	22.5	34.7	50.8	37.95
S51.1	Active carbon	11	23.9	n.d.	n.d.	n.d.
Datum System 3: Small factory boiler–13 tonne/h steam			GJt			
–	None	–	21.0	–	–	–
S22.1	Lime SDA	13	20.6	n.d.	n.d.	n.d.

n.d. No data–costs not calculated
* See Table 0.1
** Operating costs, £/equivalent kWe-year for 5694 full-load operating hours per year, assumed equal for new build and retrofit applications.
To convert from £/kWe-year to p/kWe-hour, multiply by 0.0176.

Efficiency factors, together with capital costs, and operating costs excluding the annualised capital cost element, are summarised in Table 0.5; however, because insufficient data were available at the time the processes were being evaluated, the costs have not been estimated for sea water scrubbing and active carbon adsorption (Systems 1 and 2) and for lime spray dryer absorption (System 3).

0.6 NO_x Abatement – Combustion Techniques

Combustion techniques for flue gas denitrification are, at present, capable of NO_x reductions of up to 50% individually, and up to 80% where combinations of techniques are utilised: see Table 0.6. (It should be noted that where NO_x reduction figures are reported for techniques such as low-NO_x burners, LNB, a comparison has been made with more conventional designs of burner).

Table 0.6 Estimates of Removal Efficiencies for Combustion Modifications

Technique	*Reduction in NO_x* Emission (%)	
	Dry Bottom	Wet Bottom
Low-NO_x Burners (LNB)	20–50	10–30
Reburning	30–50	30–50
Off-Stoichiometric Combustion (OSC) or Staged Combustion	10–40	10–35
Low volume heating rate (LVHR)	20–25	
Flue Gas Recirculation (FGR)	up to 15	10–25
LVHR + LNB*	36–62	
LVHR + LNB + FGR*	36–67	
LVHR + LNB + OSC*	42–77	
LVHR + LNB + Reburning*	55–81	
LNB + FGR		19–47
LNB + OSC		19–54
LNB + Reburning		43–65
LNB + OSC + Reburning		43–77

* 600 MWe new power plant; 5700 h/annum at full load.

The study appraised 21 techniques which were classified into three broad categories: burner design, furnace design/modification and furnace operation. Three of these techniques were selected for further consideration.

Foster Wheeler Dual-Register, Split-Flame Burner

A burner with internally staged combustion, primarily intended for wall-fired boilers. It has two air registers arranged in series and a perforated plate air hood and movable sleeve for controlling and optimising secondary air flow. The burner is equipped with a split-flame nozzle which forms four distinct coal streams.

Babcock and Wilcox Delayed Mixing Type Dual Register Burner

A burner with internally staged combustion, primarily intended for wall-fired boilers. It incorporates a venturi in the primary air/coal line and an adjustable plug. Two annular channels accommodate the secondary air, each having independent registers and the inner channel containing swirl vanes to ensure flame stability.

In-Furnace NO_x Reduction (IFNR)

Also referred to as reburning or fuel staging. The combustion furnace is divided into three regions: the main (fuel-lean) combustion zone, a reburning or 'deNO_xing' zone where additional fuel is injected and, finally, the combustion completion zone where additional air is injected. A number of fuel staged firing configurations can be created, the main techniques being the Mitsubishi Advanced Combustion Technology (MACT), the Babcock-Hitachi IFNR

Table 0.7 Comparison of Burners

Burner Type	Type of Air Staging	Technique	NO_x level achievable (ppm)	Applicability*
Foster Wheeler Controlled-Flow Split Flame	Internal	Combines a split-flame annular coal nozzle with dual air register	150–200	N,R
Babcock and Wilcox DRB	Internal	Relies on dual air registers	200–300	N,R
Riley Stoker Controlled Combustion Venturi	Internal	Employs a four-bladed coal spreader	250	N,R
Tangential Firing with Over-fire Air	External	Use of stoichiometry or staged combustion in tangential firing system	250–300	N,R
Low-NO_x Concentric Firing System	External	Tangential firing system with some combustion air diverted along furnace walls away from flame	150–250	N,R
Mitsubishi PM Burner	External	Tangential firing system that creates alternating fuel-rich and fuel-lean flames	120–200	N

* N = new units, R = retrofits

system, and the Kawasaki Volume Combustion (KVC) system.

Unlike flue gas treatment processes for NO_x abatement, FGD and combined SO_2-NO_x abatement processes, a merit point system was not applied to $DeNO_x$ combustion techniques. However, a comparison of all the techniques was carried out using the limited technical data available; these are summarised in Tables 0.7–0.9 inclusive for the leading techniques.

0.7 NO_x Abatement – Flue Gas Treatment

Flue gas treatment (FGT) processes for denitrification are, at present, economically competitive with combustion modifications only where high NO_x reductions of 70–90% are required. The study appraised six basic types of process, and evaluated two of them in detail. These two process types, which are outlined below, had more than 10 merit points and were regarded as potentially applicable in the UK; their appraisals are summarised in Table 0.10. Only one other process is commercially available at present–dry adsorption (8 points); this was developed as an FGD process but is now offered as a combined SO_2–NO_x process. It should be noted that the number of points in the merit point systems adopted for FGD and combined SO_2–NO_x processes are not strictly comparable with those for NO_x processes.

Table 0.8 Comparison of Furnace Designs or Modifications

Furnace Design or Modification	Flue Gas Recirculation	Over-fire Air Injection	Reburning (IFNR)	Reduced Heat in Furnace
Technique	Low combustion temperature. Low oxygen concentration	Two stage combustion	Three stage combustion	Low volumetric heating rate (LVHR)
NO_x level achievable	Up to 15% reduction (Dry bottom boiler)	10–35% reduction	120 ppm	20–25% reduction
Applicability*	N,R	N,R	N	N
Problems	Cost Flame expansion Energy penalty	Incomplete combustion Flame extension Tube erosion		Possible boiler de-rating

Table 0.9 Comparison of Furnace Operating Techniques

Furnace Operation	Low Excess Air	Biased Burner Firing	Burners Out Of Service	Derating
Technique	Low excess air combustion i.e. reduced oxygen concentration	Upper row burners operate fuel-lean; Lower row operates fuel-rich	Upper row burners are out of service injecting only air	Reduced head load in furnace thereby lowering combustion temperature
NO_x Level Achievable (% reduction)	40	30–40	30–40	20–30
Applicability*	R	R	R	N,R
Problems	Unburnt carbon CO emissions Fouling and corrosion			Derating of unit.

* N = new units, R = retrofits

Table 0.10 Appraisal of NO_x Abatement FGT Processes

Process (& Code)	Merit Points	Merits	Drawbacks	No. of Units	Equiv Mwe		%NO_x Removal	Applied to
					Range	Total		
Column	1	2	3	4	5	6	7	8
SCR (N41)	13	1,3,5,7,8	–	171	8–700	27,000	53–90	1,2
SNR (N42)	11	2,5,7,8	4	83	n.d.	n.d.	40–85	1,2

Notes

Col. 1: See Table 0.1 for scale of merit points.
Col. 2: Categories with above average merit rating (see Table 0.1).
Col. 3: Categories with below average merit rating (see Table 0.1).
Col. 4: Number of units erected or on order (all fuels).
Col. 5 & 6: 1 MWe Equivalent = 1 MWe of power produced or 3000 Nm^3/h of gas treated. (n.d. = insufficient data for SNR, but have been fitted to utility boilers)
Col. 8: Datum system for which application or process has been evaluated in detail.

Selective Catalytic Reduction (SCR)

NO_x in the flue gas is reacted with ammonia gas in the presence of a catalyst, producing nitrogen and water which are discharged with the treated gas stream.

Selective Non-catalytic Reduction (SNR)

NO_x in the flue gas is reacted with ammonia gas at high temperature, e.g. in the upper section of the boiler. No catalyst is necessary at these temperatures and the products, nitrogen and water, are discharged with the treated gas stream.

The two processes were evaluated for Datum Combustion Systems 1 and 2 only. For the smallest operating scale dealt with in this Volume (Datum System 3), it is anticipated that the lower NO_x reduction efficiencies required to meet acceptable NO_x emission levels can be readily, and more economically, attained by combustion modifications alone.

For the detailed evaluation of each of the two selected processes, a NO_x reduction level was assumed that was within the range of capability of the process giving due regard to anticipated outlet NO_x (and ammonia) concentrations. The resulting emission factors for nitrogen, and efficiency factors, are summarised in Table 0.11. It was assumed that SO_2, HCl and particulate concentrations in the flue gas would be unaffected by these processes: minor side-reactions with ammonia and particulate deposition were considered to be negligible. There was insufficient information to enable costs for either process to be presented, although the reader is referred to Section 4.6 where limited comparative deNO_x cost data are presented.

Table 0.11 Summarised Results of Detailed Evaluations of NO_x Abatement FGT

DeNO_x Process (& Code)	Merit Points*	Efficiency Factor per tonne coal	Emission Factor (kg nitrogen per tonne coal)
Datum System 1 : 4 × 500 MWe coal-fired power station boilers			
		GJe	
None	–	8.89	2.36
SCR (N41)	13	8.84	0.46
SNR (N42)	11	8.85	0.69
Datum System 2 : Large industrial boiler–450 tonne/h steam			
		GJt	
None	–	24.4	3.72
SCR (N41)	13	24.3	0.51
SNR (N42)	11	24.3	1.04

* See Table 0.1

0.8 Combined Sulphur Dioxide – Nitrogen Oxides Abatement Processes

In addition to new processes developed for the purpose, several of the FGD processes and flue gas denitrification processes have been modified to enable both SO_2 and NO_x to be removed from the flue gas simultaneously. The study appraised sixteen processes, and evaluated two of them in detail.

As with the processes for desulphurisation and for denitrification of flue gas, combined abatement processes with more than 10 merit points were regarded as being potentially commercially applicable in the UK. This criterion was met by only one process–Active Carbon/Selective Catalytic Reduction (12 merit points), and this was evaluated for Datum System 1, and (assuming the creation of an absorbent regeneration industry as already mentioned) for Datum System 2.

Like its FGD counterpart, the Lime Spray Dryer Absorption process has attractions for small as well as for large scale operation, but because of lack of information available and of operating experience, its rating is only 9 merit points. However, because of its promise for future applications to all operating scales, it was evaluated, in as much detail as the available information allowed, for Datum System 1, 2 and 3. Descriptions of the two processes selected are outlined below, and their appraisals are summarised in Table 0.12.

Table 0.12 Appraisal of Combined Abatement Processes

1	Code No.	S11.1	S31.1
2	Reagent or Process	Active Carbon/ Selective Catalytic Reduction	Lime Spray Dryer Absorber
	Appraisal		
3	Merit Points	12	9
4	Merits	2,5,6,7,8	2,5,8
5	Drawbacks	3	3,4
	Status		
6	No. of Units	7	2
7	MWe Range	0.7–370	7–32
8	Total MWe	606	39
	Performance		
9	% S Capture	90–98	90–95
10	% NO_x Abatement	60–90	20–60
	Detailed Evaluation		
11	System No.	1,2	1,2,3

Notes:

Line 1: Process Code No. denoting position of basic process in the classification.
Line 2: Principal reagent, or name of process.
Line 3: Total number of merit points (see Table 0.1 for scale of merit points).
Line 4: Categories with above average merit rating (see Table 0.1).
Line 5: Categories with below average merit rating (see Table 0.1).
Line 6: Number of combined abatement units erected or on order worldwide.
Lines 7 and 8: Range of combustion plant sizes, and total capacity installed or on order, in equivalent MWe; 1 MWe equivalent = 1 MWe of power produced or 3000 Nm^3/h of gas treated or 3 MWt.
Line 9: Range of sulphur capture performance reported.
Line 10: Range of NO_x abatement performance reported.
Line 11: Datum combustion system for which application of the process has been evaluated in detail.

Active carbon/Selective catalytic reduction (Process Code No. NS11.1)

SO_2 and NO_2 are adsorbed in the 1st-stage reactor by carbon fed from the 2nd-stage reactor. Nitric oxide is catalytically reduced to elemental nitrogen by ammonia in the 2nd-stage reactor. Carbon from the 1st stage is regenerated thermally (producing SO_2, CO_2 and elemental nitrogen) and recycled to the 2nd stage. SO_2 can be sold, or converted to sulphuric acid or elemental sulphur.

Lime Spray Dryer Absorber (SDA–Process Code No. NS31.1)

SO_2 and NO_x in hot gas are absorbed in an SDA by lime slurry containing sodium hydroxide, forming solids containing calcium sulphite, sulphate, nitrite and nitrate.

In making the detailed evaluations of the two processes selected, extents of SO_2 and NO_x abatement were assumed that were within the ranges of capability of the processes, with the most rigorous performance for Datum System 1, and the least for System 3. The resulting emission factors for sulphur and nitrogen, together with those for chlorine and particulates, are summarised in Table 0.13. Efficiency factors are also presented in the Table. There was insufficient information available at the time the evaluations were made to enable costs for either of the Combined Abatement processes to be estimated.

0.9 Fluidised Bed Combustion (FBC)

There are some 380 fluidised bed combustion plant of all types erected or on order in the western world; most are small units (e.g. boilers producing less than 20 tonne/h steam) but some are larger (up to 200 tonne/h steam) and scale-up to the size of modern utility power plant can be envisaged before the year 2000. The attractions include the ability to burn coals of poor and variable quality, the ease of removal of SO_2 by feeding limestone or dolomite to the bed, and the low emissions of NO_x.

If gas is passed upwards with sufficient velocity through a bed of particles, the particles start to move and separate from each other, and the bed takes on some of the physical characteristics of a liquid and is incipiently 'fluidised'. At higher gas velocity, e.g. 1–3 m/s, particle movement is more vigorous, and some of the gas passes upwards in the form of 'bubbles'; in this 'bubbling bed' condition, the bed resembles a boiling liquid. Further velocity increases result in progressive particle elutriation from the bed until, at the 'fast fluidised' condition (gas velocity about 10 m/s), all of the particles are transported by the gas.

Fluidisation is characterised by good mixing of solids and contact between solids and gases, uniformity of temperature throughout the bed, and high rates of heat transfer between the bed and immersed surfaces. These characteristics are favourable for: rapid and efficient combustion of coal at low temperatures (700–950°C); absorption of SO_2 in the bed by limestone or dolomite; avoidance of 'thermal NO_x' formation (i.e. oxidation of atmospheric nitrogen) which occurs readily at the high flame temperatures in most other combustion systems; and high rates of heat transfer to steam generating and superheating tubes immersed in the bed. The bed of a fluidised bed combustor has a very low carbon content, and is mainly composed of the ash of the coal (sometimes augmented by an added refractory such as sand), together with limestone or dolomite sulphur-sorbent. Cooling of burning fuel particles by surrounding bed material prevents the formation of sticky ash, so that large furnace volumes for ash cooling are unnecessary.

Provided sufficient sorbent (limestone or dolomite) is fed to the combustor, the emission of SO_2 can be reduced to any desired level. The amount of sorbent to be fed depends upon the sulphur content of the coal, the desired emission level, the reactivity of the

Table 0.13 Summarised Results of Detailed Evaluations of Combined Abatement Processes

Code No.	Combined Abatement Process	Merit Points*	Efficiency Factor per tonne coal	Emission Factors, kg/tonne coal fired			
				Sulphur	Nitrogen	Chlorine	Particulates
Datum System 1: 4 × 500 MWe coal-fired power station boilers							
			GJe				
–	None	–	8.89	16.0	2.36	2.50	0.87
NS11.1	Active C/SCR	12	8.75	0.94	0.46	0.07	0.11
NS31.1	Lime SDA	9	8.73	1.51	0.92	0.07	0.11
Datum System 2: Large industrial boiler–450 tonne/h steam							
			GJt				
–	None	–	24.4	15.0	3.72	2.60	0.98
NS11.1	Active C/SCR	12	24.1	1.69	1.92	0.08	0.13
NS31.1	Lime SDA	9	23.8	2.91	1.76	0.08	0.12
Datum System 3: Small factory boiler–13 tonne/h steam							
			GJt				
–	None	–	21.0	15.0	1.44	2.60	6.07
NS31.1	Lime SDA	9	20.5	4.49	0.82	0.09	0.13

* Total merit points awarded (see Table 0.1)

sorbent, the type of combustor and the operating conditions. The solid residues from FBC systems include coal ash, calcium oxide, sulphate and carbonate and (if dolomite is used) magnesia. The material becomes impermeable on atmospheric exposure, but some care may be needed in choice of disposal sites to avoid environmental problems from leaching of alkaline components of the freshly dumped material.

Most FBC systems operate in the bubbling bed condition. However, fast-fluidised FBCs are gaining acceptance; in these 'Circulating bed combustors', the solids transported from the furnace are captured and recycled, thus prolonging their residence time in the combustor. The main types of FBCs are described briefly below, and their status is indicated in Table 0.14. Emission factors for some applications of FBC are summarised in Table 0.15, and efficiency factors, together with some costs, are presented in Table 0.16; however, very little published cost information was available.

BUBBLING BED COMBUSTORS

Deep beds (0.6–1.5 m depth)

With pneumatic feed of crushed coal (sized up to 6 mm) to the base, or with overbed firing of coal sized up to 20 mm. NO_x emission concentrations 600–900 mg/Nm^3.

Shallow beds (less than 0.3 m depth)

With overbed feeding of washed lump coal; unsuitable for efficient sulphur capture.

Dual bed

Gases from the first bed fluidise a second bed in which further combustion and sulphur capture occurs. NO_x emission concentrations 300–500 mg/Nm^3.

Pressurised fluidised bed combustors (PFBCs)

Operate with beds of 3–4 m depth at pressures up to 20 bar, with recovery of energy in a gas turbine operating on cleaned hot flue gas. Application is for power generation, using cycles such as: the high gas temperature steam cycle (steam production in immersed tubes, with power generated in the steam and gas turbines giving enhanced power generation efficiency); low gas temperature steam cycle (additional steam production and gas-cooling in convection passes, with power generated in the steam turbine only).

Table 0.14 Fluidised Bed Combustion Plant

Type	No. of Suppliers	No. of Units*	Size range MWe**
Bubbling Bed Combustors			
Deep bed	37	251	0.4–272
Shallow bed	11	63	0.4–20
Dual bed	2	8	4–36
Pressurised	3	3	80–100
Circulating Bed Combustors			
Integral heat exchange	9	48	2–182
External heat exchange	6	10	4–185

* Number of units supplied or on order
** Equivalent MWe; 1 MWe = 2.5 tonne/h steam

Table 0.15 Emission Factors for Fluidised Bed Combustor Applications

Combustor Type	Emission Factors, kg/tonne coal fired		
	Sulphur	Nitrogen	Chlorine
Datum System 1: 4 × 500 MWe coal-fired power station boilers			
PF without abatement	16.0	2.36	2.50
PF with FGD	3.99	1.84	n.d.
AFBC deep bubbling bed	1.60	1.32	2.50
Lurgi CFB	1.60	0.68	2.50
PFBC turbocharged	4.14	0.64	n.d.
Pressurised CFB turbocharged	4.10	0.56	n.d.
Datum System 2: Large industrial boiler–450 tonne/h steam			
PF without abatement	15.0	3.72	2.60
AFBC deep bubbling bed	1.51	1.38	2.60
Ahlstrom CFB (2×225 t/h)	1.50	0.78	2.60
MSFB (2×225 t/h)	1.50	0.78	2.60
Datum System 3: Small factory boiler–13 tonne/h steam			
Chain grate without abatement	15.0	1.44	2.60
AFBC deep bubbling bed	3.0	2.3	2.6
Dual Bed	2.1	1.58	2.6

Table 0.16 Efficiency Factors and Costs for FBC Applications

Combustor Type	Efficiency Factor per Tonne coal	Capital Cost £million	Operating Cost* £/kWe-year
Datum System 1: 4 × 500 MWe coal-fired power station boilers			
PF without abatement	8.89 GJe	n.d.	n.d.
PF with FGD	7.85	1490	123.1
AFBC deep bubbling bed	8.60	1471	129.25
Lurgi CFB	8.71	n.d.	n.d.
PFBC turbocharged	8.01	n.d.	n.d.
Pressurised CFB turbocharged	8.53	n.d.	n.d.
Datum System 2: Large industrial boiler–450 tonne/h steam			
PF without abatement	24.4 GJt	n.d.	n.d.
AFBC deep bubbling bed	24.1	n.d.	n.d.
Ahlstrom CFB (2×225 t/h)	24.2	n.d.	n.d.
MSFB (2×225 t/h)	23.8	n.d.	n.d.
Datum System 3: Small factory boiler–13 tonne/h steam			
Chain grate without abatement	21.0 GJt	n.d.	n.d.
AFBC deep bubbling bed	21.0	n.d.	n.d.
Dual Bed	23.3	n.d.	n.d.

* Operating costs, £/equivalent kWe-year for 5694 full-load operating hours per year. To convert to p/kWh multiply by 0.0176.

n.d. = No data

CIRCULATING BED COMBUSTORS (CFB)

Integral heat exchange

Heat exchange surface is on the walls of the combustor shaft, and sometimes of the solids disengaging zone; e.g. Ahlstrom CFB.

External heat exchange

The main heat exchange surface is immersed in a bubbling bed of the solids being recirculated to the combustor; e.g. Lurgi CFB and Battelle Multi-Solids Fluidised Bed (MSFB).

0.10 Removal of Acid Halides

Acid halides in combustion gases from coal derive mainly from halogen compounds (chlorine, fluorine, bromine and iodine) in the coal. For UK coals, concentrations are: chlorine–between trace and 1.2%, averaging 0.32%; fluorine–between 98 and 130 ppm (average 114 ppm); bromine and iodine–present in trace concentrations. Most of the halogen compounds in coal are volatilised during combustion, forming hydrogen halides which are soluble even in highly acidic water. When a feed coal contains more than about 0.1% chlorine, most wet FGD systems operate satisfactorily only if the gas is prescrubbed with water.

Many FGD and combined abatement processes are also effective for removal of acid halides. Processes available specifically for acid halide removal are outlined below. The principal problem with all of the processes is the high solubility of the halogen compounds formed, and the lack of markets for the end-products, which often have to be disposed of at sea.

Wet scrubbing

Scrubbing with for example water; the effluent can be neutralised with lime, coal ash (containing heavy metals) filtered out (or the pH adjusted to precipitate the trace elements), and the solution concentrated or evaporated to give calcium chloride for disposal.

Spray dryer absorption (SDA)

All SDA processes for FGD or combined abatement remove virtually all of the hydrogen halides in the gas giving soluble halides in the solid product.

Dry Processes

Most of these use reaction with limestone or slaked lime injected into the gas, or in packed beds, at low temperatures (e.g. 200°C). Active carbon also removes acid halides.

There is insufficient information available for separate evaluation of the acid halides removal processes. In general, the need to remove acid halides will increase the costs of wet FGD and combined abatement processes because of the need to isolate and remove calcium chloride.

1 Introduction

1 Introduction

1.1 Background and Purpose of the Manual

This Manual has been compiled by The Fellowship of Engineering in fulfilment of a commission from the Department of the Environment to prepare a Manual describing, comparing and appraising processes for abatement of acidic emissions from fossil fuel combustion systems. It is anticipated that the readership of the Manual will include those air quality professionals wishing to have information on the acidic emissions abatement technologies available, and organisations involved in assessing the economics of emission control.

The aim has been to include only those abatement technologies applied during or after combustion; acid emission abatement processes dependent on removal of pollutant elements (sulphur, nitrogen and halogens) from the fuel before combustion have therefore been excluded, but it is hoped to include them in a later edition of the Manual. Among the technologies thus excluded are: oil desulphurisation (of particular interest for small oil-fired combustion systems); and coal gasification (which has an important potential application in gasification-combined cycle power generation systems).

In order to provide a comprehensive data base, the Manual contains:

- Brief descriptions of each type of process or technology commercially available, or showing promise of becoming commercially available by the end of the century
- Information showing the status and applicability of the processes or technologies
- Accounts of operating experience and performance
- Detailed descriptions and cost data for application of selected processes and technologies to a range of coal-fired, oil-fired and gas-fired combustion systems

In order to make it more manageable for the user, the Manual has been prepared in three volumes covering stationary combustion systems; one for each of the fuels–coal, oil and gas; and a fourth volume for mobile sources. This Volume is concerned with stationary coal-fired combustion systems.

In preparing this Manual for stationary combustion systems, the Fellowship of Engineering retained Foster Wheeler Power Products Limited with Hoy Associates Limited who acted as a collaborating subcontractor. The direction of the work by the Fellowship was organised through an overall Steering Group, and specialised Task Groups, one for each of the three fuels–coal, oil and gas. The composition of the Steering and Task Groups is given in the Acknowledgements.

1.2 Acidic Emission Inventories

The main precursors of acid deposition are sulphur dioxide, SO_2, and nitrogen oxides, NO_x (primarily nitric oxide, NO), emitted into the atmosphere from natural and man-made sources. In north-west Europe, man-made sulphur emissions contribute over 80% of total sulphur emissions into the atmosphere. Although the available data on NO_x emissions are less extensive than those for sulphur emissions, it is estimated that for north-west Europe, man-made sources of NO_x contribute between 75% and 93% of total NO_x emissions [38].

Warren Spring Laboratory (WSL) have prepared estimates of UK emissions of SO_2 and NO_x, and the quantities for 1985, and sources, are presented in Figures 1.1 and 1.2 [40]. The principal source of SO_2 emission is seen to have been power stations (71%), and in normal circumstances over 95% of the UK electric power generation from fossil fuel fired power stations is from coal-fired stations. Responsibility for the remaining 29% (about 1 million tonnes) of the 3.58 million tonnes UK SO_2 inventory was shared by oil and coal.

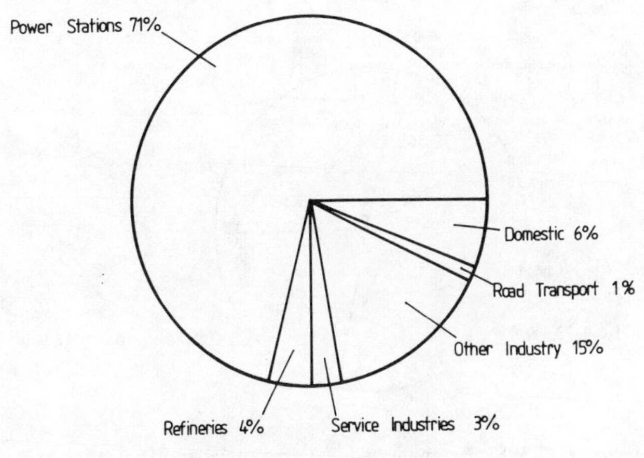

Figure 1.1 UK Man-made Sources of SO_2 Emission (1985). Total SO_2 Inventory 3.58 million tonnes

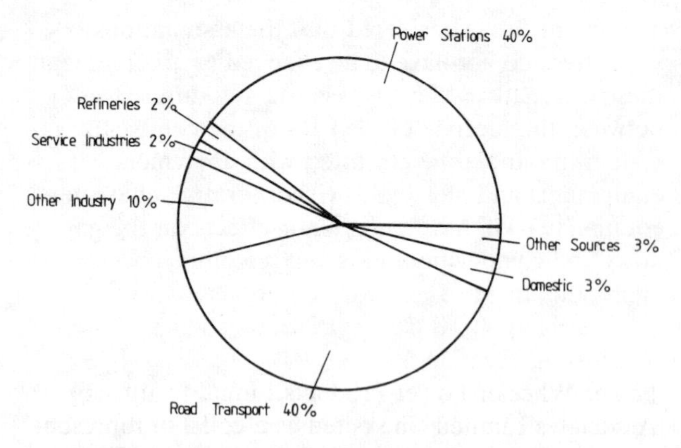

Figure 1.2 UK Man-made Sources of NO_x Emission (1985). Total NO_x Inventory (as NO_2) 1.84 million tonnes

For NO_x, of the total of 1.84 million tonnes (expressed as NO_2) road transport accounted for 40% of the emissions, power stations (predominantly coal-fired) contributed 40% (about 736,000 tonnes), and other stationary sources, coal-, oil- and gas-fired, 20% (about 368,000 tonnes). When stationary sources of emission alone are taken into consideration, it is seen that the pattern for NO_x emissions is very similar to that for SO_2, with power stations contributing 67%, refineries, service industries and other industry 23%, and domestic sources 5% of stationary-source emissions.

In 1982, WSL measured NO_x emissions from 72 selected commercial and industrial coal-, oil- and gas-fired boilers of up to 230 tonne/h steam capacity. These data were used [41] as the basis for estimating UK emissions from a limited number of sources, as shown in Figure 1.3. Of the 195,000 tonnes of NO_x (expressed as NO_2), 46,000 tonnes (about 24%) originated from combustion of coal.

On the basis of these data it can be estimated that annual emissions from coal-fired power stations are of the order of 2.43 million tonnes of SO_2 and 0.7 million tonnes of NO_x (as NO_2); and from industrial

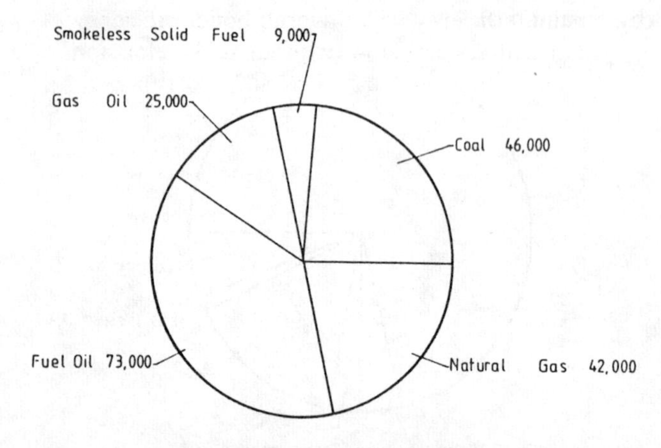

Figure 1.3 WSL Estimate of UK Commercial and Industrial NO_x Emissions (1982). Estimated Emissions, tonnes (as NO_2) from Various Fuels. Total Estimated (as NO_2) 195,000 tonnes

boilers 0.36 million tonnes SO_2 and 0.1 million tonnes of NO_x; see Table 1.1.

Table 1.1 Acid Emission Inventories

Source	Road Transport	Power Stations	Other Sources	Total
Sulphur Dioxide				
Coal	–	2.43	0.36	2.77
Oil	0.04	0.11	0.64	0.81
Total	0.04	2.54	1.00	3.58
Nitrogen Oxides				
Coal	–	0.70	0.10	0.80
Oil	0.74	0.04	0.18	0.96
Gas	–	–	0.08	0.08
Total	0.74	0.74	0.36	1.84

The continuing administration of the Alkali etc. Works Act (1863) has reduced industrial emissions of hydrogen chloride (HCl) to the point where most of the UK emissions of this pollutant now originate from the combustion of UK coal, which is unusually rich in chlorine; minor sources are principally the incineration of industrial waste containing PVC, and combustion of oil. The emissions of HCl are an order of magnitude smaller than those of SO_2 from power stations [38]. Emissions of hydrogen fluoride are much lower and are not generally regarded as significant, and those of the other acid halides–hydrogen bromide and iodine–are lower still.

1.3 Scope

The emissions inventory data in Section 1.2 indicate that coal-fired power station and industrial boilers are the greatest source of SO_2 emissions and contribute significantly to the overall NO_x inventory. The emissions of acid halides from these sources are also not insignificant, although very much smaller than SO_2 and NO_x.

ABATEMENT PROCESSES AND TECHNOLOGIES

The study in this Volume therefore covers processes and technologies, applicable to coal-fired systems, for:
- Flue gas desulphurisation
- Nitrogen oxides abatement
- Simultaneous sulphur and nitrogen oxides removal
- Acid halides removal.

As regards the last of these, many flue gas desulphurisation and combined SO_2–NO_x removal processes also remove acid halides from the gas, and processes have also been developed specifically for the removal of acid halides. Both aspects are considered in this Volume.

In addition, this Volume deals with the increasingly important coal-combustion technology of fluidised bed combustion, which gives reduced emissions of NO_x, and in which SO_2 can be removed during combustion.

APPLICATIONS

To arrive at realistic evaluations, consideration has been given to the application of selected emissions abatement processes to representative types and sizes of coal-fired combustion plant (defined as the 'datum coal-fired combustion systems' in Section 1.4).

The evaluations take into account the alternatives of:
- Incorporating abatement equipment in new combustion plant
- Retrofitting abatement equipment on existing plant.

EFFLUENTS AND BY-PRODUCTS

An important aspect of most acidic emissions abatement processes is the need to dispose of process effluents–water, other liquids, and solids.

Some effluents are potentially marketable by-products, e.g. sulphur or sulphuric acid from regenerable flue gas desulphurisation processes. Credit for their sale will be possible only when a market exists, or if no unacceptable social penalty (e.g. destruction of an existing industry) has to be paid for marketing the product.

REDUCTION IN COMBUSTION PLANT EFFICIENCY

Many of the acidic emissions abatement techniques and processes result in a diminution in the thermal efficiency of the combustion plant. In practice, this will give rise to the need to burn more fuel, and in some instances to instal additional combustion plant, to produce the desired output of useful energy. This factor has to be borne in mind when considering the impact of installing abatement equipment on national pollutant inventories, or on the degree of abatement needed.

1.4 General Assumptions for Datum Combustion Systems

Three datum coal-fired combustion systems are considered in this Volume and defined in this Section.

Although separate evaluations are made for each system for retrofitting emissions abatement equipment and for new builds incorporating emissions abatement, it is considered that the assumptions made here do not have to be changed to accommodate these alternatives. This is because any differences between the thermal efficiencies of the combustion system in old plant (retrofitted with abatement equipment) and new builds (incorporating abatement equipment) will have only minor effects on the gas flows and compositions entering the emissions abatement equipment. These considerations do not necessarily apply to the application of fluidised bed combustion (Section 6).

The datum combustion systems, selected to represent the major large coal users in the UK, are defined as follows:

System 1

2000 MWe (4 × 500 MWe) coal-fired (pulverised fuel, tangential-fired) power station. At full load: excess air level 25%; combustion efficiency 99.5%; power sent out 1860 MWe; net station efficiency 36%. Particulates emission (without acidic emissions abatement equipment) 115 mg/Nm^3 dry gas. Gas temperature leaving boiler plant to stack 150°C.

System 2

Large industrial coal-fired (pulverised fuel, opposed-fired) water-tube boiler: 450 tonne/h steam at 100 bar pressure and 480°C temperature; excess air level 25%; combustion efficiency 99%; boiler efficiency 87%. Particulates emission (without acidic emissions abatement equipment) 115 mg/Nm^3 dry gas. Gas temperature leaving boiler plant to stack 150°C.

System 3

Small factory coal-fired (small coal, chain grate) twin fire-tube boiler: 13 tonne/h of steam at 10 bar pressure and 180°C (saturated); excess air level 35%; combustion efficiency 95%; overall boiler efficiency 75%. Particulates emission (without acidic emissions abatement equipment) 10.20 kg/h in conformity with

Table 1.2 Assumed Compositions of Coal (As received basis)

Datum System	1	2 and 3
Composition, % by weight:		
Moisture	9.9	10.6
Ash	16.7	8.0
Carbon	60.3	67.7
Hydrogen	3.6	4.5
Nitrogen	1.3	1.4
Sulphur	1.6	1.5
Chlorine	0.25	0.26
Oxygen (difference)	6.4	6.0
Gross calorific value, MJ/kg	24.6	28.0
Net calorific value, MJ/kg	23.6	26.8

the Clean Air Regulations 1971 [66]. Gas temperature leaving boiler plant to stack 230°C.

Assumed coal compositions: Shown in Table 1.2. The compositions quoted were supplied by British Coal, H.Q. Technical Department, and represent the average compositions of coal supplied to the CEGB (Datum system 1, part cleaned smalls) and to industry (Datum systems 2 and 3, washed smalls).

CALCULATED DATUM FLUE GAS COMPOSITIONS

Compositions and acidic emission levels (assuming zero abatement), estimated from the definitions of the datum combustion systems and from the assumed coal composition, are presented in Table 1.3. The SO_2 and HCl contents represent the total inputs of sulphur and chlorine in the coal, i.e. the assumption has been made that none of the input sulphur has been retained by the ash of the coal. The NO_x contents have been estimated from [357, 358, 359, 360, 402, 403] supplemented by curves of baseline NO_x emissions with different firing systems [352].

DATUM EFFICIENCY AND EMISSION FACTORS

Efficiency and emission factors (assuming zero abatement) calculated from the definitions of the datum combustion systems and from the assumed coal compositions are presented in Table 1.4.

1.5 Appraisal of Processes

A procedure has been adopted to quantify the appraisals made in this Manual of processes for flue gas desulphurisation (Section 2.3), flue gas denitrification (Section 4.3) and combined SO_2-NO_x abatement (Section 5.3). In this rough appraisal procedure, 'merit points' have been awarded to each process for a number of features. The points have been awarded according to the scale:

0 – below average merit
1 – average merit
2 – above average merit
3 – outstandingly above average merit

The features to which these points have been assigned are as follows:

1. Information available: the amount of information available
2. Process simplicity
3. Operating experience: the extent of operating experience
4. Operating difficulty: availability, reliability
5. Loss of power: the equivalent percentage reduction in power sent out as a result of installing the process
   2 = Less than 2.5%
   1 = 2.5 to 7.5%
   0 = More than 7.5%
6. Reagent requirements: the general rule is
   2 = regenerative reagent processes
   1 = throw-away reagent processes
   but there are some exceptions

Table 1.3 Calculated Flue Gas Composition and Acidic Emissions

System No.	1		2		3	
Excess air, %	25		25		35	
Combustion efficiency, %	99.5		99.0		95.0	
Flue gas composition by volume:						
	Wet	Dry	Wet	Dry	Wet	Dry
H_2O%	6.5	–	6.9	–	6.4	–
CO_2%	13.9	14.8	13.6	14.4	12.1	12.9
O_2%	4.1	4.4	4.1	4.4	6.0	6.4
N_2%	75.4	80.6	75.1	80.7	75.4	80.6
SO_2 ppm	1386	1482	1146	1231	1065	1138
NO_x ppm	468	500	652	700	234	250
HCl ppm	196	209	180	193	167	178
Emissions, mg/Nm³ (dry gas; assuming no abatement):						
SO_2	4240		3520		3255	
NO_x as NO_2	1025		1440		515	
HCl	340		315		290	
Particulates	115		115		660	

Table 1.4 Calculated Efficiency and Emission Factors

System No.	1	2	3
Gross heat input, MW	1292*	469	13.1
Coal fired, tonne/h	189*	60.3	1.68
Useful energy output, GJe/h	1674*	–	–
GJt/h	–	1468	35.3
Efficiency factor, GJe/tonne	8.86	–	–
GJt/tonne	–	24.3	21.0
Ashes output, tonne/h	32.3*	5.32	0.2
C content of ashes, %	2.1	9.4	34.1
Particulates emitted, kg/h	164*	59.0	10.2
Flue gas, Thousand Nm³/h			
Wet	1524*	551	16.52
Dry	1426*	513	15.46
Water vapour	99*	38	1.06
Gas temperature, °C	150	150	230
Actual gas volume, Thousand m³/h	2362*	854	30.44
Elements in acidic emissions, kg/h (assuming no abatement)			
Sulphur	3024*	904	25.2
Nitrogen	446*	225	2.42
Chlorine	473*	156	4.37
Emission factors, kg/tonne coal (assuming no abatement)			
Sulphur	16.0	15.0	15.0
Nitrogen	2.36	3.72	1.44
Chlorine	2.50	2.60	2.60

* Per 500 MWe unit

7. Ease of end-product disposal

8. Process applicability: the general rule is
   2 = processes suitable for retrofitting (unless application is restricted by e.g. geographical considerations)
   1 = other processes

In some instances, lack of information has made it necessary to assume the number of merit points for a feature; the assumptions made were deduced from characteristics that are known of the process concerned. Process capital and operating costs could not be included in the list of features for merit points, as in most instances published cost data were unavailable or too meagre to be of value, and it was considered unsafe to assume merit ratings for these features.

The merit points system described above is based on comparisons with the average level for each of the eight features. However, because the average levels are different for the different acidic emissions being abated, the merit ratings for a flue gas desulphurisation process, for example, cannot be compared with those for flue gas denitrification or for combined abatement processes.

A process that was average in all of the eight features would, of course, have a total of 8 merit points. Very few processes achieved 1 or more points for all of the eight features. On the other hand, several processes were above average in a sufficient number of features to achieve totals exceeding 8 merit points. It was arbitrarily decided that a process with over 10 merit points was potentially applicable commercially in the UK, and its application to either two or three of the datum coal-fired combustion systems has been evaluated in detail (Sections 2.5, 4.5 or 5.5).

1.6 Notes for the Guidance of the Reader

PLANT SIZE

Plant size is expressed in a number of ways in this Manual, depending upon the particular application for the combustion system. In order to make comparisons of operating scale, the following conversion factors may be used:

1 MWe of power generated is roughly equivalent to:

3.6 GJe/h of power generated (exact conversion)
3 MWt heat input to the combustion system
10.8 GJt/h of heat input to the combustion system
3000 Nm^3/h gas treated
2.5 tonne/h of steam generated by the boiler

In those cases where the original text expresses plant size in MW or m^3/h without precise definition, and it is uncertain which units apply (e.g. MWt or MWe; Nm^3 or m^3), no further qualification of the quoted figures has been made.

DESCRIPTIONS OF EQUIPMENT

In this Manual, simplified block diagrams accompany the Outline Descriptions of processes, and the descriptions of plant in detailed evaluations of process applications are illustrated by simplified flow diagrams. A convention has been adopted that plant items shown on the diagrams are referred to in the text with their names given capital initials (e.g. Absorber, Heat Exchanger) and items not shown on the diagrams are with lower-case initials (e.g. crystalliser).

APPLICABILITY OF PROCESSES TO COAL AND OIL

All of the flue gas desulphurisation, denitrification and combined abatement processes considered in this Volume are applicable to coal- and oil-fired combustion systems. However, for coal-fired systems, many of the processes need to incorporate a gas pre-scrubbing stage to remove particulates and acid halides as well as to cool the gas. In general, the Outline Descriptions of the basic process types have been described (Sections 2.2, 4.2 and 5.2) as for application to coal-fired combustion systems, and in some instances it is also necessary to refer to the electrostatic precipitator (ESP), and to disposal of fly-ash. When the same processes are applied to oil-fired combustion systems, where the particulates and acid halides concentrations are negligible, prescrubbing and ESPs are not needed and the gas can be cooled for example by evaporative spray cooling. The acidic abatement process is in other respects the same for coal- and oil-fired systems.

In view of this, in the appraisal of the processes in this Volume (Sections 2.3, 4.3 and 5.3) actual applications have been quoted from the literature whatever the fuel, and whatever the type or purpose of the combustion system.

EMISSION ABATEMENT LEVELS ASSUMED

In the detailed evaluations of selected emission abatement processes (Sections 2.5, 4.5 and 5.5) assumptions have been made regarding the concentrations of pollutants in the gas leaving the abatement plant. These assumptions have been arbitrary, but in all instances the emissions abatement assumed has been within the range of capability of the process concerned, with the most rigorous requirement for Datum System 1 (4 × 500 MWe power station boiler). In all instances the particulates emission was assumed to be reduced to 15 mg/Nm³.

EMISSIONS AND EMISSION FACTORS

In the detailed evaluations of techniques and processes (and in Table 1.4) the emissions in kg/h, and emission factors in kg/tonne of fuel, are expressed as kg of the element (sulphur, nitrogen and chlorine) per hour or per tonne of fuel. To convert to kg of sulphur dioxide, nitric oxide or nitrogen dioxide, and hydrogen chloride, multiply the figures by the factors below.

To convert:	Multiply by
Sulphur to sulphur dioxide (SO_2)	1.998
Nitrogen to nitric oxide (NO)	2.142
Nitrogen to nitrogen dioxide (NO_2)	3.284
Chlorine to hydrogen chloride (HCl)	1.028

UNAVAILABLE INFORMATION

At the time the Manual was being prepared, some of the information needed was unavailable. In order to preserve the format for the Manual, however, Section headings etc. have been retained even where no material could be included. It is hoped that much of this information will become available for inclusion in a later edition of the Manual; in the meantime, the absence of available information is indicated by three asterisks thus: ***

EFFECT OF PLANT AVAILABILITY

In the detailed evaluations of processes in Sections 2.5, 4.5 and 5.5, an indication is given of the effect on annual average emissions and emission factors of the abatement plant being bypassed in the event of its being shut down for any reason. The effect is, of course, to increase the emissions and emission factors for those emissions that would otherwise be abated. For an abatement plant availability of A%, the average annual emissions and emission factors are increased, from values represented by E, to the higher values E′ given by:

$$E' = E.A/100 + E_o(1 - A/100)$$

where E_o represents the corresponding emissions or emission factor without abatement (given in Section 1.4). For illustration purposes, average annual emissions and emission factors for abatement plant availabilities of 100% and 90% are given for all of the processes evaluated.

2. Flue Gas Desulphurisation Processes

2. Flue Gas Desulphurisation Processes

2.1 Classification of the Processes

The general classification of flue gas desulphurisation (FGD) processes adopted in this Volume is presented in Figure 2.1. The processes are divided into seven categories according to the scheme:

First division

Distinction between wet and dry processes.

Second division

Distinction between non-regenerable and regenerable reagents; however, if regeneration of the sorbent is based on transfer of the captured sulphur to combination with another reagent, the process is classified as non-regenerable.

Third division

Depends on the method of application of the reagent:
- for wet reagents, whether the reagent is solution or slurry
- for dry reagents, whether the reagent is applied in the flue gas or (for non-regenerable processes only) in the furnace.

Fourth division

- for non-regenerable wet processes, distinction between those producing a wet end-product, or a dry end-product (excluding wet-end products that are dewatered before sale or disposal)
- for non-regenerable dry processes, distinction between reagent injection processes or absorption in a reactor
- for regenerable processes: distinction between those using thermal or chemical regeneration methods.

There are: four main wet reagent categories (Categories S10 to S40) and three main dry reagent categories (Categories S50 to S70), with the fourth division in each of these main categories denoted S11, S12, S21, S22, etc. The 'S' prefix denotes sulphur abatement technology.

Some of the processes yielding a dry end-product involve spray-drying of solutions or slurries as an integral feature of the process. Such processes are frequently described in the literature as 'dry', 'semi-dry' or 'wet-dry' processes. These terms are regarded as misleading for such processes, which are referred to in this Volume as solution-based or slurry-based (as appropriate) absorption processes with dry end-products. In the Volume, the terms 'wet' and 'dry' are used to describe the state of the reagent contacted with the gas; the terms 'semi-dry' and 'wet-dry' are not used.

PROCESS CODE NUMBERS

A number of basic process types occur within each process category. Basic process types are assigned a Code Number comprising the relevant Category Number followed by a Type Number; e.g. the limestone slurry scrubbing processes fall into Category S21 and have been assigned the Code No. S21.1.

The Code Numbers assigned to the basic process types are shown in Table 2.1.

2.2 Outline FGD Process Description

The system used to classify flue gas desulphurisation processes is illustrated in Figure 2.1; the codes of processes dealt with in this Section are listed in Table 2.1.

CATEGORY S11

Process Code S11.1 – Sea Water Scrubbing Process

Outline of Process

A simplified block diagram of the process is presented in Figure 2.2. Gas from the Boiler is cooled in a Heat Exchanger (H.E.) and by injection of sea water in the Cooler, and is then scrubbed in the Absorber with sea water. The cleaned gas is reheated in the H.E. and exhausted to the stack. The water leaving the Absorber contains sulphurous and sulphuric acid, together with hydrogen halides, trace elements and

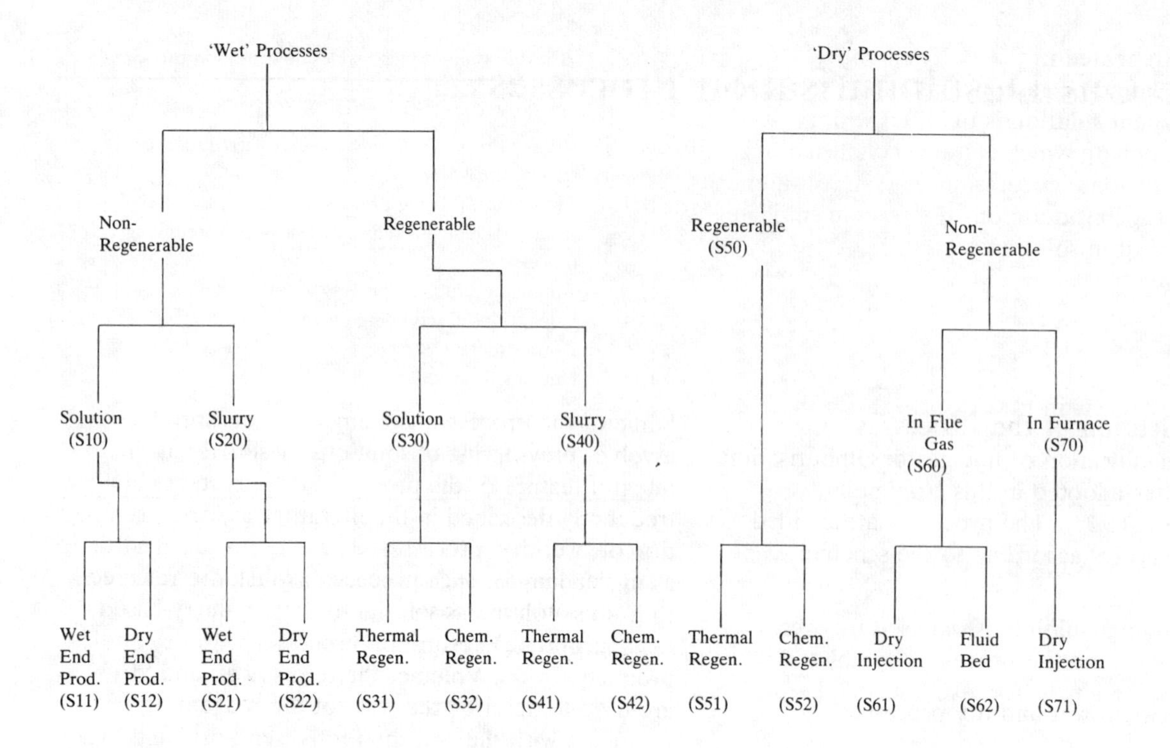

Figure 2.1 Classification of Flue Gas Desulphurisation Processes

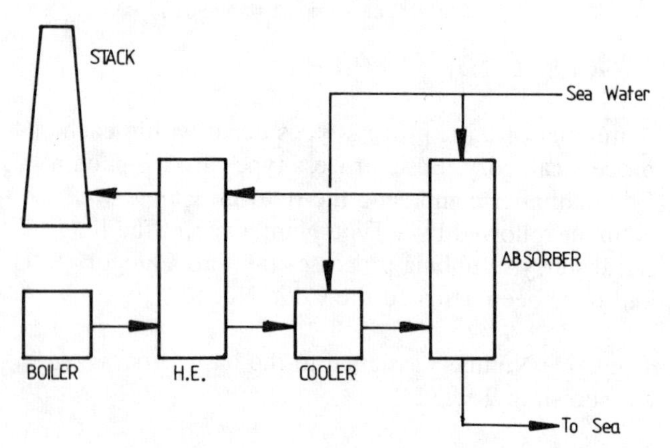

Figure 2.2 Block Diagram of Sea Water Scrubbing FGD Process

particulates, and is discharged to sea. In some circumstances the effluent can be neutralised with e.g. chalk, but it is then usually necessary to ensure that the concentration of the reaction product is below the solubility limit.

Chemistry of Process [122]

Sea water contains carbonate, bicarbonate, borate, phosphate and arsenate ions which exert a buffering effect, maintaining a pH of about 8.3 and assisting the absorption of SO_2. Sulphur oxides in the gas entering the Absorber dissolve in the water to form sulphurous acid, H_2SO_3, and sulphuric acid, H_2SO_4. Sulphurous acid oxidises fairly readily, and some of it therefore reacts with excess oxygen in the gas to form sulphuric acid. These reactions are represented by:

Absorption: $SO_2 + H_2O = H_2SO_3$
$SO_3 + H_2O = H_2SO_4$

Oxidation: $2\ H_2SO_3 + O_2 = 2\ H_2SO_4$

The water also absorbs other acidic gases, e.g. acid halides, and washes out particulates from the flue gas. Because sulphurous, sulphuric and halogen acids are themselves water pollutants, and the discharge may also possibly contain toxic trace elements leached from the particulates, the process can be applied only where the effluent can be discharged harmlessly, e.g. to the sea; this explains the use of sea water as the scrubbing agent. The process has also been applied using the River Thames as a source of scrubber water and for receiving the discharge [156]. It was here that neutralisation was required:

Neutralisation:
$$H_2SO_4 + CaCO_3 = CaSO_4 + CO_2 + H_2O$$

Process Code S11.2 – Alkali scrubbing process

Outline of Process

This basic process includes scrubbing with slurries of alkaline ash, from which alkali compounds are leached and react with sulphur oxides in solution. A simplified block diagram of the process is presented in Figure 2.3. Gas is cooled in a gas–gas Heat Exchanger (H.E.) or by injection of water, and is then scrubbed in the Absorber with a solution of alkali: caustic soda (NaOH); soda ash (impure sodium carbonate, Na_2CO_3); or a slurry of alkaline ash. The purified gas is reheated in the H.E. and/or by combustion of liquid or gaseous fuel, and exhausted to stack. The spent solution is oxidised with air to form sodium sulphate which is then crystallised out from solution; in some circumstances (where water pollution is not a consideration) the sodium sulphate can be disposed of in solution.

Chemistry of Process

Sulphur dioxide reacts in the Absorber with the alkali to form sodium sulphite, Na_2SO_3:

$$2\ NaOH + SO_2 = Na_2SO_3 + H_2O$$

$$Na_2CO_3 + SO_2 = Na_2SO_3 + CO_2$$

Some of the sodium sulphite is oxidised by excess oxygen in the gas, and the process is completed in the Oxidiser, using air, to form sodium sulphate, Na_2SO_4:

$$2\ Na_2SO_3 + O_2 = 2\ Na_2SO_4$$

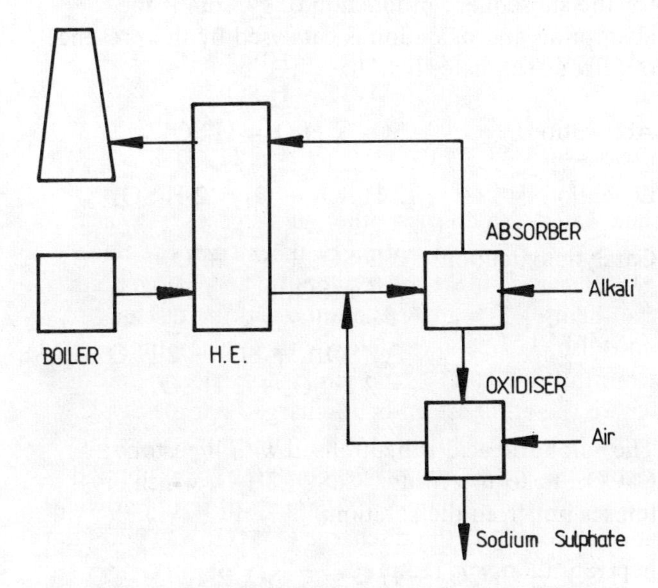

Figure 2.3 Block Diagram of Alkali Scrubbing FGD Process

Process Code S11.3 – Ammonia Scrubbing Process

Outline of Process

A simplified block diagram of the process (which is not to be confused with the regenerable ammonia scrubbing process outlined under Process Code S31.3) is presented in Figure 2.4. Gas is cooled in a gas–gas Heat Exchanger (H.E.) or by injection of water, and is then scrubbed in the Absorber with a solution of ammonium hydroxide (NH_4OH) obtained by injecting ammonia gas into the Absorber. The purified gas is reheated in the H.E. or by combustion of liquid or gaseous fuel, and exhausted to stack. The spent solution containing ammonium sulphite, $(NH_4)_2SO_3$,

Table 2.1 Classification of flue gas desulphurisation processes

	(a) Processes using wet reagents
Category S10: Non-regenerable solution-based wet processes	
Category S11: Wet end-product	
S11.1	Sea water scrubbing process
S11.2	Alkali (including alkaline fly-ash slurry) scrubbing process
S11.3	Ammonia scrubbing process
S11.4	Sulphuric acid scrubbing process
S11.5	Dowa process
S11.6	Dual alkali process
Category S12: Dry end-product	
S12.1	Alkali scrubbing/spray drying process
S12.2	Walther process
Category S20: Non-regenerable slurry-based wet processes	
Category S21: Wet end-product	
S21.1	Limestone or lime (including high-calcium fly-ash) slurry scrubbing process
Category S22: Dry end-product	
S22.1	Lime slurry scrubbing/spray drying process
Category S30: Regenerable solution-based wet reagent processes	
Category S31: Thermal regeneration	
S31.1	Wellman-Lord process
S31.2	Flakt-Boliden process
S31.3	Catalytic Inc./IFP or ammonia scrubbing process
Category S32: Chemical regeneration	
S32.1	Citrate process
S32.2	Conosox process
S32.3	Ispra Mark 13A process
S32.4	Aqueous carbonate process
Category S40: Regenerable slurry-based wet reagent processes	
Category S41: Thermal regeneration	
S41.1	Magnesia slurry scrubbing process
Category S42: Chemical regeneration	
S42.1	Sulf-X process
	(b) Processes using dry reagents
Category S50: Regenerable dry reagent processes	
Category S51: Thermal regeneration	
S51.1	Active carbon adsorption process
Category S52: Chemical regeneration	
S52.1	Wet active carbon adsorption process
S52.2	Copper oxide process
S52.3	Catalytic oxidation process
Category S60: Non-regenerable dry reagent applied to flue gas	
Category S61: Dry injection	
S61.1	Hydrated lime injection process
S61.2	Alkali injection process
Category S62: Dry reactor	
S62.1	Lurgi CFB lime absorber process
Category S70: Non-regenerable dry reagent applied in furnace	
Category S71: Dry injection	
S71.1	Sorbent direct injection process (limestone, lime, sodium bicarbonate or soda ash direct injection to furnace)
Category S72: Fluidised bed combustion (with limestone or dolomite injection; see also Section 6)	
S72.1	Atmospheric pressure, circulating and pressurised fluidised bed combustion

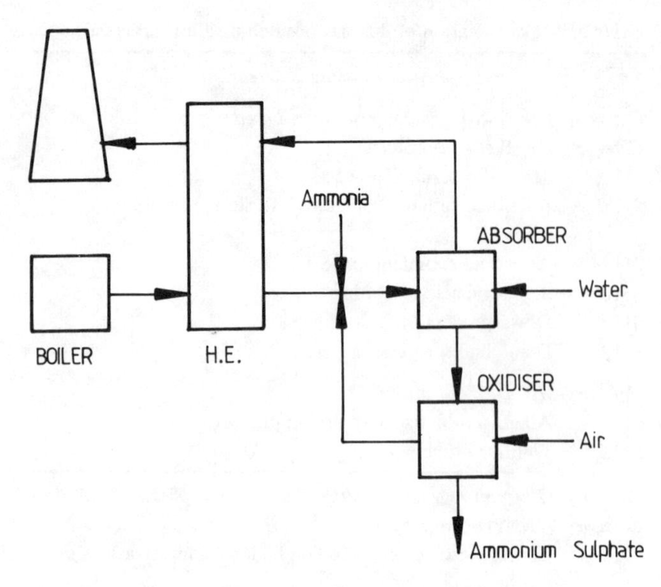

Figure 2.4 Block Diagram of Ammonia Scrubbing FGD Process

is filtered and the sulphite can be recovered for use in the manufacture of caprolactam. However, some ammonium sulphite is oxidised in the flue gas, and the oxidation may be completed in the Oxidiser, using air, to form ammonium sulphate, $(NH_4)_2SO_4$, as a marketable by-product.

Chemistry of Process

In the Absorber, sulphur dioxide reacts with the ammonium hydroxide to form ammonium sulphite:

$$2\ NH_4OH + SO_2 = (NH_4)_2SO_3 + H_2O$$

Oxidation of the sulphate leads to the formation of ammonium sulphate:

$$2\ (NH_4)_2SO_3 + O_2 = 2\ (NH_4)_2SO_4$$

Process Code S11.4 – Sulphuric Acid Scrubbing Process

Outline of Process [172]

A simplified block diagram of the process is presented in Figure 2.5. Gas from the Boiler is passed through a Prescrubber to remove acid halides and particulates, and then into the Absorber where it is scrubbed with a recirculated stream of dilute sulphuric acid, which absorbs the SO_2. The gas then passes via a Reheater to the stack. The acid is oxidised with air in the Oxidiser to ensure conversion of absorbed SO_2 to sulphuric acid. A side stream of the acid is pumped to the Neutraliser, where it is converted to calcium sulphate di-hydrate (gypsum) which is crystallised out for sale, for the manufacture of plaster board, or safe disposal. The liquor from the gypsum separation process is pumped to a waste water plant.

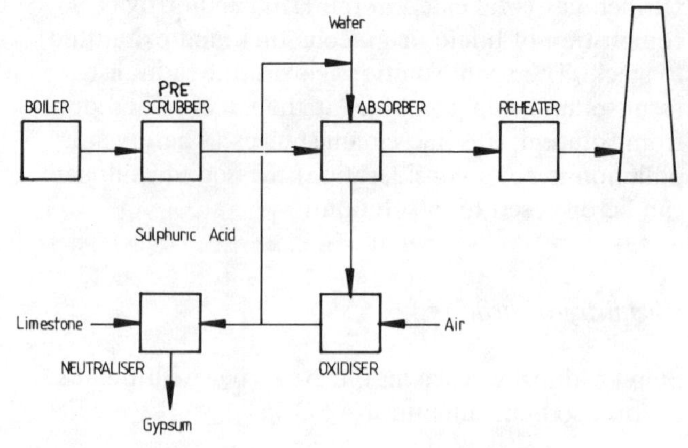

Figure 2.5 Block Diagram of Sulphuric Acid Scrubbing FGD Process

Chemistry of Process [172]

The absorption and oxidation of SO_2 to give sulphuric acid is favoured by the use of the minimum practicable sulphuric acid concentration, but this concentration can be increased to levels making it easier for the subsequent production of gypsum if the absorption and oxidation is catalysed by the presence of ferrous sulphate, $FeSO_4$:

Absorption: $SO_2 + H_2O = H_2SO_3$

Oxidation: $2\ H_2SO_3 + O_2 = 2\ H_2SO_4$

Catalytic oxidation: $2\ FeSO_4 + SO_2 + O_2 = Fe_2(SO_4)_3$

$Fe_2(SO_4)_3 + SO_2 + 2\ H_2O = 2\ FeSO_4 + 2\ H_2SO_4$

The sulphuric acid is neutralised with limestone, $CaCO_3$, to form gypsum, $CaSO_4.2H_2O$, which crystallises out from the solution:

$$H_2SO_4 + CaCO_3 + H_2O = CaSO_4.2H_2O + CO_2$$

Process Code S11.5 – Dowa Process

Outline of Process [95]

A simplified block diagram of the process is presented in Figure 2.6. Gas leaving the Boiler passes to an Absorber, where it is scrubbed with a solution of basic aluminium sulphate, which absorbs SO_2. The gas then flows via a Reheater to the stack. The solution from the Absorber is treated with air in the oxidiser, where aluminium sulphite formed in the Absorber is oxidised to the sulphate. The solution then passes to the Neutraliser, where it is treated with limestone, which precipitates gypsum, regenerating the basic aluminium sulphate. The gypsum is separated from the solution in a Thickener and Filter. The concentration of unwanted soluble compounds (e.g. chlorides and magnesium salts) is controlled by treating a side stream of the solution with an excess of limestone in the Alumina Recovery tank. This precipitates aluminium hydroxide; the precipitate is concentrated in a Thickener and returned to the system via a Redissolver tank where it is mixed with water. The gypsum joins that formed in the Neutraliser and is subsequently removed.

Although the actual sorbent, basic aluminium sulphate, is regenerated in the Neutraliser, the Dowa process is not classed as a regenerable process because the captured sulphur is transferred to another reagent.

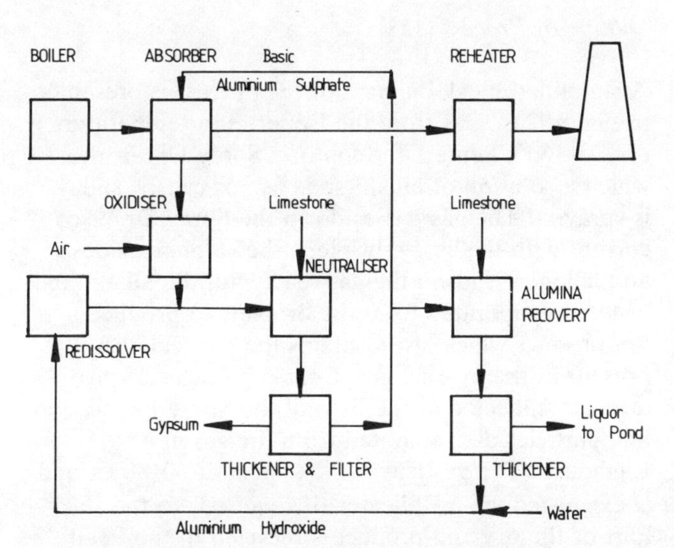

Figure 2.6 Block Diagram of Dowa FGD Process

Chemistry of Process [95]

In the Absorber, SO_2 reacts with basic aluminium sulphate solution, $Al_2(SO_4)_3.Al_2O_3$, to form the sulphate-sulphite:

$$Al_2(SO_4)_3.Al_2O_3 + 3\ SO_2 = Al_2(SO_4)_3.Al_2(SO_3)_3$$

The reaction occurs at a pH of 3.0–3.5. In the Oxidiser, the sulphite is oxidised by air to the sulphate. Reaction is rapid, as the SO_2 is present in the solution in the ionic state:

$$2\ Al_2(SO_4)_3.Al_2(SO_3)_3 + 3\ O_2 = 4\ Al_2(SO_4)_3$$

Treatment with limestone in the Neutraliser then regenerates the basic aluminium sulphate, precipitates calcium sulphate dihydrate (gypsum, $CaSO_4.2\ H_2O$) and evolves carbon dioxide:

$$2\ Al_2(SO_4)_3 + 3\ CaCO_3 + 6\ H_2O = Al_2(SO_3)_3.Al_2O_3 + 3\ CO_2 + 3\ CaSO_4.2\ H_2O$$

If the treatment is carried out with a large excess of limestone, the reaction products are aluminium hydroxide, $Al_2(OH)_3$, which is precipitated at high pH, and gypsum:

$$Al_2(SO_4)_3 + 3\ CaCO_3 + 9\ H_2O = 2\ Al(OH)_3 + 3\ CaSO_4.2\ H_2O + 3\ CO_2$$

Process Code S11.6 – Dual (or Double) Alkali Process

Outline of Process [95]

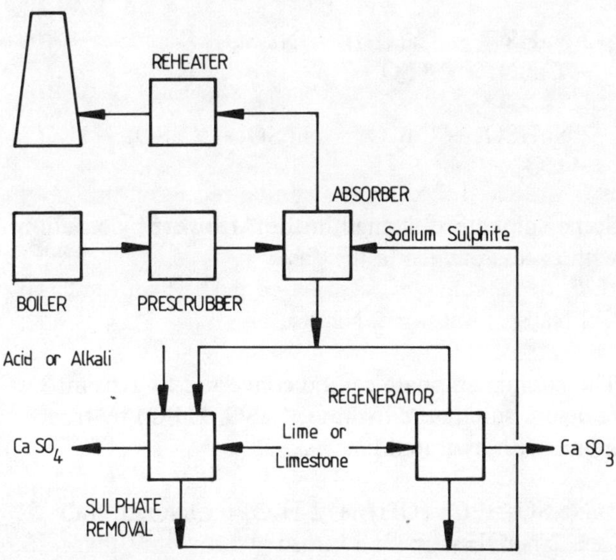

Figure 2.7 Block Diagram of Dual Alkali FGD Process

A simplified block diagram of the process is presented in Figure 2.7. The gas passes through a Prescrubber to remove particulates and acid halides, and is then treated in the Absorber with sodium sulphite solution to absorb sulphur dioxide. After reheating, the gas is exhausted to stack. The alkali solution contains sodium bisulphite and some sulphate. It is passed to the Regenerator, where it is regenerated with hydrated lime or limestone to produce solid calcium sulphite. This is separated from the regenerated sodium sulphite which is returned to the Absorber. A slip-stream of the unregenerated alkali solution is treated with either sulphuric acid (to reduce the pH) or alkali (to increase the pH) before treatment with lime; this is to allow calcium sulphate to be precipitated separately from sulphite without having to operate at very low sulphite concentrations. The calcium sulphite and sulphate are disposed of.

Although the actual sorbent, calcium sulphite, is regenerated in the Regenerator, the Dual Alkali process is not classed as a regenerable process because the captured sulphur is transferred to another reagent.

Chemistry of Process [122]

In the Absorber, sodium sulphite, Na_2SO_3, reacts with SO_2 to form sodium bisulphite, $NaHSO_3$:

$$Na_2SO_3 + SO_2 + H_2O = 2\ NaHSO_3$$

The ingoing solution has a pH of about 9, but the bisulphite solution leaving the Absorber has a pH of only 6.

The sulphite is regenerated in the Regenerator with hydrated lime, $Ca(OH)_2$, or limestone, $CaCO_3$; the insoluble calcium sulphite, $CaSO_3$, is separated from the solution, which is returned to the Absorber:

$$2\ NaHSO_3 + Ca(OH)_2 = Na_2SO_3 + CaSO_3 + 2\ H_2O$$

$$2\ NaHSO_3 + CaCO_3 = Na_2SO_3 + CaSO_3 + H_2O + CO_2$$

Some sulphate is formed in the Absorber by reaction with excess oxygen in the gas:

$$2\ Na_2SO_3 + O_2 = 2\ Na_2SO_4$$

The sodium sulphate can be converted to gypsum (calcium sulphate dihydrate, $CaSO_4.2H_2O$) by treatment, with hydrated lime:

$$Na_2SO_4 + Ca\ (OH)_2 + 2\ H_2O = CaSO_4.2\ H_2O + 2\ NaOH$$

but co-precipitation of the gypsum with the calcium sulphite formed in the normal regeneration reaction occurs only when the sulphite concentration is very low. One solution is to add sulphuric acid:

$$Na_2SO_4 + 2\ CaSO_3 + H_2SO_4 = 2\ CaSO_4 + 2\ NaHSO_3$$

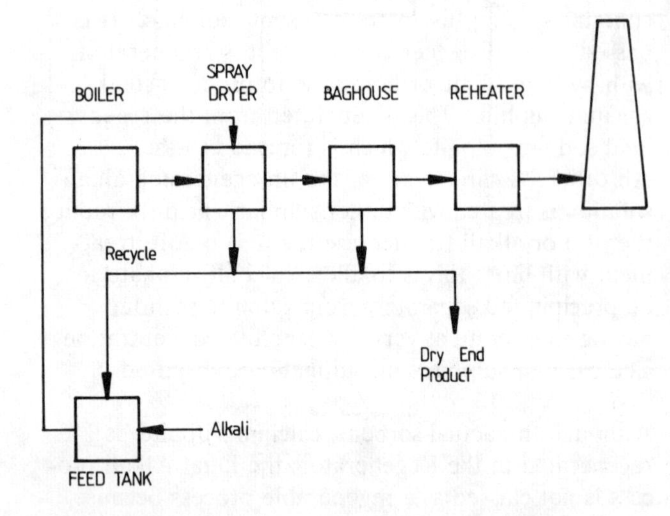

Figure 2.8 Block Diagram of Alkali Scrubbing/Spray Drying FGD Process

but more sulphuric acid is needed to reduce the pH to the value of 2–3 needed for the precipitation. Another solution is to increase the pH to 13–14 by adding alkali, whereupon calcium sulphate is precipitated by lime, and the sodium sulphite is regenerated:

$$Na_2SO_4 + Ca(OH)_2 = CaSO_4 + 2\ NaOH$$

$$NaOH + NaHSO_3 = Na_2SO_3 + H_2O$$

CATEGORY S12

Process Code S12.1 – Alkali Scrubbing/Spray Drying Process

Outline of Process [122]

A simplified block diagram of the process is presented in Figure 2.8. Gas from the Boiler, at a temperature of 120–160°C, enters the top of a Spray Dryer into which a solution of alkali (soda ash or caustic soda) is sprayed. The gas passes down the Spray Dryer co-current with the spray droplets; the sulphur oxides and halogen acids in the gas react with the alkali, and water is evaporated from the droplets to produce a dry product which also contains the particulates present in the ingoing gas. Coarse particles of product are collected at the base of the Spray Dryer, and fine particles are separated from the gas in a Baghouse. The gas leaves the system at 60–80°C, and is exhausted via a Reheater (if required) to the stack. Part of the dry end-product is recycled to the Feed Tank to increase the conversion of alkali. An elaboration of the process regenerates the alkali (see under Process Code S32.4).

The dry end-product contains sodium sulphite, sulphate and halides, together with some sodium carbonate. It is usual to recycle part of the dry end-product to the Feed Tank to increase the alkali utilisation. The acidic components of the gas are absorbed with very high efficiency, and reaction is faster than with hydrated lime (see Process Code S21.1) so that lower liquid/gas flow ratios can be used.

Chemistry of Process [122]

The sulphur oxides, SO_2 and SO_3, and acid halides, HF, HCl and HBr (represented below as 'HHa') react with the alkali solution as follows:

With caustic soda (NaOH)

$2\ NaOH + SO_2 = Na_2SO_3 + H_2O$ (forming sodium sulphite)

$2\ NaOH + SO_3 = Na_2SO_4 + H_2O$ (forming sodium sulphate)

$NaOH + HHa = NaHa + H_2O$ (forming sodium halides)

With soda ash (sodium carbonate, Na_2CO_3) the reactions are similar, but with liberation of CO_2 in place of H_2O:

$Na_2CO_3 + SO_2 = Na_2SO_3 + CO_2$ (forming sodium sulphite)

$Na_2CO_3 + SO_3 = Na_2SO_4 + CO_2$ (forming sodium sulphate)

$Na_2CO_3 + 2\ HHa = 2\ NaHa + CO_2 + H_2O$ (forming sodium halides)

Reaction is rapid whilst the droplets are wet, and the gas temperature in the Spray Dryer is maintained slightly above the saturation temperature (e.g. within 20°C) to delay the drying out of the droplets sufficiently for efficient absorption.

Process Code S12.2 – Walther Process

Outline of Process [Buckau-Walther Group promotional literature]

A simplified block diagram of the process, which is not to be confused with the ammonia scrubbing process (see Process Code S11.3) is presented in Figure 2.9. Part of the gas from the Boiler Economiser flows via a Hot Electrostatic Precipitator (Hot ESP) into a Spray Dryer where it is contacted with ammonia gas and a solution of ammonium sulphate pumped from the Oxidiser; both the gas and the solution enter at the base of the Spray Dryer. The ammonia absorbs some SO_2 and SO_3 and as the solution dries out it produces solid ammonium sulphate which is captured in an Electrostatic Precipitator (ESP) or Baghouse as the potentially marketable fertiliser end-product. The material collected in the ESP will also contain particulates and acid halides captured from the gas. The gas is then mixed with the remaining cooler, unpurified gas leaving the Economiser via the boiler, Air Heater and ESP. The reunited gas stream is cooled in a Heat Exchanger (H.E.) then flows to two spray towers–a Scrubber and Rescrubber–denoted S1 and S2 in the block diagram. The gas passes up the spray towers counter-current to a flow of water which removes unreacted ammonia from the gas, and absorbs more SO_2. The purified gas is reheated in the H.E. and exhausted to stack. The solution leaving the Scrubber is oxidised with air in an Oxidiser and is then passed to the Spray Dryer.

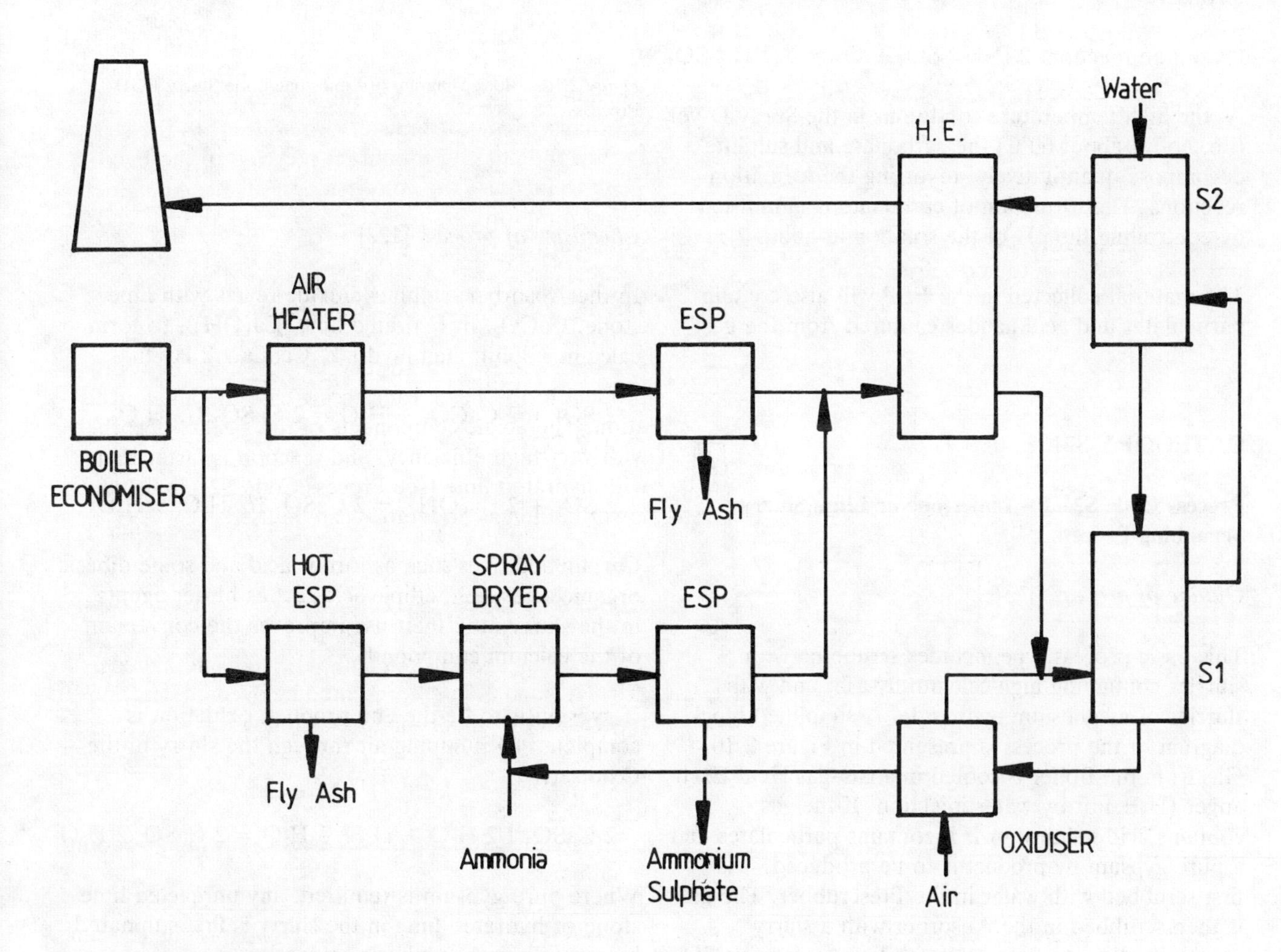

Figure 2.9 Block Diagram of Walther FGD Process

Chemistry of Process [Buckau-Walther Group promotional literature]

The ammonia solution absorbs SO_2, SO_3 and some CO_2 from the gas in the Scrubber (S1), Rescrubber (S2) and Spray Dryer to form ammonium sulphite, sulphate and carbonate:

Sulphite formation:

$$2\ NH_4OH + SO_2 = (NH_4)_2SO_3 + H_2O$$

Sulphate formation:

$$2\ NH_4OH + SO_3 = (NH_4)_2SO_4 + H_2O$$

Carbonate formation:

$$2\ NH_4OH + CO_2 = (NH_4)_2CO_3 + H_2O$$

In the Scrubber (S1) and Rescrubber (S2), much of the SO_2 absorption results from reaction with the ammonium carbonate solution:

Carbonate neutralisation:

$$(NH_4)_2CO_3 + SO_2 = (NH_4)_2SO_3 + CO_2$$

Sulphite is oxidised to sulphate by reaction with atmospheric oxygen in the Oxidiser, and the sulphate, dried in the Spray Dryer and removed as a solid in the ESP or Baghouse, is potentially marketable as a fertiliser:

Oxidation reaction: $2\ (NH_4)_2SO_3 + O_2 = 2\ (NH_4)_2SO_4$

At the high temperature conditions in the Spray Dryer (i.e. above about 60°C) the carbonate and sulphite decompose quantitatively, reversing the formation reactions. The formation of carbonate is minimised by controlling the pH of the solution at about 7.

The material collected in the ESP will also contain particulates and acid halides captured from the gas.

CATEGORY S21

Process Code S21.1 – Limestone or Lime Slurry Scrubbing Process

Outline of process [173]

This basic process type includes scrubbing with slurries containing high-calcium fly-ash, and with slurries of magnesium hydroxide. A simplified block diagram of the process is presented in Figure 2.10. Gas from the Boiler is cooled in a gas–gas Heat Exchanger (H.E.) or by water injection. If the gas contains acid halides, or if it contains particulates and a pure gypsum by-product is to be produced, it is first scrubbed with water in the Prescrubber. The gas is then scrubbed in the Absorber with a slurry containing limestone or hydrated lime to remove SO_2. The purified gas is reheated in the H.E. and/or by combustion of oil or gaseous fuel, and exhausted to stack. The spent slurry, containing calcium sulphite hemihydrate, is difficult to deal with as it is not easily thickened or dewatered. It is either pumped to settling lagoons, or oxidised with air in the Oxidiser to form gypsum which, after dewatering, is disposed of or sold as a by-product for the manufacture of plasterboard.

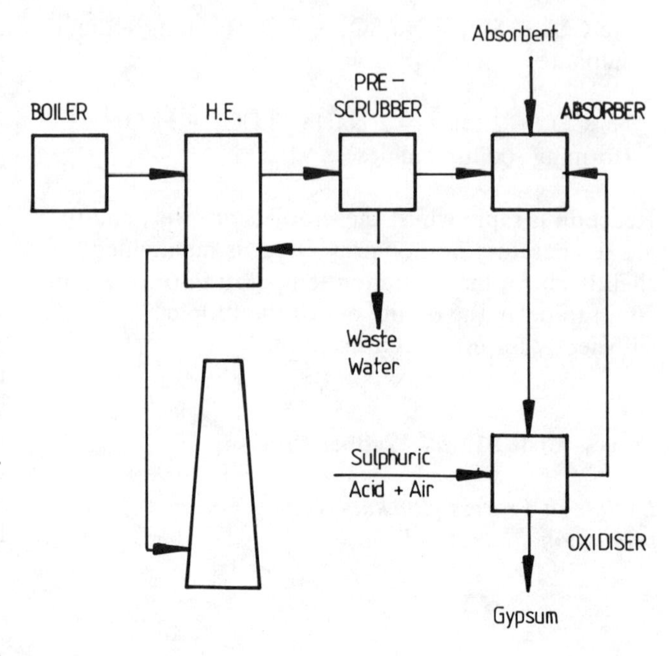

Figure 2.10 Block Diagram of Limestone/Lime Slurry FGD Process

Chemistry of process [122]

In the Absorber, sulphur dioxide reacts with limestone, $CaCO_3$, or hydrated lime, $Ca(OH)_2$, to form calcium sulphite hemihydrate, $CaSO_3.1/2\ H_2O$:

$$2\ SO_2 + 2\ CaCO_3 + H_2O = 2\ CaSO_3.1/2\ H_2O + 2\ CO_2$$

$$2\ SO_2 + 2\ Ca(OH)_2 = 2\ CaSO_3.1/2\ H_2O + H_2O$$

Certain additives such as formic acid and some dibasic organic acids, e.g. adipic acid, act as buffer agents in the slurry, and their use improves the conversion of the calcium compounds.

If gypsum is to be the end product, oxidation is completed by pumping air through the slurry in the Oxidiser:

$$2\ CaSO_3.1/2\ H_2O + O_2 + 3\ H_2O = 2\ CaSO_4.2\ H_2O$$

Where pure gypsum is required, any unreacted limestone or hydrated lime in the slurry is first sulphated by reaction with sulphuric acid:

$$CaCO_3 + H_2O + H_2SO_4 = CaSO_4.2\ H_2O + CO_2$$

$$Ca(OH)_2 + H_2SO_4 = CaSO_4.2\ H_2O$$

When magnesium hydroxide is used as the absorbent, the absorption, oxidation and neutralisation reactions are similar to those described above for calcium hydroxide, and the end-products are magnesium sulphite or magnesium sulphate.

CATEGORY S22

Process Code S22.1 – Lime Slurry Scrubbing/Spray Drying Process

Outline of Process [203]

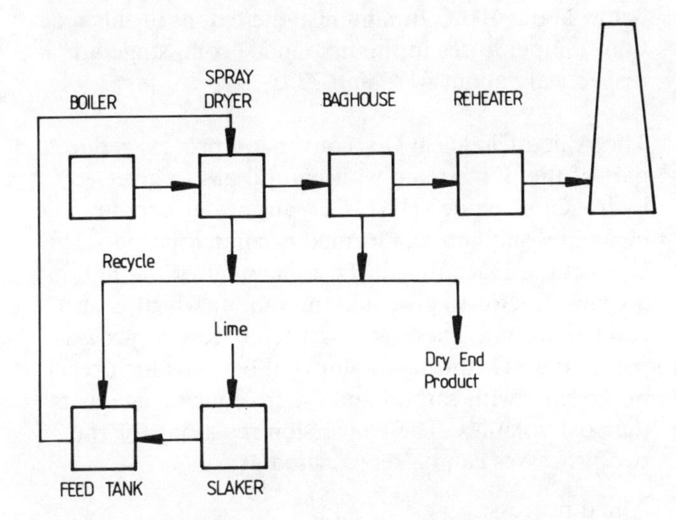

Figure 2.11 Block Diagram of Lime Slurry Scrubbing/Spray Drying FGD Process

A simplified block diagram of the process is presented in Figure 2.11. Gas from the Boiler, at a temperature of 120–160°C, enters the top of a Spray Dryer into which a slurry of lime is sprayed. The lime, which is hydrated and slurried in the Slaker, is pumped via a Feed Tank to the slurry atomiser nozzles. The gas passes down the Spray Dryer co-current with the slurry spray droplets; the sulphur oxides and halogen acids in the gas react with the lime, and water is evaporated from the slurry droplets to produce a dry product which also contains the particulates present in the ingoing gas. Coarse particles of product are collected at the base of the Spray Dryer, and fine particles are separated from the gas in a Baghouse or Electrostatic Precipitator. The gas leaves the system at 60–80°C, and is exhausted via a Reheater (if required) to the stack.

The dry end product contains calcium sulphite, sulphate and halides, together with some calcium carbonate, unreacted lime and coal ash. It is usual to recycle part of the dry end product to the slurry preparation stage to increase the lime utilisation. The acid halides and SO_3 are absorbed with very high efficiency (e.g. 99% removal).

The percentage capture of SO_2 depends on spray droplet size, particle size of the hydrated lime, and the Ca/S molar ratio used; typically, captures of 75% are obtained at a Ca/S = 1, and 95% at Ca/S = 2.

Chemistry of Process [122]

The sulphur oxides, SO_2 and SO_3, and acid halides, HF, HCl and HBr (represented below as 'HHa') react in the Spray Dryer with the lime in the slurry as follows:

$Ca(OH)_2 + SO_2 = CaSO_3 + H_2O$ (forming calcium sulphite)

$Ca(OH)_2 + SO_3 = CaSO_4 + H_2O$ (forming calcium sulphate)

$Ca(OH)_2 + 2HHa = CaHa_2 + 2H_2O$ (forming calcium halides)

Reaction occurs in two phases: the first while the droplets are still wet; and the second after dryout, when porous particles are formed. Reaction is rapid during the wet first phase, the rate being controlled by the rate of dissolution of lime through a calcium sulphite layer. For this reason, the gas temperature in the Spray Dryer is maintained slightly above the saturation temperature (e.g. within 20°C) to delay the drying out of the droplets sufficiently for efficient absorption. In the dry second phase, reaction is controlled by pore diffusion in the particle. The gas residence time in the Spray Dryer is around 10 seconds, with most of the reaction occurring within about 2 seconds.

CATEGORY S31

Process Code S31.1 – Wellman-Lord Process

Outline of Process [265]

A simplified block diagram of the process is presented in Figure 2.12. Gas flows, via a Prescrubber to remove particulates and acid halides, to the Absorber where it is scrubbed by a concentrated solution of sodium sulphite to form the bisulphite. The purified gas passes via entrainment separators to a steam-heated Reheater before being exhausted to stack. The sodium bisulphite solution is pumped to the Regenerator. Here it is converted to sodium sulphite, with evolution of sulphur dioxide, by thermal decomposition, using steam in a double-effect evaporator/crystalliser. Some sulphate formation occurs in the

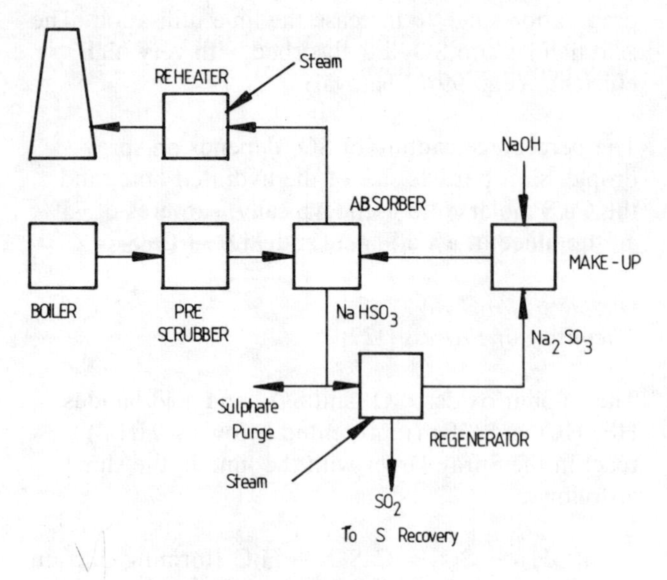

Figure 2.12 Block Diagram of Wellman-Lord FGD Process

Absorber, and a side stream is treated by fractional crystallisation at 0°C to remove the sulphate and thiosulphate; the loss of sodium is made up by feeding fresh alkali. Caustic soda has to be used as the make-up alkali in the UK, as commercial soda ash cannot be used because of its chloride content.

The concentrated SO_2 stream leaving the Regenerator is cooled to condense water vapour, and is then compressed for further processing. It can be catalytically oxidised for the manufacture of sulphuric acid, but it is more usual to reduce it to elemental sulphur which is potentially marketable for a number of uses, but which can also be disposed of safely.

Chemistry of the Process [122, 137, 146]

Sodium sulphite solution reacts with sulphur dioxide in the Absorber to form the bisulphite:

$$Na_2SO_3 + SO_2 + H_2O = 2\ NaHSO_3$$

In the Absorber there is no phase separation, and the SO_2 partial pressure is first order with respect to bisulphite ion concentration, $[HSO_3^-]$. Sulphur capture efficiencies of about 95% can be achieved. The reaction in the Regenerator is the reverse of that in the Absorber, but with the difference that sulphite crystallises out. The SO_2 partial pressure is then proportional to the product of the square of the sodium and bisulphite ion concentrations: $[Na^+]^2.[HSO_3^-]^2$.

Evaporation produces a small increase in concentration and therefore a large increase in SO_2 partial pressure, resulting in efficient stripping of SO_2 from the solution and conversion of the solution to sulphite.

Some sodium sulphate is formed in the Absorber by reaction of the sulphite with excess oxygen in the gas:

$$2\ Na_2SO_3 + O_2 = 2\ Na_2SO_4$$

Sodium sulphate is also formed, together with sodium thiosulphate, $Na_2S_2O_3$ by disproportionation in the Regenerator:

$$6\ NaHSO_3 = 2\ Na_2SO_4 + Na_2S_2O_3 + 2\ SO_2 + 3\ H_2O$$

The evaporator/crystalliser temperature has to be kept below about 100°C to minimise the extent of this reaction; temperatures in the first and second stage are respectively about 94°C and 77°C.

The Allied Chemical Co. conversion process reduces part of the SO_2 stream with natural gas to give hydrogen sulphide (H_2S), CO_2 and water vapour. The elemental sulphur also formed is condensed out. The H_2S is then reacted with the remainder of the SO_2 in a Claus reactor to give sulphur vapour which is also condensed out. The Foster Wheeler 'Resox' process passes the SO_2 through a hot coal bed in which reaction occurs with carbon and other reducing agents in the coal volatiles. The conversion reactions for the two processes can be represented as:

Allied process:

$$3\ SO_2 + 2\ CH_4 = 2\ H_2S + 2\ CO_2 + 2\ H_2O + S$$

$$SO_2 + 2\ H_2S = 3\ S + 2\ H_2O$$

Resox process:

$$SO_2 + C = S + CO_2$$

Process Code S31.2 – Fläkt-Boliden Process

Outline of Process [137]

This process is one form of the Citrate process; the other form is described in Process Code S32.1. A simplified block diagram of the process is presented in Figure 2.13. The gas is first passed through a Prescrubber to remove particulates and acid halides, and is then passed into an Absorber where it is treated with sodium citrate solution. The solution, containing bisulphite ion, is pumped via a heat exchanger to the Regenerator, where SO_2 is removed by steam stripping. The regenerated solution is returned to the Absorber.

A side stream of the solution from the Absorber is treated by fractional crystallisation to remove the sulphate, and the loss of sodium is made up by feeding fresh alkali. Caustic soda has to be used as the make-up alkali in the UK, as commercial soda ash cannot be used because of its chloride content.

The concentrated SO_2 stream leaving the Regenerator is cooled to condense water vapour, and is then compressed for further processing. It can be catalytically oxidised for the manufacture of sulphuric acid, but it is more usual to reduce it in a sulphur recovery unit to elemental sulphur which is potentially marketable for a number of uses, but which can also be disposed of safely.

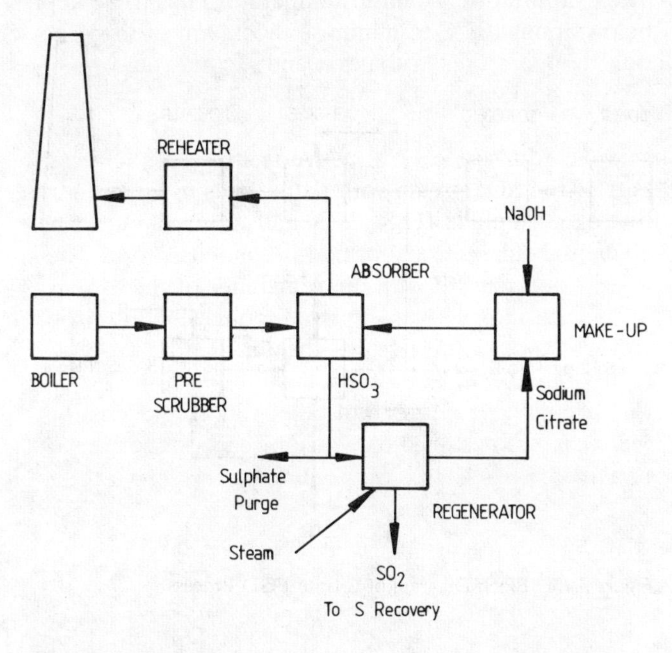

Figure 2.13 Block Diagram of Fläkt-Boliden FGD Process

Chemistry of the Process [122]

Sulphur dioxide dissolves in water, and the solution ionises to give bisulphite and hydrogen ions:

$$SO_2 + H_2O = HSO_3^- + H^+$$

The sodium citrate removes the hydrogen ion, driving the ionisation reaction towards the right. It therefore acts as a buffering agent enhancing the solubility of SO_2 in the water and maintaining the pH of the solution within the range 3.5 to 5. The reaction is reversed in the Regenerator, giving a concentrated SO_2 stream for further processing.

Some sodium sulphate is formed in the Absorber by reaction of the bisulphite with excess oxygen in the gas.

The SO_2 conversion processes are similar to those described in S31.1 above.

Process Code S31.3 – Catalytic Inc./Institut Francais du Petrole (IFP) or Ammonia Scrubbing Process

Outline of Process [137]

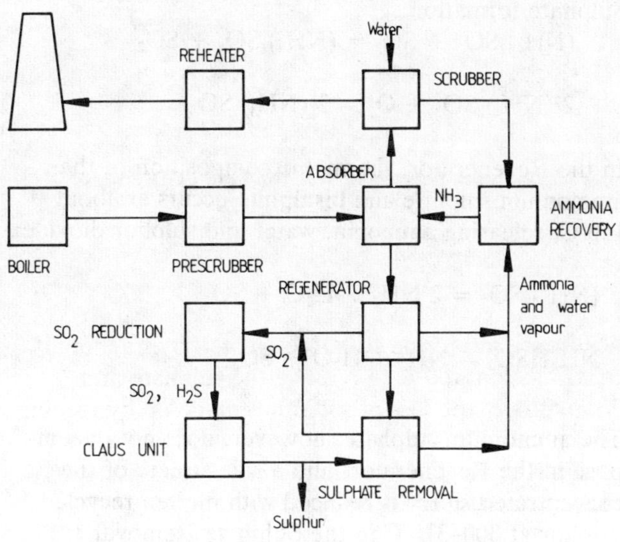

Figure 2.14 Block Diagram of IFP/Catalytic FGD Process

A simplified block diagram of the process (which should not be confused with the throwaway ammonia process outlined in Process Code S11.3) is presented in Figure 2.14. After passing through a Prescrubber to remove particulates, acid halides and some SO_3, the gas from the Boiler is scrubbed with ammonium hydroxide solution in a multi-stage Absorber, and then with water in a separate Scrubber. The cleaned gas passes through the Reheater before it is exhausted to stack. The spent liquor, containing ammonium sulphite and bisulphite, is evaporated in the Regenerator to about 40% of its bulk at 150°C and is then thermally decomposed to produce SO_2, ammonia and water. Some ammonium sulphate is formed in the Absorber by absorption of SO_3 from the gas, and by oxidation of the ammonium sulphite with excess oxygen in the gas. To deal with this, a side stream of the concentrated slurry from the Regenerator is reduced with recycled molten sulphur in the Sulphate Removal system. This also produces SO_2, ammonia and water. The ammonia and water are returned to the Absorber, and the SO_2 is partially reduced with a reducing gas containing hydrogen and converted to elemental sulphur in a liquid-phase Claus Unit.

Chemistry of Process [137]

The absorption reactions occurring in the Absorber, which can remove more than 90% of the sulphur oxides in the gas, are:

Sulphite formation:
$$2\ NH_4OH + SO_2 = (NH_4)_2SO_3 + H_2O$$

Bisulphite formation:
$$(NH_4)_2SO_3 + SO_2 + H_2O = 2\ NH_4HSO_3$$

Sulphate formation:
$$(NH_4)_2SO_3 + SO_3 = (NH_4)_2SO_4 + SO_2$$

$$2\ (NH_4)_2SO_3 + O_2 = 2\ (NH_4)_2SO_4$$

In the Regenerator, thermal decomposition of the ammonium sulphite and bisulphite occurs at about 150°C, releasing ammonia, water and sulphur dioxide:

$$(NH_4)_2SO_3 = 2\ NH_3 + H_2O + SO_2$$

$$NH_4HSO_3 = NH_3 + H_2O + SO_2$$

The ammonium sulphate, however, does not decompose in the Regenerator, and a side stream of the concentrated slurry is reduced with molten recycled sulphur at 300–370°C in the Sulphate Removal system:

$$(NH_4)_2SO_4 = NH_4HSO_4 + NH_3$$

$$2\ NH_4HSO_4 + S = 3\ SO_2 + 2\ NH_3 + 2\ H_2O$$

$$(NH_4)_2SO_4 = SO_3 + 2\ NH_3 + H_2O$$

The ammonia and water from the Regenerator and Sulphate Removal system are returned, via the Ammonia Recovery system, to the Absorber, and the sulphur oxides are partially reduced in the SO_2 Reduction system with a reducing gas containing hydrogen to give a gas containing two volumes of H_2S per volume of unreduced SO_2. This reacts in a liquid phase Claus Unit to give elemental sulphur of more than 99% purity:

Reduction: $SO_2 + 3\ H_2 = H_2S + 2\ H_2O$

Sulphur Recovery: $2\ H_2S + SO_2 = 3\ S + 2\ H_2O$

CATEGORY S32

Process Code S32.1 – Citrate Process

Outline of Process [6, 122, 146]

Another form of this process, using thermal regeneration, is described in Process Code S31.2. A simplified block diagram of the process is presented in Figure 2.15. The gas is first passed through a Prescrubber to remove particulates and acid halides, and is then passed into an Absorber where it is treated with sodium citrate solution. The solution, containing bisulphite ions, is pumped to the Regenerator, where SO_2 is reacted with hydrogen sulphide (H_2S) in a liquid-phase Claus reaction yielding elemental sulphur. The regenerated solution is returned to the Absorber.

A side stream of the solution from the Absorber is treated by fractional crystallisation to remove the sulphate, and the loss of sodium is made up by feeding fresh alkali. Caustic soda has to be used as the make-up alkali in the UK, as commercial soda ash cannot be used because of its chloride content.

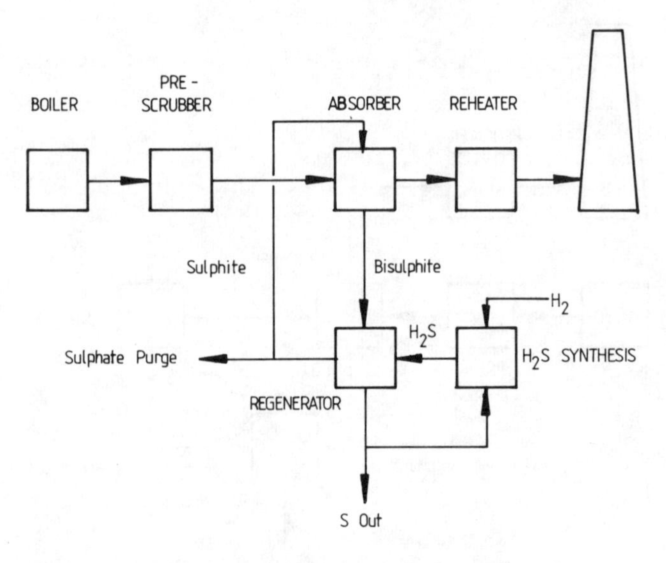

Figure 2.15 Block Diagram of Citrate FGD Process

Chemistry of the Process [122]

Sulphur dioxide dissolves in water, and the solution ionises to give bisulphite and hydrogen ions:

$$SO_2 + H_2O = HSO_3^- + H^+$$

The sodium citrate removes the hydrogen ion, driving the ionisation reaction towards the right. It therefore acts as a buffering agent for enhancing the solubility of SO_2 in the water, maintaining the pH of the solution within the range 3.5 to 5. Sulphur capture efficiencies of 90% can be attained.

In the Regenerator, H_2S reacts with bisulphite and hydrogen ions to give elemental sulphur in a two-step process: a fast reaction giving thiosulphate ions; and a slow rate-controlling reaction giving sulphur:

Fast reaction:
$$2\ H_2S + 4\ HSO_3^- = 3\ S_2O_3^{--} + 2\ H^+ + 3\ H_2O$$

Slow reaction:

$$6\ H_2S + 3\ S_2O_3^{--} + 6\ H^+ = 3/2\ S_8 + 9\ H_2O$$

Overall reaction: $2\ H_2S + HSO_3^- + H^+ = 3/8\ S_8 + 3\ H_2O$

The sulphur recovered has a purity of 99%. The H_2S is prepared by hydrogenation of part of the sulphur produced, using hydrogen derived from reforming natural gas.

Some sodium sulphate is formed in the Absorber by reaction of the bisulphite with excess oxygen in the gas.

Process Code S32.2 – Conosox Process

Outline of Process [95]

A simplified block diagram of the process is presented in Figure 2.16. Gas from the Boiler is treated in the Absorber with a solution of potassium carbonate containing some potassium hydrosulphide, and then passes via the Reheater to stack. The solution absorbs SO_2 in the Absorber, and is then pumped to the Reducer where it is regenerated at about 230°C and 45 bar pressure. The reducing agent is carbon monoxide generated in the Partial Oxidiser from the reaction of oil with oxygen. The spent reducing gas from the Reducer, containing H_2S, passes to the Sulphur Recovery system, incorporating an H_2S absorber and stripper, together with a Claus unit to treat the concentrated H_2S stream. The regenerated absorbent solution from the Reducer contains some potassium sulphate which is separated from the solution in the Crystalliser before the solution is returned to the Absorber.

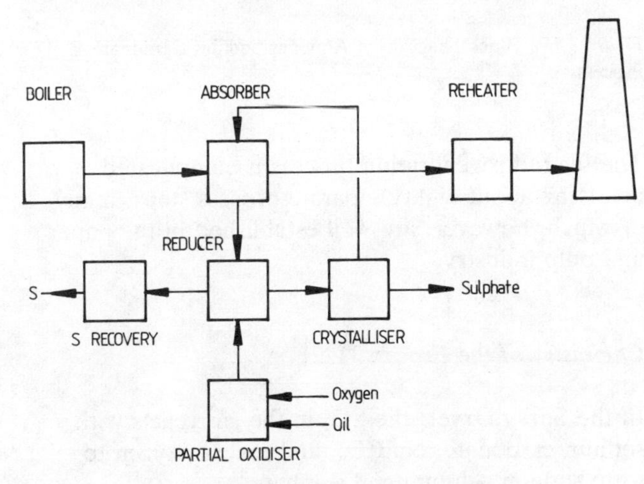

Figure 2.16 Block Diagram of Conosox FGD Process

Chemistry of Process [95]

The reactions occurring in the Absorber between SO_2, potassium carbonate, K_2CO_3, and potassium bisulphide, KHS, form the sulphite, bisulphite and thiosulphate:

Sulphite formation:

$$K_2CO_3 + SO_2 = K_2SO_3 + CO_2$$

Bisulphite formation:

$$K_2SO_3 + SO_2 + H_2O = 2\ KHSO_3$$

Thiosulphate formation:

$$2\ KHS + 4\ KHSO_3 = 3\ K_2S_2O_3 + 3\ H_2O$$

The thiosulphate is highly soluble, allowing the solution to contain absorbed sulphur in a highly concentrated form without risk of scale formation. The thiosulphate also reduces the equilibrium partial pressure of SO_2 in the gas, and inhibits the formation of potassium sulphate. Nevertheless, some sulphate is formed (equivalent to about 1% of the total sulphur captured) by oxidation of the sulphite, and this has to be separated from the regenerated solution in the Crystalliser:

$$2\ K_2SO_3 + O_2 = 2\ K_2SO_4$$

In the Reducer the carbonate and bisulphide are regenerated by reduction of thiosulphate with carbon monoxide at a temperature of about 230°C and pressure of 45 bar:

$$3\ K_2S_2O_3 + 12\ CO + 5\ H_2O = 2\ K_2CO_3 + 2\ KHS + 4\ H_2S + 10\ CO_2$$

The CO is generated by partial oxidation of fuel oil in the Partial Oxidiser, where the reactions include the water gas shift reaction:

$$CO_2 + H_2 = CO + H_2O$$

The gas leaving the Reducer, containing H_2S and CO_2 passes to an absorption tower forming part of the Sulphur Recovery system. Here the H_2S and some of the CO_2 is absorbed in a mixed solution of polyethylene glycol dimethyl ethers. This solution passes to a stripper giving a concentrated H_2S–CO_2 gas (approximately equal volumes) which is treated in a Claus unit to produce elemental sulphur for marketing or safe disposal:

Oxidation: $2\ H_2S + 3\ O_2 = 2\ SO_2 + 2\ H_2O$

Claus Reaction: $2\ H_2S + SO_2 = 3\ S + 2\ H_2O$

Process Code S32.3 – Ispra Mark 13A Process

Outline of Process [272]

A simplified block diagram of the process is presented in Figure 2.17. Gas from the boiler is passed in turn through a Concentrator, a Reactor and a final

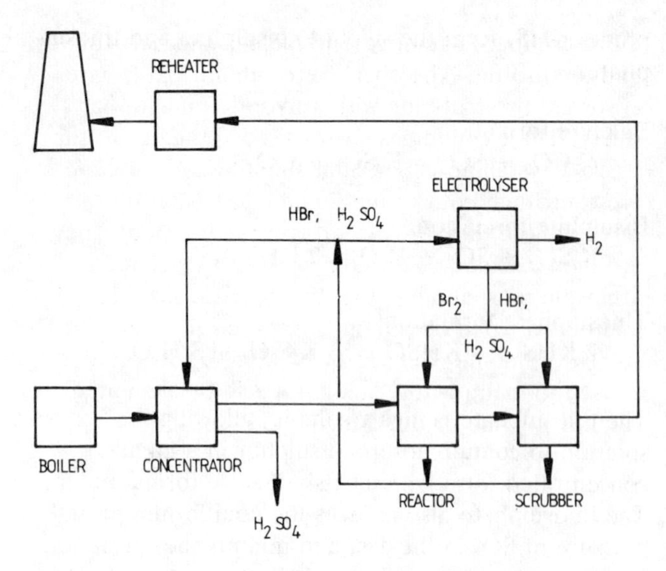

Figure 2.17 Block Diagram of Ispra Mark 13A FGD Process

Scrubber. In the Reactor and Scrubber, the gas is contacted with an aqueous solution of hydrobromic and sulphuric acids containing a small amount of bromine. The bromine reacts with the sulphur dioxide to form sulphuric acid. Part of the solution is pumped to the Concentrator, where the sensible heat of the gas is used to evaporate water and drive off bromine, leaving sulphuric acid. The remainder of the solution is pumped to an electrolyser where hydrobromic acid is converted to bromine and hydrogen. The cleaned gas flows through a Reheater before being exhausted to stack.

Chemistry of Process [272]

The principal reaction occurring in the Reactor (temperature 60–70°C) and final Scrubber (temperature 40–50°C) is the oxidation of SO_2 to sulphuric acid (H_2SO_4) by bromine, which is reduced to hydrobromic acid (HBr):

$$SO_2 + 2\ H_2O + Br_2 = H_2SO_4 + 2\ HBr$$

The bromine is regenerated from the hydrobromic acid by electrolysis in the Electrolyser:

$$2\ HBr \xrightarrow{\text{Electrolysis}} Br_2 + H_2$$

Hydrogen is therefore one of the byproducts of the process. The other by-product, sulphuric acid at a concentration of 80% by weight, is obtained from the Concentrator which is operated at a temperature of 140°C. This process, which is developmental, is claimed to give over 90% sulphur capture.

Process Code S32.4–Aqueous Sodium Carbonate Process

Outline of Process [6, 43, 137, 146]

A simplified block diagram of the process is presented in Figure 2.18. After clean-up in a Cyclone, the gas is treated with aqueous sodium carbonate in a Spray Dryer FGD unit. It then passes via an Electrostatic Precipitator to stack. The solids from the Spray Dryer and Electrostatic Precipitator, containing sodium sulphite and sulphate, are converted to sodium sulphide by treatment in the Reactor with coal or coke in a molten salt bath at 900–1050°C. The melt is quenched and dissolved in water in the Quench vessel, and undissolved solids are separated from the solution. The gas from the Reactor contains carbon dioxide; it is cooled to remove volatilised chlorides, and these, together with carry-over solids, are separated from the gas. The sodium sulphide solution and the cleaned Reactor off-gas pass to the Carbonation tower, where the sodium sulphide is converted to sodium carbonate and hydrogen sulphide. The sodium carbonate is returned to the Spray Dryer, and the hydrogen sulphide is treated in a Claus Unit to produce elemental sulphur.

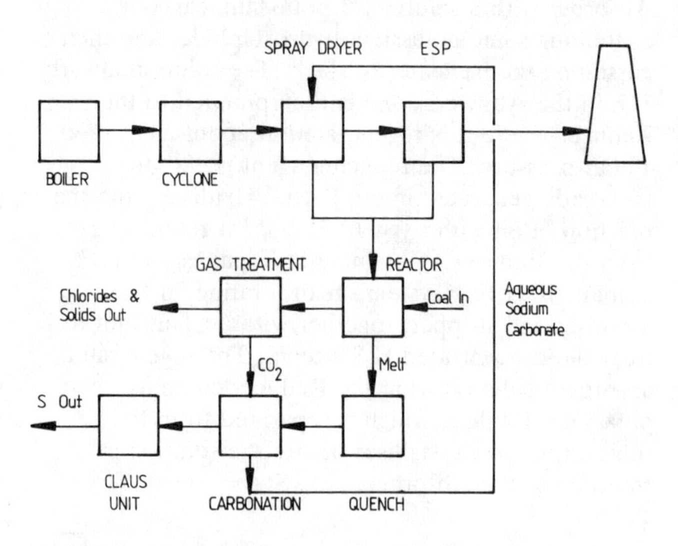

Figure 2.18 Block Diagram of Aqueous Sodium Carbonate FGD Process

The overall regeneration process is complicated, involving about eighty separate process steps, many of which, however, are well established in the paper and pulp industry.

Chemistry of the Process [122]

In the Spray Dryer, the SO_2 in the gas reacts with sodium carbonate, Na_2CO_3, and excess oxygen to form sodium sulphite and sulphate:

Sulphite formation:

$$SO_2 + Na_2CO_3 = Na_2SO_3 + CO_2$$

Sulphate formation: $2\ Na_2SO_3 + O_2 = 2\ Na_2SO_4$

In the Reactor, the sodium sulphite and sulphate are

both reduced at 900–1050°C to sodium sulphide (Na_2S) by carbon, supplied as coal or coke:

$$2\ Na_2SO_3 + 3\ C = 2\ Na_2S + 3\ CO_2$$

$$Na_2SO_4 + 2\ C = Na_2S + 2\ CO_2$$

Further carbon dioxide is generated in the Reactor by combustion of the fuel. The melt is quenched and dissolved in water in the Quench vessel, and undissolved solids are removed. The sodium sulphide solution is then passed to the Carbonation vessel. Here it is treated with CO_2-rich gases from the Reactor (after condensation and seperation of volatilised chlorides from the gas, and removal of other solids carried over from the fuel) to form carbonate, bicarbonate and H_2S:

$$Na_2S + CO_2 + H_2O = Na_2CO_3 + H_2S$$

$$Na_2CO_3 + CO_2 + H_2O = 2\ NaHCO_3$$

The bicarbonate solution is heated, releasing the carbon dioxide and restoring the sodium carbonate for return to the Spray Dryer. The gas from the Carbonation tower, containing H_2S, is sent to a Claus Unit, where part of the H_2S is oxidised to SO_2 which then reacts with the remainder of the H_2S:

$$2\ H_2S + 3\ O_2 = 2\ H_2O + 2\ SO_2$$

$$2\ H_2S + SO_2 = 3\ S + 2\ H_2O$$

CATEGORY S41

Process Code S41.1 – Magnesia Slurry Scrubbing Process

Outline of Process [6, 122, 146]

A simplified block diagram of the process is presented in Figure 2.19. Gas from the Boiler is passed through

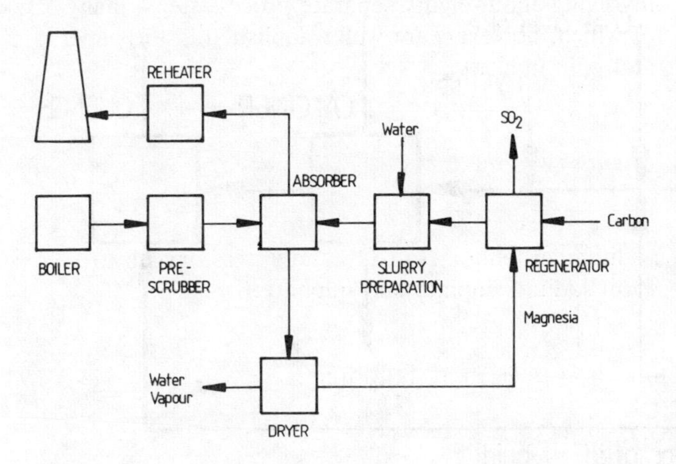

Figure 2.19 Block Diagram of Magnesia Slurry Scrubbing FGD Process

a Prescrubber to remove acid halides and particulates, and then to the Absorber, were sulphur oxides are absorbed by scrubbing with a hydrated magnesia slurry. The cleaned gas passes via a Reheater to the stack. Crystalline magnesium sulphite and sulphate are centrifuged out of the liquid and treated in the Dryer where they lose water of crystallisation. They are then calcined in the Regenerator in a reducing atmosphere at about 760–870°C to yield magnesia; addition of coke or carbon is required to reduce the sulphate. Sulphur dioxide evolved in the Regenerator is used to manufacture sulphuric acid or elemental sulphur for sale or disposal.

Chemistry of the Process [122]

Hydrated magnesia, $Mg(OH)_2$, reacts with sulphur dioxide in the Absorber to form magnesium sulphite trihydrate and hexahydrate. Some of the magnesium sulphite is converted to magnesium sulphate heptahydrate:

Sulphite formation:

$$Mg(OH)_2 + SO_2 + 2\ H_2O = MgSO_3.3\ H_2O$$

$$Mg(OH)_2 + SO_2 + 5\ H_2O = MgSO_3.6\ H_2O$$

Sulphate formation:

$$2\ MgSO_3 + O_2 + 14\ H_2O = 2\ MgSO_4.7\ H_2O$$

The sulphites and sulphate crystallise out, and are transferred to the Dryer where they lose their water of crystallisation. The anhydrous solids are then calcined in the Regenerator, to which carbon is added to reduce the sulphate. The regeneration reactions, which produce magnesia, MgO, and SO_2, occur at 760–870°C:

Sulphite regeneration: $MgSO_3 = MgO + SO_2$

Sulphate regeneration:

$$2\ MgSO_4 + C = 2\ MgO + 2\ SO_2 + CO_2$$

The magnesia is then hydrated in the Slurry Preparation stage:

$$MgO + H_2O = Mg(OH)_2$$

CATEGORY S42

Process Code S42.1 – Sulf-X Process

Outline of Process [175]

A simplified block diagram of the process is presented in Figure 2.20. Gas from the Boiler first passes through a Prescrubber to remove acid halides and particulates, and then into the Absorber where it is scrubbed with a slurry containing a mixture of iron sulphides in a sodium sulphide solution. The reactions with sulphur dioxide produce a range of insoluble iron/sulphur compounds, some soluble iron sulphate, and trace amounts of elemental sulphur. Some reduction of NO_x to elemental sulphur also occurs, but the liquid/gas flow ratio for efficient NO_x abatement is much higher than for SO_2 removal. The solid reaction products are separated from the solution in the Dewatering and Filtration system, dried in a steam-heated Dryer, and roasted at 650–750°C in an indirectly fired Calciner to which coal or coke is fed as a reducing agent, regenerating the iron sulphides and releasing sulphur. Part of the solution from the Filtration stage is treated in a Sulphate Separation system to crystallise out sodium sulphate which is transferred to the calciner. The solids from the Calciner pass to the Quench, where they are mixed with liquid from the Filtration and Sulphate Separation stages to reform the sulphide slurry sent to the Absorber. Sulphur vapour in the tail gas from the Calciner is condensed for marketing or safe disposal, and the gas is returned to the Prescrubber inlet.

Chemistry of Process [95]

The sorbent slurry is a complex mixture of iron compounds in a sodium sulphide solution, with a pH controlled at 6.0–6.4 by the buffering action of $Fe(OH)_2$. The main absorbing compound is FeS. The reactions occurring in the Absorber include:

Ionisation of FeS:

$$FeS + H_2O = Fe^{++} + HS^- + OH^-$$

Ionisation of SO_2:

$$SO_2 + H_2O = H^+ + HSO_3^-$$

Thiosulphate formation:

$$2\ HS^- + 2\ HSO_3^- + O_2 = 2\ S_2O_3^{--} + H_2O$$

Sulphur release:

$$S_2O_3^{--} + H^+ = HSO_3^- + S$$

Sulphate formation:

$$2\ HSO_3^- + O_2 = 2\ SO_4^{--} + H^+$$

The overall reaction can be expressed by the following simplification (note that this equation does not balance):

Overall reaction:

$$FeS + SO_2 + O_2 = Fe^{++} + SO_4^{--} + H^+$$

These reactions result in the formation of complex mixtures of iron sulphides, Fe_xS_y, together with $FeSO_4$, $Fe(OH)_2$ and Na_2SO_4. The key controlling factors are maintenance of the correct pH, the

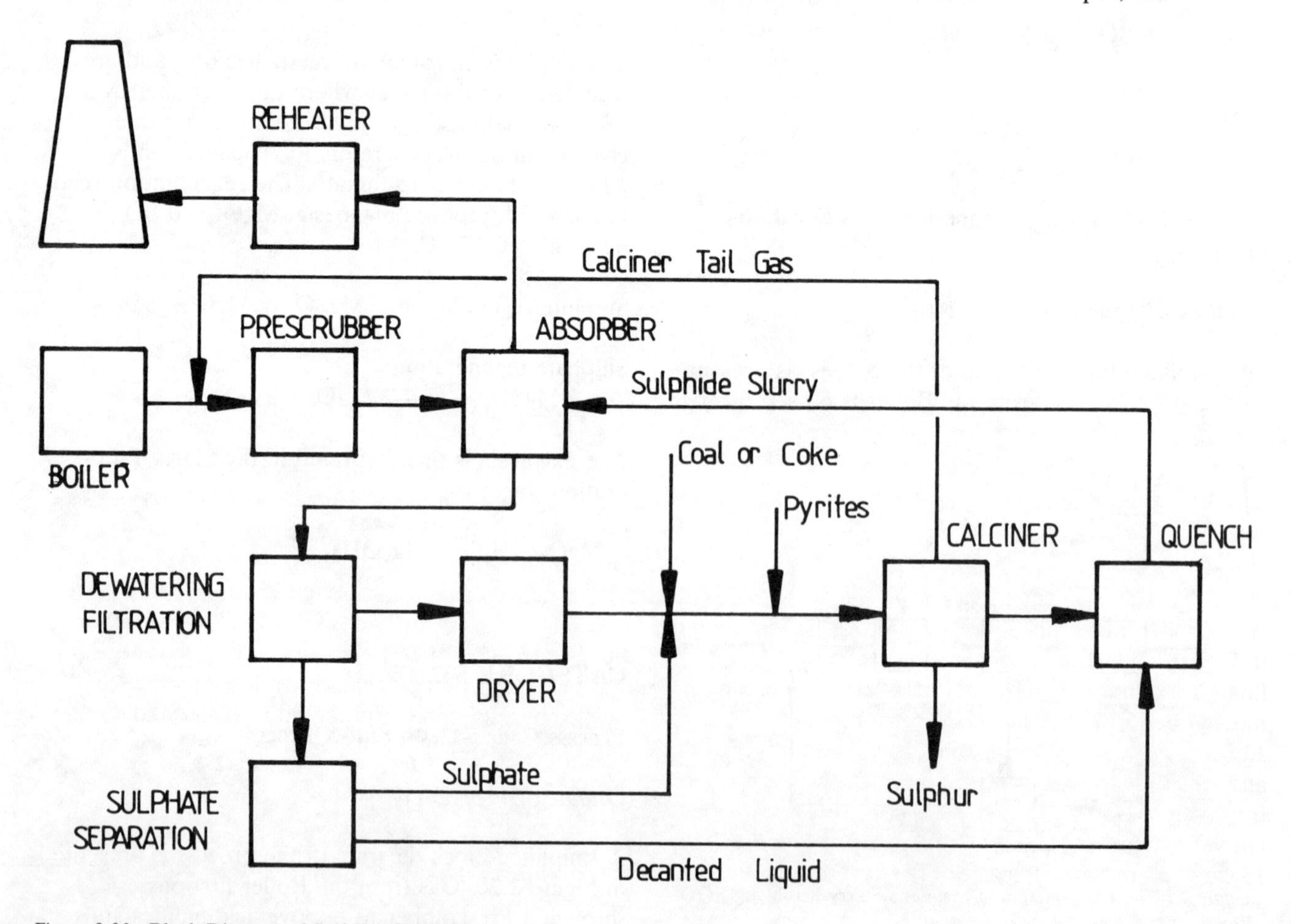

Figure 2.20 Block Diagram of Sulf-X FGD Process

correct Na_2S concentration, and the buffering action of the $Fe(OH)_2$.

The main regeneration reaction is the thermal decomposition of the iron sulphides to give the sorbent FeS and sulphur; e.g. for FeS_2:

$$FeS_2 = FeS + S$$

Sodium sulphate is crystallised out from the solution, and the solid is sent to the Calciner where it is reduced by carbon to the sulphide. The ferrous sulphate is subsequently reduced in solution:

$$Na_2SO_4 + 2\,C = Na_2S + 2\,CO_2$$

$$FeSO_4 + Na_2S = FeS + Na_2SO_4$$

There are some losses of iron from the system, and these are made up by feeding pyrites, FeS_2, to the Calciner.

CATEGORY S51

Process Code S51.1 – Active Carbon Adsorption Process

Outline of Process [6, 137]

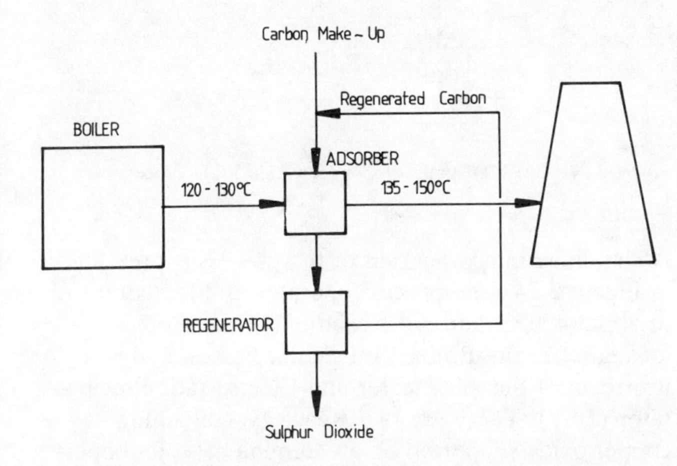

Figure 2.21 Block Diagram of Carbon Adsorption FGD Process

A simplified block diagram of the process is presented in Figure 2.21. Gas from the Boiler at 120–130°C flows horizontally through the Adsorber containing a moving bed of either activated carbon pellets sized 13 mm or of coke. The carbon adsorbs sulphur dioxide and water, which react with oxygen in the gas to form sulphuric acid. It also adsorbs nitrogen dioxide and acid halides. There is a temperature rise of 15–20°C across the Adsorber. The carbon bed acts as a panel bed filter and captures particulates from the gas. The cleaned gas then exhausts to stack. The carbon is screened to remove captured particulates, and then moves into the Regenerator, where it is heated in an inert atmosphere to 400–600°C by hot gas or by contact with hot sand. The sulphuric acid is decomposed, and the SO_3 released is reduced by carbon to SO_2 which can be used for manufacturing sulphuric acid for sale, or elemental sulphur for either sale or disposal. The adsorbed nitrogen dioxide reacts with carbon in the Regenerator to release elemental nitrogen and carbon dioxide, and the adsorbed acid halides are desorbed. The acid halides in the Regenerator off-gas are removed before treatment of the SO_2. The carbon lost in the reduction reactions is replaced by fresh activated carbon.

Chemistry of Process [6, 137]

In the Adsorber, SO_2 and water vapour are adsorbed by the activated carbon, and they react on the carbon surface with oxygen in the gas. The reaction can be represented by:

Oxidation: $2\,SO_2 + O_2 + 2\,H_2O = 2\,H_2SO_4$

The sulphuric acid formed remains adsorbed in the pores of the carbon. The reaction is exothermic, and there is a 15–20°C rise in gas temperature. Nitrogen dioxide and acid halides are also adsorbed from the gas.

At the higher temperature (400–650°C) in the Regenerator, acid halides are desorbed. The adsorbed sulphuric acid decomposes to sulphur trioxide and water, and the trioxide is reduced by the carbon to form the dioxide:

Decomposition: $H_2SO_4 = SO_3 + H_2O$

Reduction: $2\,SO_3 + C = 2\,SO_2 + CO_2$

The carbon also reduces adsorbed NO_2:

NO_2 Reduction: $2\,NO_2 + 2\,C = N_2 + 2\,CO_2$

A carbon make-up is required to replace that consumed in the reduction reactions (0.09 kg/kg SO_2 removed, and 0.26 kg/kg NO_2 removed). The loss of carbon during reduction increases porosity, and hence the internal surface of the carbon remaining, which is therefore further activated. However, the increase in porosity weakens the carbon particles, so that there are carbon break-down losses which also have to be replaced, resulting in a carbon make-up rate that is typically 30–100% greater than the theoretical.

Active carbon can also catalyse the reduction of nitric oxide, and a modification of the process can be used for the simultaneous abatement of SO_2 and NO_x. This is described in Section 5.

CATEGORY S52

Process Code S52.1 – Wet Active Carbon Adsorption Process

Outline of Process [157]

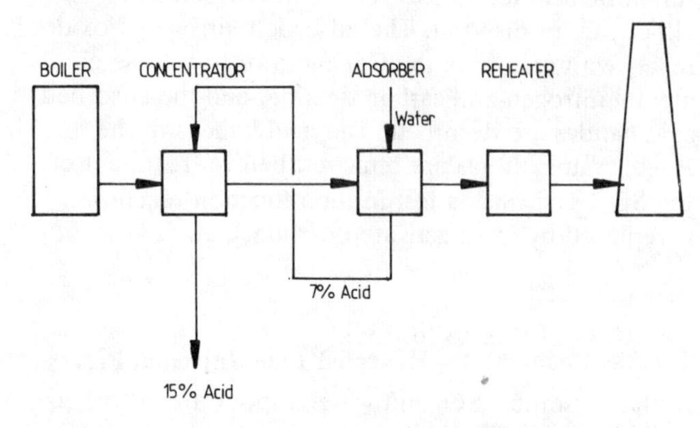

Figure 2.22 Block Diagram of Wet Active Carbon Adsorption (Lurgi Sulfacid) FGD Process

A simplified block diagram of one form of the process, the Lurgi Sulfacid process, is presented in Figure 2.22. Gas from the Boiler flows through the Concentrator, where it is cooled by dilute (7%) sulphuric acid from the Adsorber. It then passes through the Adsorber containing active carbon, which adsorbs sulphur dioxide. Water is admitted to the Adsorber intermittently, without interrupting the gas flow, and this removes sulphuric acid formed in the pores of the carbon. The dilute acid formed is concentrated somewhat (to 15%) by evaporation in the Concentrator. In another form of the process, the Hitachi process (see block diagram presented in Figure 2.23), up to five Adsorbers are operated in parallel: four on stream at any time, and one being washed with dilute acid; a side stream of the acid is concentrated in a submerged combustion evaporation Concentrator.

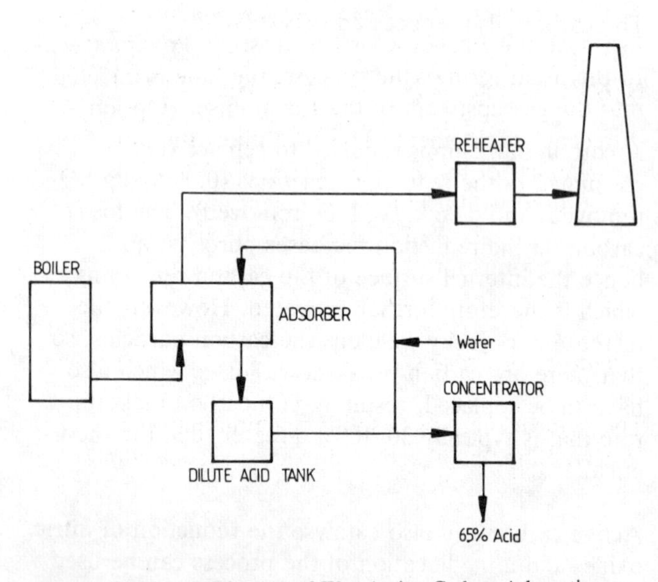

Figure 2.23 Block Diagram of Wet Active Carbon Adsorption (Hitachi) FGD Process

Chemistry of Process [6, 137]

In the Adsorber, SO_2 and water vapour are adsorbed by the activated carbon, and they react on the carbon surface with oxygen in the gas. The reaction can be represented by:

$$2\ SO_2 + O_2 + 2\ H_2O = 2\ H_2SO_4$$

Washing with water or dilute acid removes the sulphuric acid from the carbon. The adsorption and stripping occur at relatively low temperatures, and the gas has to be reheated. There is not the loss of carbon associated with the dry active carbon adsorption process (see Process Code S51.1).

Process Code S52.2 – Copper Oxide Process

Outline of Process [6, 137]

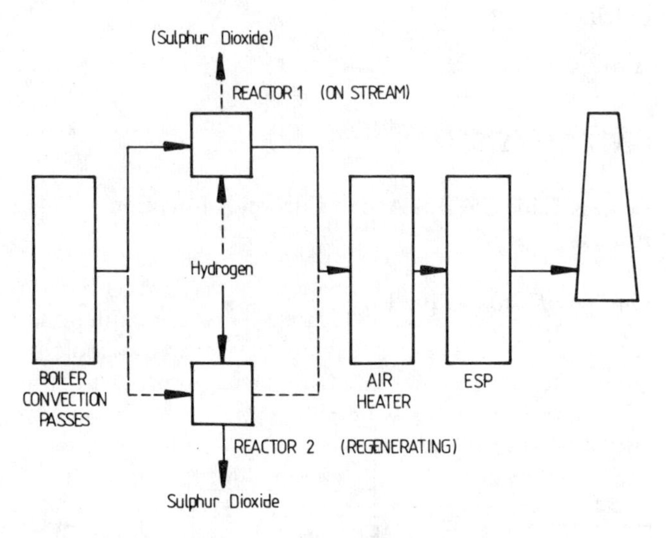

Figure 2.24 Block Diagram of Copper Oxide FGD Process

A simplified block diagram of the process is presented in Figure 2.24. The process operates at a temperature of about 400°C, and the Reactors are therefore located after the Boiler Convection Passes, and upstream of the Air Heater and Electrostatic Precipitator (ESP). There are two Reactors, containing copper oxide supported on an alumina base, and operated cyclically: one Reactor is on stream, with the gas flowing over (not through) the copper oxide, whilst the other is being regenerated by passing hydrogen over the sulphated copper oxide, reducing it to metallic copper and giving an SO_2-rich off-gas. When the Reactor is put back on stream, the copper is oxidised by oxygen in the gas, reforming copper oxide for further reaction.

Chemistry of Process [6, 122, 137]

The reaction occurring in the absorption phase of the cycle is:

$$2\ CuO + 2\ SO_2 + O_2 = 2\ CuSO_4$$

In the regeneration phase, hydrogen reduces the copper sulphate to metallic copper. The copper is re-oxidised when the absorption phase is resumed. The reactions are:

$$CuSO_4 + 2\ H_2 = Cu + SO_2 + 2\ H_2O$$

$$2\ Cu + O_2 = 2\ CuO$$

All of the reactions occur at about 400°C, and the absence of temperature cycling between the absorption and regeneration phases avoids thermal stresses on the copper oxide.

The oxide and sulphate can catalyse the reduction of NO_x with ammonia, so that a modification of the process can be used for the simultaneous abatement of SO_2 and NO_x. This is described in Section 5.

Process Code S52.3 – Catalytic Oxidation Process

Outline of Process [157]

A simplified block diagram of the process is presented in Figure 2.25. Gas at about 500°C is passed through a High-Temperature Electrostatic Precipitator (H.T. ESP) and then through the Reactor containing a bed of vanadium pentoxide which catalyses the oxidation of SO_2 to SO_3. The gas leaving the Reactor is cooled to not less than about 230°C by passage through the Economiser and Air Heater, and is then scrubbed in the Absorber with dilute acid at about 110°C before being exhausted via a Mist Eliminator to the stack. The SO_3 forms sulphuric acid, and this is continuously withdrawn.

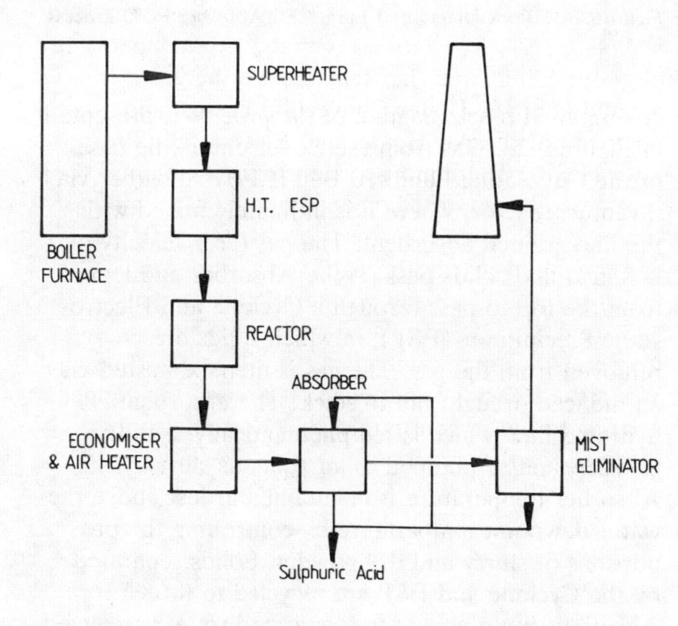

Figure 2.25 Block Diagram of Catalytic Oxidation FGD Process

Chemistry of Process

The oxidation of SO_2 by oxygen in the flue gas in the Reactor is catalysed by vanadium pentoxide, as in the familiar contact process for the manufacture of the acid. Subsequent absorption in dilute acid forms sulphuric acid. The reactions are simply represented by:

Oxidation: $2\ SO_2 + O_2 = 2\ SO_3$

Acid formation: $SO_3 + H_2O = H_2SO_4$

CATEGORY S61

Process Code S61.1 – Hydrated Lime Injection Process

Outline of Process [137]

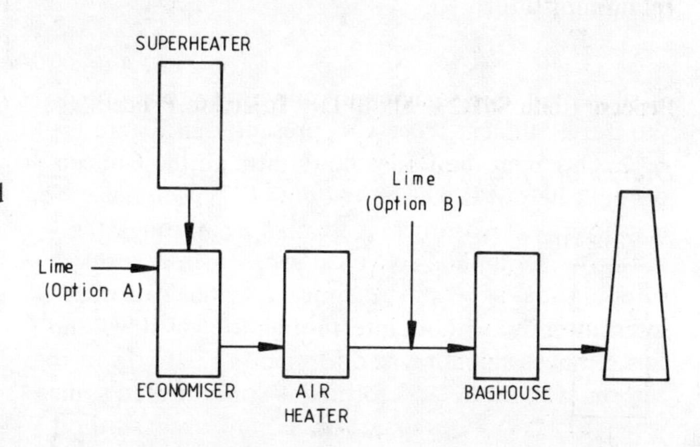

Figure 2.26 Block Diagram of Hydrated Lime Injection FGD Process

A simplified block diagram of the process is presented in Figure 2.26. Hydrated lime is injected into the gas and absorbs sulphur dioxide to form calcium sulphite, which is removed, together with unreacted lime and flyash, in the Baghouse or Electrostatic Precipitator. In the usual form of the process, the lime is injected into the gas upstream of the Economiser (Option A). A recent development [291] is to inject the lime downstream of the air heater (Option B). Option A is also used with magnesia or magnesium hydroxide to reduce SO_3 concentrations, and hence acid dewpoint temperatures.

Chemistry of Process

The heterogeneous gas-solids reaction between hydrated lime and sulphur dioxide gives calcium sulphite, and is represented as:

$$Ca(OH)_2 + SO_2 = CaSO_3 + H_2O$$

A large excess of lime over the stoichiometric quantity

is needed for efficient sulphur capture. The lime utilisation is improved by adoption of Option B–injection into the gas downstream of the Air Heater [291].

The improvement obtained with Option B is greatest when the gas temperature at the point of injection is only 75–80°C (i.e. only slightly above the water dewpoint) and when the hydrate contains 3–4% moisture; the moisture content is critical.

For Option B the lime should be 'dry hydrated', i.e. hydrated with the theoretical quantity of water to give a dry product. This has a different morphology from the lime prepared by 'wet slaking' with excess water: it contains more lattice defects which increase its reactivity. The actual form depends on the water temperature and on the additives in the water, which concentrate on the surface of the dry product. Thus, sodium compounds concentrate on the surface as NaOH, which is deliquescent and helps the lime to retain moisture.

Process Code S61.2 – Alkali Dry Injection Process

Outline of Process [95]

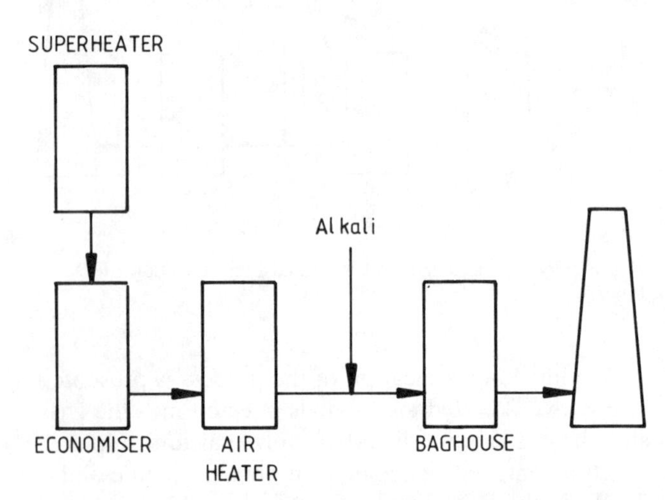

Figure 2.27 Block Diagram of Alkali Injection FGD Process

A block diagram of the process is presented in Figure 2.27. Powdered alkali (sodium carbonate or bicarbonate) is injected into the gas downstream of the Air Heater. The temperature should be greater than 135°C when sodium bicarbonate is used. The reaction occurs in the duct and in the Baghouse used to remove the solids from the gas, before the gas is exhausted to stack.

Chemistry of Process [95]

The alkali reacts with sulphur dioxide and oxygen in the gas to form sodium sulphate:

$$2\ Na_2CO_3 + 2\ SO_2 + O_2 = 2\ Na_2SO_4 + 2\ CO_2$$

If sodium bicarbonate is used, it first decomposes to form sodium carbonate:

$$2\ NaHCO_3 = Na_2CO_3 + H_2O + CO_2$$

This decomposition is slow at temperatures below 135°C, giving less time for the subsequent absorption of SO_2. However, at higher temperatures, nahcolite (a naturally occurring sodium bicarbonate found in the US) is a more effective injection reagent than sodium carbonate. This appears to be due to the porosity generated by the decomposition reaction. The effectiveness of the reagents is increased by reduction in particle size, e.g. to below 50 micron.

CATEGORY S62

Process Code S62.1 – Lurgi Circulating Fluidised Bed (CFB) Lime Absorber Process

Outline of Process [293]

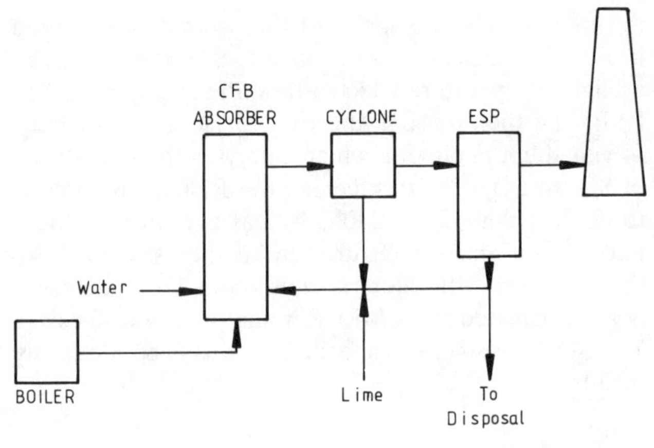

Figure 2.28 Block Diagram of Lurgi CFB Absorber FGD Process

A simplified block diagram of the process is presented in Figure 2.28. Gas from the Boiler enters the base of the Circulating Fluidised Bed (CFB) Absorber via a venturi reactor, where it is intimately mixed with the fine-grained absorbent. The gas (at a velocity of 2–8 m/s) and solids pass up the Absorber and leave from the top to pass through a Cyclone and Electrostatic Precipitator (ESP), in which solids are removed from the gas. The gas is then exhausted via an induced draught fan to stack. The absorbent is hydrated lime which is fed pneumatically as a dry powder, and/or pumped as an aqueous slurry. The Absorber temperature is maintained at just above the water dewpoint temperature by controlling the proportions of slurry and dry powder. Solids separated by the Cyclone and ESP are recycled to the CFB Absorber, but a part of the separated solids is rejected for disposal.

Chemistry of Process

The heterogeneous gas–solids reaction between hydrated lime and sulphur dioxide gives calcium sulphite, and is represented as:

$$Ca(OH)_2 + SO_2 = CaSO_3 + H_2O$$

With a residence time of solids in circulation of up to 20 minutes, sulphur capture of over 95% can be achieved with an input calcium to sulphur molar ratio of about 1.7. The best performance is to be expected when the reaction temperature is only 75–80°C (i.e. only slightly above the water dewpoint) and when the hydrate contains 3–4% moisture; the moisture content is critical.

The lime should be 'dry hydrated', i.e. hydrated with the theoretical quantity of water to give a dry product. This has a different morphology from the lime prepared by 'wet slaking' with excess water: it contains more lattice defects which increase its reactivity. The actual form depends on the water temperature and on the additives in the water, which concentrate on the surface of the dry product. Thus, sodium compounds concentrate on the surface as NaOH, which is deliquescent and helps the lime to retain moisture.

CATEGORY S71

Process Code S71.1 – Sorbent Direct Injection Process

Outline of Process [137]

A simplified block diagram of the process is presented in Figure 2.29. The finely ground sorbent (usually limestone, but hydrated lime, soda ash or sodium bicarbonate can also be used) is injected into the Boiler Furnace with the fuel. The sorbent calcines, losing CO_2, and absorbs SO_2 to form sulphites and sulphates. When the process is applied to systems using modified burners for NO_x abatement (see Sections 3 and 5) the sorbent can also react with H_2S in the reducing regions of the flame to form the sulphide, some of which subsequently oxidises to the sulphates. The reacted and unreacted sorbent is removed, with fly-ash, in the Electrostatic Precipitator (ESP) or Baghouse.

Chemistry of Process [122]

Limestone is the sorbent usually used, and in the furnace it undergoes: calcination; sulphite formation; sulphate formation; and (with burners modified for NO_x abatement) sulphide formation:

Calcination: $CaCO_3 = CaO + CO_2$

Sulphite formation: $CaCO_3 + SO_2 = CaSO_3 + CO_2$

$$CaO + SO_2 = CaSO_3$$

Sulphate formation:

$$2\ CaCO_3 + 2\ SO_2 + O_2 = 2\ CaSO_4 + 2\ CO_2$$

$$2\ CaO + 2\ SO_2 + O_2 = 2\ CaSO_4$$

$$2\ CaSO_3 + O_2 = 2\ CaSO_4$$

$$CaS + 2\ O_2 = CaSO_4$$

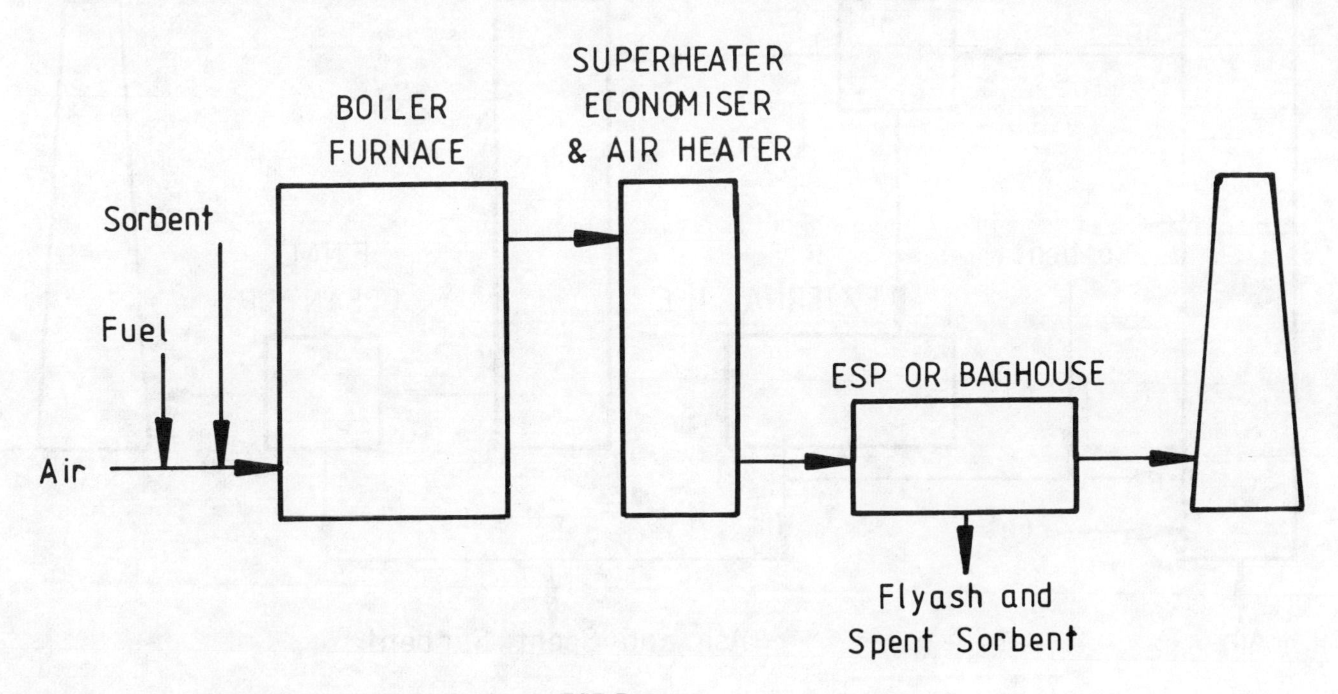

Figure 2.29 Block Diagram of Sorbent Direct Injection FGD Process

Sulphide formation:

$$CaCO_3 + H_2S = CaS + H_2O + CO_2$$

$$CaO + H_2S = CaS + H_2O$$

$$4\ CaSO_3 = 3\ CaSO_4 + CaS$$

Similar reactions occur with the other sorbents that can be used.

A large excess of sorbent is usually required because:

The flame temperature is in the region of 1700–2000°C i.e. well above the calcium sulphate decomposition temperature of 1200°C or calcium sulphite temperature of 450°C.

Calcination at the flame temperature results in formation of an inactive ('dead-burned') lime resulting from melting of a calcium carbonate/calcium oxide eutectic. This gives a non-porous structure which inhibits diffusion of SO_2 into the solid particles.

In consequence the sulphur capture is less than 50% even with twice the stoichiometric ratio of limestone. However, with the burners modified for NO_x abatement, the peak flame temperatures are generally lower and may be located away from the main sulphur capture zone. Furthermore, the possibility exists of sulphur capture by sulphide formation. Such burners therefore allow the attainment of more efficient sulphur capture.

CATEGORY S72

Process Code S72.1 – Fluidised Bed Combustion

Outline of Process [468]

Simplified block diagrams of the three main forms of Fluidised Bed Combustors (FBCs) are presented in: Figure 2.30 for Atmospheric Pressure Fluidised Bed Combustors (AFBC); Figure 2.31 for Circulating Fluidised Bed Combustors (CFBC); and Figure 2.32 for Pressurised Fluidised Bed Combustors (PFBC). The subject is dealt with in detail in Section 6, and only the sulphur capture aspects of FBC are considered here. Fuel is burned in the FBC in the presence of a sulphur sorbent (usually limestone or

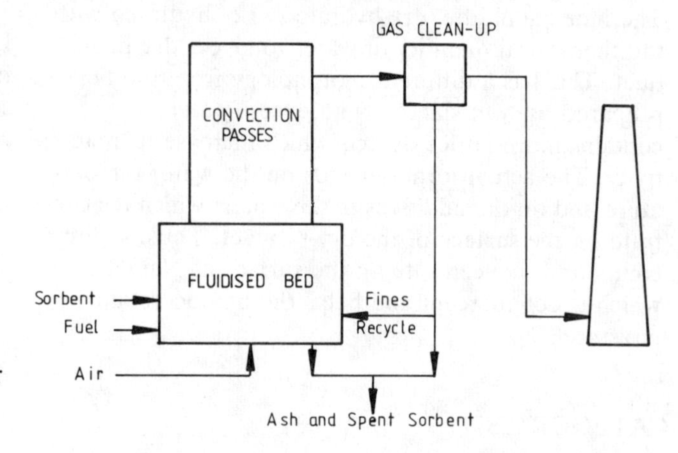

Figure 2.30 Block Diagram of Atmospheric Pressure Fluidised Bed Combustor

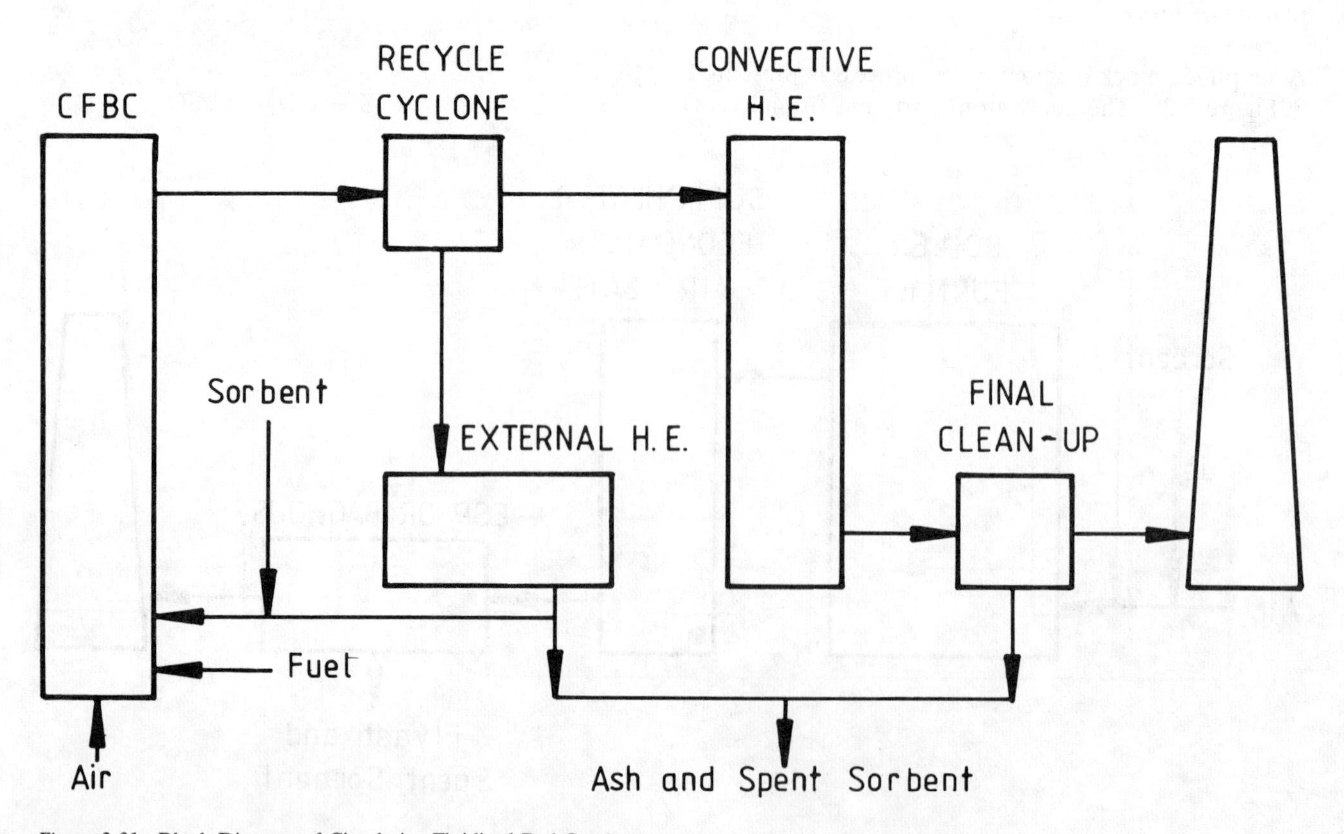

Figure 2.31 Block Diagram of Circulating Fluidised Bed Combustor

dolomite) which reacts with the sulphur oxides, either directly or following calcination of the calcium carbonate, to give calcium sulphate. An excess of sorbent over the stoichiometric requirement is fed to the bed, and an equivalent quantity of spent sorbent is removed, together with fuel ash, from the bed and gas clean-up system (AFBC and PFBC) or from the External Heat Exchanger (H.E.) and Final Clean-Up (CFBC). The sorbent requirements can be reduced by: use of reactive limestone or dolomite; careful choice of operating conditions; recycle of sorbent fines carried over from the bed; and either partial hydration or chemical regeneration of sulphated sorbent (at the expense of complexity and thermal efficiency).

Chemistry of Process [468]

The sorbent reactions occurring in the bed are: calcination of $CaCO_3$ and half calcination of dolomite; and sulphation of $CaCO_3$ and of CaO:

Calcination of limestone: $CaCO_3 = CaO + CO_2$

Half calcination of dolomite:

$$CaCO_3.MgCO_3 = [CaCO_3 + MgO] + CO_2$$

Full calcination of dolomite:

$$[CaCO_3 + MgO] = [CaO + MgO] + CO_2$$

Sulphation:

$$2\ CaCO_3 + 2\ SO_2 + O_2 = 2\ CaSO_4 + 2\ CO_2$$

$$2\ CaO + 2\ SO_2 + O_2 = 2\ CaSO_4$$

The sulphation reaction is probably a two-stage reaction:

$$2\ SO_2 + O_2 = 2\ SO_3$$

$$CaO + SO_3 = CaSO_4$$

with the oxidation of SO_2 occurring at the sorbent surface.

The key to sorbent behaviour is calcination, which creates porosity, giving access to the interior of the sorbent particle. The rate of calcination increases with increase in temperature; rapid calcination, such as occurs in AFBCs, results in the formation of small CaO crystallites which present a large internal surface area (i.e. high reactivity), but with access only through small pores. Reaction with SO_2 occurs rapidly, but because the reaction product is bulky, it blocks the pore entrances and reaction soon ceases, giving low fractional utilisation of the sorbent. For these reasons there is an optimum temperature for sulphur capture at around 850°C: when sorbent utilisation, increased by increased calcination and rate of sulphation, starts to fall by increased blockage of pore entrances.

Because calcium sulphate is more bulky than the carbonate or oxide, it tends to form a shell, up to about 50 micron thick, on the surface of the lime. The fractional utilisation is therefore increased by the use of fine particles. Impact of the particle with surfaces, e.g. in cyclones, can cause this shell to flake off,

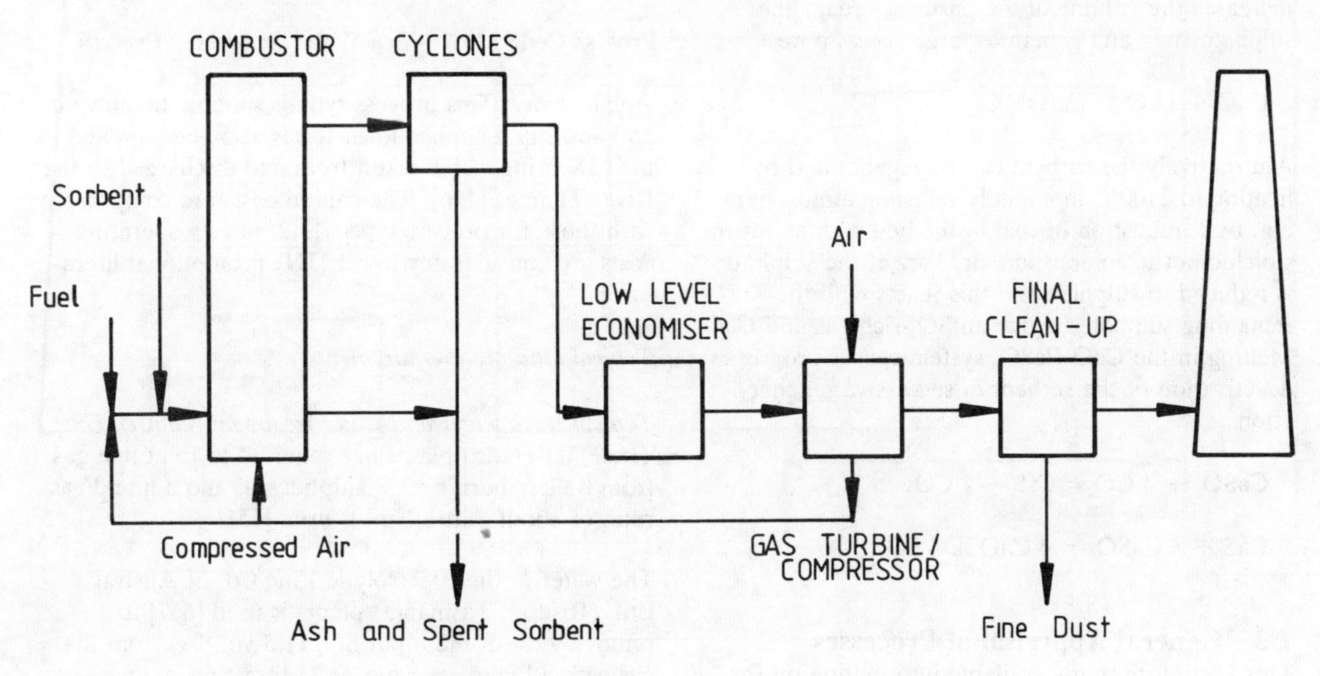

Figure 2.32 Block Diagram of Pressurised Fluidised Bed Combustor

revealing unreacted sorbent surface for further sulphur capture.

The rate of calcination falls with increase in CO_2 partial pressure, and therefore only partial calcination occurs in PFBCs. Slow calcination reduces the initial rate of sulphation, but allows the CaO crystallites to sinter, forming larger pores which are less easily blocked; fractional utilisation of the sorbent is thereby increased [536]. No optimum temperature for sulphur capture has been detected in PFBCs.

Dolomites 'half-calcine' at low temperatures, creating porosity even without calcination of the $CaCO_3$ component, allowing direct sulphation of $CaCO_3$, so that the pore-blockage problem is less acute, and the optimum temperature is less well defined.

In general, only the calcium component of dolomites undergoes sulphation, as the temperature in FBCs (in the range 750–950°C) is unfavourable for the formation of magnesium sulphate. However, 'half-calcined' dolomite can form a double sulphate, $CaSO_4.3\ MgSO_4$ in significant concentrations in PFBCs at temperatures below 800°C.

Some designs of FBC allow for staged combustion, giving reducing zones within the bed. This has been found to improve sorbent utilisation [539], probably because of formation of calcium sulphide (which is not bulky and does not block pore entrances) and its subsequent oxidation:

$$CaO + H_2S = CaS + H_2O$$

$$CaS + 2\ O_2 = CaSO_4$$

Sorbent utilisation can be increased by partial hydration of spent sorbent with water or steam. This increases the volume of the particle, breaks the sulphate shell and generates large access pores:

$$CaO + H_2O = Ca(OH)_2$$

Alternatively the sorbent can be regenerated by heating to 1100°C in a mildly reducing atmosphere, e.g. by combustion of coal in the bed with about the stoichiometric combustion air. Part of the sulphate is reduced to sulphide and this reacts with the remaining sulphate to give an SO_2-rich gas and CaO. Melting in the CaO-$CaSO_4$ system causes progressive deactivation of the sorbent in successive regenerations:

$$CaSO_4 + 4\ CO = CaS + 4\ CO_2$$

$$CaS + 3\ CaSO_4 = 4\ CaO + 4\ SO_2$$

2.3 General Appraisal of Processes

This Section presents available information on the characteristics of the FGD process types: status, applicability, space and typical land area requirements, fresh water and water treatment requirements, reagent consumption, end-product or waste materials disposal requirements, typical power consumptions and reductions in combustion plant efficiency, operating experience and process developments.

Where available, published data are presented, but in some instances the information on the process characteristics listed above is also (or alternatively) derived from published estimates for specific cases (all at 100% full load), referred to as Cases 1 to 6. The background data for these estimates are presented in Table 2.2.

Table 2.2 Background Data on Published Estimates

Case		1	2	3	4	5	6
Reference		95	95	6	277	280	*
MWe		2×500	2×500	800	3×660	500	2000
Coal: Sulphur	wt.%	4.0	0.48	3.6	2.0	4.0	2.0
Chlorine	wt.%	0.1	0.07	0.1	0.4	–	–

*Davy McKee promotional literature
– Indicates data unavailable

The system used to classify flue gas desulphurisation processes is illustrated in Figure 2.1; the codes of processes dealt with in this Section are listed in Table 2.1.

CATEGORY S11

The status of Category S11 processes is indicated in Table 2.3, which presents information on the size of plants built, year of completion and percentage sulphur capture.

Process Code S11.1 – Sea Water Scrubbing Process

Applicability: This process type is suitable for retrofit and new build applications. It has also been applied in the UK using water taken from and discharged to the River Thames [156]. The capital costs are comparable with those for process type S11.2, but its operating costs are considerably lower [IHI promotional literature].

Typical land area requirements: ***

Typical fresh water and water treatment requirements: The Fläkt-Hydro plants are required to treat flue gas from boilers burning 3% sulphur oil, and a liquid/gas ratio of about 8 litre/Nm^3 is used [651].

The water in the Electrolytic Zinc Co. of Australia Ltd. (Risdon, Tasmania) plants is used [667] to remove 99% of the input SO_2 (1.5 vol. %) from the gas with a liquid/gas ratio of 27 litre/Nm^3.

Consumption of reagents: The CEGB plants neutralised the waste water with chalk or with waste sludge from water softening plant [156].

The Fläkt-Hydro plants are located at the Porsgrunn Chemical Plant of Norsk Hydro, and the effluent is neutralised with wastes from a magnesium production plant, and neutralised with waste $Ca(OC1)_2$ [651].

The Risdon plants return the effluent to the sea at a pH of 2.5.

End-product or waste disposal requirements: The CEGB plants used river water and were required to discharge water of suitably low sulphur compounds concentration [156]; the concentration was below the solubility limits.

Typical pressure losses: At the Fläkt-Hydro plants: 25–35 mbar [651].

Typical power requirements: For the Fläkt-Hydro plants, 1.1–1.6% of the equivalent power generation.

Typical reductions in combustion plant efficiency: ***

Operating experience: The CEGB plants [156] and Fläkt-Hydro plants [651] have given some problems with corrosion, but the availability of the Risdon plant [667] is very high.

The process is similar to the 'Inert Gas System' installed by IHI on 15 ships, handling 11,000–36,000 m^3/h of gas (aggregate of 302,000 m^3/h) with 94–99% sulphur capture and 84–92% dust removal [IHI promotional literature].

Process developments: Little development is likely, owing to the limited applicability of the process.

Appraisal:

1.	Information available	1
2.	Process simplicity	3
3.	Operating experience	1
4.	Operating difficulty	1
5.	Loss of power	1*
6.	Reagent requirements	2 (1 if water is neutralised)
7.	Ease of end-product disposal	1
8.	Process applicability	1 (geographical limitation)
	Total	11

*Assumed in absence of data

Table 2.3 Status of Processes Using Non-Regenerable Solution-Based Wet Reagents: Wet End-Products

Code No.	Vendor	No. of Plants	Plant size Range MWe	Total MWe	Dates	% S Capture	Ref.
S11.1	A/S Ardal & Sundall	–	–	–	–	80–90	651
	CEGB	4	120–228	754	1933–62	90–95	156, 650
	El'ytic Zinc Co.	2	–	194*	1948–51	99	651
	Fläkt/Hydro	4	10–20	70	1975–	96–99	651
	IHI	4	272–370*	1272*	1972–75	–	**
S11.2	Andersen 2000	170	0.5–50	–	–	95–99	135
	B&W Co.	3	550	1650	–	86.5	135, 125
	FMC	5a	–	–	–	90–95	135
	IHI	65	6–480*	2285*	1968–75	–	**
	Thyssen/CEA	4	125–295	670	–	–	125
	UOP Air Correction Div.	2	330–550	880	–	98	125, 655
S11.3	IHI	9	13–217*	1458*	1966–81	–	**
	Simon-Carves	1	17	17	1957	90	156
S11.4	Chiyoda 101	14b	14–1050*	4426*	1972–75	–	95,**
S11.5	Dowa Mining Co.	10	4–150*	761*	1972–80	90–98	95
S11.6	Airpol	70	–	–	–	95–99	135
	Andersen 2000	2	0.5–250	–	–	95–97	135
	FMC	12	265–420c	685c	1979–86	90–95	95c, 135
	Thyssen/CEA	2	85–1167*	1252*	1975–79	–	95,**
	Envirotech/GEESI	1	575	575	1979	90	95, 667
	Showa Denko	5	62–150	416	1973–74	–	95
	Kawasaki/Kureha	5	150–450	1950	1974–77	–	95

*Gas flow, thousand Nm^3/h
**Vendor's promotional literature
a Includes some calcium slurry scrubbing installations
b Total for 9 boilers; remaining 5 installations comprised other types of combustion plant
c For 2 utility installations
– Indicates information unavailable

Process Code S11.2 – Alkali Scrubbing Process

Applicability: This process type is suitable for retrofit and new build applications. Depending upon operating conditions, sulphur capture can exceed 95%.

Typical land area requirements: For a 550 MWe boiler, 0.56% sulphur coal, 98% sulphur capture [655]: 0.4 hectare for FGD plant; 61 hectare pond for waste disposal (estimated 10-year capacity).

Typical fresh water and water treatment requirements: Water requirements are 0.3–0.7 m^3/MWh [135]. For small units, e.g. 12 MWt, a water consumption of 69 litres/GJt is estimated [160].

Consumption of reagents: Alkali: sulphur molar ratio of 1.5 [160].

End-product or waste disposal requirements: Some large units oxidise the end-product to form sodium sulphate for marketing; the quantity would be about equivalent to the sulphur captured. Others, including small units, e.g. 12 MWt, dispose of waste material as a sludge (containing sulphate, sulphite, bisulphite and particulates with about 85% water) having a volume of about 33 litres/GJt [160]. This would generally be difficult to dispose of in the UK in an environmentally acceptable manner.

Typical pressure losses: About 15 mbar [655].

Typical power requirements: For small units, e.g. 12 MWt, about 4 kW/MWt [160].

For a 550 MWe boiler, 0.56% sulphur coal, 98% sulphur capture [655]: about 7 kW/MWe.

Typical reductions in combustion plant efficiency: ***

Operating experience: Availability quite high. The main problems appear to be plugging of absorber spray nozzles and scale on absorber walls. Other problems (e.g. with dampers) have been solved [655].

Process developments: ***

Appraisal:

1.	Information available	1
2.	Process simplicity	2
3.	Operating experience	1
4.	Operating difficulty	1
5.	Loss of power	1*
6.	Reagent requirements	1
7.	Ease of end-product disposal	0
8.	Process applicability	2
	Total	9

*Assumed in absence of data

Process Code S11.3–Ammonia Scrubbing Process

Applicability: This process type is suitable for retrofit and new build applications.

Typical land area requirements: ***

Typical fresh water and water treatment requirements: ***

Consumption of reagents: ***

End-product or waste disposal requirements: ***

Typical pressure losses: ***

Typical power requirements: ***

Typical reductions in combustion plant efficiency: ***

Operating experience: The CEGB Plant [156] operated successfully, but was shut down after about two years because of the difficulty of obtaining by-product ammonia, the impurity of the product and the objectional odour of the plume.

Process developments: The Walther process (Section 2.2 and 2.3) overcomes some of the objections to the simple ammonia scrubbing process.

Appraisal:

1.	Information available	0
2.	Process simplicity	2
3.	Operating experience	0
4.	Operating difficulty	1
5.	Loss of power	1*
6.	Reagent requirements	1
7.	Ease of end-product disposal	1
8.	Process applicability	2
	Total	8

*Assumed in absence of data

Process Code S11.4 – Sulphuric Acid Scrubbing Process

Applicability: This process type is suitable for retrofit and new build applications.

Typical land area requirements: ***

Typical fresh water and water treatment requirements: For a 500 MWe boiler, 1% sulphur oil [172], 0.08 m^3/MWh fresh water.

Consumption of reagents [172]: Limestone (to neutralise the sulphuric acid produced) about equivalent to the sulphur captured; catalyst (ferrous sulphate) about 0.13 kg/MWh.

End-product or waste disposal requirements: ***

Typical pressure losses: 55 mbar [172].

Typical power requirements [172]: For a 250 MWe boiler, 1% sulphur oil, 90% sulphur capture: 2.0% of the gross power produced.

Typical reductions in combustion plant efficiency: ***

Operating experience: A run of 50 days was reported in 1974 [172].

Process developments: ***

Appraisal:

1.	Information available	0
2.	Process simplicity	2
3.	Operating experience	1
4.	Operating difficulty	0*
5.	Loss of power	1*
6.	Reagent requirements	2
7.	Ease of end-product disposal	1
8.	Process applicability	2
	Total	9

*Assumed in absence of data

Process Code S11.5–Dowa Process

Applicability: This process type is suitable for retrofit and new build applications.

Typical land area requirements: ***

Typical fresh water and water treatment requirements: See Table 2.4 for estimates.

Consumption of reagents: See Table 2.4 for estimates.

Table 2.4 Published Estimated Requirements

Case No. (see Table 2.2)		1
Reference		95
MWe		2×500
Coal: Sulphur	wt. %	4.0
Chlorine	wt. %	0.1
Sulphur capture	%	90
Typical fresh water and water treatment requirements per MWh:		
Raw water		0.37 m^3
Steam (28 bar, 300°C)		0.10 tonne
Consumption of reagents, kg/MWh:		
Limestone at a Ca/S molar ratio of 1.0		53.3
Alum make-up at 0.0034 mol/mol limestone		0.8
End-product or waste disposal requirements, m^3/MWh:		
Waste gypsum sludge volume		0.09
Fly-ash volume		0.06
Typical pressure losses, mbar:		11
Typical power requirements, MWe:		14.8
Typical reductions in combustion plant efficiency, %:		
Loss in power sent out		4.6

End-product or waste disposal requirements: See Table 2.4 for estimates.

Typical pressure losses: See Table 2.4 for estimates.

Typical power requirements: See Table 2.4 for estimates.

Typical reductions in combustion plant efficiency: See Table 2.4 for estimates.

Operating experience: Applied in nine Japanese installations (acid plants, smelters and one oil-fired boiler), and tested by EPRI at a U.S. coal-fired boiler [95].

Process developments: ***

Appraisal:

1.	Information available	1
2.	Process simplicity	1
3.	Operating experience	1
4.	Operating difficulty	0*
5.	Loss of power	1
6.	Reagent requirements	1
7.	Ease of end-product disposal	1
8.	Process applicability	2
	Total	8

*Assumed in absence of data

Process Code S11.6 – Dual Alkali Process

Applicability: This process type is suitable for retrofit and new build applications.

Typical land area requirements [654]: For a 265 MWe boiler, 3.35% sulphur coal, 90% sulphur capture: 0.29 hectare for the FGD plant.

Typical fresh water and water treatment requirements: See Table 2.5 for estimates.

Consumption of reagents: See Table 2.5 for estimates.

End-product or waste disposal requirements: See Table 2.5 for estimates.

Typical pressure losses: See Table 2.5 for estimates.

Typical power requirements: See Table 2.5 for estimates.

Typical reductions in combustion plant efficiency: See Table 2.5 for estimates.

Operating experience: The system using lime as the regenerating reagent has been applied to about 50 industrial and utility installations in the U.S. and Japan, with many others under construction or planned. The system using limestone has been applied to full-scale installations only in Japan [95].

A summary of problems reported by Battelle in their visit reports [653, 654, 664, 667, 683] is as follows (total number of plants visited, numbers experiencing problems in various plant areas, and those with problems remaining unsolved):

No. visited	5
Total equiv. MWe	1564
Total streams	14

Plant Area	Number of Problems Total	Unsolved	Solved
Prescrubber	1	0	1
Absorber	4	2	2
Demisters	2	1	1
Flues or stack	1	1	0
Dampers	3	3	0
Sorbent preparation	0	0	0
Waste/product system	2	1	1
Other	2	2	0
Totals	15	10	5

Process developments: The system using limestone as the regenerating reagent is now being offered (by FMC only) in the U.S. [95], where there are four installations [125] in the range 265–390 MWe (total 1310 MWe) and three [664] each handling 160,000 Nm³/h.

Appraisal:

1.	Information available	2
2.	Process simplicity	1
3.	Operating experience	1
4.	Operating difficulty	0
5.	Loss of power	1
6.	Reagent requirements	1
7.	Ease of end-product disposal	1
8.	Process applicability	2
	Total	9

CATEGORY S12

The status of Category S12 processes is indicated in Table 2.6, which presents information on the size of plants built, year of completion, and percentage sulphur capture.

Process Code S12.1 – Alkali Scrubbing/Spray Drying Process

Applicability: This process type is suitable for retrofit or new build applications. Depending upon operating conditions, up to 95% sulphur capture can be achieved. However, the reagent is expensive unless naturally occurring deposits (such as nahcolite in the U.S.) are available; and the waste product is highly water-soluble, making environmentally acceptable disposal difficult.

Table 2.5 Published Estimated Requirements

Case No. (see Table 2.2)		1	1	3
Reference		95	95	6
MWe		2×500	2×500	800
Coal: Sulphur	wt.%	4.0	4.0	3.6
Chlorine	wt.%	0.1	0.1	0.1
Sulphur Capture,	%	90	90	95
Regeneration by:		Limestone	Lime	Lime
Typical fresh water and water treatment requirements per MWh:				
Raw water,	m³	0.27	0.28	–
Steam (29 bar, 300°C)	tonne	0.10	0.10	–
Consumption of reagents, kg/MWh:				
NaOH		1.7	1.0	1.0
Regenerant		55.9	31.1	28.5
Sludge fixative lime		4.2	4.5	–
End-product or waste disposal requirements:				
Sludge volume	m³/MWh	0.13	0.13	0.09
Fly-ash	wt.%	30	30	–
Moisture	wt.%	30	30	–
Typical pressure losses, mbar:		15	15	–
Typical power requirements, MWe:		14.3	12.9	–
Typical reductions in combustion plant efficiency, %:				
Loss in power sent out		4.5	4.9	–

– Indicates information unavailable

Table 2.6 Status of Processes Using Non-Regenerable Solution-Based Wet Reagents: Dry End-Products

Code No.	Vendor	No. of Plants	Plant size Range MWe	Total MWe	Dates	% S Capture	Ref.
S12.1	Fläkt	2	3.5–110	114	1980–81	73–90	655, 671
	United McGill	11a	–	–	–	90	135
	Rockwell/Wheelabrator Frye	1	410	410	1981	70	223
S12.2	Walther	4	64–440	980	1984–88	Over 92	29, 659

a Includes some lime spray dryer plants
– Indicates information unavailable

Typical land area requirements: ***

Typical fresh water and water treatment requirements: ***

Consumption of reagents: Reagent is supplied at up to twice stoichiometric requirements, but the supply rate is usually lower. The Fläkt plant (20% of 550 MWe, 0.4% sulphur coal, 73% sulphur capture) operates with a Na/inlet S molar ratio of 1.2 or Ca/S molar ratio of 1.6 [671].

The Rockwell International/Wheelabrator-Frye plant (410 MWe, 0.87% sulphur coal) achieved 70% sulphur capture with a stoichiometric ratio of 1 mol/mol input S [684].

End-product or waste disposal requirements: ***

Typical pressure losses: 40 mbar [223].

Typical power requirements: For a 410 MWe boiler, 0.87% sulphur coal, about 70% sulphur capture: 0.5–0.6% of generation [684].

Typical reductions in combustion plant efficiency: ***

Operating experience: A summary of problems reported by Battelle in their visit reports [655, 666, 671, 684] is as follows (total number of plants visited, numbers experiencing problems in various plant areas, and those with problems remaining unsolved):

No. visited	2
Total equiv. MWe	520
Total streams	2

Plant area	Number of Problems Total	Unsolved	Solved
Absorber	2	1	1
Flues or stack	0	0	0
Dampers	0	0	0
Sorbent preparation	0	0	0
Waste/product system	0	0	0
Other	1	1	0
Totals	3	2	1

Process developments: ***

Appraisal:

1.	Information available	1
2.	Process simplicity	3
3.	Operating experience	0
4.	Operating difficulty	0
5.	Loss of power	1*
6.	Reagent requirements	1
7.	Ease of end-product disposal	0
8.	Process applicability	2
	Total	8

*Assumed in absence of data

Process Code S12.2 – Walther Process

Applicability: This process type is suitable for retrofit and new build applications.

Typical land area requirements: For handling 50% of the flue gas from a 475 MWe boiler, 1.1% sulphur coal, over 92% sulphur capture, the FGD plant requires 0.2 hectare [659].

Typical fresh water and water treatment requirements: The water requirement is estimated at about 0.17 m^3/MWh [Walther promotional literature].

Consumption of reagents: The consumption of ammonia is about equivalent to the sulphur captured [Walther promotional literature].

End-product or waste disposal requirements: The end product, ammonium sulphate, is potentially marketable as a fertiliser.

Typical pressure losses: 20 mbar can be assumed [Walther promotional literature].

Typical power requirements: ***

Typical reductions in combustion plant efficiency: ***

Operating experience: No operating data are yet available.

Process developments: ***

Appraisal:

1.	Information available	1
2.	Process simplicity	2
3.	Operating experience	0
4.	Operating difficulty	0*
5.	Loss of power	1*
6.	Reagent requirements	1
7.	Ease of end-product disposal	1
8.	Process applicability	2
	Total	8

*Assumed in absence of data

CATEGORY S21

The status of Category S21 processes is indicated in Table 2.7, which presents information on the size of plants built, year of completion, and percentage sulphur capture. The list includes plants operating with limestone and lime slurries, with alkaline fly-ash added in some instances; plants operating with formic, adipic or succinic acid or DBA additives, or with magnesia-based, sodium thiosulphate or calcium chloride additives; and plants operating with and without forced oxidation of the calcium sulphite hemihydrate produced.

Process Code S21.1 – Limestone or Lime Slurry Scrubbing Processes

Applicability: This process type is suitable for retrofit and new build applications. Depending on the operating conditions, 90–95% sulphur capture is readily attained.

Typical land area requirements: Values reported are very variable, probably partly because they do not all encompass areas for raw material and product processing and handling (excluding areas for waste disposal). They are:

Table 2.7 Status of Processes Using Non-Regenerable Slurry-Based Wet Reagents: Wet End-Products

Code No.	Vendor	No. of Plants	Plant size Range MWe	Total MWe	Dates	% S Capture	Ref.
S21.1	AAF	7	65–495	1955	1975–81	–	125, 685
	Babcock-Hitachi	1	250	250	1976	93	678
	Babcock Power Ltd.	23	25–1628*	15,938*	1974–87	90–95	135,**
	B&W Co.	16	55–875	7320	1977–83	90–99	125, 654, 680
	Bahco	3	102*a	102*a	1969–	98	652
	Bechtel	2	778	1555	1983	–	125
	Bischoff	33	10–1500*	29,727*	1971–88	Over 95	135,**
	Chiyoda 121	7	23–200	700	1978–84	97–99	95, 680
	Combustion Engineering	31	72–890	12020b	1973–81b	75–90b	135
	Deutsche Babcock	31	150–470	8970	1985–88	90–95	29, 135
	Environeering/Riley Stoker	3	180–415	775	–	50–85	125, 662
	E.S.T.S.	2	309–600	909	1985–88	–	**
	Fläkt/Babcock	2	96–181	277	1987–88	–	29
	GEESI	24	175–1150	14305	1973–	70–85	125, 675
	IHI	33	10–2400*	13055*	1971–87	90–95	678, **
	Kawasaki HI	2	156–250	406	1983–84	98	679
	Kobe Steel	1	720*	720	–	90	684
	Lisop Oy	4	–	–	–	Over 80	135
	Mitsubishi	77	12–2450*	41,060*	1972–87	90–98	135, **
	Peabody P.S.	10	179–625	4525	–	–	125
	Pullman Kellogg	7	400–917	3995	1979–84	92–98	125, 661
	Research Cottrell	27	115–750	10,305	1973–88	95	135
	Saarberg-Holter	25	1–600	5760	1974–88	95	95, 135
		+ 7	4–131*	350	1972–82		
	Steinmuller	25	72–707c	6100c	1981–88c	85–95	135, **
	TVA	1	550	550	1977	80	125, 665
	Texcel	25	5 Min.	–	–	Over 95	135
	Thyssen/CEA	39	10–740	6430d	1975–88d	Over 95	135
	UOP Air Corr'n Div.	6	195–720	2990	–	80–85	125
	Waagner-Biro	3	160–330	690	1986	–	**
	Wheelabrator	1	750	750	–	–	125

* Gas flow, Thousand Nm^3/h
** Vendor's promotional literature
a For 1 installation
b For 24 installations [125, 653, 669, 677]
c For 20 installations [29]
d For 24 plants in the range 77–500 MWe [29, 125, 653]
– Indicates information unavailable

Vendor	MWe	Hectare	Reference
B & W	55	0.22a	680
	364	0.21b	685
	618	4.0c	654
Environeering	2×180	0.40	662
GEESI	600	1.0	670
MHI/Thyssen	185	0.8	659
Peabody PS	440–447	0.46	685
UOP	532	0.19	656

Notes: a: for FGD system; b: for absorbers and thickeners; c: for FGD system, sludge treatment and lime handling.

Typical fresh water and water treatment requirements: Water requirements are typically 0.1–0.4 m^3/MWh; the quantity appears to depend on the SO_2 concentration, though one vendor (Research-Cottrell) is reported to require only 0.03–0.05 m^3/MWh for inlet SO_2 concentrations as high as 10,000 mg/Nm^3 [135]. Cooling and reheating by gas–gas heat exchange reduces the requirement for water.

Waste water from systems with forced oxidation in Japan has 30–100 ppm chemical oxygen demand (COD), and contains 500–600 ppm fluorides, up to 500 ppm iron, 1000 ppm magnesium and 8000 ppm chlorides; it has to be treated with lime to precipitate the heavy metals and fluorides, followed by ion exchange to reduce the COD [184].

Further information in the form of published estimates is presented in Table 2.8 for limestone as absorbent, and Table 2.9 for lime as absorbent.

Consumption of reagents: The requirements for lime or limestone can be reduced if high-calcium fly-ash can be used [175], or by the use of fine grinding of the limestone, e.g. to about 95% finer than 874 micron [215]. Requirements for other reagents, e.g. organic acid or magnesia-based additives, depend on details of process variations, and are dealt with in Section 2.5.

Further information in the form of published estimates is presented in Table 2.8 for limestone as absorbent, and Table 2.9 for lime as absorbent.

End-product or waste disposal requirements: Estimates are presented in Table 2.8 for limestone as absorbent, and in Table 2.9 for lime as absorbent.

Typical pressure losses: Gas pressure drops across the FGD system are typically 15–20 mbar for units with spray towers; the Saarberg-Holter (25–34 mbar) and Chiyoda Thoroughbred 121 (40–50 mbar) pressure drops are higher, however. The higher pressure loss with the Saarberg-Holter process results from the use of venturi absorbers [192]. The use of venturi prescrubbers also increases the total pressure drop of plant with spray tower absorbers to about 50 mbar [175].

Further information in the form of published estimates is presented in Table 2.8 for limestone as absorbent, and Table 2.9 for lime as absorbent.

Typical power requirements: Estimates are presented in Table 2.8 for limestone as absorbent, and in Table 2.9 for lime as absorbent.

Typical reductions in combustion plant efficiency: Estimates are presented in Table 2.8 for limestone as absorbent, and in Table 2.9 for lime as absorbent.

Operating experience: This is the most extensively adopted FGD system, with over 400 installations worldwide indicated in Table 2.7 above. Early operating difficulties resulted from poor design and inadequate appreciation of details of the process chemistry; these have now been largely overcome, and operating reliability is high. At the TVA Paradise station, the major problems concerned: plugging of mist eliminators; collapse of eliminators due to temperature excursions; and heat exchanger failures due to poor design. Other problems included corrosion of outlet ducts and other components, and the use of unsuitable types of component. All of these problems were overcome by modifications to the plant [230].

The older units in the U.S. are operated without forced oxidation of the calcium sulphite, but most units in Japan and the F.R.G., and an increasing number of new units in the U.S., have forced oxidation in order to reduce end-product processing and disposal costs, and the environmental impact of waste disposal; for example, forced oxidation results in 40% reduction in area for disposal compared with ponding of natural-oxidation sludge [184]. However, retrofitting of forced oxidation to existing lime or limestone slurry FGD plant would incur high capital cost, as it involves changes in the process chemistry, water balance, pH control and product purity. Examples of FGD plant with forced oxidation are offered by Chiyoda, Combustion Engineering, GEESI, IHI, Pullman-Kellogg, Research Cottrell, Saarberg-Holter.

Testing of the Chiyoda Thoroughbred 121 system on a 23 MW pilot plant in the U.S. demonstrated efficient and reliable operation when applied to a coal-fired boiler; four further units (including two units for 200 MWe utility boilers) are under construction [95]. Two 200 MWe units tested for over one year in Japan were found to give essentially 100% reliability [188].

Table 2.8 Published Estimated Requirements Limestone Absorbent

Case No.			1a	1b	1c	3a	3b
Reference			95	95	95	6	6
MWe			2×500	2×500	2×500	800	800
Coal:	Sulphur	wt.%	4.0	4.0	4.0	3.6	3.6
	Chlorine	wt.%	0.1	0.1	0.1	0.1	0.1
Forced Oxidation			No	Yes	Yes*	No	Yes*
Sulphur capture		%	90	90	90	90	95
Molar ratio Ca/captured S			1.15	1.15	1.01	–	–
Typical fresh water and water treatment requirements per MWh:							
Cooling Tower blowdown		m^3	0.32	0.42	0.30	–	–
Steam		tonne	0.10	0.10	0.10	–	–
Typical sorbent requirements, kg/MWh:							
Limestone			60.8	60.8	53.7	44.9	45.0
Fixative lime			2.3	Nil	Nil	–	Nil
Formic acid			Nil	Nil	0.06	Nil	Nil
End-product or waste disposal requirements:							
Volume		m^3/MWh	0.13	0.15	0.17	0.11	0.13
Fly-ash		wt.%	30	30	30	–	–
Water		wt.%	30	20	20	–	–
Typical pressure losses, mbar:			15	15	40	–	–
Typical power requirements, MWe:			20.6	20.5	18.8	–	–
Typical reductions in combustion plant efficiency, %:							
Loss of power sent out			5.2	4.9	5.1	–	–

* Chiyoda Thoroughbred 121 process
– Indicates information unavailable

Table 2.9 Published Estimated Requirements Lime Absorbent

Case No.			1d	1e	3c
Reference			95	95	6
MWe			2×500	2×500	800
Coal:	Sulphur	wt.%	4.0	4.0	3.6
	Chlorine	wt.%	0.1	0.1	0.1
Forced Oxidation			No	Yes*	No
Sulphur capture		%	90	90	95
Molar ratio Ca/captured S			1.10	1.02	–
Typical fresh water and water treatment requirements per MWh:					
Raw water		m^3	0.12	0.10	–
Cooling tower blowdown		m^3	0.18	0.16	–
Steam (29 bar, 300°C)		tonne	0.10	0.10	–
Typical sorbent requirements, kg/MWh:					
Lime			34.3	31.8	22.8
End-product or waste disposal requirements:					
Volume		m^3/MWh	0.12	0.15	0.11
Fly-ash		wt.%	30	30	–
Water		wt.%	30	20	–
Typical pressure losses, mbar:			15	25	–
Typical power requirements, MWe:			17.4	17.0	–
Typical reductions in combustion plant efficiency, %:					
Loss of power sent out		%	4.9	4.9	–

* Saarberg-Holter process
– Indicates information unavailable

A summary of problems reported by Battelle in their visit reports [650, 652–657, 659, 661–665, 668, 670, 674–682, 684, 685] is shown in Table 2.10 (total number of plants visited, numbers experiencing problems in various plant areas, and those with problems remaining unsolved).

Process developments: The use of additives, e.g. formic acid, adipic acid and 'dibasic acid, DBA' (a mixture of glutaric, adipic and succinic acids) allows improved performance and reduced scale formation with negligible increase in capital cost [95, 182] and without significant environmental impact [186]; the liquid/gas ratio needed in the absorber can be greatly reduced, e.g. by 50%, by the addition of 2 g/l of formic acid to the slurry [187]. Alternative additives are: sodium thiosulphate at a concentration of 100–700 ppm in the slurry which reduces scale-formation and produces a sulphite sludge containing 85% solids [184]; magnesia (Thiosorbic lime process [181]) added at the rate of 4–8% by weight of the solids. The magnesia forms soluble magnesium sulphite which is then the principal absorbing reagent, forming magnesium bisulphite and sulphate which react with lime to regenerate the magnesium sulphite.

Simultaneous absorption and forced oxidation, as in e.g. the Chiyoda Thoroughbred 121 process, improves sulphur capture by removing sulphite and bisulphite from the system. The development of spray towers in single units for 1000 MWe installations is held to be feasible [173]. The pressure drop associated with gas–gas heat exchangers (about 10 mbar), or the expenditure of fuel for reheating, can be avoided by exhausting the purified gas to atmosphere in the power station cooling tower with a suitable acid-resisting lining [187].

Appraisal:

1.	Information available	2
2.	Process simplicity	2
3.	Operating experience	3
4.	Operating difficulty	0
5.	Loss of power	1
6.	Reagent requirements	1
7.	Ease of end-product disposal	1 (0 if no forced oxidation)
8.	Process applicability	2
	Total	12

CATEGORY S22

The status of Category S22 processes is indicated in Table 2.11, which presents information on the size of plants built, year of completion, and percentage sulphur capture.

Table 2.10 Summary of Problems Reported by Battelle

Absorbent Forced Oxidation	LST No	LST Yes	LST Yes SH	L No	L Yes	L Yes CT121
No. of plants visited	12	2	1	15	7	1
Total equiv. MWe	7634	591	177	7687	6587	87
Total No. of streams	48	3	1	51	19	1
No. of problems in:						
Prescrubber	0	0	0	4	2	0
Unsolved	0	0	0	3	1	0
Absorber	3	0	0	9	3	1
Unsolved	2	0	0	8	3	0
Demister	2	0	0	8	3	0
Unsolved	1	0	0	6	3	0
Flue or stack	8	0	0	4	4	0
Unsolved	5	0	0	4	4	0
Damper	5	0	0	3	3	0
Unsolved	5	0	0	3	3	0
Sorbent prep.	2	1	1	3	1	0
Unsolved	2	1	1	2	1	0
Waste/prod. system	2	0	0	6	3	0
Unsolved	2	0	0	5	2	0
Other areas	3	1	1	9	4	0
Unsolved	3	1	1	8	3	0
Total of problems	25	2	2	46	23	1
Total unsolved	20	2	2	39	20	0

LST: Limestone; L: Lime;
SH: Saarberg-Holter; CT121: Chiyoda Thoroughbred 121

Table 2.11 Status of Processes Using Non-Regenerable Slurry-Based Wet Reagents: Dry End-Products

Code No.	Vendor	No. of Plants	Plant size Range MWe	Total MWe	Dates	% S Capture	Ref.
S22.1	B&W	2	447–575	1017	1982–85	85–87	95, 222
	Combustion Engineering	6	100–850	–	–	Over 90	135
	DB Gas Cleaning	10	–	–	–	95 Max	135
	Ecolaire	1	79*	79*	1981	65–90	682
	Fläkt (Drypac)	23	133–3, 152a	12, 188a	1982–88	70–90	29, ** 125, 135, 655, 671, 681
	Fläkt/Niro/Joy	31	43–4300a	33, 294a	1980–89	Over 95	135, **
	GEESI	1	44	44	1980	80	95
	Mikropul	2	3–100	–	–	90–95	135
	Research-Cottrell	9	10–60	–	–	90	135
	Rockwell/	2	276–440	716	1981–85	70	95,
	Wheelabrator	+2	50–113*	163*	1980–83	70	222, 654, 676
	Texcel	5	4 minimum	–	–	Over 95	135

* Steam production, tonne/h
** Vendor's promotional literature
– Indicates information unavailable
a Gas flow, thousands Nm^3/h

Process Code S22.1 – Lime Slurry Scrubbing/Spray Drying Process

Applicability: This process type is suitable for retrofit and new build applications. All of the utility installations in the U.S. have been for combustion systems firing low-sulphur coal, for which this process type is more economical than wet slurry scrubbing systems [210], though the process is technically capable of application to high-sulphur coals [222] and has been applied to industrial boilers firing coal of up to 5% [198]. Depending upon operating conditions, any sulphur capture up to about 90% is readily attained. The process is capable of dealing with variable loads [197].

Typical land area requirements: Values depend upon the size of the plant as follows:

Vendor	Mwe	Steam tonne/h	Gas Nm^3/h	Hectare	Reference
Ecolaire		79	93,000	0.06	682
Fläkt		106	128,000	0.09	685
Joy/Niro	440		2,507,000	0.49	678
	316		1,539,000	0.37	684
		181	175,000	0.10	683
Research-Cottrell	60		243,000	0.08	684

Typical fresh water and water treatment requirements: Water requirements depend upon the gas temperature and other site-specific factors, and can be in the range of 0.01–0.2 m^3/MWh [95, 135]. Water for slaking should be of higher quality than needed for limestone slurry processes, and dilution water for slurry preparation should be of low hardness to avoid blinding of the lime surface [222].

Further information is presented in Table 2.12 in the form of published estimates.

Consumption of reagents: The consumption of lime is between 1.3 and 1.85 times stoichiometric. However, in some instances the ash of the coal provides a significant amount of lime, and the fresh added lime requirement is thereby reduced. The lime is slaked on site; high-calcium, soft-burned, low-inerts pebble lime should be used; spontaneous slaking during storage should be avoided; and care is needed to produce a reactive, finely sized material.

Further information is presented in Table 2.12 in the form of published estimates.

End-product or waste disposal requirements: Published estimates are presented in Table 2.12.

Typical pressure losses: Observed pressure losses are in the range 18 mbar [678] to 37 mbar [654].

Further information is presented in Table 2.12 in the form of published estimates.

Typical power requirements: Measurements of power consumption for a number of installations in the range 60–440 MWe, firing low-sulphur coals (0.3–0.8% S) were equivalent to 0.5–0.9% of the gross power generated [678, 684].

Further information is presented in Table 2.12 in the form of published estimates.

Table 2.12 Published Estimated Requirements

Case No. (Table 2.2)		2
Reference		95
MWe		2 × 500
Coal: Sulphur	wt. %	0.48
Chlorine	wt. %	0.07
Sulphur capture	%	70
Typical fresh water and water treatment requirements:		
Raw water	m^3/MWh	0.19
Consumption of reagents:		
Pebble lime, Ca/S molar ratio 0.9	kg/MWh	4.9
End-product or waste disposal requirements:		
Waste volume (20% moisture, 63% fly-ash)	m^3/MWh	0.04
Typical pressure loss, mbar:		30
Typical power requirements MW:		12.8
Typical reduction in combustion plant efficiency:		
Loss of power sent out	%	1.2

Typical reductions in combustion plant efficiency: Published estimates are presented in Table 2.12.

Operating experience [222]: Early problems included: solids deposition form incomplete drying due to design faults, and to excessive water content of the slurry (ameliorated by recycle of the dry product, which reduces the water content of the slurry and increases the surface area); mechanical problems with rotary atomisers and plugging of nozzle atomisers; lime slaking and slurry piping problems. These problems can be overcome by better design and operating practices.

Corrosion due to condensation is not common but can occur at part-load conditions, and insulation of potentially cold spots has been found desirable; it is advantageous to reheat the gas to avoid such problems. Particulate removal (electrostatic precipitators, or the more commonly used fabric filters, both of which give enhanced sulphur capture performance [211,212]) have not generally given problems.

A summary of problems reported by Battelle in their visit reports [652, 654, 663, 668, 670, 671, 676, 678, 680–685] is as follows (total number of plants visited, numbers experiencing problems in various plant areas, and those with problems remaining unsolved):

No. visited	14
Total equiv. MWe	1733
Total streams	22

Plant area	Number of Problems		
	Total	Unsolved	Solved
Absorber	9	8	1
Flue or stack	0	0	0
Dampers	0	0	0
Sorbent preparation	3	2	1
Waste/product system	7	6	1
Other areas	3	3	0
Totals	22	19	3

Process developments: A development is the 'dry scrubbing' process in which the exit temperature of gas from the scrubber is as close as possible to the adiabatic saturation temperature; the 'dry' particles of the end-product then retain sufficient surface moisture to allow rapid absorption of SO_2 to continue [210] as in the processes based on dry injection into the flue gas (Option B described in Section 2.2).

Other developments, still in the laboratory or pilot plant stage, include: 'in-duct' scrubbing (i.e. spraying slurry into the horizontal duct, obviating the need for a separate spray-dryer vessel) [190]; use of calcium chloride as an activation agent, particularly for limestone slurries [195]; and the use of magnesia-based slurries [196].

Appraisal:

1.	Information available	2
2.	Process simplicity	3
3.	Operating experience	2
4.	Operating difficulty	0
5.	Loss of power	2
6.	Reagent requirements	1
7.	Ease of end-product disposal	1
8.	Process applicability	2
	Total	13

CATEGORY S31

The status of Category S31 processes is indicated in Table 2.13, which presents information on the size of plants built, year of completion, and percentage sulphur capture.

Process Code S31.1 – Wellman-Lord Process

Applicability: This process type is suitable for new build and retrofit applications. It has been installed on utility and industrial coal-fired and oil-fired boilers as well as on Claus plant and sulphuric acid plant. Depending upon operating conditions, it is capable of giving 90–99% sulphur capture. It is capable of dealing with a variable load and SO_2 input.

Land area requirements: Data reported by Battelle are presented in Table 2.14. Published estimates are presented in Table 2.15.

Typical fresh water and water treatment requirements: Published estimates are presented in Table 2.15.

Consumption of reagents: Data reported by Battelle are presented in Table 2.14. Published estimates are presented in Table 2.15.

End-product or waste disposal requirements: Published estimates are presented in Table 2.15.

Typical power requirements: Data reported by Battelle are presented in Table 2.14. Published estimates are presented in Table 2.15.

Typical reductions in combustion plant efficiency: Published estimates are presented in Table 2.15.

Operating experience: The operational reliability is quoted as being in the range 95–98%; one plant showed a reliability of 96% for the absorption area and 92% for the regeneration area. Prescrubbing is essential for coal-fired plants, to remove fly-ash, HCl, SO_3 and trace impurities [276].

Table 2.13 Status of Processes Using Regenerable Solution-Based Wet Reagents: Thermal Regeneration

Code No.	Vendor	No. of Plants	Plant size Range MWe	Total MWe	Dates	% S Capture	Ref.
S31.1	Davy McKee	35	17–4640*	22276*	1970–88	90–99	**, 266
S31.2	Fläkt R'ch/Boliden AB	3	0.1–5*	–	1976–77	–	95
S31.3	Air Products/Catalytic/IFP	2 (Pilot plant)	2–25	27	–	95	6, 137

* Thousand m^3/h
** Vendor's promotional literature
– Indicates information unavailable

For plants in which elemental sulphur is recovered, the sulphur condenser was a source of trouble: mainly sulphur deposition, leaks and valve failures. Another plant gave trouble from acidic conditions and unreliable steam supply [146].

A summary of problems reported by Battelle in their visit reports [658, 659] is given below (total number of plants visited, numbers experiencing problems in various plant areas, and those with problems remaining unsolved).

No. visited	2
Total equiv. MWe	1958
Total number of streams	7

Plant area	Number of Problems		
	Total	Unsolved	Solved
Prescrubber	1	1	0
Absorber	1	1	0
Demisters	2	0	2
Flues or stack	0	0	0
Dampers	0	0	0
Sorbent preparation	1	1	0
Product system	1	1	0
Other	2	2	0
Totals	8	6	2

Process developments: The formation of sulphate is now being reduced by 75% at some plants by modified operational techniques and the use of an anti-oxidant. This may allow the sulphate crystallisation part of the system to be dispensed with in future plants, with sulphate concentrations being controlled by a small purge [276].

Appraisal:

1.	Information available	2
2.	Process simplicity	1
3.	Operating experience	2
4.	Operating difficulty	0
5.	Loss of power	0
6.	Reagent requirements	2
7.	Ease of end-product disposal	2
8.	Process applicability	2
	Total	11

Process Code S31.2 – Fläkt-Boliden Process

Applicability: This process type is suitable for retrofit and new build applications.

Typical land area requirements: ***

Table 2.14 Requirements Reported by Battelle

Reference	658	658	659
MWe	711	1068	3×60
Fuel S content, %	0.8	0.8	5.8a
Sulphur capture, %	90	84	90
End-product	S (by Allied Process) or acid		Sulphuric Acid
Typical land area requirements, hectare:			
Absorbers	0.43	0.58	–
Regeneration	0.21*		–
Crystallisation	0.15*		–
Sulphur production	0.14*		–
Acid production (alternative)	0.16*		–
Total area (FGD plant)	1.51–1.53*		2.4
Steam consumption:			
Tonne/tonne S removed	7.8*		–
Tonne/MWh (3.8 bar)	–		0.20–0.23
(42 bar from acid plant)	–		0.06
Liquid/gas ratio, litre/m³:			
Venturi pre-scrubber	0.9*		3.1
Absorber: liquid feed	0.1*		0.06
recirculation	0.24		0.24
Consumption of reagents:			
Soda ash, kg/kg S removed	0.19*		Nil
NaOH, kg/MWh	Nil*		about 4
Natural gas, m³/kg S removed	0.39*		Nil
Pressure drop, mbar	70–90	less than 70	75
Power consumption, MWe:	27	–	11

Note a: Mean of fuels supplied (petroleum coke, refinery fuel gas and oil)
* For (or shared by) both the 711 MWe and the 1068 MWe plant
– Indicates information unavailable

Typical fresh water and water treatment requirements: Published estimates are presented in Table 2.16.

Consumption of reagents: Published estimates are presented in Table 2.16.

End-Product or waste disposal requirements: Published estimates are presented in Table 2.16.

Typical power requirements: Published estimates are presented in Table 2.16.

Typical reductions in combustion plant efficiency: Published estimates are presented in Table 2.16.

Operating experience: The process has been operated only on pilot plants and a few small commercial plants.

Process developments: ***

Appraisal:

1.	Information available	1
2.	Process simplicity	1
3.	Operating experience	0
4.	Operating difficulty	0*
5.	Loss of power	0
6.	Reagent requirements	2
7.	Ease of end-product disposal	2
8.	Process applicability	2
	Total	8

*Assumed in absence of data

Process Code S31.3 – Catalytic Inc./IFP Process

Applicability: This process type is suitable for retrofit and new build applications. It is capable of giving about 95% sulphur capture [6].

Table 2.15 Published Estimated Requirements

Case No. (Table 2.2)		1	3	4	6
Reference		95	6	265	**
MWe		2×500	800	3×660	2000
Coal: Sulphur	wt. %	4.0	3.6	2.0	2.0
Chlorine	wt. %	0.1	0.1	0.4	–
S Capture	%	90	95	90	91
S Recovery process		Claus	Resox	Claus	Claus
Typical land area requirements, hectare:					
Absorption plant		–	–	0.9	–
Chemical plant		–	–	1.1	–
Limestone stock (6 weeks supply)		–	–	0.6	–
Rail siding		–	–	2.3	–
Total (Approx.)		–	–	5	–
Typical fresh water and water treatment requirements per MWh:					
Process water	m^3	0.26	–	–	–
Cooling water	m^3	7.8	–	–	3.0
High pressure steam	tonne	0.10	–	(0.11)	
Low pressure steam	tonne	0.20	–	(0.14)	0.08
Consumption of reagents per MWh (See Note a):					
NaOH (solid)	kg	1.1	2.0	Nil	Nil
(47% liquor)	kg	Nil	Nil	2.6	1.2
Natural gas	Nm^3	7.0	Nil	2.2	2.3
Coal or coke	kg	Nil	15.3	Nil	Nil
Limestone (for prescrubber water treatment)	kg	–	0.6	2.7	Nil
Lime (for ditto)	kg	–	Nil	Nil	1.25
End-product or waste disposal requirements per MWh:					
Sulphur	kg	15.5	12.8	5.5	5.8
Sodium sulphate:					
solution	m^3	0.002	Nil	Nil	Nil
solid	kg	Nil	3.4	2.2	1.0 (b)
Calcium chloride (25% soln.)	m^3	–	–	–	0.01
Slurry	m^3	0.06	0.004	0.06	–
Typical power requirements, MWe:		28.8	–	50	38
Typical reductions in combustion plant efficiency:					
Loss in power sent out	%	11	–	7.1	–

**Vendor's promotional literature
– Indicates information unavailable
Note a: Additive (NA_4 EDTA) is also required at a rate of 0.006 kg/MWh [135]
Note b: Composition [vendor's promotional literature]: Na_2SO_4 55%, Na_2SO_3 17%, $Na_2S_2O_5$ 26%, $Na_2S_2O_3$ 2%

Table 2.16 Published Estimated Requirements

Case No. (Table 2.2)		1	5
Reference		95	268
MWe		2×500	500
Coal: Sulphur	wt. %	4.0	4.0
Chlorine	wt. %	0.1	–
Sulphur capture	%	90	90
Typical fresh water and water treatment requirements per MWh:			
Process water	m^3	0.25	0.25
Cooling water	m^3	12.9	10.0
L.P. steam (4.5 bar)	tonne	0.01	0.02
H.P. steam (25 bar)	tonne	0.34	0.34
Consumption of reagents per MWh:			
NaOH	kg	0.27	1.2
Citric acid	kg	0.05	0.04
Natural gas	m^3	7.2	–
End-product or waste disposal requirements per MWh:			
Sulphur	kg	15.8	21.8
Sodium sulphate decahydrate	kg	–	3.4
Fly-ash + waste solids, 20% moisture	m^3	0.06	–
Typical power requirements, MWe:		19.9	9.1
Typical reductions in combustion plant efficiency:			
Loss in power sent out	%	9.1	–

– Indicates information unavailable

Table 2.17 Published Estimated Requirements

Case No. (Table 2.2)		3
Reference		6
MWe		800
Coal: Sulphur	wt. %	3.6
Chlorine	wt. %	0.1
Sulphur capture	%	95
Consumption of reagents, kg/MWh:		
Ammonia		0.19
Limestone		0.57
Coal and coke		22
End-product or waste disposal requirements per MWh:		
Elemental sulphur (Over 99.9% pure)	kg	13.8
Dry solids	kg	2.0
Slurry waste	m^3	0.003

Typical land area requirements: ***

Typical fresh water and water treatment requirements: ***

Consumption of reagents: Published estimates are presented in Table 2.17.

End-product or waste disposal requirements: Published estimates are presented in Table 2.17.

Typical power requirements: ***

Typical reductions in combustion plant efficiency: ***

Operating experience: Experience is limited to operation of two pilot plants [6].

Process developments: A modification of the process can simultaneously give about 70% removal of NO_x [6, 137].

Appraisal:

1.	Information available	1
2.	Process simplicity	1
3.	Operating experience	0
4.	Operating difficulty	0*
5.	Loss of power	0*
6.	Reagent requirements	2
7.	Ease of end-product disposal	2
8.	Process applicability	2
	Total	8

*Assumed in absence of data

CATEGORY S32

The status of Category S32 processes is indicated in Table 2.18, which presents information on the size of plants built, year of completion, applicability (i.e. suitability for retrofit); and percentage sulphur capture.

Table 2.18 Status of Processes Using Regenerable Solution-Based Wet Reagents: Chemical Regeneration

Code No.	Vendor	No. of Plants	Plant size Range MWe	Total MWe	Dates	% S Capture	Ref.
S32.1	–	2	1–60	61	1979	90	6, 653
S32.2	Conoco Coal Dev. Co.	2 (Pilot plants)	0.35–7	7.35	–	–	95
S32.3	–	–	–	–	–	–	–
S32.4	Rockwell	1	100	100	1983	92	95, 269

– Indicates information unavailable

Process Code S32.1 – Citrate Process

Applicability: This process type is suitable for new build and retrofit applications. A sulphur capture of over 90% can be attained [6].

Typical land area requirements: ***

Typical fresh water and water treatment requirements: ***

Consumption of reagents: Data reported by Battelle [653] are presented in Table 2.19. Published estimated requirements are presented in Table 2.20.

End-product or waste disposal requirements: Published estimated requirements are presented in Table 2.20.

Pressure losses: Data reported by Battelle [653] are presented in Table 2.19.

Power consumption: Data reported by Battelle [653] are presented in Table 2.19.

Typical reduction in combustion plant efficiency: ***

Operating experience: ***

Process developments: ***

Appraisal:

1.	Information available	1
2.	Process simplicity	1
3.	Operating experience	0
4.	Operating difficulty	0*
5.	Loss of power	0*
6.	Reagent requirements	2
7.	Ease of end-product disposal	2
8.	Process applicability	2
	Total	8

*Assumed in absence of data

Table 2.19 Data Reported by Battelle

Reference		653
MWe		60
Coal: Sulphur	wt. %	3.05
Gas: SO_2 content	ppmv	2000
Sulphur capture	%	90
Liquid/gas ratio, litre/m³:		
Prescrubber		3.8
Absorber		1.0
Consumption of reagents, kg/tonne recovered sulphur:		
Citric acid		14.4
Caustic soda		46.0
Pressure loss, mbar:		15
Power consumption MWe:		1.6

Process Code S32.2 – Conosox Process

Applicability: This process type is suitable for new build and retrofit applications.

Typical land area requirements: ***

Typical fresh water and water treatment requirements: Published estimated requirements are presented in Table 2.21.

Consumption of reagents: Published estimated requirements are presented in Table 2.21.

End-product or waste disposal requirements: Published estimated requirements are presented in Table 2.21.

Table 2.20 Published Estimated Requirements

Case No. (Table 2.2)		3
Reference		6
MWe		800
Coal: Sulphur	wt. %	3.6
Chlorine	wt. %	0.1
Sulphur capture	%	95
Consumption of reagents kg/MWh:		
Sodium citrate		0.19
Sodium hydroxide		0.6
Sodium thiosulphate		0.9
Limestone		0.57
Coal and coke		13
End-product or waste disposal requirements per MWh:		
Sulphur (99% pure)	kg	13.5
Sodium sulphate decahydrate	kg	4.7
Waste solids and slurry	m³	0.004

Table 2.21 Published Estimated Requirements

Case No. (Table 2.2)		1
Reference		95
MWe		2×500
Coal: Sulphur	wt. %	4.0
Chlorine	wt. %	0.1
Sulphur capture	%	90
Typical fresh water and water treatment requirements, m³/MWh:		
Raw water		0.03
Cooling water		5.7
Cooling tower blowdown		0.3
Consumption of reagents per MWh:		
Potassium hydroxide	kg	1.0
Liquid oxygen	kg	35
Fuel oil	m³	0.027
Methane	Nm³	1.33
End-product or waste disposal requirements per MWh:		
Sulphur	kg	17.1
Steam (5.8 bar)	tonne	0.034
Fly-ash to landfill (20% moisture)	m³	0.06
Typical power requirements, MWe:		9.2
Typical reductions in combustion plant efficiency:		
Loss in power sent out	%	0.03

Typical power requirements: Published estimated requirements are presented in Table 2.21.

Typical reductions in combustion plant efficiency: Published estimated requirements are presented in Table 2.21. It can be noted that the reduction in power sent out is smaller than the power consumed by the process owing to the credit for steam generated at 5.8 bar.

Operating experience [95]*:* This process type has been operated on the pilot plant scale only.

Process developments [95]*:* Requires demonstration on the 100 MWe operating scale.

Appraisal:

1.	Information available	1
2.	Process simplicity	1
3.	Operating experience	0
4.	Operating difficulty	0*
5.	Loss of power	0
6.	Reagent requirements	1
7.	Ease of end-product disposal	2
8.	Process applicability	2
	Total	7

*Assumed in absence of data

Process Code S32.3 – Ispra Mark 13A Process

Applicability: ***

Typical land area requirements: ***

Typical fresh water and water treatment requirements: ***

Consumption of reagents: ***

End-product or waste disposal requirements: Published estimates are presented in Table 2.22.

Typical power requirements: Published estimates are presented in Table 2.22.

Table 2.22 Published Estimated Requirements

Reference		272
MWe		400
Coal: sulphur	wt. %	1.0
Sulphur capture	%	93
End-product or waste disposal requirements per MWh:		
Hydrogen	Nm^3	2.1
Sulphuric acid (80%)	kg	5.4
Typical power requirements, MWe:		5.2

Typical reductions in combustion plant efficiency: ***

Operating experience: The process has been operated only on the bench scale (about 10 m^3/h gas) at the Ispra (Italy) Joint Research Centre of the Commission of the European Communities. The Commission has invited proposals for construction and operation of a 20,000 m^3/h pilot plant.

Process developments [273]*:* To make the process more compatible with the gas temperature from power stations without sacrificing the ability to concentrate the sulphuric acid produced, the final stages of concentration could be effected by using a small side stream of the flue gas at 380°C. Part of the gas leaving the air heater would then pass to the first-stage concentrator, and the remainder would flow to the reactor via a recuperator which would be used for reheating the purified gas.

Appraisal:

1.	Information available	1
2.	Process simplicity	1
3.	Operating experience	0
4.	Operating difficulty	0*
5.	Loss of power	0*
6.	Reagent requirements	2
7.	Ease of end-product disposal	1
8.	Process applicability	2
	Total	7

*Assumed in absence of data

Process Code S32.4 – Aqueous Sodium Carbonate Process

Applicability: This process type is suitable for new build and retrofit applications.

Typical land area requirements: ***

Typical fresh water and water treatment requirements: Published estimates are presented in Table 2.23.

Consumption of reagents: Published estimates are presented in Table 2.23.

End-product or waste disposal requirements: Published estimates are presented in Table 2.23.

Typical power requirements: Published estimates are presented in Table 2.23.

Typical reductions in combustion plant efficiency: Published estimates are presented in Table 2.23.

The 100 MW unit at the Huntley station used 4–5% of the station power [269].

Operating experience [269]: A 100 MW spray dryer with a 60 MW regeneration system, 2.5–3.5% sulphur coal, has been operated. The spray dryer gave

Table 2.23 Published Estimated Requirements

Case (Table 2.2)		1	3
Reference		95	6
MWe		2×500	800
Coal: Sulphur	wt. %	4.0	3.6
Chlorine	wt. %	0.1	0.1
Sulphur capture	%	90	95
Typical fresh water and water treatment requirements, m³/MWh:			
Process water, filtered, softened		0.28	–
Cooling water, filtered		19	–
Consumption of reagents, kg/MWh:			
Sodium hydroxide		1.0	0.53
Coke		29	18.7
Diatomaceous earth filter precoat		0.25	–
End-product or waste disposal requirements per MWh:			
Sulphur	kg	16.4	15.5
Dry solids	kg	Nil	3.9
Wet waste	m³	0.06	0.002
Typical power requirements, MWe:		37.6	–
Typical reductions in combustion plant efficiency:			
Loss in power sent out	%	3.6	–

– Indicates information unavailable

deposition problems. Regenerator problems arose because it was rated for only 60% of the absorber capacity, and because some items (e.g. reducer feed pumps, equipment for handling molten sulphur) operated at the low end of their design range. The reducer lost its refractory lining.

Process developments [269]: Design changes to increase operating efficiency would have included installation of a mechanical particulates collector upstream of the spray dryer to collect 85% of the particulates (the remainder would be taken out with the filter cake in the regeneration unit), and of a fabric filter to collect the dry product from the spray dryer. However, research on the project terminated in 1984 because of lack of funding [681].

Appraisal:

1.	Information available	1
2.	Process simplicity	1
3.	Operating experience	0
4.	Operating difficulty	0*
5.	Loss of power	1
6.	Reagent requirements	1
7.	Ease of end-product disposal	2
8.	Process applicability	2
	Total	8

*Assumed in absence of data

CATEGORY S41

The status of Category S41 processes is indicated in Table 2.24, which presents information on the size of plants built, year of completion, and percentage sulphur capture.

Process Code S41.1 – Magnesia Slurry Scrubbing Process

Applicability: This process type is suitable for new build and retrofit applications. Depending upon operating conditions, it is capable of attaining 98% or more sulphur capture [279].

Typical land area requirements: ***

Typical fresh water and water treatment requirements: Published estimates are presented in Table 2.25.

Consumption of reagents: Published estimates are presented in Table 2.25.

End-product or waste disposal requirements: Published estimates are presented in Table 2.25.

Typical pressure losses: Published estimates are presented in Table 2.25.

Typical power requirements: Published estimates are presented in Table 2.25.

Typical reductions in combustion plant efficiency: Published estimates are presented in Table 2.25.

Operating experience [95, 269, 278, 279]: The earliest plants, built by Chemico, gave generally satisfactory service, but are no longer operational as the boilers have been shut down.

The absorption systems of the United Engineers & Constructors (UE&C) plants operate with lower pH in the slurry than the Chemico plants (6.3 vs. 8.5); the process is tolerant of variations in the pH of the slurry. They had an average availability of over 99% under base load conditions. Early difficulties were mainly concerned with: slaking of the magnesia; erosion of centrifuge scrapers; and sulphite carry-over from the dryer cyclones. Modifications to correct these faults have been reported.

Table 2.24 Status of Processes Using Regenerable Slurry-Based Wet Reagents: Thermal Regeneration

Code No.	Vendor	No. of Plants	Plant size Range, MWe	Total MWe	Dates	% S Capture	Ref.
S41.1	United Engrs & Constructors	4	120–360	980	1974–82	90–98	95, 279
	Chemico	2	155–190	345	1972–73	–	95
	–	2	–	160	–	–	95

– Indicates information unavailable

The UE&C plants use off-site regeneration systems, with oil- or gas-fired fluidised bed calciners (compared with rotary calciners in the Chemico plants). The regeneration system availability has risen from 40% to over 80%; initial problems included: temperature control difficulties in the calciners leading to 'dead-burning' of the MgO; failure of the waste heat boiler to achieve its design performance; and mechanical problems with a booster fan.

The last two plants shown in Table 2.24 are operating in Japan: one on smelter gas, the other on oil-fired boiler and Claus gases; their availability is up to 95% in spite of some residual mechanical problems.

Table 2.25 Published Estimated Requirements

Case No. (Table 2.2)		1	3
Reference		95	6
MWe		2×500	800
Coal: Sulphur	wt. %	4.0	3.6
Chlorine	wt. %	0.1	0.1
S Capture	%	90	95
Typical fresh water and water treatment requirements, m^3/MWh:			
Raw water		0.36	–
Consumption of reagents, kg/MWh:			
Magnesia		3.2	0.47
Limestone		–	0.57
Fuel oil		0.45	–
Coal and coke		–	37.1
End-product or waste disposal requirements per MWh:			
Sulphuric acid (93%)	kg	48	Nil
Sulphur	kg	Nil	14.5
Waste material	m^3	0.06	0.006
Typical pressure losses, mbar:		20	–
Typical power requirements, MWe:		19.7	–
Typical reductions in combustion plant efficiency:			
Loss of power sent out	%	5.2	–

– Information unavailable

A summary of problems reported by Battelle [674] from a visit to a power station with UE&C FGD plants is as follows (problems in various plant areas, and remaining unsolved):

Total equiv MWe	700	
Number of absorption streams	5	

	Problems	
Plant area	Unsolved	Solved
Prescrubber	No	Yes
Absorber	No	No
Demisters	Yes	No
Flues or stack	No	Yes
Dampers	No	No
Sorbent preparation	No	Yes
Product system	No	No
Other	No	Yes

Process developments: Design modifications to the regeneration system for the UE&C plants were expected [279] to increase availability to 85%.

Appraisal:

1.	Information available	2
2.	Process simplicity	1
3.	Operating experience	0
4.	Operating difficulty	1
5.	Loss of power	1
6.	Reagent requirements	2
7.	Ease of end-product disposal	2
8.	Process applicability	2
	Total	11

CATEGORY S42

The status of Category S42 processes is indicated in Table 2.26, which presents information on the size of plants built, year of completion, and percentage sulphur capture.

Process Code S42.1 – Sulf-X Process

Applicability: This process type is suitable for new build and retrofit applications. It is capable of giving very high (99%) sulphur capture and some abatement of NO_x emissions, but the high liquid/gas ratio needed for high NO_x emission abatement would increase the absorber size and cost unduly.

Typical land area requirements: ***

Typical fresh water and water treatment requirements: Published estimates are presented in Table 2.27.

Consumption of reagents: Published estimates are presented in Table 2.27.

End-product or waste disposal requirements: Published estimates are presented in Table 2.27.

Typical pressure losses: Published estimates are presented in Table 2.27.

Typical power requirements: Published estimates are presented in Table 2.27.

Typical reductions in combustion plant efficiency: Published estimates are presented in Table 2.27.

Operating experience [277]: Experience limited to 1.5 MWt scale, with some stages of the regeneration system omitted. Initial difficulties, claimed to have been overcome, included mechanical transfer of the filter cake to and through the dryer, and oxidation of the wet cake in the calciner. Operation ceased in 1984, and for contractual reasons is not expected to be resumed [281].

Table 2.26 Status of Processes Using Regenerable Slurry-Based Wet Reagents: Chemical Regeneration

Code No.	Vendor	No. of Plants	Plant size Range MWe	Total MWe	Dates	% S Capture	Ref.
S42.1	P'burgh Env. Systems Inc. (PENSYS)	3	1–1.5	3.5	1977–82	99	95, 277

Process developments: Forecast in 1983 [277]: tests of alternative, less expensive reducing agents for the calciner; long-term reliability verification; scale-up to the 10–60 MWe scale.

Appraisal:

1.	Information available	1
2.	Process simplicity	0
3.	Operating experience	0
4.	Operating difficulty	0*
5.	Loss of power	1
6.	Reagent requirements	2
7.	Ease of end-product disposal	2
8.	Process applicability	2
	Total	8

*Assumed in absence of data

Table 2.27 Published Estimated Requirements

Case No. (Table 2.2)		1
Reference		95
MWe		2×500
Coal: Sulphur	wt. %	4.0
Chlorine	wt. %	0.1
S Capture	%	90
Typical fresh water and water treatment requirements per MWh:		
Raw water	m^3	0.06
Cooling water blowdown	m^3	0.27
High pressure steam	tonne	0.09
Low pressure steam	tonne	0.004
Consumption of reagents per MWh:		
Sodium sulphide	kg	0.52
Pyrite	kg	5.1
Coke	kg	9.6
Fuel oil	litre	1.4
End-product or waste disposal requirements per MWh:		
Sulphur	kg	16.2
Waste to landfill	m^3	0.06
Fly-ash	wt. %	72
Scrubber blowdown solids	wt. %	8
Moisture	wt. %	20
Typical pressure losses, mbar:		40
Typical power requirements, MWe:		25.6
Typical reductions in combustion plant efficiency:		
Loss in power sent out	%	5.5

CATEGORY S51

The status of Category S51 processes is indicated in Table 2.28, which presents information on the size of plants built, year of completion, and percentage sulphur capture.

Process Code S51.1 – Active Carbon Adsorption Process

Applicability: This process type is suitable for new build and retrofit applications, and is capable of attaining greater than 98% sulphur capture; it also removes hydrogen chloride and fluoride. A modification of the process (see Section 5) provides simultaneous SO_2 and NO_x emission abatement.

Typical land area requirements: ***

Typical fresh water and water treatment requirements: ***

Consumption of reagents: Published estimates are presented in Table 2.29.

End-product or waste disposal requirements: Published estimates are presented in Table 2.29.

Table 2.29 Published Estimated Requirements

Case No. (Table 2.2)		3
Reference		6
MWe		800
Coal: Sulphur	wt. %	3.6
Chlorine	wt. %	0.1
Sulphur Capture	%	95
Sulphur recovery process		Resox
Consumption of reagents, kg/MWh:		
Coal or coke		41
Active carbon		3.2
End-product or waste disposal requirements, kg/MWh:		
Sulphur		14.5
Dry solids		5.6

Table 2.28 Status of Processes Using Regenerable Dry Reagents: Thermal Regeneration

Code No.	Vendor	No. of Plants	Plant size Range*	Total *	Dates	% S Capture	Ref.
S51.1	BF/Uhde/Mitsui	5	1–1100	1505	1973–87	80–98	424, 435
	Sumitomo H.I.	1	300	300	–	90	424
	Chemiabau Reinluft	4	–	–	–	–	157

* Thousand Nm^3/h
– Indicates information unavailable

The Reinluft process uses granular low-temperature coke (prepared by devolatilising peat, lignite or certain bituminous coals at temperatures below 700°C), sized 2.5–3.5 mm, in place of the specially prepared active carbon used by the Bergbau-Forschung process [157].

Typical power requirements: ***

Typical reductions in combustion plant efficiency: ***

Operating experience [157]: The combined Bergbau Forschung/Foster Wheeler process was tested in a 20 MWe unit at the Scholtz station (Gulf Power Co.). The major problems were: hot spots in the adsorber; poor reliability of the carbon/sand separator; plugging of the Resox system condenser with sulphur and carbon. The 35 MWe prototype at Lunen, which used a Claus sulphur recovery system, gave better performance in the adsorption and sulphur recovery systems.

In the Reinluft process, the coke make-up reaches its peak reactivity in 3–10 cycles. Some operating difficulties were found in the operation of the four plants: fires in the coke bed; corrosion of equipment by sulphuric acid. A new design has been developed to overcome these problems.

Process developments [157]: Westvaco are developing a fluidised bed adsorption and regeneration process in which adsorbed SO_2 is catalytically oxidised to sulphuric acid on the carbon, and regeneration of the carbon is by reaction with hydrogen. A 560 Nm^3/h pilot plant has operated for 350 hours with one short interruption from sulphur-plugging. Hydrogen consumption is 3–4 mol/mol SO_2.

Appraisal:

1.	Information available	1
2.	Process simplicity	2
3.	Operating experience	0
4.	Operating difficulty	0*
5.	Loss of power	2*
6.	Reagent requirements	2
7.	Ease of end-product disposal	2
8.	Process applicability	2
	Total	11

*Assumed in absence of information

CATEGORY S52

The status of Category S52 processes is indicated in Table 2.30, which presents information on the size of plants built, year of completion, and percentage sulphur capture.

Process Code S52.1 – Wet Active Carbon Adsorption Process

Applicability: This process type is suitable for new build and retrofit applications.

Typical land area requirements: ***

Typical fresh water and water treatment requirements: ***

Consumption of reagents [157]: The carbon consumption for the Hitachi process is about 2% p.a.

End-product or waste disposal requirements [157]: The Lurgi process produces 7% sulphuric acid which is concentrated to 15% in the gas cooler. The Hitachi process produces 20% acid which is concentrated to 65% using a submerged combustor.

Typical power requirements: ***

Typical reductions in combustion plant efficiency: ***

Operating experience [157]: The Hitachi process encountered minor problems initially, but its general performance is claimed to be very successful. In one plant the dilute (17%) acid was not concentrated, but used to produce gypsum from limestone.

Process developments [157]: A continuous process, with water washing of the carbon simultaneously with adsorption, is at the laboratory stage of development.

Appraisal:

1.	Information available	1
2.	Process simplicity	1
3.	Operating experience	0
4.	Operating difficulty	1
5.	Loss of power	2*

Table 2.30 Status of Processes Using Regenerable Dry Reagents: Chemical Regeneration

Code No.	Vendor	No. of Plants	Plant size Range MWe	Total MWe	Dates	% S Capture	Ref.
S52.1	Lurgi	2	30–170*	200*	–	Over 90	157
	Hitachi	2	–	–	–	80	157
S52.2	Shell/	1	0.6	0.6	1974	–	137
	UOP	+1	125*	125*	1983	90	282
S52.3	Monsanto	2	15–110	125	1960–72	Over 90	158, 137

* Thousand Nm^3/h
– Indicates information unavailable

6.	Reagent requirements	2
7.	Ease of end-product disposal	1
8.	Process applicability	2
	Total	10

*Assumed in absence of data

Process Code S52.2 – Copper Oxide Process

Applicability: This process type is suitable only for new build applications.

Typical Land area requirements: ***

Typical fresh water and water treatment requirements: ***

Consumption of reagents: The hydrogen consumption [282] is 0.2 kg/kg sulphur captured. Additional information derived from published estimates is presented in Table 2.31.

End-product or waste disposal requirements: Published estimates are presented in Table 2.31.

Typical pressure losses: A plant handling 125,000 Nm^3/h, 2500 ppmv SO_2, 90% sulphur capture has a pressure drop of 20 mbar [282].

Table 2.31 Published Estimated Requirements

Case No. (Table 2.2)		3
Reference		6
MWe		800
Coal: Sulphur	wt. %	3.6
Chlorine	wt. %	0.1
Sulphur Capture	%	90
Sulphur recovery process		Claus
Consumption of reagents kg/MWh:		
Coal or coke		3
End-product or waste disposal requirements kg/MWh:		
Sulphur		13.7
Dry solids		5.0

Typical power requirements: ***

Typical reductions in combustion plant efficiency: ***

Operating experience [137]: The 0.6 MW unit was operated on a side stream from a 400 MWe coal-fired boiler at the Big Bend station, Florida, for two years. The copper oxide underwent 13,000 cycles without loss of desulphurisation reactivity.

The 125,000 Nm^3/h plant mentioned above [282] has been in operation for over 2 years; initial problems in the recovered SO_2 concentration unit were resolved.

Process developments: Cost reductions for units larger than about 50 MWe (about 175,000 Nm^3/h) would result from using 2–3 reactors on acceptance to 1 reactor on regeneration [282].

A fluidised bed version of the process is being developed [137].

A modification of the process can give simultaneous removal of SO_2 and NO_x (see Section 5).

The process is no longer being licensed by Shell.

Appraisal:

1.	Information available	1
2.	Process simplicity	1
3.	Operating experience	0
4.	Operating difficulty	0*
5.	Loss of power	2*
6.	Reagent requirements	1
7.	Ease of end-product disposal	2
8.	Process applicability	1
	Total	8

*Assumed in absence of information

Process Code S52.3 – Catalytic Oxidation Process

Applicability: This process type is suitable for new build applications; with reheat of the gas upstream of the FGD system it is also suitable for retrofit applications. It is capable of attaining over 90% sulphur capture.

Typical land area requirements: ***

Typical fresh water and water treatment requirements: ***

Consumption of reagents: There is a catalyst loss of about 2.5% each time the catalyst is cleaned (about four times p.a.) [158].

End-product or waste disposal requirements: The 110 MW plant was expected [158] to produce 26 kg/MWh of 78% sulphuric acid.

Typical power requirements: ***

Typical reductions in combustion plant efficiency: ***

Operating experience: A plant of 110 MWt capacity was installed on a coal-fired boiler, with natural gas burners to achieve the gas temperature needed for the catalytic oxidation reaction. The plant was operated for 444 hours, but was then shut down to replace the natural gas burners with oil burners. Troubles encountered were: plugging of catalyst with soot from the oil burners (corrected by using an external burner giving better combustion); failure of the burner refractory lining; corrosion and plugging in the acid system. These problems stem mainly from attempting to operate the process as a retrofit system. The system has not been operated since 1975 [137, 157, 158].

Process developments: Development of this process is not being considered at present.

Appraisal:

1.	Information available	1
2.	Process simplicity	2
3.	Operating experience	0
4.	Operating difficulty	0
5.	Loss of power	1*
6.	Reagent requirements	2
7.	Ease of end-product disposal	1
8.	Process applicability	1
	Total	8

*Assumed in absence of information

CATEGORY S61

The status of Category S61 processes is indicated in Table 2.32, which presents information on the size of plants built, year of completion, and percentage sulphur capture.

Process Code S61.1 – Hydrated Lime Injection Process

Applicability: ***

Typical land area requirements: ***

Typical fresh water and water treatment requirements: ***

Consumption of reagents: In field tests on a 1 MW scale by the Conoco Coal Research Division of Conoco Inc., hydrated lime, sized 80% below 43 micron, with a surface area of 23 m^2/g and a moisture content of 1.5%, was used at a calcium/sulphur molar ratio up to 2.7, giving sulphur capture of about 80%. The additive, sodium hydroxide, was supplied as an aqueous solution at $NaOH/Ca(OH)_2$ mass ratio of 0.1.

End-product or waste disposal requirements: ***

Typical power requirements: ***

Typical reductions in combustion plant efficiency: ***

Operating experience: ***

Process developments: ***

Appraisal:

1.	Information available	0
2.	Process simplicity	2
3.	Operating experience	0
4.	Operating difficulty	0*
5.	Loss of power	2*
6.	Reagent requirements	1
7.	Ease of end-product disposal	1
8.	Process applicability	2
	Total	8

*Assumed in absence of information

Process Code S61.2 – Alkali Injection Process

Applicability: ***

Typical land area requirements: ***

Typical fresh water and water treatment requirements: The requirement for water is negligible [95].

Consumption of reagents: For Nahcolite a feed rate equivalent to 1.2 times the stoichiometric rate is estimated to give about 75% sulphur capture [95].

End-product or waste disposal requirements: ***

Typical power requirements: For a 1000 MWe installation the power consumption is estimated to be about 7 MWe [95].

Typical reductions in combustion plant efficiency: ***

Operating experience: ***

Process developments: ***

Appraisal:

1.	Information available	0
2.	Process simplicity	2
3.	Operating experience	0
4.	Operating difficulty	0*
5.	Loss of power	2*
6.	Reagent requirements	1
7.	Ease of end-product disposal	0
8.	Process applicability	2
	Total	7

*Assumed in absence of information

Table 2.32 Status of Processes Using Non-Regenerable Dry Reagents Applied to Flue Gas: Dry Injection

Code No.	Vendor	No. of Plants	Plant size Range MWe	Total MWe	Dates	% S Capture	Ref.
S61.1	Conoco Coal Research	1	1	1	–	80	–
S61.2	–	–	–	–	–	–	–

– Indicates information unavailable

Table 2.33 Status of Processes Using Non-Regenerable Dry Reagents Applied to Flue Gas: Dry Reactor

Code No.	Vendor	No. of Plants	Plant size Range*	Total *	Dates	% S Capture	Ref.
S62.1	Lurgi	–	100–300	3,000	–	94–98	**

* Thousand Nm^3/h ** Vendor's promotional literature
– Indicates information unavailable

CATEGORY S62

The status of Category S62 processes is indicated in Table 2.33, which presents information on the size of plants built, year of completion, and percentage sulphur capture.

Process Code S62.1 – Lurgi CFB Lime Absorber Process

Applicability: This process type is suitable for new build and retrofit applications.

Typical land area requirements: ***

Typical fresh water and water treatment requirements: ***

Consumption of reagents: Hydrated lime is typically used at a Ca/S molar ratio of 1.7 [Vendor's promotional literature].

End-product or waste disposal requirements: ***

Typical power requirements: A plant handling 250,000 Nm^3/h flue gas consumed 520 kW [Vendor's promotional literature].

Typical reductions in combustion plant efficiency: ***

Operating experience [Vendor's promotional literature]: A plant handling 250,000 Nm^3/h started operation in 1984. Problems encountered included: deposits at the base of the CFB at the ESP entry; coarse fly-ash entering the CFB; dust build-up in the second field of the ESP; wear on the suspension nozzle; and instability at part load. All of these were overcome by small design changes.

Process developments: ***

Appraisal:

1.	Information available	1
2.	Process simplicity	2
3.	Operating experience	0
4.	Operating difficulty	0
5.	Loss of power	2*
6.	Reagent requirements	1
7.	Ease of end-product disposal	1
8.	Process applicability	2
	Total	9

*Assumed in absence of information

CATEGORY S71

The status of Category S71 processes is indicated in Table 2.34, which presents information on the size of plants built, year of completion, and percentage sulphur capture.

Process Code S71.1 – Sorbent Direct Injection Process

Applicability: This process type is suitable for retrofit and new build applications.

Typical land area requirements: ***

Typical fresh water and water treatment requirements: ***

Consumption of reagents: The sorbent considered is usually limestone. A Ca/input S molar ratio of about 2 is needed to give 50–70% sulphur capture [296] though estimates [302] have been based on the assumption that this would give only 35% sulphur capture (about 65% if $Ca(OH)_2$ is used) and that the stone would be ground to 50% below 18 micron, 100% below 100 micron.

Table 2.34 Status of Processes Using Non-Regenerable Dry Reagents Applied in Furnace: Dry Injection

Code No.	Vendor	No. of Plants	Plant size Range MWe	Total MWe	Dates	% S Capture	Ref.
S71.1	Tampella	1	–	250	1986	50–70	290

– Indicates information unavailable

Further information is presented in Table 2.35 in the form of published estimates.

End-product or waste disposal requirements: Published estimates are presented in Table 2.35.

Typical power requirements: Published estimates are presented in Table 2.35.

Typical reductions in combustion plant efficiency: Published estimates are presented in Table 2.35.

Table 2.35 Published Estimated Requirements

Reference		302
MWe		4×500
Coal: Sulphur	Wt. %	2.0
Ash	Wt. %	20
Sulphur capture	%	35
Consumption of reagents, kg/MWh:		
limestone		46
End-product or waste disposal requirements, kg/MWh:		
Additional		47
Typical power requirements, MWe:		11
Typical reductions in combustion plant efficiency:		
Reduction in power sent out	%	0.6

Operating experience: There is very little published information on operating experience. The process has not yet been applied on a commercial scale [302].

Process developments: The system incorporating Limestone Injection into Multistage Burners (LIMB) combines NO_x abatement with enhanced sulphur capture performance [296]. LIMB is described in Section 5.

Appraisal:

1.	Information available	0
2.	Process simplicity	2
3.	Operating experience	0
4.	Operating difficulty	0*
5.	Loss of power	2
6.	Reagent requirements	1
7.	Ease of end-product disposal	1
8.	Process applicability	2
	Total	8

*Assumed in absence of information

Process Code S72.1 – Fluidised Bed Combustion Processes

The appraisal of fluidised bed combustion processes is given in Section 6.

2.4 Processes for Detailed Study

The selection of processes for detailed study in this Volume has been based upon their suitability for application in the UK for the three datum combustion systems (Section 1.3) considered:

- 2000 MWe (4 × 500 MWe) power station (Datum system 1)
- Large (450 tonne steam/h) industrial boiler (Datum system 2)
- Small (13 tonne steam/h) factory boiler (Datum system 3)

In principle, all of the processes listed in Section 2.1 can be applied to all combustion plant, but the attraction of many processes diminishes with factors such as decrease in plant operating scale, and increases in FGD process complexity, reagent costs, and end-product disposal difficulty.

Appraisal of processes

To evaluate some of these factors, a rough appraisal of each process type has been made in Section 2.3 by assigning 'merit points' for a number of features; merit points have been awarded according to the scale:

0 Below average merit
1 Average merit
2 Above average merit
3 Outstandingly above average merit

The features to which these points have been assigned are described in Section 1.5; they are briefly:

1. Information available
2. Process simplicity
3. Operating experience–extent and difficulties encountered
4. Operating difficulty–availability, reliability
5. Loss of power sent out–by installation of the FGD process
6. Reagent requirements–quantities
7. Ease of end-product disposal
8. Process applicability–e.g. for retrofit

All of the processes listed in Section 2.1 and outlined in Section 2.2 (except fluidised bed combustion processes) are appraised in Section 2.3. The merit points assigned to the process types for each of the above features are summarised in Table 2.36. It should be noted that the number of points in the merit point system adopted in other Sections of the Manual are not strictly comparable with those considered here.

Processes suitable for the UK

The principal purpose of assigning merit points to each of the processes was to aid in the selection of processes that could be considered suitable for application in the UK. It was arbitrarily assumed that suitable FGD processes would be those having more

Table 2.36 Summary of FGD Process Appraisals

Code No.	Name	Merit Points for Feature No: 1	2	3	4	5	6	7	8	Total Points
S11.1	Sea water scrubbing	1	3	1	1	1	2	1	1	11
S11.2	Alkali scrubbing	1	2	1	1	1	1	0	2	9
S11.3	Ammonia scrubbing	0	2	0	1	1	1	1	2	8
S11.4	Sulphuric acid scrubbing	0	2	1	1	1	2	1	2	8
S11.5	Dowa	1	1	1	0	1	1	1	2	8
S11.6	Dual alkali	2	1	1	0	1	1	1	2	9
S12.1	Alkali spray drying	1	3	0	0	1	1	0	2	8
S12.2	Walther process	1	2	0	0	1	1	1	2	8
S21.1	Limestone slurry scrubbing	2	2	3	0	1	1	1	2	12
S22.1	Lime spray drying	2	3	2	0	2	1	1	2	13
S31.1	Wellman-Lord	2	1	2	0	0	2	2	2	11
S31.2	Fläkt-Boliden	1	1	0	0	0	2	2	2	8
S31.3	Catalytic/IFP	1	1	0	0	0	2	2	2	8
S32.1	Citrate	1	1	0	0	0	2	2	2	8
S32.2	Conosox	1	1	0	0	0	1	2	2	7
S32.3	Ispra Mk. 13A	1	1	0	0	0	2	1	2	7
S32.4	Aqueous carbonate	1	1	0	0	1	1	2	2	8
S41.1	Magnesia scrubbing	2	1	0	1	1	2	2	2	11
S42.1	Sulf-X	1	0	0	0	1	2	2	2	8
S51.1	Active carbon	1	2	0	0	2	2	2	2	11
S52.1	Wet active carbon	1	1	0	1	2	2	1	2	10
S52.2	Copper oxide	1	1	0	0	2	1	2	1	8
S52.3	Catalytic oxidation	1	2	0	0	1	2	1	1	8
S61.1	Hydrated lime injection	0	2	0	0	2	1	1	2	8
S61.2	Alkali injection	0	2	0	0	2	1	0	2	7
S62.1	Lurgi circulating bed	1	2	0	0	2	1	1	2	9
S71.1	Sorbent injection	0	2	0	0	2	1	1	2	8
S72.1	FBC	Appraised in Section 6								

Features: 1. Information available
2. Process simplicity
3. Operating experience
4. Operating difficulty
5. Loss of power
6. Reagent requirements
7. Ease of end-product disposal
8. Process applicability

than 10 merit points. In addition, the fluidised bed combustion processes (see Section 6) are suitable for application in the UK; they have not been appraised in Section 2.3.

Selection of processes for detailed study

All of the FGD process types are shown in Table 2.37 with an indication of which of the Datum combustion systems are considered, from the above criteria, to be suitable for application of the process in the UK.

Only two basic process types are considered to be suitable for the smallest operating scale dealt within this Volume (Datum System 3) as well as for larger operating scales. These are:

- Lime slurry scrubbing/spray dryer process (Process Code S22.1; merit rating 13 points). The relative simplicity of this process, the relative cheapness of the reagent (lime, manufactured from limestone which is abundant in the UK) and the production of an easily handled dry end-product, make it suitable for application to all three Datum combustion systems. This process is evaluated in Section 2.5.

- Atmospheric pressure fluidised bed combustion (one of the basic process types under Process Code S72.1). This is dealt with in detail in Section 6.

For utilities and large industrial boilers, the simplest processes for consideration in the UK are:

- Sea water scrubbing process (Process Code S11.1; merit rating 11 points). This process is noteworthy for its extreme simplicity, but it is, of course, applicable only to coastal locations. The process is evaluated in Section 2.5.

- Limestone or lime slurry scrubbing processes (Process Code S21.1; merit rating 12 points). This uses cheap reagents and is a very well-established process, with more FGD plants of this basic type

Table 2.37 Applications of FGD Processes Considered in Detail

Code No.	Name	Application to Datum System 1	2	3	Section
S11.1	Sea water scrubbing	Yes	Yes	–	2.5
S11.2	Alkali scrubbing	–	–	–	–
S11.3	Ammonia scrubbing	–	–	–	–
S11.4	Sulphuric acid scrubbing	–	–	–	–
S11.5	Dowa	–	–	–	–
S11.6	Dual alkali	–	–	–	–
S12.1	Alkali spray drying	–	–	–	–
S12.2	Walther process	–	–	–	–
S21.1	Limestone slurry scrubbing	Yes	Yes	–	2.5
S22.1	Lime spray drying	Yes	Yes	Yes	2.5
S31.1	Wellman-Lord	Yes	Yes*	–	2.5
S31.2	Fläkt-Boliden	–	–	–	–
S31.3	Catalytic/IFP	–	–	–	–
S32.1	Citrate	–	–	–	–
S32.2	Conosox	–	–	–	–
S32.3	Ispra Mk. 13A	–	–	–	–
S32.4	Aqueous carbonate	–	–	–	–
S41.1	Magnesia scrubbing	Yes	Yes*	–	2.5
S42.1	Sulf-X	–	–	–	–
S51.1	Active carbon	Yes	Yes*	–	2.5
S52.1	Wet active carbon	–	–	–	–
S52.2	Copper oxide	–	–	–	–
S52.3	Catalytic oxidation	–	–	–	–
S61.1	Hydrated lime injection	–	–	–	–
S61.2	Alkali injection	–	–	–	–
S62.1	Lurgi circulating bed	–	–	–	–
S71.1	Sorbent injection	–	–	–	–
S72.1	Atmospheric pressure FBC	Yes	Yes	Yes	6.9
do.	Circulating FBC	Yes	Yes	–	6.9
do.	Pressurised FBC	Yes	Yes	–	6.9

*In general, only if a centralised reagent reprocessing plant were available

throughout the world than of all other processes put together. For UK applications, only those processes embodying forced oxidation and producing gypsum (marketable, or easily disposed of according to circumstances) would be considered. Because of its importance, three examples of this basic process type are evaluated in Section 2.5, namely the IHI Gypsum process, the Chiyoda Thoroughbred 121 process and the Saarberg-Hölter process.

- All of the forms of fluidised bed combustion: atmospheric pressure, circulating bed and pressurised FBC (Process Code 72.1, appraised and evaluated in Section 6).

Finally, the processes that are, in principle, the most complex are the regenerable reagent processes; their main attractions are: the reduced demand for reagent and the reduced production of wastes; and the potential for producing marketable products. Because of their major chemical processing content, they are of interest mainly for the largest operating scale, e.g. Datum System 1. However, if future legislation imposes limits on sulphur oxides emissions in the UK, it is likely that a reagent reprocessing industry will be created, and if this occurs, the regenerative processes would be applicable to the large industrial combustion systems also, e.g. Datum System 2. The processes to be considered are:

- Wellman-Lord process (Process Code S31.1; 11 merit points)–a well-established process that has been evaluated by the CEGB [265]. This is evaluated in Section 2.5.

- Magnesia slurry scrubbing process (Process Code S41.1; 11 merit points) evaluated in Section 2.5. This is an example of a process that has been operated with spent reagent sent to an off-site reprocessing plant [279].

- Active carbon adsorption process (Process Code S51.1; 11 merit points) evaluated in Section 2.5.

2.5 Evaluation of Selected FGD Processes

Evaluation of Sea Water Scrubbing FGD Process (Process Code S11.1)

See Section 2.2 for: outline of the basic process; its chemistry; block diagram.

See Section 2.3 for a list of manufacturers offering this type of equipment.

See Section 2.3 for general appraisal of the basic process.

See Section 2.4 for the reason for choosing the Sea Water Scrubbing FGD Process.

This basic process type is considered (Section 2.4) to be suitable for application in the UK only to utility boilers and large industrial boilers, and hence it is evaluated here only for Datum Combustion Systems 1 and 2.

Process Description: Figure 2.33 shows a simplified flow diagram for application of the process to a pulverised fuel fired water tube boiler.

Gas from downstream of the electrostatic precipitator or baghouse, boosted by a Fan, is cooled in a Heat Exchanger, and is then scrubbed with sea water in the Scrubber to remove sulphur oxides, acid halides and particulates. The purified gas then flows via a Demister to the Heat Exchanger, where it is reheated to above the dewpoint, and is then exhausted to the stack.

Status and Operating Experience: The status of the sea water scrubbing process is indicated in Section 2.3, where it is seen that more than fourteen plants have been built in the UK (using river water), Scandinavia, Japan and USA. The Japanese plants built by IHI are similar to their 'Inert gas systems' installed on 15 ships.

Information on operating experience quoted in Section 2.3 is limited. The CEGB plants [156] and Fläkt-Hydro plants [651] have given some trouble with corrosion, but the Risdon, Tasmania plant has enjoyed a high availability [667].

Variations and Development Potential: The main potential variations in this process are:

- Design of the Absorber: a number of alternatives can be used: spray towers (the usual choice, giving the lowest gas pressure losses); venturi scrubbers with spray banks (higher efficiency but higher pressure loss); and turbulent contact absorbers (TCAs) with spray banks (high efficiency, but also with high pressure loss).

- Method of reheat: Heat exchange (in a heat exchanger) from input gas; combustion of liquid or gaseous fuel in part or all of the purified gas; injection of air heated in a steam-heated air heater; bypassing of uncooled flue gas around the scrubber.

- Treatment of effluent: Neutralisation of effluent with e.g. chalk, as an alternative to no treatment before discharging to sea. Where the effluent is treated, operating conditions should preferably be chosen to ensure that the calcium sulphite and sulphate formed are in sufficiently low concentration to remain in solution.

The process has the merit of extreme simplicity of design and operation, especially if it is unnecessary to treat the effluent water before discharge to sea.

The process can be applied only where the effluent (treated or untreated) can be discharged to sea; its application is therefore likely to be limited, and little development can be expected.

Process Requirements of Each Application Considered: These are shown in Table 2.38 for the two applications considered: Datum Combustion Systems 1 and 2.

It is assumed that for System 1:

- The SO_2 content of the gas is to be reduced to 400 mg/Nm^3

- The effluent is treated with chalk before discharge

It is assumed that for System 2:

- The SO_2 content of the gas is to be reduced to 650 mg/Nm^3

- The effluent is not treated before discharge to sea

It is further assumed that for both systems:

- The NO_x content is unaffected

- The HCl content of the gas is reduced by 80%

- The particulates content is reduced to 15 mg/Nm^3

- The sea water input is sufficiently high to give an equivalent sulphate ion concentration in the effluent of 1400 ppmw.

By-products and Effluents: The effluents from the process consist of sulphate, sulphite, and chloride ions in solution together with calcium ions (System 1), and particulates.

The composition of the effluent water is summarised in Table 2.39 for the two Datum Combustion Systems considered.

Efficiency and Emission Factors: The efficiency and emission factors for the process are summarised in Table 2.40 for the two applications considered: Datum Combustion Systems 1 and 2.

In calculating the efficiency factors, it is assumed that at full load:

- The performance without the incorporation of the FGD plant would be as shown in Table 1.4

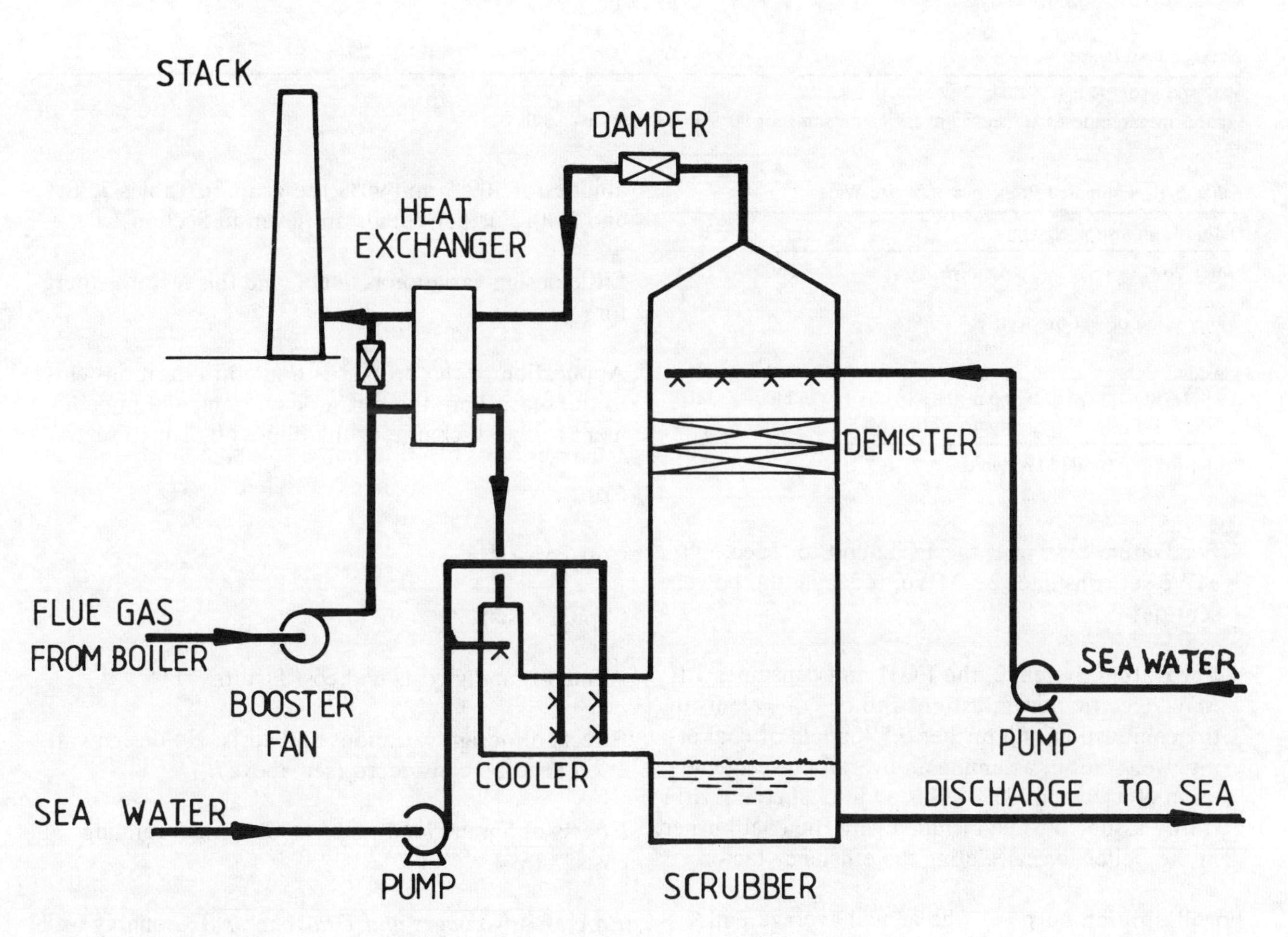

Figure 2.33 Sea Water Once-Through Process

Table 2.38 Process Requirements

Datum Combustion System		1	2
Inlet Gas at full load			
Volume flow	'000 Nm^3/h	1524*	551
Dry gas	'000 Nm^3/h	1426*	513
Water vapour	'000 Nm^3/h	99*	38
Actual volume flow	'000 m^3/h	2362*	854
Temperature	°C	150	150
Particulates content	mg/Nm^3 (dry)	115	115
SO_2 content	mg/Nm^3 (dry)	4240	3520
NO_x content	mg/Nm^3 (dry)	1025	1440
HCl content	mg/Nm^3 (dry)	340	315
Exit Gas at full load			
Volume flow	'000 Nm^3/h	1469*	528
Dry gas	'000 Nm^3/h	1423*	512
Water vapour	'000 Nm^3/h	45*	16
Actual volume flow	'000 m^3/h	2007*	721
Temperature	°C	100	100
Particulates content	mg/Nm^3 (dry)	15 (25)	15 (25)
SO_2 content	mg/Nm^3 (dry)	400 (785)	650 (935)
NO_x content	mg/Nm^3 (dry)	1025 (1025)	1440 (1440)
HCl content	mg/Nm^3 (dry)	68 (95)	63 (88)
Reaction temperature	°C	25	25
Particulates removal		Simultaneous	Simultaneous
Reagent (effluent treatment)		Chalk	None
Requirements at full load			
Sea water	tonne/h	5815*	1565
Chalk	tonne/h	8.81*	Nil
Electric Power	MWe	2.5*	1.0
Manpower	men/shift	***	***
Average load factor	%	***	***

*Per 500 MWe boiler
Figures in parentheses are annual average emissions for 90% FGD plant availability

Table 2.39 Estimated Properties of Waste Water

Datum Combustion System		1	2
Water flow,	tonne/h	23500*	1585
Composition of wet product:			
Coal ash	ppm	25	30
Added Ca^{++}	ppm	585	–
SO_4^{--}	ppm	1400	1400
Cl^-	ppm	65	80

*Total from four 500 MWe units

- For Datum System 1, the FGD unit for each 500 MWe set consumes 2.5 MWe, reducing the power sent out

- For Datum System 2, the FGD unit consumes 1.0 MWe electric power, assumed to be equivalent to the combustion of a further 0.5 tonne/h of coal at a power station, assuming an overall power generation efficiency of 33%. This additional coal is arbitrarily assumed to be included with the coal burned in the boiler for calculating the efficiency factor.

For illustration purposes, the annual average emissions and emission factors for FGD plant availabilities of 100% and 90% are given in Tables 2.38 and 2.40. Further details are given in Section 1.6.

Little design variation is likely, and this factor is therefore ignored.

Application of the process is limited to locations close to the sea, where there is no environmental impediment to the discharge of the dilute effluent to sea.

Costs:

Cost basis: ***

Capital costs: ***

Annual running costs and cost factors: ***

Effects of design variations and costs: No design variation need be considered (see above).

Effects of annual load patterns on annual running costs: ***

Process Advantages and Drawbacks: Reasonably well-established process, with a pedigree dating from

Table 2.40 Efficiency and Emission Factors

Datum Combustion System		1	2
Coal heat input (gross)	MWt	1292*	469
Coal fired	tonne/h	189*	60.3
FGD plant power consumed	MWe	2.5*	1.0
Equivalent coal input	tonne/h	–	0.5+
Useful energy from system	GJe/h	1665*	–
	GJt/h	–	1468
Total equiv. coal input	tonne/h	189*	60.8(a)
Efficiency factor	GJe/tonne	8.81	–
	GJt/tonne	–	24.1(a)
Emissions			
Sulphur in SO_2	kg/h	285* (559*)	167 (241)
Nitrogen in NO_x	kg/h	445* (445*)	225 (225)
Chlorine in HCl	kg/h	94* (132*)	31 (44)
Particulates	kg/h	21* (35*)	8 (13)
Emission factors (per tonne coal)			(a)
Sulphur	kg/tonne	1.51 (2.96)	2.75 (3.96)
Nitrogen	kg/tonne	2.35 (2.35)	3.70 (3.70)
Chlorine	kg/tonne	0.50 (0.70)	0.51 (0.72)
Particulates	kg/tonne	0.11 (0.19)	0.13 (0.21)

* For each 500 MWe boiler FGD plant
\+ Calculated assuming overall power generation efficiency = 0.33
(a) Based on coal fired to boiler plus coal equivalent to electric power consumed
Figures in parentheses are annual average emissions for 90% FGD plant availability

1933; simple chemistry; simple equipment; land not required for stocking reagent (unless neutralisation with chalk is required) or for holding or disposing of product; high sulphur capture efficiency (over 95% capture can be attained); potential for environmentally acceptable effluent disposal, with adjustment of pH if required.

Disadvantages are that the design is limited to applications at certain coastal locations; careful design and control needed to avoid scaling problems.

Evaluation of IHI Limestone (or Gypsum) FGD Process (Process Code S21.1 – Limestone or Lime Slurry Scrubbing Processes)

See Section 2.2 for: outline of the basic process; its chemistry; block diagram.

See Section 2.3 for a list of manufacturers offering this type of equipment.

See Section 2.3 for general appraisal of the basic process.

See Section 2.4 for the reason for choosing the Limestone/Lime Slurry Scrubbing Process.

This basic process type is considered (Section 2.4) to be suitable for application in the UK only to utility boilers and large industrial boilers, and hence it is evaluated here only for Datum Combustion Systems 1 and 2. The process presented here, which is of the type producing gypsum as an end product, is offered by Ishikawajima-Harima Heavy Industries (IHI) Company Limited (see Appendix 2 for details of this manufacturer).

See also below for evaluations of the Chiyoda Thoroughbred 121 Limestone-gypsum process and the Saarberg-Hölter-Lurgi lime-gypsum process for two other versions of this basic process type.

Process Description: Figure 2.34 shows a simplified flow diagram for application of the process to a pulverised fuel fired water tube boiler.

The system comprises two main sections: the dust and sulphur oxides removal section; and the gypsum production section.

For Datum System 1 it is assumed that pure marketable gypsum by-product is needed, so that it is necessary:

- To remove as much as practicable of the residual particulates content of the gas leaving the electrostatic precipitator
- To convert unreacted limestone in the product to gypsum

For Datum System 2 it is assumed that impure gypsum will be produced as a waste product, and that these measures will not be needed.

For dust and sulphur oxides removal, gas from downstream of the electrostatic precipitator or baghouse, boosted by a Fan, is cooled in the Heat Exchanger, scrubbed with water circulated through a Prescrubber and its Demister (Datum System 1 only)

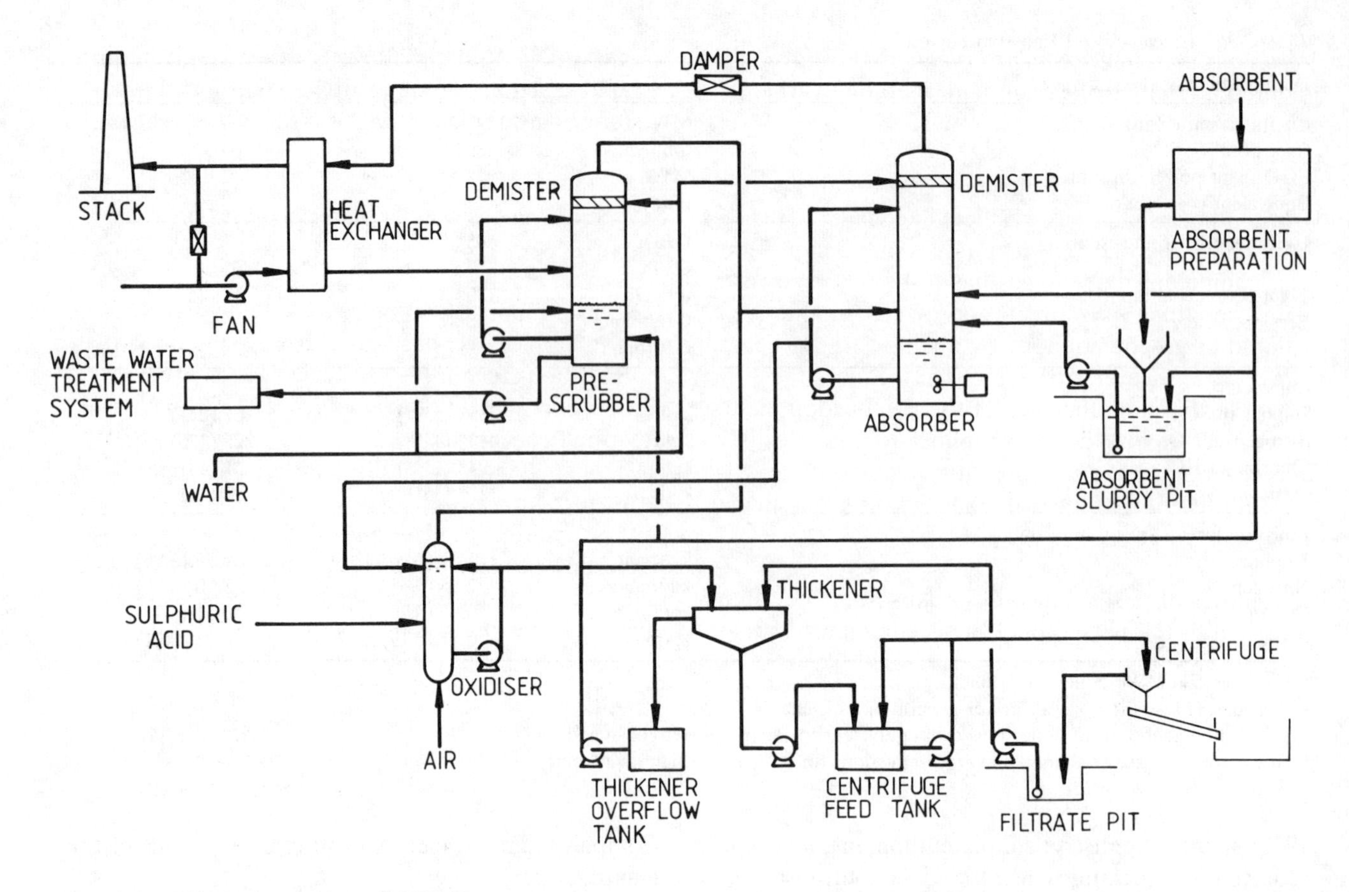

Figure 2.34 IHI Limestone–Gypsum FGD Process

to remove acid halides and particulates, and is then scrubbed with limestone slurry circulated through the Absorber to remove sulphur oxide. The purified gas then flows via a Demister to the Heat Exchanger, where it is reheated by exchange with incoming gas, before flowing via a damper to the Stack.

The prescrubber water, containing suspended solids and dissolved halide, is circulated to a Waste Water Treatment System. Limestone slurry is prepared from raw limestone, which is ground and mixed with water in the Limestone Slurry Preparation System. The slurry is circulated through the Absorber; a proportion of the circulating slurry is bled off to the Gypsum Production Section.

In the Gypsum Production Section the slurry is circulated through the Oxidiser, in which calcium sulphite hemihydrate reacts with the oxygen in air blown through the slurry. Sulphuric acid is added to convert any unreacted limestone or lime to gypsum also (Datum System 1 only), and to adjust the pH of the gas entering the Absorber. Part of the slurry circulating through the Oxidiser is bled off to a Thickener, where the gypsum is concentrated. The overflow from the Thickener flows via the Thickener Overflow Tank, from which it is pumped to the Prescrubber and Absorber, and to the Limestone Slurry Preparation System. The thickened underflow is pumped via a pump tank to the Centrifuges, from which gypsum, containing less than 10% water, is removed. The centrifuge filtrate is pumped back to the Thickener.

About 8% of the nitrogen oxides content of the gas is removed by the Absorber. A modification of the process (Section 4) allows for simultaneous removal of the sulphur and nitrogen oxides. As acid halides react with limestone or hydrated lime to form soluble calcium halides, the prescrubbing stage reduces the loss of calcium to solution.

Status and Operating Experience: The status of the IHI process is indicated in Section 2.3, where it is seen that since 1971, thirty-three plants handling between 10,000 and 2,400,000 Nm³/h of gas (total 13,055,000 Nm³/h) have been built or are under construction. Of these, nine plants have been erected for utility coal-fired and oil-fired boilers for generating capacities between 26 MWe and 700 MWe, and a further four (156 to 500 MWe coal-fired utility units) are under construction.

Information on operating experience is limited. Operating experience for a plant for a 700 MWe coal-fired boiler was found [678] to be generally satisfactory. However, it has been reported [189] that for the same plant, scaling occurred in the demister, but the scale formation was corrected by changing the spray pattern on the demister. Small extents of scaling occurred on other surfaces in contact with slurry–this is a hazard in the presence of calcium sulphate, which is slightly soluble and crystallises out.

Variations and Development Potential: IHI have adopted a number of design and operating variations, including:

- Cooling of the gas by injection of a water spray instead of using a gas–gas heat exchanger. This simplifies the plant, as it removes one major item of equipment, but the gas has to be reheated (to restore plume buoyancy), e.g. by combustion of liquid or gaseous fuel, or by mixing with air heated in a steam-heater air heater. This reduces the overall thermal efficiency of the combustion plant, but gives greater design simplification and flexibility as regards stack gas temperature. This variation would therefore be worth consideration for smaller plants.

- Removal of particulates in the Absorber, i.e. elimination of the prescrubber. This eliminates a major plant item, but increases the quantity of limestone or hydrated lime required and gives an impure gypsum product. The latter would not be a serious drawback in the majority of instances where no market for by-product gypsum is available.

- Use of hydrated lime in place of limestone. This entails calcination of limestone (usually by others) and slaking of lime (usually on site, but careful control is required to avoid problems). The advantages include a reduced requirement for fine grinding (hydrated lime is a fine powder) and a slightly higher solubility in water and higher reactivity (giving faster reaction).

- Design of the Absorber: a number of alternatives can be used: spray towers (the usual choice, giving the lowest gas pressure losses); venturi scrubbers with spray banks (higher efficiency but higher pressure loss); and turbulent contact absorbers (TCAs) with spray banks (maximum efficiency, but also with high pressure loss).

A number of design and operating variations also exist between manufacturers. The most important of these are:

- Elimination of the oxidation stage. This eliminates a major plant item, but gives rise to the need for disposing of calcium sulphite hemihydrate (see Section 2.2) to a settling lagoon. This material is difficult to concentrate and dewater, so that its disposal gives major problems.

- Combining absorption and oxidation in one vessel, eliminating a process stage, as in the Chiyoda Thoroughbred 121 process (see below), the Saarberg-Hölter-Lurgi process (see below) and the Babcock-Hitachi process. This also reduces the requirements for lime or limestone.

- Use of additives such as formic acid and adipic acid. These act as buffering agents which result in more rapid reaction between sulphur oxides and limestone, but they increase the cost for the reagent slurry.

In essence, this is a well-established and proven design.

The basic simplicity of the process allows for scaling-up of the plant over very wide ranges, and adoption of modifications to suit the particular needs of the combustion plant. For example, IHI [173] have described conceptual designs for plant to handle the gas from a 1000 MWe utility boiler in a single Absorber spray tower. The sulphur capture efficiency of the Absorber can be controlled by varying the liquid flow/gas flow ratio (L/G ratio) in the Absorber, and by controlling the pH of the slurry; maximum efficiency of capture is at a pH of about 5.3.

Process Requirements for Each Application Considered: These are shown in Table 2.41 for the two applications considered: Datum Combustion Systems 1 and 2.

It is assumed that for System 1:

- The SO_2 content of the gas is to be reduced to 400 mg/Nm^3

- The end product is gypsum which will be sold as a byproduct; in order to achieve a pure product, the gas is prescrubbed to remove particulates and HCl, and the prescrubber water is neutralised with NaOH. Unreacted limestone is sulphated by addition of sulphuric acid.

It is assumed that for System 2:

- The SO_2 content of the gas is to be reduced to 650 mg/Nm^3

- The end product is gypsum which will be disposed of to a landfill. High purity is not required, and the gas is not prescrubbed. The HCl in the gas therefore reacts with some of the limestone. Only minor quantities of sulphuric acid will be required to adjust the pH for oxidation. Caustic soda will not be required.

It is further assumed that for both combustion systems:

- The same fresh sorbent/sulphur ratio is used, with differences in the percentage capture of sulphur oxides being met by the use of different Liquid/Gas ratios.

- The nitrogen oxides content of the gas is reduced by 8%.

- The HCl content of the gas is reduced by 80%.

- The particulates emission is reduced to 15 mg/Nm³.

By-products and Effluents: The effluents from the process consist of gypsum and waste water requiring treatment. The gypsum has to be disposed of, e.g. to a landfill, unless:

- It is produced in sufficient purity–achieved without difficulty, as for Datum System 1 above, by prescrubbing to remove particulates, and conversion of unreacted limestone or lime by addition of sulphuric acid; however, these measures are probably worthwhile only on the largest scale.

- There is a market for it, e.g. for manufacturing plasterboard–this will apply in only a small number of instances.

For System 2, where it is assumed that the gypsum produced is disposed of to a landfill, some care has to be taken in the selection of a suitable site for disposal, as calcium sulphate is slightly soluble in water, and the sulphate anion is harmful to concrete.

The rates of production and impurities contents of gypsum produced are summarised in Table 2.42 for the two Datum Combustion Systems considered.

Water treatment requirements are dependent on the composition of the waste water leaving the prescrubber and other parts of the plant, and on the local waste water disposal consent requirements. The main constituents of waste water to a water treatment plant will be particulates, together with calcium and sodium cations, and sulphite, sulphate, bicarbonate, chloride and fluoride anions. Some coal ash constituents may be leached out.

Table 2.41 Process Requirements

Datum Combustion System		1	2
Inlet Gas at full load			
Volume flow	'000 Nm³/h	1524*	551
Dry gas	'000 Nm³/h	1425*	513
Water vapour	'000 Nm³/h	99*	38
Actual volume flow	'000 m³/h	2362*	854
Temperature	°C	150	150
Particulates content	mg/Nm³ (dry)	115	115
SO_2 content	mg/Nm³ (dry)	4250	3520
NO_x content	mg/Nm³ (dry)	1025	1440
HCl content	mg/Nm³ (dry)	340	315
Exit Gas at full load			
Volume flow	'000 Nm³/h	1684*	608
Dry gas	'000 Nm³/h	1423*	512
Water vapour	'000 Nm³/h	261*	96
Actual volume flow	'000 m³/h	2239*	808
Temperature	°C	90	90
Particulates content	mg/Nm³ (dry)	15 (25)	15 (25)
SO_2 content	mg/Nm³ (dry)	400 (785)	650 (935)
NO_x content	mg/Nm³ (dry)	945 (953)	1325 (1337)
HCl content	mg/Nm³ (dry)	68 (95)	63 (88)
Reaction temperature	°C	55	55
Particulates removal		Separate	Simult.
Reagent		Limestone	Limestone
Particle size		90% below 75 micron	90% below 75 micron
Concentration in slurry	wt. %	18	18
Ca/S Molar Ratio		1.05	1.05
Requirements at full load			
Limestone (97% pure)	tonne/h	10.2*	3.05
Sulphuric acid (98% pure)	tonne/h	1.36*	Minor
Caustic soda (47% liquor)	tonne/h	0.9*	Nil
Water	tonne/h	68*	14
Electric Power	MWe	12.3*	3.3*
Manpower	men/shift	3+	1++
Average load factor	%	***	***

* Per 500 MWe boiler
+ Total for four boiler FGD units
++ Plus additional labour for materials handling
Figures in parentheses are annual average emissions for 90% FGD plant availability

Table 2.42 Estimated Properties of Gypsum Produced

Datum Combustion System		1	2
Rate of production:			
Wet	tonne/h	75.8*	5.36
Dry	tonne/h	68.2*	4.82
Composition of wet product:			
Moisture	wt. %	10.0	10.0
Coal ash	wt. %	Nil	1.0
Calcium chloride	wt. %	Nil	11.1
Calcium carbonate	wt. %	Nil	2.4
Other impurities	wt. %	1.6	1.7
Gypsum	wt. %	88.4	73.8
Purity of dry product	wt. %	98.2	82.0

*Total from four 500 MWe boiler FGD plants

Efficiency and Emission Factors: The efficiency and emission factors for the process are summarised in Table 2.43 for the two applications considered: Datum Combustion Systems 1 and 2.

In calculating the efficiency factors, it is assumed that at full load:

- The performance without the incorporation of the FGD plant would be as shown in Table 1.4.

- For Datum System 1, the FGD units for each 500 MWe set consumes 12.3 MWe, reducing the power sent out.

- For Datum System 2, the FGD unit consumes:

- 3.3 MWe electrical power, equivalent to the combustion of a 1.3 tonne/h of coal at a power station, assuming an overall power efficiency of 33%. This additional coal is arbitrarily assumed to be included with the coal burned in the boiler for calculating the efficiency and emission factors.

For illustration purposes, the annual average emissions and emission factors for FGD plant availabilities of 100% and 90% are given in Tables 2.41 and 2.43. Further details are given in Section 1.6.

Effect of Load Variations: ***

Effect of Design Variations: ***

Limitations: ***

Costs:

Cost basis: See Section 2.6

Capital costs: See Section 2.6–for retrofit and for new build applications.

Annual running costs and cost factors: See Section 2.6–for retrofit and for new build applications.

Effects of Design Variations on Costs: ***

Effects of Annual Load Patterns on Annual Running Costs: ***

Process Advantages and Drawbacks: Well-established process; simple chemistry; high sulphur capture efficiency (over 95% capture can be attained); low reagent/sulphur stoichiometric ratio; waste product potentially marketable, and reasonably benign if disposal is required.

The disadvantages are that a large land area is required (reduced if gypsum is produced); fine grinding needed (for limestone but not for lime); careful design and control needed to avoid scaling problems.

Table 2.43 Efficiency and Emission Factors

Datum Combustion System		1	2
Coal heat input (gross)	MWt	1292*	469
Coal fired	tonne/h	189*	60.3
FGD plant power consumed	MWe	12.3*	3.3
Equivalent coal input	tonne/h	–	1.3+
Useful energy from system	GJe/h	1630*	–
	GJt/h	–	1468
Total equiv. coal input	tonne/h	189*	61.6(a)
Efficiency factor	GJe/tonne	8.62	–
	GJt/tonne	–	23.8(a)
Emissions			
Sulphur in SO_2	kg/h	285* (559*)	167 (241)
Nitrogen in NO_x	kg/h	410* (414*)	207 (209)
Chlorine in HCl	kg/h	94* (132*)	31 (44)
Particulates	kg/h	21* (35*)	8 (13)
Emission factors (per tonne coal)			(a)
Sulphur	kg/tonne	1.51 (2.96)	2.71 (3.91)
Nitrogen	kg/tonne	2.17 (2.19)	3.36 (3.39)
Chlorine	kg/tonne	0.50 (0.70)	0.50 (0.71)
Particulates	kg/tonne	0.11 (0.19)	0.13 (0.21)

* For each 500 MWe boiler FGD plant
+ Calculated assuming overall power generation efficiency = 0.33
(a) Based on coal fired to boiler plus coal equivalent to electric power consumed
Figures in parentheses are annual average emissions for 90% FGD plant availability

Evaluation of Chiyoda Thoroughbred 121 FGD Process (Process Code S21.1 – Limestone or Lime Slurry Scrubbing Processes)

See Section 2.2 for: outline of the basic process; its chemistry; block diagram.

See Section 2.3 for a list of manufacturers offering this type of equipment.

See Section 2.3 for general appraisal of the basic process.

See Section 2.4 for the reason for choosing the Limestone/Lime Slurry Scrubbing Process.

This basic process type is considered (Section 2.4) to be suitable for application in the UK only to utility boilers and large industrial boilers, and hence it is evaluated here only for Datum Combustion Systems 1 and 2. The process presented here, which is of the type producing gypsum as an end product, is offered by Chiyoda Chemical Engineering and Construction Company Ltd, Yokohama, Japan (see Appendix 2 for details of this manufacturer).

See also the IHI limestone-gypsum process (above) and Saarberg-Hölter-Lurgi lime-gypsum process (below) for two other versions of this basic process type.

Process Description: Figure 2.35 shows a simplified flow diagram for application of the process to a pulverised fuel fired water tube boiler.

The system comprises two main sections: the dust and sulphur oxides removal section; and the gypsum dewatering section.

For Datum System 1 it is assumed that a pure marketable gypsum by-product is needed, so that it is necessary to remove as much as practicable of the acid halides and residual particulates by prescrubbing the gas before it enters the sulphur dioxide absorber.

For Datum System 2 it is assumed that the gypsum will be produced as a waste product; however, as the process has a limestone conversion to gypsum approaching 99% it is considered preferable to remove acid halides from the gas by prescrubbing, and this will also remove most of the particulates, so that the gypsum produced will also be of high purity.

For dust and sulphur oxides removal, gas from downstream of the electrostatic precipitator or baghouse, boosted by a Fan, is cooled in the Heat Exchanger, and is then scrubbed with water, in a Venturi Prescrubber and its Mist Eliminator, to remove acid halides and particulates. The gas enters a Jet Bubbling Reactor (JBR) in which the following processes occur simultaneously: absorption of sulphur oxides, by reaction with a limestone slurry, to form calcium sulphite and sulphate; oxidation of calcium sulphite to sulphate by reaction with oxygen in air pumped into the JBR; and precipitation of the oxidation product (gypsum). In the JBR, the gas is bubbled through the slurry, via a large number of pipes dipping into the slurry. Most of the limestone dissolves in the slurry owing to the existence of a

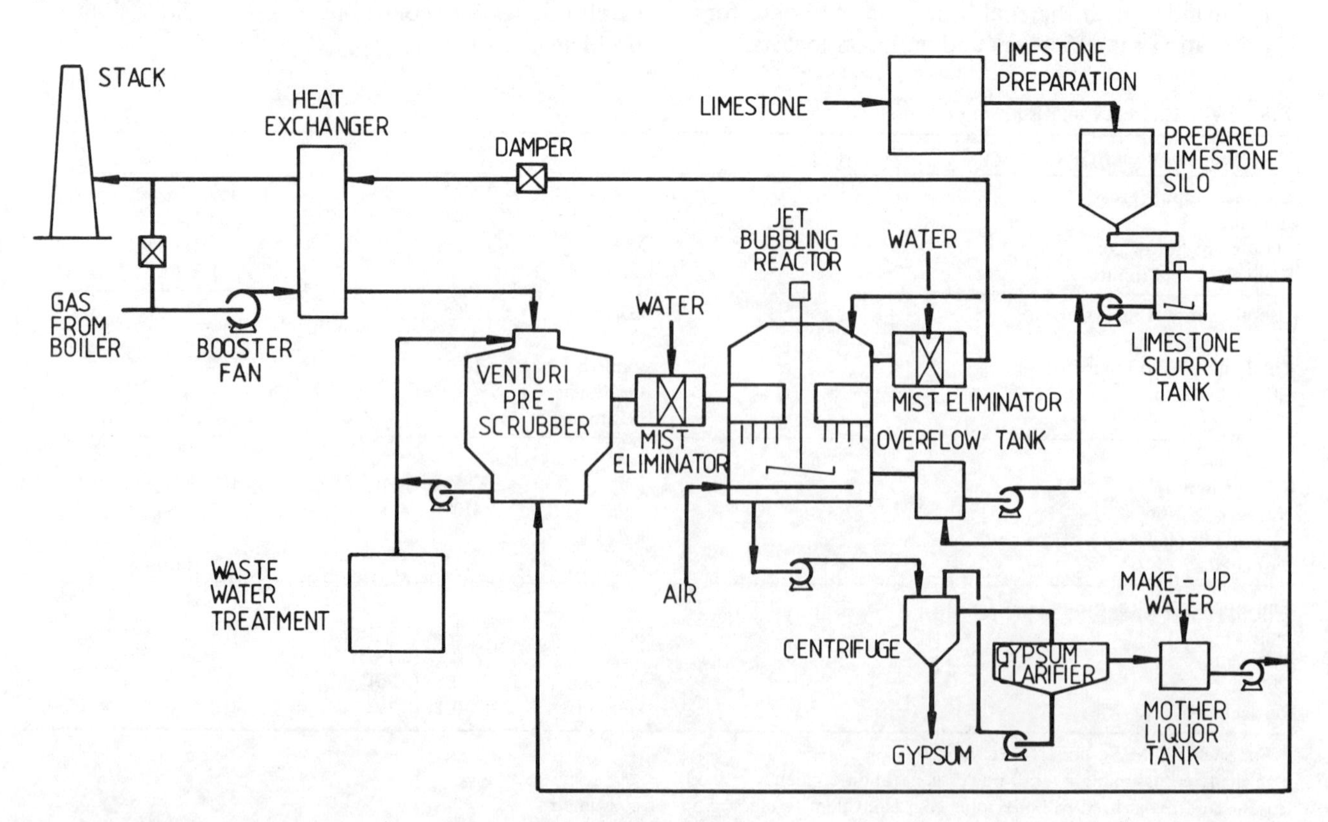

Figure 2.35 Chiyoda Thoroughbred 121 Limestone-Gypsum FGD Process

low pH (in the range 3.5–5.0). The oxidising air is admitted via sparge pipes at the base of the JBR. The purified gas leaving the JBR, together with nitrogen and unreacted oxygen from the oxidising air, flows via a Mist Eliminator and Damper to the Heat Exchanger, where it is reheated by exchange with incoming gas, before flowing to the Stack.

The prescrubber water, containing suspended solids and dissolved halides, is circulated to a Waste Water Treatment System. Limestone slurry is prepared from raw limestone, which is ground and mixed with water in the Limestone Slurry Preparation System. The slurry is pumped to the JBR. Gypsum slurry is pumped to the Gypsum Dewatering Section.

In the Gypsum Dewatering Section the slurry is pumped with a Gypsum Slurry Pump to a Basket Centrifuge which yields a product containing about 10% water, and a filtrate which flows into a Gypsum Clarifier. The underflow from the Gypsum Clarifier is returned to the Centrifuge, and the overflow is pumped to the Limestone Slurry Tank, the JBR and the Prescrubber as required.

Status and Operating Experience: The status of the Chiyoda process is indicated in Section 2.3, where it is seen that since 1978, seven units handling gas from plants of size equivalent to between 23 and 200 MWe (total 700 MWe) have been built. Of these, three units (two 200 MWe and one 23 MWe) have been erected for utility coal-fired boilers, three (52 to 87 MWe equivalent) for coal- and oil-fired boilers, and one (75 MWe equivalent) for a Claus sulphur plant.

The limited information on operating experience available indicates that reliability and availability are very high. For the two 200 MWe installations, it is reported [263] that for a period of nearly 11 months the reliabilities were 99.9 and 100% respectively. Another installation at an oil-fired boiler plant (87 MWe equivalent) was also reported [685] to have a reliability of 100%.

Variations and Development Potential: These can include:

- Cooling of the gas by injection of a water spray instead of using a gas–gas heat exchanger. This simplifies the plant as it removes one major item of equipment, but the purified gas has to be reheated to 90–150°C, e.g. by burning liquid or gaseous fuel in it, or by injecting air from a steam-heated air heater, in order to restore plume buoyancy. This reduces the overall thermal efficiency of the combustion plant, but gives greater design simplification and flexibility as regards stack gas temperature. This variation would therefore be worth consideration for smaller plants.

- Reheating, as described above, in conjunction with use of a heat exchanger in locations where greater reheat is needed for plume buoyancy.

- Removal of particulates in the JBR, i.e. elimination of the prescrubber. This has the attraction of eliminating a major plant item, but it would increase the quantity of limestone required and would give an impure gypsum product. The latter would not be a serious drawback in the majority of instances where no market for by-product gypsum is available.

Chiyoda claim that the adoption of the JBR qualifies this as a second-generation limestone-gypsum FGD process, in that it eliminates the need for slurry recycle pumps, slurry spray nozzles, separate reaction vessels for absorption and oxidation reactions, thickeners and other pre-watering equipment, and ancillary equipment such as pumps, agitators, rakes and process instrumentation. The low pH allows the limestone to dissolve, resulting in more rapid reaction with sulphur oxides and freedom from scaling and plugging in the mist eliminator. Further, rapid dissolution of limestone allows a coarser material (e.g. 90% below 75 micron) to be used.

Chiyoda have not built JBRs that can handle gas flows greater than those from the equivalent of 200 MWe boiler plants, but the high reliability, and the consequent absence of need for spare units, suggest that boiler plant of larger size could be equipped with single absorption trains.

Process Requirements for Each Application Considered: These are shown in Table 2.44 for the two application considered: Datum Combustion Systems 1 and 2.

It is assumed that for System 1:

- The SO_2 content of the gas is to be reduced to 400 mg/Nm^3

- The end product is gypsum which will be sold as a byproduct; in order to achieve a pure product, the gas is prescrubbed to remove particulates and HCl, and the prescrubber water is neutralised with NaOH. The conversion of limestone to gypsum is about 99%, and the percentage of unreacted limestone in the product is therefore acceptable.

- There are two prescrubber/JBR systems for each 500 MWe unit (total of eight systems)

It is assumed that for System 2:

- The SO_2 content of the gas is to be reduced to 650 mg/Nm^3

- The end product is gypsum which will be disposed of to a landfill. Although high purity is not required, the gas is prescrubbed in order to avoid conversion of limestone to calcium chloride.

- There is one prescrubber/JBR system

It is further assumed that for both combustion systems:

- The same molar ratio of (fresh sorbent)/(captured sulphur) is used (1.01 mol Ca per mol S absorbed), giving a limestone utilisation of 99%.

- The NO_x content of the gas is unaffected.

- The Prescrubber removes 80% of the HCl from the gas, and no absorption of HCl occurs in the JBR.

- The Prescrubber reduces the particulates content of the gas from 115 to 15 mg/Nm³, and no further reduction occurs in the JBR.

By-products and Effluents: The effluents from the process consist of gypsum and waste water requiring treatment. The gypsum has to be disposed of, e.g. to a landfill, unless:

- It is produced in sufficient purity–achieved without difficulty by prescrubbing to remove particulates and HCl.

- There is a market for it, e.g. for manufacturing plasterboard–this will apply in only a small number of instances.

For System 2, where it is assumed that the gypsum produced is disposed of to a landfill, some care has to be taken in the selection of a suitable site for disposal, as calcium sulphate is slightly soluble in water, and the sulphate anion is harmful to concrete.

The rates of production and impurities contents of

Table 2.44 Process Requirements

Datum Combustion System		1	2
Inlet Gas at full load			
Volume flow	'000 Nm³/h	1524*	551
Dry gas	'000 Nm³/h	1426*	513
Water vapour	'000 Nm³/h	99*	38
Actual volume flow	'000 m³/h	2362*	854
Temperature	°C	150	150
Particulates content	mg/Nm³ (dry)	115	115
SO_2 content	mg/Nm³ (dry)	4240	3520
NO_x content	mg/Nm³ (dry)	1025	1440
HCl content	mg/Nm³ (dry)	340	315
Exit Gas at full load			
Volume flow	'000 Nm³/h	1644*	592
Dry gas	'000 Nm³/h	1423*	512
Water vapour	'000 Nm³/h	221*	80
Actual volume flow	'000 m³/h	2186*	787
Temperature	°C	90	90
Particulates content	mg/Nm³ (dry)	15 (25)	15 (25)
SO_2 content	mg/Nm³ (dry)	400 (784)	650 (937)
NO_x content	mg/Nm³ (dry)	1025 (1025)	1440 (1440)
HCl content	mg/Nm³ (dry)	68 (95)	63 (88)
Reaction temperature	°C	53	53
Particulates removal		Separate	Separate
Reagent		Limestone	Limestone
Purity	Wt. %	97	97
Particle size		90% below 150 micron	90% below 150 micron
Concentration in slurry	Wt. %	21	21
Ca/(Captured S) Molar Ratio		1.01	1.01
Requirements at full load			
Limestone (97% pure)	tonne/h	8.1*	2.0
Caustic soda (47% liquor)	tonne/h	0.9*	0.3
Water	tonne/h	111*	40
Electric Power	MWe	6.0*	2.1
Manpower	men/shift	***	***
Average load factor	%	***	***

*Per 500 MWe boiler
Figures in parentheses are annual average emissions for 90% FGD plant availability

gypsum produced are summarised in Table 2.45 for the two Datum Combustion Systems considered.

Table 2.45 Estimated Properties of Gypsum Produced

Datum Combustion System		1	2
Rate of production:			
Wet	tonne/h	60.8*	3.4
Dry	tonne/h	54.8*	3.1
Composition of wet product:			
Moisture	wt. %	10.0	10.0
Inerts	wt. %	1.7	1.7
Calcium carbonate	wt. %	0.5	0.5
Gypsum	wt. %	87.8	87.8
Purity of dry product	wt. %	97.6	97.6

*Total from four 500 MWe boiler FGD plants

The requirements for water treatment are dependent on the composition of the waste water leaving the prescrubber and other parts of the plant, and on the local waste water disposal consent requirements. The main constituents of waste water to a water treatment plant will be particulates, together with calcium and sodium cations, and sulphite, sulphate, bicarbonate, chloride and fluoride anions. Some coal ash constituents may be leached out.

Efficiency and Emission Factors: The efficiency and emission factors for the process are summarised in Table 2.46 for the two applications considered: Datum Combustion Systems 1 and 2.

In calculating the efficiency factors, it is assumed that at full load:

- The performance without the incorporation of the FGD plant would be as shown in Table 1.4.
- For Datum System 1, the two FGD units for each 500 MWe set consume a total of 6.0 MWe electric power.
- For Datum System 2, the power consumption of 2.1MWe by the FGD plant is equivalent to the combustion of a further 0.8 tonne/h of coal at a power station, assuming an overall power generation efficiency of 33%. This additional coal is arbitrarily assumed to be included with the coal burned in the boiler for calculating the efficiency factor.

For illustration purposes, the annual average emissions and emission factors for FGD plant availabilities of 100% and 90% are given in Tables 2.44 and 2.46. Further details are given in Section 1.6.

Effect of load variations: ***

Effect of design variations: ***

Limitations: ***

Costs:

Cost basis: See Section 2.6.

Capital costs: For retrofit and for new build applications–see Section 2.6.

Annual running costs and cost factors: See Section 2.6

Effects of Design Variations on Costs: ***

Effects of Annual Load Patterns on Annual Running Costs: ***

Table 2.46 Efficiency and Emission Factors

Datum Combustion System		1	2
Coal heat input (gross)	MWt	1292*	469
Coal fired	tonne/h	189*	60.3
FGD plant power consumed	MWe	6.0*	2.1
Equivalent coal input	tonne/h	–	0.8+
Useful energy from system	GJe/h	1652*	–
	GJt/h	–	1468
Total equiv. coal input	tonne/h	189*	61.1(a)
Efficiency factor	GJe/tonne	8.74	–
	GJt/tonne	–	24.0(a)
Emissions			
Sulphur in SO_2	kg/h	285* (559*)	167 (241)
Nitrogen in NO_x	kg/h	445* (445*)	225 (225)
Chlorine in HCl (assumed)	kg/h	94* (132*)	31 (44)
Particulates	kg/h	21* (35*)	8 (13)
Emission factors (per tonne coal)			(a)
Sulphur	kg/tonne	1.51 (2.96)	2.73 (3.94)
Nitrogen	kg/tonne	2.35 (2.35)	3.68 (3.68)
Chlorine (assumed)	kg/tonne	0.50 (0.70)	0.51 (0.72)
Particulates	kg/tonne	0.11 (0.19)	0.13 (0.21)

* For each 500 MWe boiler FGD plant
\+ Calculated assuming overall power generation efficiency = 0.33
(a) Based on coal fired to boiler plus coal equivalent to electric power consumed

Process Advantages and Drawbacks: Process well-established in Japan, and successfully tested in USA on 23 MWe scale; very high reliability; simple chemistry; high sulphur capture efficiency (over 95% capture can be attained); very low reagent/sulphur stoichiometric ratio; land area required is smaller than for many other limestone-gypsum processes; waste product potentially marketable, and reasonably benign if disposal is required.

The disadvantages are that fine grinding is needed (but coarser than for other limestone-gypsum processes). No large scale experience–may be scale up problems unless multiple parallel units are employed.

Evaluation of Saarberg-Hölter FGD Process (Process Code S21.1–Limestone or Lime Slurry Scrubbing Processes)

See Section 2.2 for: outline of the basic process; its chemistry; block diagram.

See Section 2.3 for a list of manufacturers offering this type of equipment.

See Section 2.3 for general appraisal of the basic process.

See Section 2.4 for the reason for choosing the Limestone/Lime Slurry Scrubbing Process.

This basic process type is considered (Section 2.4) to be suitable for application in the UK only to utility boilers and large industrial boilers, and hence it is evaluated here only for Datum Combustion Systems 1 and 2. The process presented here is of the type using lime or limestone slurry with an organic acid (formic acid) additive and producing gypsum as an end product. The process is offered in Europe by Saarberg-Hölter-Lurgi GmbH, and in the United States by the Davy McKee Corporation (see Appendix 2 for details of these manufacturers).

See also the IHI limestone-gypsum process (above) and the Chiyoda Thoroughbred 121 limestone-gypsum process (above) for two other versions of the same basic process.

Process Description: Figure 2.36 shows a simplified flow diagram for application of the process to a pulverised fuel fired water tube boiler.

The system comprises two main sections: the dust and sulphur oxides removal section, and the gypsum dewatering section.

For Datum System 1 it is assumed that pure marketable gypsum by-product is needed, so that it is necessary to remove as much as practicable of the acid halides and residual particulates by prescrubbing the gas before it enters the sulphur dioxide absorber.

For Datum System 2 it is assumed that the gypsum will be produced as a waste product; however, as the process has a lime conversion to gypsum of about 98%, it is considered preferable to remove acid

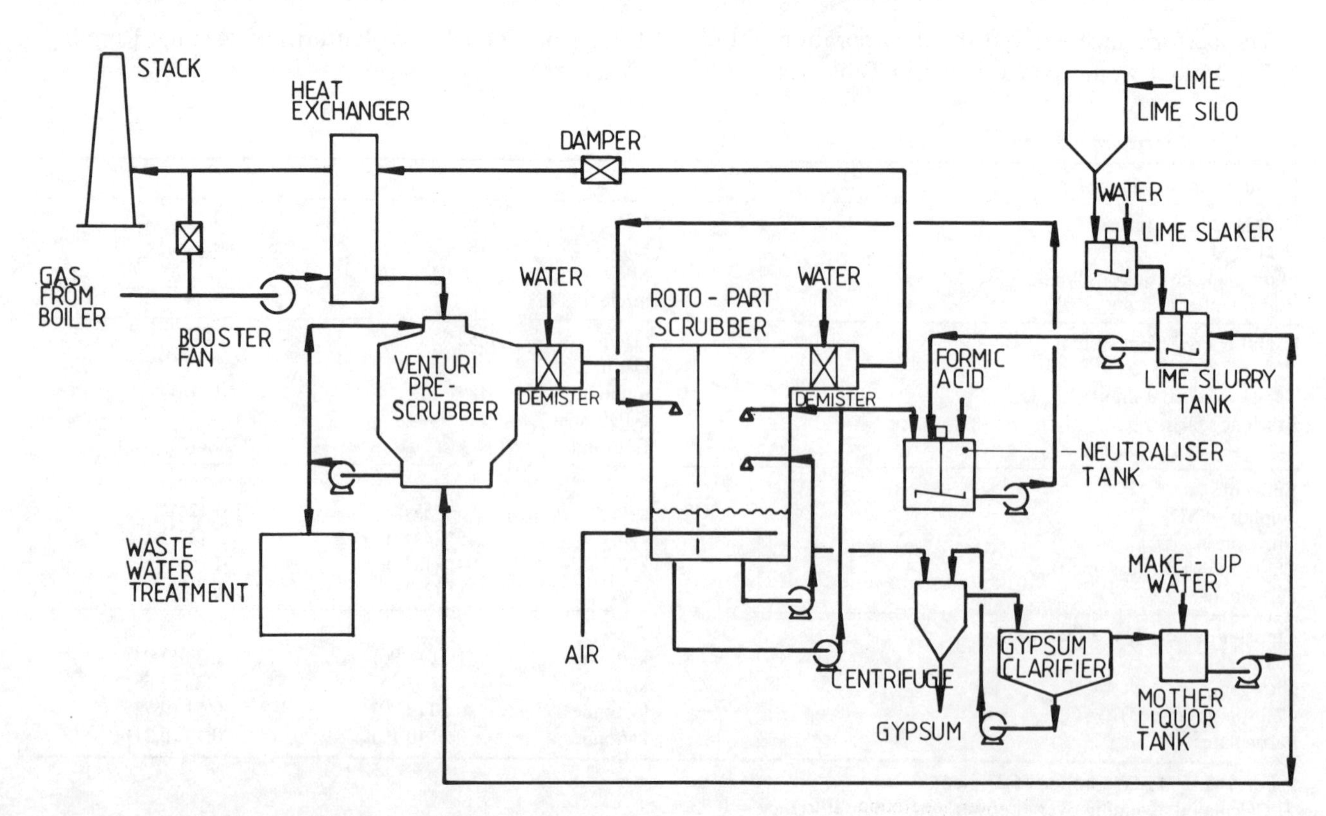

Figure 2.36 Saarberg-Hölter-Lurgi Lime-Gypsum FGD Process

halides from the gas by prescrubbing, and this will also remove most of the particulates, so that the gypsum produced will also be of high purity.

For dust and sulphur oxides removal, gas from downstream of the electrostatic precipitator or baghouse, boosted by a fan, is cooled in the Heat Exchanger, and is then scrubbed with water, in a Venturi Prescrubber and its Demister, to remove acid halides and particulates. The gas enters a 'Rotopart Scrubber' which is divided into two sections, each provided with sprays through which the Absorbent is circulated. In the first stage, the incoming gas flows downwards, co-current with the absorbent spray; in the second section, the gas flow is upwards counter-current to the spray. The following processes occur simultaneously in the Scrubber: absorption of sulphur oxides, by reaction with the lime or limestone slurry, to form calcium sulphite, bisulphite and sulphate; oxidation of calcium sulphite and bisulphite to sulphate by reaction with oxygen in air pumped into the Rotopart Scrubber; and precipitation of the oxidation product (gypsum). Most of the lime or limestone dissolves in the slurry due to the existence of a low pH (in the range 3.5–5.0) resulting from the formation of soluble calcium bisulphite. A liquid/gas flow ratio of about 3 litre/Nm^3 is required with lime as reagent, or 9–10 litre/Nm^3 with limestone (compared with up to about 18 litre/Nm^3 with limestone slurry scrubbers without integrated oxidation). The oxidising air is admitted via sparge pipes at the base of the Rotopart Scrubber. The purified gas leaving the Rotopart Scrubber, together with nitrogen and unreacted oxygen from the oxidising air, flows via a Demister and Damper to the Heat Exchanger, where it is reheated by exchange with incoming gas, before flowing to the Stack.

The prescrubber water, containing suspended solids and dissolved halides, is circulated to a Waste Water Treatment System.

Lime slurry is prepared from raw lime (assumed to be calcined off-site), which is stored in the Lime Silo, slaked in the Lime Slaker with water, and slurried in the Lime Slurry Tank with recycled mother liquor from the Gypsum Separation System. The slurry is pumped to the Neutraliser Tank, where it is mixed with a bleed from the first section of the Rotopart Scrubber. The formic acid additive is also added in the Neutraliser Tank, and the absorbent is pumped from the Neutraliser Tank to the first section of the Rotopart Scrubber. Part of the absorbent in the base of the first section of the Scrubber is pumped to sprays in the second section. The absorbent in both sections is continuously treated with oxidising air admitted through sparge pipes, and this precipitates gypsum as a slurry. A bleed of slurry circulated in the second section is pumped to the Gypsum Dewatering Section.

In the Gypsum Dewater Section the slurry is pumped with a Gypsum Slurry Pump to a Basket Centrifuge which yields a product containing about 10% water, and a filtrate which flows into a Gypsum Clarifier. The underflow from the Gypsum Clarifier is returned to the Centrifuge, and the overflow is pumped to the Mother Liquor Tank, and thence to the Lime Slurry Tank and the Prescrubber as required.

Status and Operating Experience: The status of the Saarberg-Hölter process is indicated in Section 2.3, where it is seen that since 1974, twenty-five units handling gas from plants of size equivalent to between 1 and 600 MWe (total 5760 MWe) have been built or are under construction; and a further seven units have been built handling 4000–131,000 Nm^3/h gas (total 350,000 Nm^3/h) from incinerators, Claus plants and boilers. One unit installed in the 230 MWe station of Sarrbergwerke at Völklingen in the FRG has been erected within the cooling tower, into which the exhaust gases are discharged, thus dispensing with the need for reheating and for a stack.

The limited information on operating experience available indicates that reliability and availability are very high. A plant treating 25% of the gas from a 700 MWe power station at Weiher, FRG, was reported in 1983 [192] to have operated for more than 12,500 hours with a reliability of 98%; most of the downtime was in the early history of this unit, involving balancing of the flue gas fan and modification of the solution pumps. Other problems reported [652] for the same plant were caking of lime in the Lime Silo, and difficulty in finding the right cloth for the vacuum filter (replacing the Centrifuge: see below).

Variations and Development Potential: Design and operating variations can include:

- Cooling of the gas by injection of a water spray instead of using a gas–gas heat exchanger. This simplifies the plant, as it removes one major item of equipment, but the purified gas has to be reheated to 90–150°C, e.g. by burning liquid or gaseous fuel in it, or by injecting air from a steam-heated air heater, in order to restore plume buoyancy. This reduces the overall thermal efficiency of the combustion plant, but gives greater design simplification and flexibility as regards stack gas temperature. This variation would therefore be worth consideration for smaller plants.

- Reheating, as described above, in conjunction with a heat exchanger in locations where greater reheat is needed for plume buoyancy.

- Removal of particulates and acid chlorides in the Rotopart Scrubber, i.e. elimination of the Prescrubber. This has the attraction of eliminating a major plant item, but it would increase the quantity of lime or limestone required and would give an impure gypsum product. The latter would not be a serious drawback in the majority of instances where no market for by-product gypsum is available.

- Use of limestone as the absorbent in place of lime. This reduces reagent costs, but increases the liquid/gas flow ratio needed (to about 9 litre/Nm^3 gas) and also increases the energy requirement for grinding.

- When lime is used as the absorbent, the sprays in the second section of the Rotopart Scrubber can be dispensed with. In this case, the base of the Rotopart Scrubber becomes little more than a narrow trough, and oxidation takes place in a separate Oxidiser.

- Replacement of the Centrifuge for dewatering the gypsum by a Thickener and Rotary Vacuum Filter. The Thickener also replaces the Neutraliser Tank, and the formic acid and lime slurry are added to the slurry flowing to the Thickener.

The adoption of the Rotopart Scrubber, with simultaneous absorption of SO_2 and oxidation to gypsum, can eliminate the need for: slurry circulation pumps, or alternatively for separate reaction vessels for absorption and oxidation reactions; thickeners and other pre-dewatering equipment; and ancillary equipment such as agitators, rakes and some process instrumentation. The low pH allows the lime or limestone to dissolve, resulting in more rapid reaction with sulphur oxides and freedom from scaling and plugging in the mist eliminator.

Potential for process development: ***

Process Requirements for Each Application Considered: These are shown in Table 2.47 for the two applications considered: Datum Combustion Systems 1 and 2.

It is assumed that for System 1:

- The SO_2 content of the gas is to be reduced to 400 mg/Nm^3.

- The end product is gypsum which will be sold as a by-product; in order to achieve a pure product, the gas is prescrubbed to remove particulates and HCl, and the prescrubber water is neutralised with NaOH. The conversion of lime to gypsum is about 98%, and the percentage of unreacted slaked lime in the product is therefore acceptable.

- There is one prescrubber/Rotopart Scrubber system for each 500 MWe unit (total of four systems)

It is assumed that for System 2:

- The SO_2 content of the gas is to be reduced to 650 mg/Nm^3.

- The end product is gypsum which will be disposed of to a land fill. Although high purity is not required, the gas is prescrubbed in order to avoid conversion of lime or limestone to calcium chloride.

- There is one prescrubber/Rotopart Scrubber system.

It is further assumed that for both combustion systems:

- The absorbent is hydrated lime, prepared by slaking quicklime on site.

- The same molar ratio of (fresh sorbent)/(captured sulphur) is used (1.02 mol Ca per mol S absorbed), giving a lime or limestone utilisation of 98%.

- The NO_x content of the gas is unaffected.

- The Prescrubber removes 80% of the HCl from the gas, and no absorption of HCl occurs in the Rotopart Scrubber.

- The Prescrubber reduces the particulates content of the gas from 115 to 15 mg/Nm^3, and no further reduction occurs in the Rotopart Scrubber.

By-products and Effluents: The effluents from the process consist of gypsum and waste water requiring treatment. The gypsum has to be disposed of, e.g. to a landfill, unless:

- It is produced in sufficient purity–achieved without difficulty by prescrubbing to remove particulates and HCl.

- There is a market for it, e.g. for manufacturing plasterboard–this will apply in only a small number of instances.

For System 2, where it is assumed that the gypsum produced is disposed of to a landfill, some care has to be taken in the selection of a suitable site for disposal, as calcium sulphate is slightly soluble in water, and the sulphate anion is harmful to concrete.

The rates of production and impurities contents of gypsum produced are summarised in Table 2.48 for the two Datum Combustion Systems considered.

Water treatment requirements are dependent on the composition of the waste water leaving the prescrubber and other parts of the plant, and on the local waste water disposal consent requirements. The main constituents of waste water to a water treatment plant will be particulates, together with calcium and

Table 2.47 Process Requirements

Datum Combustion System		1	2
Inlet Gas at full load			
Volume flow	'000 Nm^3/h	1524*	551
Dry gas	'000 Nm^3/h	1426*	513
Water vapour	'000 Nm^3/h	99*	38
Actual volume flow	'000 m^3/h	2362*	854
Temperature	°C	150	150
Particulates content	mg/Nm^3 (dry)	115	115
SO_2 content	mg/Nm^3 (dry)	4240	3520
NO_x content	mg/Nm^3 (dry)	1025	1440
HCl content	mg/Nm^3 (dry)	340	315
Exit Gas at full load			
Volume flow	'000 Nm^3/h	1614*	581
Dry gas	'000 Nm^3/h	1423*	512
Water vapour	'000 Nm^3/h	191*	69
Actual volume flow	'000 m^3/h	2146*	773
Temperature	°C	90	90
Particulates content	mg/Nm^3 (dry)	15 (25)	15 (25)
SO_2 content	mg/Nm^3 (dry)	400 (784)	650 (937)
NO_x content	mg/Nm^3 (dry)	1025 (1025)	1440 (1440)
HCl content	mg/Nm^3 (dry)	68 (95)	63 (88)
Reaction temperature	°C	55	55
Particulates removal		Separate	Separate
Reagent		Lime	Lime
Purity	Wt. %	94.8	94.8
Particle size		90% below 150 micron	90% below 150 micron
Concentration in slurry	Wt. %	15	15
Ca/(Captured S) Molar Ratio		1.02	1.02
Additive		Formic acid	Formic acid
Concentration in slurry	Wt. %	0.08	0.08
Liquid/gas ratio	litre/m^3	5.2	5.2
Requirements at full load			
Lime (94.8% pure)	tonne/h	5.15*	1.39
Formic acid (87% soln.)	kg/h	23*	6
Antifoulant	kg/h	33*	8
Caustic soda (47% liquor)	tonne/h	0.91*	0.30
Process air (0.5 bar)	tonne/h	9*	2
Water	tonne/h	26*	6
Electric power	MWe	5.0*	1.7
Manpower	men/shift	***	***
Average load factor	%	***	***

*Per 500 MWe boiler
Figures in parentheses are annual averages for 90% FGD plant availability

sodium cations, and sulphite, sulphate, bicarbonate, chloride and fluoride anions. Some coal ash constituents may be leached out.

Efficiency and Emission Factors: The efficiency and emission factors for the process are summarised in Table 2.49 for the two applications considered: Datum Combustion Systems 1 and 2.

In calculating the efficiency factors, it is assumed that at full load:

- The performance without the incorporation of the FGD plant would be as shown in Table 1.4.

- For Datum System 1, the FGD unit for each 500 MWe set consumes 5 MWe of electrical power.

Table 2.48 Estimated Properties of Gypsum Produced

Datum Combustion System		1	2
Rate of production:			
Wet	tonne/h	67.1*	4.5
Dry	tonne/h	60.4*	4.1
Composition of wet product:			
Moisture	wt. %	10.0	10.0
Calcium hydroxide	wt. %	0.5	0.5
Calcium formate	wt. %	0.2	0.2
Other impurities	wt. %	1.6	1.6
Gypsum	wt. %	87.7	87.7
Purity of dry product	wt. %	97.4	97.4

*Total from four 500 MWe boiler FGD plants

- For Datum System 2, the power consumption of 1.7 MWe by the FGD plant is equivalent to the combustion of a further 0.7 tonne/h of coal at a power station, assuming an overall power generation efficiency of 33%. This additional coal is arbitrarily assumed to be included with the coal burned in the boiler for calculating the efficiency factor.

For illustration purposes, the annual average emissions and emission factors for FGD plant availabilities of 100% and 90% are given in Tables 2.47 and 2.49. Further details are given in Section 1.6.

Effect of load variations: ***

Effect of design variations: ***

Limitations: ***

Costs:

Cost basis: *** See Section 2.6.

Capital costs: *** For retrofit and for new build applications–see Section 2.6.

Annual running costs and cost factors: See Section 2.6.

Effects of design variations on costs: ***

Effects of annual load patterns on annual running costs: ***

Process Advantages and Drawbacks: Process well established in the FRG; high reliability; land area required is smaller than for many other lime-gypsum processes (and has been demonstrated to be virtually zero when the FGD plant has been built inside a cooling tower); fine grinding not needed if lime is used as absorbent (calcium hydroxide is a fine powder); absorbent dissolves under operating conditions; simple chemistry; high sulphur capture efficiency (over 95% capture can be attained); very low reagent/sulphur stoichiometric ratio (98% utilisation of lime or limestone); product is easily dewatered; product potentially marketable, and reasonably benign if disposal is required.

The disadvantages are the tendency for the aeration system to become plugged by gypsum build-up; low pH necessitates use of corrosion-resistant materials.

By-products and Effluents from Limestone Slurry Scrubbing

Liquid Effluents: The quantity and composition of the liquid effluents depend on whether the scrubbing liquid circuit is operated on a closed loop or an open loop basis. Closed loop operation means that all liquid effluents are recycled after treatment to remove salts and suspended particulate matter that would otherwise interfere with the functioning of the scrubbing process and the correct operation of the Mist Eliminators.

Open loop operation means that liquid is discharged continuously in substantial quantities from the scrubbing plant, to be replaced by fresh water from a river, piped mains supply or a waste water discharge from some other process (e.g. boiler blowdown).

Table 2.49 Efficiency and Emission Factors

Datum Combustion System		1	2
Coal heat input (gross)	MWt	1292*	469
Coal fired	tonne/h	189*	60.3
FGD plant power consumed	MWe	5*	1.7
Equivalent coal input	tonne/h	–	0.7+
Useful energy from system	GJe/h	1656*	–
	GJt/h	–	1468
Total equiv. coal input	tonne/h	189*	61.0(a)
Efficiency factor	GJe/tonne	8.76	–
	GJt/tonne	–	24.07(a)
Emissions			
Sulphur in SO_2	kg/h	285* (559*)	167 (241)
Nitrogen in NO_x	kg/h	445* (445*)	225 (225)
Chlorine in HCl	kg/h	94* (132*)	31 (44)
Particulates	kg/h	21* (35*)	8 (13)
Emission factors (per tonne coal)			(a)
Sulphur	kg/tonne	1.51 (2.96)	2.74 (3.95)
Nitrogen	kg/tonne	2.35 (2.35)	3.69 (3.69)
Chlorine	kg/tonne	0.50 (0.70)	0.51 (0.72)
Particulates	kg/tonne	0.11 (0.19)	0.13 (0.21)

* For each 500 MWe boiler FGD plant
+ Calculated assuming overall power generation efficiency = 0.33
(a) Based on coal fired to boiler plus coal equivalent to electric power consumed
Figures in parentheses are annual average emissions for 90% FGD plant availability

In closed loop operation, liquid effluent discharge rate is very low, or zero, the process being an overall consumer of water. This is because the treated flue gases have a higher dew-point than the gases entering the FGD system, and because some moisture is removed with the solid product or waste residue. Maintenance of a correct water balance in closed loop operation is often very difficult.

In open loop operation (except for sea-water or estuary-water scrubbing) a high proportion of the scrubbing liquid is likewise recycled after treatment, because this reduces both the water input to the process and the amount of contaminated effluent that has to be discharged. Because of the difficulty of maintaining a water balance in closed loop operation, open loop is preferable, except where water is in severely limited supply. Many inland U.S. plants employ closed loop operation.

The liquid leaving the prescrubber contains a large fraction of the HCl and particulate matter entering with the flue gas. In the treatment of flue gases from combustion of most UK coals, the HCl content is considerably higher than for American coals, the chlorine content of UK coals ranging up to about 0.8%. The quantity of chloride washed out in wet FGD processes is seen as posing some problems in UK operation, both in causing corrosion of scrubbing plant and in its disposal [11, 109, 144]. Because of the high solubility of nearly all chlorides, the chloride ion builds up in the prescrubber liquid circuit and the bleed stream removed therefore contains high concentrations of chloride. It is unlikely that an effluent containing such high concentrations could be disposed of without some treatment on site, unless the liquid can be discharged to the sea.

Workers at the UK Central Electricity Research Laboratories of CEGB, in association with manufacturers of FGD process plant, have addressed this problem. One solution proposed [447] is to withdraw a bleed stream from the recirculating prescrubber liquid with pH in the range 1.4–2.0 (at which level HCl but not SO_2 is absorbed), neutralise this bleed stream with lime slurry, screen out the flyash content and then evaporate the filtrate to dryness. The condensate will then be available for recycling to the process or for other use, and the solid calcium chloride, together with any fluoride also removed in the prescrubber, is transported from the power station for disposal (see below).

A pH of 1.4 with a $CaCl_2$ concentration of 30% corresponds to a HCl concentration of 0.002% w/w which, at 60°C is in equilibrium with an HCl vapour concentration of only 1.6 vpm. When burning coal with a chlorine content of 0.3%, the concentration of HCl in dry flue gas is 253 vppm, hence 99.4% of the HCl will be removed in the prescrubber, for a liquor pH of 1.4 and 30% concentration of calcium chloride at 60°C. This is important for the purity of a gypsum product intended for marketing.

The low pH values encountered in the prescrubber enhance the solubility of some trace elements present in the very fine particles of ash that escape collection in the electrostatic precipitator. Hence the calcium chloride produced will contain such elements as arsenic, chromium, copper, mercury, nickel, lead, selenium, tellurium, vanadium and zinc [144]. In processes where the calcium chloride is disposed of as a concentrated solution, most of these elements may be precipitated from solution by adjusting the pH to between 9 and 10 using caustic soda, after initial neutralisation with lime. If it is necessary to reduce the concentration of trace elements in $CaCl_2$ (dry form) before disposal, the $CaCl_2$ solution may be similarly treated before evaporation to dryness.

The amounts of trace elements present in FGD liquid effluent are very variable, depending on the composition of the coal and on the furnace design and operating conditions. Values measured at the Scholven F power station in Germany and quoted in reference [144] are shown in Table 2.50, but these should not be regarded as average, or even typical values, because of the very wide variation in the trace element contents of coals [35].

To maintain the $CaCl_2$ concentration in the prescrubber circuit at 30% w/w when firing a 0.3% chlorine coal, the rate of flow of liquid in the bleed stream withdrawn for evaporation is 3.9 tonne/h per 500 MWe set, and it will contain 3 tonne/h of water. For coals of other chlorine contents the rate of flow is increased or decreased in proportion. For a 0.3% chlorine coal, the energy required for evaporation to dryness would be of the order of 0.25% of the boiler heat input, and there would therefore be significant savings if it were possible to dispose of the $CaCl_2$ in solution.

The removal of most of the residual fly-ash in the prescrubber means that the liquid effluent arising from the SO_2 absorption circuit is relatively free from trace elements, as well as from chlorides. The principal source of contaminants in the effluent from the SO_2 absorption circuit are impurities present in the limestone or lime used as sorbent. Apart from silica and alumina compounds, magnesium and iron are the most common impurities. All are present in very variable amounts and, in some applications of the slurry scrubbing process, magnesium is deliberately introduced to improve absorption of SO_2 and to reduce scaling.

Part of the alumina and most of the iron and magnesium dissolve in the scrubbing liquor. The silica and undissolved aluminosilicates are carried through to form harmless impurities in the gypsum product. There is a risk of iron being precipitated as ferric hydroxide in the oxidiser, resulting in undesirable discolouration of the product, but magnesium and aluminium sulphates remain in solution although the aluminium sulphate may be partly hydrolysed to aluminium hydroxide which finishes up as an impurity in the gypsum. The removal of magnesium from the product is achieved by thorough washing of the gypsum after separation from the supernatant liquor.

The principal substances present in the effluent derived from the scrubbing circuit are therefore calcium sulphate at saturation SO_2 level (about 0.2% by weight) and iron, magnesium and possibly aluminium sulphates, together with trace amounts of other elements derived from the sorbent, and any chloride and fluoride escaping capture in the prescrubber.

The quantity of liquid effluent arising from the SO_2 absorption stage varies according to the rate of build-up of impurities in the scrubbing circuit and hence with the limestone composition. Typically, the rate is of the order of 60 tonne/h, but it may not be necessary to discharge the whole of this–it may be environmentally or economically attractive to recycle part of the effluent (or the whole of it) following treatment to remove soluble salts and suspended particulate matter.

Table 2.50 Heavy Metals in FGD Effluent (Scholven F Power Station): [319]

Metal	Concentration, micro-grammes/litre	
	From FGD Plant	After Treatment
As	170	15
Be	85	0.2 max
Bi	10 max	3 max
Cd	40	0.9 max
Co	60	2 max
Cr	510	98
Cu	660	21
Hg	140	13
Ni	280	17
Pb	520	10 max
Sb	20 max	10 max
Se	1120	26
Te	30 max	10
Tl	2 max	1 max
V	250	17
Zn	4300	50 max

The processes used for water treatment consist of several steps. In the first step, sulphite, sulphates and any fluorides are precipitated as their calcium salts by the addition of lime. In the second step, the pH is raised to 9–10 by the addition of caustic soda, resulting in the precipitation of metal oxides and hydroxides. Many of these come down in the form of flocculant precipitates and their removal is often assisted by adding ferric chloride or a polyelectrolyte coagulant. Settling of the precipitate is effected in a thickener, the underflow from which forms a sludge that may be further dewatered in a filter press for ease of handling and disposal.

The overflow from the thickener is treated for separation of oil, if necessary, and further purified by ion exchange using weak anion exchange resins to remove thiosulphate and other reducing anions. The pH of the effluent is then adjusted before discharge.

For discharge to surface watercourses there are usually restrictions on concentrations of suspended and dissolved solids, chemical oxygen demand and pH, as well as on concentrations of specific toxic elements.

If it is proposed to recycle the effluent, a lesser degree of purification would probably be required.

Solid Residues and Products: These include gypsum and waste materials for disposal, mainly calcium chloride and residues from water treatment.

Kyte and Cooper [144] have discussed the requirements for gypsum quality to satisfy existing market applications, and have estimated the rate of production from UK power stations equipped with FGD plants in relation to the size of existing UK market for gypsum. Table 2.51 taken from reference [144] shows some typical guideline specifications used in the UK, USA, Germany and Japan. In Table 2.52, the quality requirements for gypsum in cement and in wallboard manufacture are compared with the quality of gypsum produced in Japanese power stations [675–677]. The FGD products are seen to be well within the requirements for both applications, apart from that for purity (gypsum content) in the wallboard application, where some samples fell outside the limit.

Table 2.51 Guidelines Issued by Users for Gypsum Purity [144]

	National Gypsum Co.	Georgia Pacific Co.	U.S. Gypsum Co.	Westroc	German/ Japanese	UK
Gypsum content min %	94	90	–	95	95	95
Calcium sulphite max. %	0.5	–	–	–	0.25	0.25
Sodium ion max ppm	500	200	75	80	600	500
Chloride max ppm	800	200	75	80	100	100
Magnesium max ppm	500	–	50	50	1000	1000
Free water, max %	15	10	12	10	10	8
pH	6–8	3–9	6.5–8	6–9	5–9	5–8
Particle size, microns	–	–	20–40	–	–	16–63

Table 2.52 By-Product Gypsum Quality in relation to requirements

Property	General Requirements For Cement	For wallboard	Quality ex FGD Takasaqo Takehara	Matsushima
Surface H_2O %	10–12 (max)	10–12 (max)	6–8	6
Purity %	90 (min)	95 (min)	90–97	99
Size, microns	50 (min)	50 (min)	50–150	50
Cryst. shape	–	–	pillar and plate shaped	
pH	–	5–7	6–8	6.8
Composition: %				
$CaCO_3$	2 (max)	0.5–2 (max)	0.5–1	0.1
CaO (Combined)	–	–	32.4	32.3
SO_3	–	44 (min)	45.3	46.2
SO_2	–	–	0.05	0.07
MgO	–	0.08 (max)	0.03	0.01g (-Mg)
Na_2O	–	0.04 (max)	0.03 (-Na)	0.003 (-Na)
Carbon	–	–	0.1 (max)	0
Ash	2 (max)	2 (max)	1.5–2 (max)	0.2
Cl	–	0.03 (max)	0.03 (max)	0.002 (min)
Wet tensile strength Kg/cm^2	–	8 (min)	12–13	13
Adhesion to paper	–	acceptable	acceptable	acceptable

Table 2.53 Gypsum Composition

Component %	From Scholven F.P.S.	Natural Gypsum
$CaSO_4.2\ H_2O$	98–99	78–85
$CaSO_3.1/2\ H_2O$	0–0.7	0
Cl	0.01	0.002
Na_2O	0.01	0.03
MgO	0.1	0.5–3.0
Fe	0.05	–
Silicate	–	10–20
Moisture	7–10	1 (max)

A comparison of the quality of gypsum from Scholven F power station with that of natural gypsum is given in Table 2.53. In contrast to some of the Japanese samples, purity here is high–far superior to that of the natural gypsum with which it is compared (which would not meet the Japanese requirements for either cement or wallboard manufacture).

For wallboard manufacture, large crystal size is wanted, and this can be achieved more easily by carrying out the forced oxidation process in a separate vessel. Hence, if it is intended to market the gypsum for wallboard manufacture, processes such as the IHI process, which use a separate vessel for oxidation, are superior to those which carry out oxidation in the absorber.

In the production of plaster of paris, which is the basic material for making wallboard and other plaster products such as stucco, the gypsum ($CaSO_4.2\ H_2O$) is carefully calcined to form the alpha- and beta- crystalline forms of calcium sulphate hemihydrate ($CaSO_4.1/2\ H_2O$).

For some uses the alpha form is preferred and for others the beta form. Proportions of each are controlled by using appropriate conditions for calcination. The plaster sets rapidly on mixing with water by rehydrating to the dihydrate: setting may be delayed by incorporating small amounts of an additive, e.g. borax or certain organic materials.

If calcination is carried beyond the hemihydrate stage, anhydrous calcium sulphate is produced and, unless 'hard burnt', this also rehydrates on mixing with water. Keene's cement and a number of other proprietary plasters are of this type, containing various additives to control setting rate and the properties to the hardened material.

The application of gypsum in the cement industry is mainly as a setting retardant and the quality of gypsum is less critical than for wallboard manufacture. The rate of addition represents only a small percentage of the cement, so the market for this use is not large.

The production of gypsum from a 200 MWe power station burning 2% sulphur coal has been estimated to be 15% of the existing UK market for gypsum [144]. Hence, if a large number of stations were to be equipped with gypsum-producing FGD plant it would probably be necessary to dispose of some of the gypsum produced. The disposal of unwanted, substandard gypsum is referred to as 'stacking', and for disposal in this way the quality of the product would not be of such importance. There would not, for example, be a need to convert excess limestone to the sulphate, but pH control in the oxidiser may still be necessary in order to achieve a satisfactory degree of oxidation.

It is possible that other outlets may be found for gypsum, if large amounts become available from flue gas desulphurisation. Research has shown that it may be used to stabilise other wastes which have to be tipped, and for this purpose a small amount of $Ca(OH)_2$ may also have to be added [328, 330]. In one patented technique [325] for the use of FGD gypsum as a stabilising agent for the treatment of wastes, such as radioactive or toxic materials, the waste is mixed with FGD gypsum and a smaller amount of water-dispensable melamine formaldehyde resin.

It is however, unlikely that the waste stabilisation application will command anything like the same market value as its use as plaster.

It is reasonable to enquire whether, in the event of the product having to be dumped, the oxidation stage could be omitted. The majority of lime/limestone slurry scrubbing processes installed in the USA are, in fact, of this type and do not use forced oxidation.

However, many of the more recently installed plants do employ forced oxidation, and some of the older ones have had it installed, even though it is not proposed to market the gypsum produced.

The reasons why oxidation is being adopted in the USA is because experience has shown that the calcium sulphite/sulphate sludge produced in the absence of forced oxidation is very troublesome to handle and to dewater, and many plant failures have occurred. On top of this, the sludges tend to be thixotropic, making them less than ideal for landfill.

In order to create stable tips, the dewatered sludges have to be carefully blended with the correct proportions of fly-ash and lime. Following tipping, care has to be taken to avoid contamination of water courses or subterranean water reserves with leachates from the tipped material. Monitoring of ground water in the vicinity of the tipping area on a routine basis is insisted on by some authorities.

The amount of calcium chloride produced by a 2000 MWe station on 65% load factor, burning 0.3% chlorine coal, would be about 5125 tonnes per annum. There is virtually no market for calcium chloride and, being highly soluble, inland tipping is environmentally unacceptable. The most satisfactory method for disposal would be tipping at sea, or discharge of $CaCl_2$ solution to the sea. In any such disposal, the concentrations of other elements such as heavy metals and fluorine would be of concern, and may require controlling.

Water treatment sludges will be produced in relatively small quantities, the amounts depending largely on the purity of the sorbent used. Tipping, after fixation with a suitable stabilising agent (possibly gypsum plus lime) should result in no serious environmental risks.

Evaluation of Fläkt-Niro FGD Process (Process Code S22.1 – Lime Slurry Spray Dryer Processes)

See Section 2.2 for: outline of the basic process; its chemistry; block diagram.

See Section 2.3 for a list of manufacturers offering this type of equipment.

See Section 2.3 for general appraisal of the basic process.

See Section 2.4 for the reason for choosing the Fläkt-Niro process to illustrate the Code S22.1 basic process.

This basic process type is considered (Section 2.4) to be suitable for application in the UK to utility boilers, large industrial boilers and small factory boilers, and hence it is evaluated here for Datum Combustion Systems 1, 2 and 3. The process presented here is of the type using lime slurry in a Spray Dryer Absorber, producing a dry end product. The process is offered in Europe by Fläkt Industri AB (Växjö, Sweden) in collaboration with A/S Niro Atomiser (Copenhagen, Denmark), and in the USA by Joy-Niro. See Appendix 2 for details of these manufacturers.

Process Description: Figure 2.37 shows a simplified flow diagram for application of the process to a coal fired boiler. It is assumed that for all applications, the solid end-product will be disposed of to a landfill.

For dust and sulphur oxides removal, gas from downstream of the electrostatic precipitator or baghouse enters a Spray Dryer Absorber (SDA) fed with a spray of calcium hydroxide slurry. Operating conditions are chosen to give a gas temperature leaving the SDA that is only 10–20°C above the saturation temperature, so as to delay the dry-out of the slurry; this is because absorption of sulphur oxides (and of acid halides) is more rapid when water is present, and these gas constituents are therefore removed more efficiently when dry-out is delayed. The residence time of the slurry droplets in the SDA is about 10 s. The gas then passes to an electrostatic precipitator (ESP), or more usually a Baghouse, to remove the gas-borne reaction products which have not been collected in the hopper base of the SDA. A significant proportion (10–20%) of the total sulphur oxides capture occurs in the ESP or Baghouse. The collection hoppers of the SDA and of the ESP or Baghouse are electrically heated to prevent solids build-up resulting from condensation. The cleaned gas then passes, via a fan, to the Stack. If necessary, it is first reheated to 90–150°C by combustion of a gaseous or liquid fuel, or by mixing with air heated by steam.

Part of the collected solids is returned via a Recycle Product Silo to the Feed Tank, where it is mixed with fresh lime slurry. The fresh lime is slaked with water in the Slaker, and the calcium hydroxide is slurried with water in the Lime Slurry Tank before being pumped to the Feed Tank. The dry waste end product contains fuel ash, calcium sulphite, sulphate and chloride, and unreacted calcium hydroxide. The presence of the fuel ash gives the material pozzolanic properties, which render the material self-hardening and stable on exposure to air.

Status and Operating Experience: The status of the Fläkt-Niro process is indicated in Section 2.3, where it is seen that Fläkt, Niro, and Joy have installed, or are constructing, over 50 units.

Availability of operating experience is generally high. The principal problems encountered are: solids build-up on atomiser wheels [668, 681, 683, 684]; solids build-up on walls [681, 684, 685]; difficulties with solids extraction from the SDA and Baghouse [678, 683]; instrumentation inadequacies [683]; and corrosion in the Baghouse [683].

Variations and Development Potential: Design and operating variations can include:

- Number, type and location of slurry atomisers in the SDA.
- Alternatives of Electrostatic Precipitator and Baghouse for removal of gas-borne end-product.
- Removal of fly-ash from the boiler flue gas either upstream of the SDA, or in the ESP or Baghouse downstream of the SDA.

The simplicity of the process is its chief characteristic, making it suitable for application on even small-scale plant (e.g. System 3).

Process Requirements for Each Application Considered: These are shown in Table 2.54 for the three applications considered: Datum Combustion Systems 1, 2 and 3.

It is assumed that the SO_2 content of the gas is to be reduced to the following levels:

- System 1: 400 mg/Nm^3 (dry), equivalent to 90.6% capture of SO_2
- System 2: 1000 mg/Nm^3 (dry), equivalent to 71.6% capture of SO_2
- System 3: 1000 mg/Nm^3 (dry), equivalent to 69.3% capture of SO_2

It is further assumed that for all three systems:

- The gas leaving the Baghouse has a particulates content of 15 mg/Nm^3 (dry).
- The gas to the stack is reheated to 90°C by addition of air from a steam-heated air-heater.
- The same molar ratio of (fresh sorbent)/(captured sulphur) is used (1.3 mol Ca per mol S absorbed), giving a lime utilisation of about 75%.

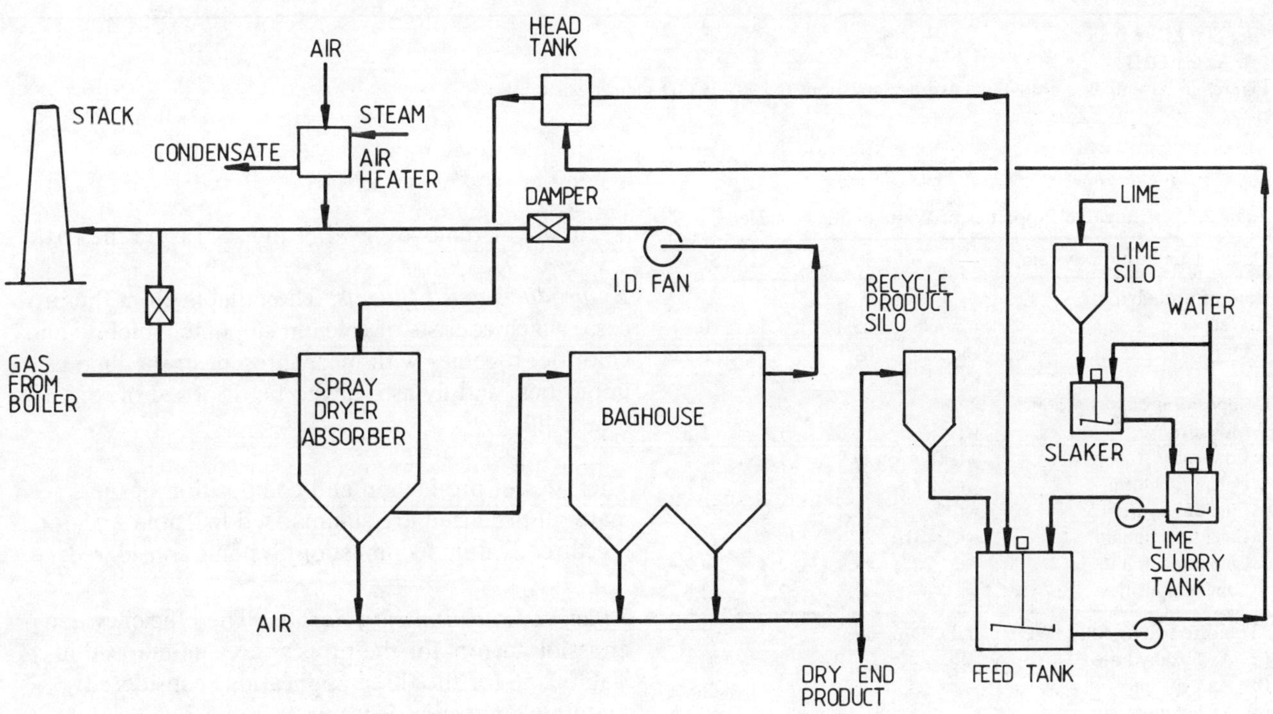

Figure 2.37 Lime Spray Drying FGD Process

Table 2.54 Process Requirements

Datum Combustion System		1	2	3
Inlet Gas at full load				
Volume flow	'000 Nm^3/h	1524*	551	16.52
Dry gas	'000 Nm^3/h	1426*	513	15.46
Water vap.	'000 Nm^3/h	99*	38	1.06
Actual volume flow	'000 m^3/h	2362*	854	30.44
Temperature	°C	150	150	230
Particulates content	mg/Nm^3 (dry)	115	115	660
SO_2 content	mg/Nm^3 (dry)	4240	3520	3255
NO_x content	mg/Nm^3 (dry)	1025	1440	515
HCl content	mg/Nm^3 (dry)	340	315	290
Exit Gas at full load				
Volume flow	'000 Nm^3/h	1604*	580	18.30
Dry gas	'000 Nm^3/h	1423*	512	15.42
Water vap.	'000 Nm^3/h	181*	68	2.88
Actual volume flow	'000 m^3/h	2015*	729	23.66
Temperature	°C	70	70	80
Particulates content	mg/Nm^3 (dry)	15 (25)	15 (25)	15 (80)
SO_2 content	mg/Nm^3 (dry)	400 (784)	1000 (1252)	1000 (1226)
NO_x content	mg/Nm^3 (dry)	1025 (1025)	1440 (1440)	515 (515)
HCl content	mg/Nm^3 (dry)	68 (95)	63 (88)	58 (81)
Reaction temperature	°C	50	50	60
Particulates removal		Simultaneous	Simultaneous	Simultaneous
Reagent		Lime	Lime	Slaked Lime
Purity	Wt. %	94.8	94.8	96.0
Concentration in slurry	Wt. %	5	13(a)	10(a)
CA/(Captured S) Molar Ratio		1.3	1.3	1.3
Requirements at full load				
Quicklime (94.8%)	tonne/h	7.25*	2.17	–
Slaked Lime (96%)	tonne/h	–	–	0.076
Water (slaking)	tonne/h	2.2*	0.47	–
Water (slurrying)	tonne/h	64.3*	23.4	1.45
Electric power	MWe	10.7*	2.9	0.1
Manpower	men/shift	***	***	***
Average load factor	%	***	***	***

*Per 500 MWe boiler
(a) As $Ca(OH)_2$
Figures in parentheses are annual average emissions for 90% FGD plant availability

Table 2.55 Estimated Properties of Waste Solids Produced

Datum Combustion System		1	2	3
Rate of production:				
Wet	tonne/h	65.3*	4.56	0.12
Dry	tonne/h	64.1*	4.48	0.11
Composition of wet product:				
Moisture	wt. %	1.8	1.8	1.8
Coal ash	wt. %	0.9	1.1	8.4
Calcium chloride	wt. % (a)	7.2	8.5	9.3
Calcium hydroxide	wt. %	14.4	23.9	12.2
Calcium sulphate	wt. % (b)	23.5	19.9	21.0
Calcium sulphite	wt. % (c)	49.9	42.3	44.7
Other impurities	wt. %	2.3	2.5	2.6

*Total from four 500 MWe boiler FGD plants
(a) As hexahydrate
(b) As gypsum
(c) As hemihydrate

– The SDA removes 80% of the HCl from the gas.

By-products and Effluents: The effluent from the process, which consists of calcium sulphate, sulphite and chloride, together with inerts present in the lime as impurities, and fly-ash, has to be disposed of, e.g. to a land fill.

The rates of production and composition of the material produced are summarised in Table 2.55 for the three Datum Combustion Systems considered.

Efficiency and Emission Factors: The efficiency and emission factors for the process are summarised in Table 2.56 for the three applications considered: Datum Combustion Systems 1, 2 and 3.

In calculating the efficiency factors, it is assumed that at full load:

- The performance without the incorporation of the FGD plant would be as shown in Table 1.4.

- For Datum System 1, the FGD units for each 500 MWe set consumes 10.7 MWe of electric power, reducing the power sent out.

- For Datum System 2, the FGD unit consumes 2.8 MWe of electric power, equivalent to the combustion of a further 1.1 tonne/h of coal at a power station, assuming an overall power generation efficiency of 33%. This additional coal is arbitrarily assumed to be included with the coal burned in the boiler for calculating the efficiency factor.

- For Datum System 3, the FGD unit consumes 0.1 MWe of electric power, equivalent to the combustion of a further 0.04 tonne/h of coal at a power station, assuming an overall power generation efficiency of 33%. This additional coal is arbitrarily assumed to be included with the coal burned in the boiler for calculating the efficiency factor.

In all three cases the gas leaving the SDA is at a temperature sufficiently high to make reheating unnecessary.

Effect of plant availability: For illustration purposes, the annual average emissions and emission factors for FGD plant availabilities of 100% and 90% are given in Tables 2.54 and 2.56. Further details are given in Section 1.6.

Effect of load variations: ***

Effect of design variations: ***

Limitations: ***

Costs:

Cost basis: See Section 2.6.

Capital costs: See Section 2.6.

Annual running costs and cost factors: See Section 2.6–for retrofit and for new build applications.

Effects of design variations on costs: ***

Effects of annual load patterns on annual running costs: ***

Process Advantages and Drawbacks: Basic process well established in USA and Europe; simple process allowing application to small scale without high labour demands; fairly high reliability; land area required is small; fine grinding not needed as lime is used as absorbent (calcium hydroxide is a fine powder); simple chemistry; fairly high sulphur capture efficiency (over 90% capture) can be attained; waste product can be easily and safely disposed of.

Table 2.56 Efficiency and Emission Factors

Datum Combustion System		1	2	3
Coal heat input (gross)	MWt	1292*	469	13.1
Coal fired	tonne/h	189*	60.3	1.68
FGD plant power consumed	MWe	10.7*	2.8	0.1
Equivalent coal input	tonne/h	–	1.1+	0.04+
Useful energy from system	GJe/h	1635*	–	–
	GJt/h	–	1468	35.3
Total equiv. coal input	tonne/h	189*	61.4(a)	1.72(a)
Efficiency factor	GJe/tonne	8.65	–	–
	GJt/tonne	–	23.9(a)	20.6(a)
Emissions				
Sulphur in SO_2	kg/h	285* (559*)	257 (322)	7.72 (9.47)
Nitrogen in NO_x	kg/h	446* (446*)	225 (225)	2.42 (2.42)
Chlorine in HCl	kg/h	94* (132*)	31 (44)	0.87 (1.22)
Particulates	kg/h	21* (35*)	8 (13)	0.23 (1.23)
Emission factors (per tonne coal)			(a)	(a)
Sulphur	kg/tonne	1.51 (2.96)	4.19 (5.24)	4.49 (5.51)
Nitrogen	kg/tonne	2.36 (2.36)	3.66 (3.66)	1.41 (1.41)
Chlorine	kg/tonne	0.50 (0.70)	0.50 (0.72)	0.51 (0.71)
Particulates	kg/tonne	0.11 (0.19)	0.13 (0.21)	0.13 (0.72)

* For each 500 MWe boiler FGD plant
+ Calculated assuming overall power generation efficiency = 0.33
(a) Based on coal fired to boiler plus coal equivalent to electric power consumed
Figures in parentheses are annual average emissions for 90% FGD plant availability

The disadvantages are the tendency for deposits to form on SDA walls; absorbent more costly than limestone, and higher stoichiometric ratios needed.

Effluents and Residues from Lime Slurry Spray-Dry

Liquid Effluents: The outstanding feature of spray-dry FGD processes is that, when operated correctly, they do not produce any liquid effluent directly consequent upon the desulphurisation process. A boiler plant using the process will generate only those liquid effluents normally associated with coal-fired boiler plants, i.e. boiler water blowdown, site run-off from rainfall and hose-down operations, sewage, etc.

There remains a possibility that, in cases of solids build-up in the dryer, the use of high-pressure water jets to clear the obstruction could produce quantities of strongly alkaline water containing suspended solids and calcium salts in solution. The disposal of these could present problems of differing magnitude depending on the volume of water that it was necessary to use to clear the blockage and on the situation of the plant. Although a more disagreeable operation and one introducing a risk of damage to the plant, removal of deposits in the dry state might be preferable if disposal of contaminated water is likely to be difficult. It is present practice on some facilities to recycle the contaminated water to the lime slaker.

Solid Residues: As discharged from the baghouse or electrostatic precipitator, the solid residues from a lime slurry spray-dry FGD process are in the form of a dry, free-flowing powder with properties resembling those of fly-ash [337] and composed of small, irregular shaped clusters of reaction products [338]. The composition of the residues varies, depending on the ash content of the coal, the quantity of lime necessary to achieve the required SO_2 emission level, and the amounts of sulphur dioxide and water reacting with lime.

Compounds present, in addition to coal ash, according to Donnelly [310] are: calcium sulphate hemihydrate and dihydrate, calcium hydroxide and calcium carbonate. For most UK coals, calcium chloride will also be present in appreciable amounts.

A considerable effort has been applied, mainly by Niro Atomizer, in the study of environmental effects of the dumping of the residues, and into possible uses for them. In the absence of opportunities for marketing the material, the probable method of disposal is as landfill. Research has shown [310] that the moistened residues, when properly tipped and compacted, undergo fixation reactions which yield a filled material of high strength and low permeability to water. It is recommended, as an extra precaution, that the run-off should be collected and treated as necessary before it is released to natural drainage. It can also be recycled to the lime slaker line. Moistening of the residues can be carried out in various types of mixer, such as a plug mill or double-bladed type mixer, and sufficient water should be added to lubricate the grains of dry material to facilitate compaction. Care must be taken not to use too much water as this would produce a sticky material that is difficult to handle.

The properties of the compacted mass that determines its suitability for landfill are its density, its compressive strength, its permeability, and the composition of any leachate generated. The density of the compacted mass is an indicator of the effectiveness of compaction, and the optimum amount of water addition is that which gives the maximum density upon compaction, which generally corresponds to the maximum compressive strength. The optimum water addition is selected on the basis of laboratory tests for a given waste product.

An unconfined compressive strength of 172 kN/m^2 (25 p.s.i.) is considered a minimum for supporting the machinery used in tipping operations, and values of this property for a wide range of spray-dry residues have been shown to lie between 414 and 6,900 kN/m^2 after 28 days curing at 23°C [310]. The lower values were obtained with coals that produce ashes of low pozzolanic activity, but the strengths of the compacted residues were greater than those obtained from the coal ashes without the spray-dry reaction products.

Tests showed very low permeability coefficients (of the order of 10^{-6} cm/s) after 14 days of curing. The nature of the coal ash was found to have only a minor influence on permeability. It was noted also that the actual migration of water through a properly constructed landfill is substantially lower than indicated by laboratory determinations of permeability coefficients. Water penetrates the landfill only when there is standing water on it.

Leachates showed initially high pH, up to about 12, but this gradually falls, becoming eventually acid with pH in the range 5 to 6. Leaching of heavy metals from the ash was retarded by the alkaline reaction, and concentrations of several minor elements in the extracted leachate were at least an order of magnitude lower than the limits specified by the U.S. Environmental Protection Agency.

Chloride was not measured in the Niro study, and, since residues from application of the process at a UK power station might contain as much as 5% of the highly soluble calcium chloride (measured as the hexahydrate) [307], some leaching of $CaCl_2$ from tipped residues is to be anticipated. This reinforces the case for collecting run-off from a tipping area and releasing it for processing in such a way that chloride limits are not exceeded in a receiving watercourse.

Possible uses for lime spray-dry residues have been identified. They are: acid mine drainage control and co-disposal as a fixation agent for hazardous wastes [310]; pelletising and curing of the moistened residues to form a synthetic aggregate for use in civil engineering and construction projects [308, 310]; use, with cement addition, in the production of bricks and building blocks, grouting materials, 'stabilisate' (e.g. sub-base material for roads, etc.) or use in concrete [337]; the production of blocks for the construction of artificial reefs for coastal protection (using similar processes to those used for making building blocks [309]; incorporation with Portland cement to produce mortars [338]; and use in cement manufacture as a setting retarder [320].

Of particular concern, where the material is likely to come into contact with steel as, for example, in reinforced concrete, is the possible effect of chloride and sulphate ions on the protective passive layer of steel which is normally formed in contact with cement. Tests on several products resulting from combustion of a number of coals with spray-dry FGD showed that there was generally no chloride (or sulphate) attack. Only one of the products tested showed an increase in corrosion, and that only in half of the six tests carried out with it [338].

A broad view of the conclusions of all the test work indicates that, on the basis of field exposure of laboratory-produced artefacts and synthetic aggregates, satisfactory quality products can be made from spray-dry residues by using mixes of the right proportions. The amount of cement that has to be added to the mix ranges from 0 to 5 % for 'stabilisate' and grouting material; 15 to 20% for building blocks; and of the order of 70% as a cement constituent of a concrete mix (i.e. excluding fines and aggregate).

As with most residues from coal combustion, the main problems arise from the variable nature of the product due to variations in coal composition, lime slurry feed rate to meet emission limits, and other process variables. This variation is likely to make marketing more difficult, and will tend to depress market values for the material below those commanded by prime quality materials normally purchased.

Evaluation of Wellman-Lord FGD Process (Process Code S31.1)

See Section 2.2 for: outline of the basic process; its chemistry; block diagram.

See Section 2.3 for a list of manufacturers offering this type of equipment.

See Section 2.3 for general appraisal of the basic process.

See Section 2.4 for the reason for choosing the Wellman-Lord (Code S31.1) basic process.

This basic process type is considered (Section 2.4) to be suitable for application in the UK to utility boilers, and to large industrial boilers having access to reagent reprocessing facilities, and hence it is evaluated here only for Datum Combustion Systems 1 and 2. The process is offered by the Davy-McKee Corporation. See Appendix 2 for details of this manufacturer.

Process Description: Figures 2.38 and 2.39 show simplified flow diagrams for application of the process to a coal-fired boiler. It is assumed that the sulphur dioxide produced in the process is treated to recover elemental sulphur.

For dust and sulphur oxides removal, gas from downstream of the electrostatic precipitator or baghouse, boosted by a fan, is scrubbed with water in a Venturi Prescrubber and its Mist Eliminator, to cool the gas and to remove acid halides and particulates. The gas enters the Absorber, which is of the valve-tray design, where it is contacted counter-currently with sodium sulphite solution, which absorbs sulphur oxides from the gas to form sodium bisulphite, together with a small proportion of sodium sulphate. The gas passes through a mist eliminator at the top of the Absorber, and is then heated by mixing with air heated by steam in the Air Heater, before being exhausted to the Stack.

The prescrubber water contains about 2% solids together with dissolved halides. It is neutralised with caustic soda and circulated to a Waste Water Treatment System.

The absorbent solution from the base of the Absorber is pumped to the Absorber Product Tank; a side stream is pumped to the Sulphate Purge Tank. The further treatment of these two streams is described below.

The absorbent regeneration system is shown in the simplified flow diagram, Figure 2.39. The solution from the Absorber Product Tank is pumped, via a Solution Preheater, to the First Effect Evaporator of a Double Effect Evaporator system. In the First Effect, the solution is heated with steam, with the condensate from the steam being used to preheat the ingoing solution in the Solution Preheater. Part of the solution circulating through the First Effect Evaporator is bled to the Second Effect Evaporator, where it is heated at a lower pressure with vapour from the First Effect Evaporator and from the Sulphate Crystalliser (see below). Condensate and vapour from the Second Effect Evaporator pass through a water-cooled condenser. From here, condensate is treated with steam in a Condensate Stripper to strip out dissolved sulphur dioxide. The gas, with water vapour, passes through a second

Condenser; condensate from this returns to the Condensate Stripper, and the stripped SO_2 is pumped by the SO_2 Compressor to the SO_2 Recovery Plant.

The sulphate purge treatment is also shown in Figure 2.39. The solution from the Sulphate Purge Tank is pumped via a Sulphate Purge Preheater to the Crystalliser, where it is concentrated by evaporation using heat supplied by steam; the steam condensate passes to the Sulphate Purge Preheater to preheat the ingoing sulphate purge. The vapour from the Sulphate Crystalliser passes to the steam jacket of the Second Effect Evaporator (see above). Concentrated slurry from the Crystalliser passes to a Centrifuge, in which sodium sulphate crystals are separated from the liquor. The liquor passes via the Crystalliser Liquor Tank to the Absorber Product Tank for regeneration of its bisulphite content.

As seen in Figure 2.39, part of the solution leaving the First Effect Evaporator is pumped to the Mother Liquor Tank. From here it passes in two streams: one to the Crystalliser Liquor Tank (and hence back to the Absorber Product Tank); and the other to the Dissolving Tank. The solution from the Second Effect Evaporator, containing regenerated sodium sulphite, passes directly to the Dissolving Tank, which also receives the stripped condensate from the Condensate Stripper. The losses from the system of sodium sulphate are made up by feeding sodium hydroxide to the solution in the Dissolving Tank; the sodium hydroxide forms sodium sulphite by reaction with SO_2 in the Absorber. The solution is pumped from the Dissolving Tank to the Absorber Feed Tank (Figure 2.38) and the start of a new cycle.

Status and Operating Experience: The status of the Wellman-Lord process is indicated in Section 2.3, where it is seen that thirty-five installations have been, or are being, erected in the size range 17,000–4,636,000 m^3/h (total 22,276,000 m^3/h). These plants (twelve in USA, eighteen in Japan, four in the FRG and one in Austria) have been installed on: power station and other boilers burning coal, coke, lignite and oil; Claus plants; and sulphuric acid plants.

Operating experience shows (see Section 2.3) that availability and reliability are generally high. The principal problems that have been encountered appear not to have been process-related, but to have been caused by damage to the FGD plant resulting from difficulties in operation of the boiler plant, inexperienced FGD plant operation, and unsuitable materials of construction [658, 659]. Reliability has been quoted [276] as greater than 95%, and often over 98%.

Variations and Development Potential: Design and operating variations can include:

- Reheating of the effluent gas by heat exchange with ingoing unpurified gas.

- Adoption of mechanical vapour recompression in the evaporator system.

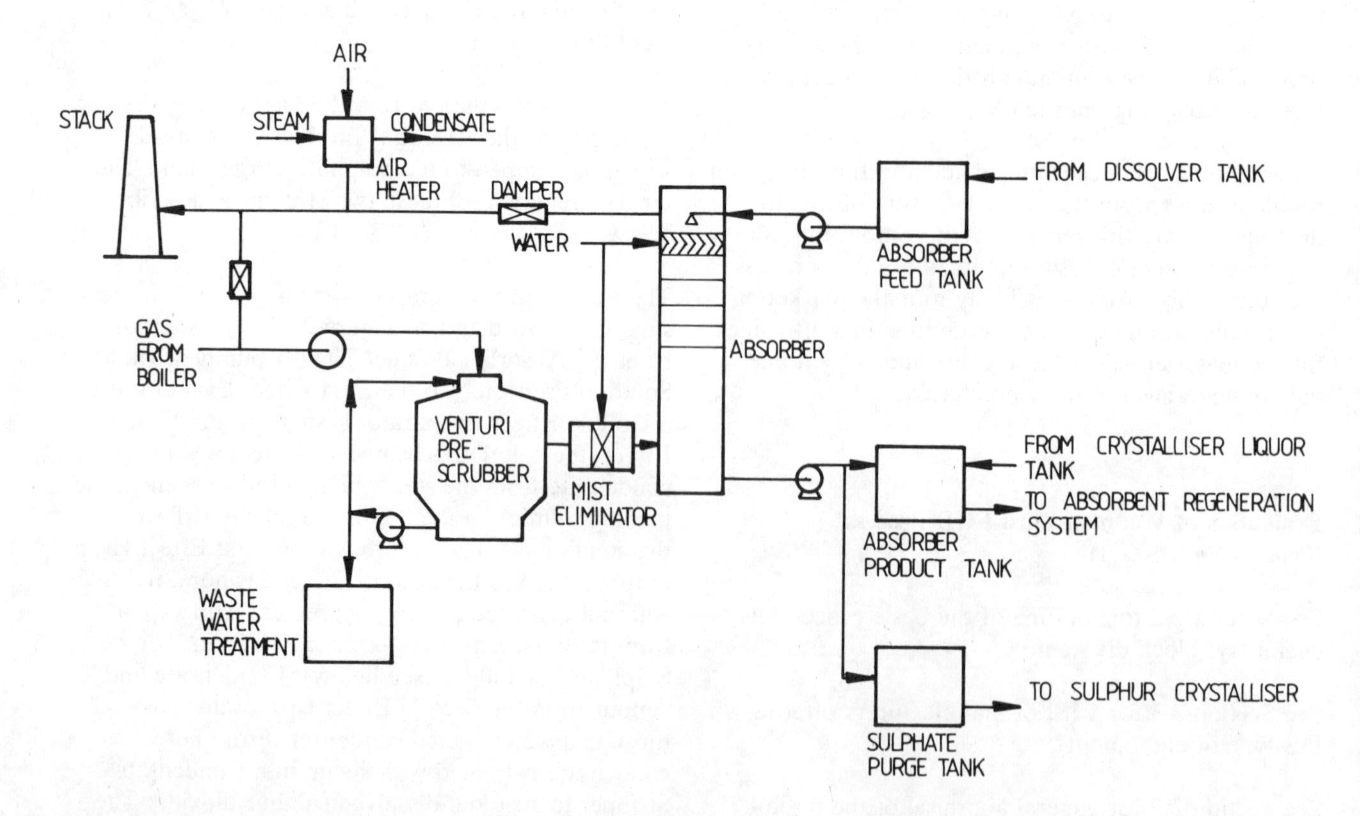

Figure 2.38 Wellman-Lord FGD Process: SO_2 and Dust Removal System

- Refrigerated vacuum crystallisation of the sulphate purge (this was employed on the earlier plants, but is less easy to operate than the steam-heated evaporation crystallisation) [276].

- Use of an oxidation inhibitor to reduce the extent of oxidation of absorbent to sulphate by about 75% [276].

- Alternatives for sulphur recovery: as elemental sulphur by a variety of processes (Alliance process, Claus process, Foster-Wheeler 'Resox' process); as sulphuric acid (by conventional catalytic oxidation of SO_2); or as liquid SO_2.

Although complex, the process can achieve high sulphur capture without excessive reagent make-up requirements or production of waste products. The sulphur produced can be marketed as sulphuric acid, liquid SO_2 or as elemental sulphur; or it can be disposed of safely as elemental sulphur. High operating flexibility is obtainable by installing large surge capacity for regenerated and unregenerated reagent.

Potential for process development: ***

Process Requirements for Each Application Considered: These are shown in Table 2.57 for the two applications considered: Datum Combustion Systems 1 and 2.

It is assumed that the SO_2 content of the gas is to be reduced to the following levels:

- System 1: 400 mg/Nm^3 (dry) at the inlet to the reheater, equivalent to 90.6% capture of SO_2

- System 2: 650 mg/Nm^3 (dry) at the inlet to the reheater, equivalent to 81.5% capture of SO_2

It is further assumed that for both systems:

- The particulates content of the gas is reduced in the Prescrubber from 115 to 15 mg/Nm^3, and no further reduction occurs in the Absorber.

- The NO_x content of the gas is unaffected.

- The HCl content of the gas is reduced in the Prescrubber by 80%, and no further reduction occurs in the Absorber.

- The gas to the stack is reheated to 90°C by addition of air from a steam-heated air-heater

By-products and Effluents: The by-products from the system are crystallised sodium sulphate and elemental sulphur. The production of sodium sulphate can be reduced by the use of an anti-oxidant. The sulphur can be marketed, but can also be safely disposed of.

The rates of production and composition of the material produced are summarised in Table 2.58 for the two Datum Combustion Systems considered.

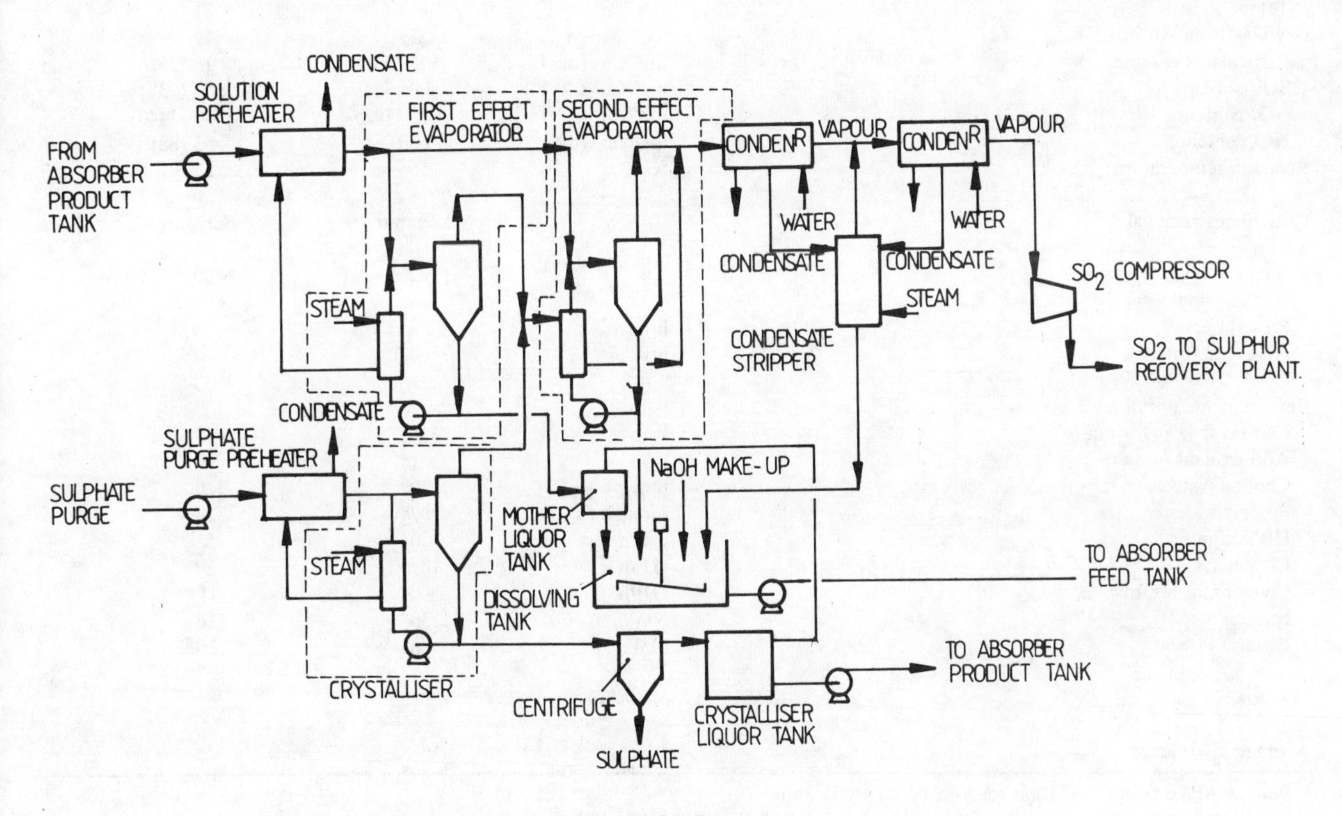

Figure 2.39 Absorbent Regeneration System of Wellman-Lord FGD Process

Efficiency and Emission Factors: The efficiency and emission factors for the process are summarised in Table 2.59 for the two applications considered: Datum Combustion Systems 1 and 2.

In calculating the efficiency factors, it is assumed that at full load:

- For Datum System 1, the FGD unit for each 500 MWe set consumes: 259 GJt/h of steam, reducing the power generated by 4.5 MWe; and 6.4 MWe of electric power, reducing the power sent out.

- For Datum System 2, the FGD unit consumes: 80 GJt/h of steam, reducing the useful energy sent out; and 2.3 We of electric power, equivalent to the combustion of a further 0.9 tonne/h of coal at a power station, assuming an overall power generation efficiency of 33%. This additional coal is arbitrarily assumed to be included with the coal burned in the boiler for calculating the efficiency factor.

Effect of plant availability: For illustration purposes, the annual average emission factors for FGD plant availabilities of 100% and 90% are given in Tables 2.57 and 2.59. Further details are given in Section 1.6.

Effect of load variations: ***

Effect of design variations: ***

Table 2.57 Process Requirements

Datum Combustion System		1	2
Inlet Gas at full load			
Volume flow	'000 Nm^3/h	1524*	551
Dry gas	'000 Nm^3/h	1426*	513
Water vapour	'000 Nm^3/h	99*	38
Actual volume flow	'000 m^3/h	2362*	823
Temperature	°C	150	150
Particulates content	mg/Nm^3 (dry)	115	115
SO_2 content	mg/Nm^3 (dry)	4240	3520
NO_x content	mg/Nm^3 (dry)	1025	1440
HCl content	mg/Nm^3 (dry)	340	315
Exit Gas to stack at full load			
Volume flow	'000 Nm^3/h	2258*	789
Dry gas	'000 Nm^3/h	1982*	706
Water vapour	'000 Nm^3/h	276*	83
Actual volume flow	'000 m^3/h	2928*	1021
Temperature	°C	81	81
Dry Gas from Absorber	'000 Nm^3/h	1423*	512
Particulates content	mg/Nm^3 (dry)	15 (25)	15 (25)
SO_2 content	mg/Nm^3 (dry)	400 (784)	650 (937)
NO_x content	mg/Nm^3 (dry)	1025 (1025)	1440 (1440)
HCl content	mg/Nm^3 (dry)	68 (95)	63 (88)
Reaction temperature	°C	56	56
Particulates removal		Separate	Separate
Reagent		NaOH	NaOH
Liquid/gas flow ratio			
Prescrubber	litre/m^3	3	3
Absorber (total)	litre/m^3	0.35	0.35
Requirements at full load			
Caustic soda (47% liquor)	kg/h	366*	98
Anti-oxidant	kg/h	1*	0.3
Cooling water	tonne/h	2670*	720
Process water	tonne/h	87*	24
HP steam	GJt/h	150*	53
LP steam	GJt/h	109*	29
Condensate return	GJt/h	−18*	−13
Natural gas	Nm^3/h	1130*	310
Electric Power	MWe	6.4*	0.9
Manpower	men/shift	***	***
Average load factor	%	***	***

* Per 500 MWe boiler + Total for four boiler FGD units
Figures in parentheses are annual average emissions for 90% FGD plant availability

Limitations: ***

Costs:

Cost basis: *** See Section 2.6.

Capital costs: *** See Section 2.6–for retrofit and for new build applications.

Annual running costs and cost factors: *** See Section 2.6–for retrofit and for new build applications.

Effects of design variations on costs: ***

Effects of annual load patterns on annual running costs: ***

Process Advantages and Drawbacks: Process well established in USA, Japan and Europe; fairly high reliability; simple chemistry; high sulphur capture efficiency (over 90% capture) can be attained; operating flexibility; no requirement for large quantities of reagent; no large quantities of waste products; by-product potentially marketable or can be easily and safely disposed of.

The disadvantages are complexity of process; need for experienced labour.

Table 2.58 Estimated Rates of Output Solids Production

Datum Combustion System		1	2
Rate of production:			
Waste sulphate/sulphite	tonne/h	1.30*	0.09
Sulphur	tonne/h	10.65*	0.72
Composition of waste product:			
Na_2SO_3	wt. %	47	47
Na_2SO_4	wt. %	53	53

*Total from four 500 MWe boiler FGD plants

Evaluation of Magnesia Slurry Scrubbing FGD Process (Process Code S41.1)

See Section 2.2 for: outline of the basic process; its chemistry; block diagram.

See Section 2.3 for a list of manufacturers offering this type of equipment.

See Section 2.3 for general appraisal of the basic process.

See Section 2.4 for the reason for choosing the Magnesia Slurry Scrubbing (Code S41.1) basic process.

This basic process type is considered (Section 2.4) to be suitable for application in the UK to utility boilers, and to large industrial boilers having access to reagent reprocessing facilities, and hence it is evaluated here only for Datum Combustion Systems 1 and 2. The process is offered by United Engineers and Constructors. See Appendix 2 for details of this manufacturer.

Table 2.59 Efficiency and Emission Factors

Datum Combustion System		1	2
Coal heat input (gross)	MWt	1292*	469
Coal fired	tonne/h	189*	60.2
FGD plant power consumed	MWe	6.4*	2.3
Equivalent coal input	tonne/h	–	0.9+
Steam for FGD plant	GJt/h	259*	82
Condensate heat recovery	GJt/h	11*	13
Loss of power generation	MWe	4.5*	–
Useful energy from system	GJe/h	1635*	–
	GJe/h	–	1399
Total equiv. coal input	tonne/h	189*	61.1(a)
Efficiency factor	GJe/tonne	8.65	–
	GJt/tonne	–	22.9(a)
Emissions			
Sulphur in SO_2	kg/h	285* (559*)	167 (241)
Nitrogen in NO_x	kg/h	445* (445*)	225 (225)
Chlorine in HCl	kg/h	94* (132*)	31 (44)
Particulates	kg/h	21* (35*)	8 (13)
Emission factors (per tonne coal)			(a)
Sulphur	kg/tonne	1.51 (2.96)	2.73 (3.94)
Nitrogen	kg/tonne	2.35 (2.35)	3.68 (3.68)
Chlorine	kg/tonne	0.50 (0.70)	0.51 (0.72)
Particulates	kg/tonne	0.11 (0.19)	0.13 (0.21)

* For each 500 MWe boiler FGD plant
+ Calculated assuming overall power generation efficiency = 0.33
(a) Based on coal fired to boiler plus coal equivalent to electric power consumed
Figures in parentheses are annual average emissions for 90% FGD plant availability

Process Description: Figure 2.40 shows a simplified flow diagram for application of the process to a coal fired boiler. It is assumed that the sulphur dioxide produced in the process is treated to recover sulphuric acid.

For dust and sulphur oxides removal, gas from downstream of the electrostatic precipitator or baghouse, boosted by a Fan, is cooled in a gas–gas Heat Exchanger, and scrubbed with water in a Venturi Prescrubber and its Mist Eliminator to remove acid halides and particulates. The gas enters the Absorber, which is of the spray grid design, where it is contacted counter-currently with magnesium hydroxide slurry, which absorbs sulphur oxides from the gas to form magnesium sulphite trihydrate $MgSO_3.3H_2O$, together with some magnesium sulphate heptahydrate $MgSO_4.7H_2O$. The gas passes through a mist eliminator at the top of the Absorber, and is then reheated by heat exchange with the ingoing gas in the Heat Exchanger. The absorbent slurry from the base of the Absorber flows to the Absorber Product Tank.

The prescrubber water, containing suspended solids and dissolved halides, is circulated to a Waste Water Treatment System.

Regenerated magnesia pellets from the Recycle MgO Silo are ground in a Mill, and together with make-up magnesia, slaked with water in the MgO Slaker. The resultant magnesium hydroxide is slurried with water in the Feed Tank.

For absorbent regeneration and sulphur recovery, the slurry from the Absorber Product Tank is pumped to a Centrifuge, where the crystallised reaction products are separated from liquid. The liquid centrate is returned to the slurry Feed Tank, and the solid is dried and decomposed to anhydrous sulphite and sulphate in a Rotary Dryer. The magnesia is then regenerated from the solid by heating in a Fluidised Bed (FB) Calciner fired with fuel oil. Calcination decomposes magnesium sulphite and sulphate to form pellets of magnesia, with release of sulphur dioxide. The off gas from the calciner passes via a Cyclone dust collector to the Cooler, where it preheats the combustion air to the Calciner. The gas is then passed via a Hot Electrostatic Precipitator (ESP) to the Sulphuric Acid Plant, the tail gas from which is mixed with the untreated boiler flue gas entering the FGD system. Magnesia pellets from the FB Calciner, and magnesia carryover captured by the Cyclone and Hot ESP, are cooled in a water-cooled Rotary Cooler and fed to the Recycle MgO Silo for reuse.

Status and Operating Experience: The status of the Magnesia Slurry process is indicated in Section 2.3, where it is seen that eight installations have been built. Of these, six were installed on coal- or oil-fired utility boilers in the USA: four by United Engineers and Constructors in the range 120–360 MWe (total 980 MWe), and two by Chemico (155 MWe and 190 MWe, totalling 345 MWe). The remaining two units are installed in Japan: one treating smelter gas, the other treating gases from oil-fired boiler and Claus

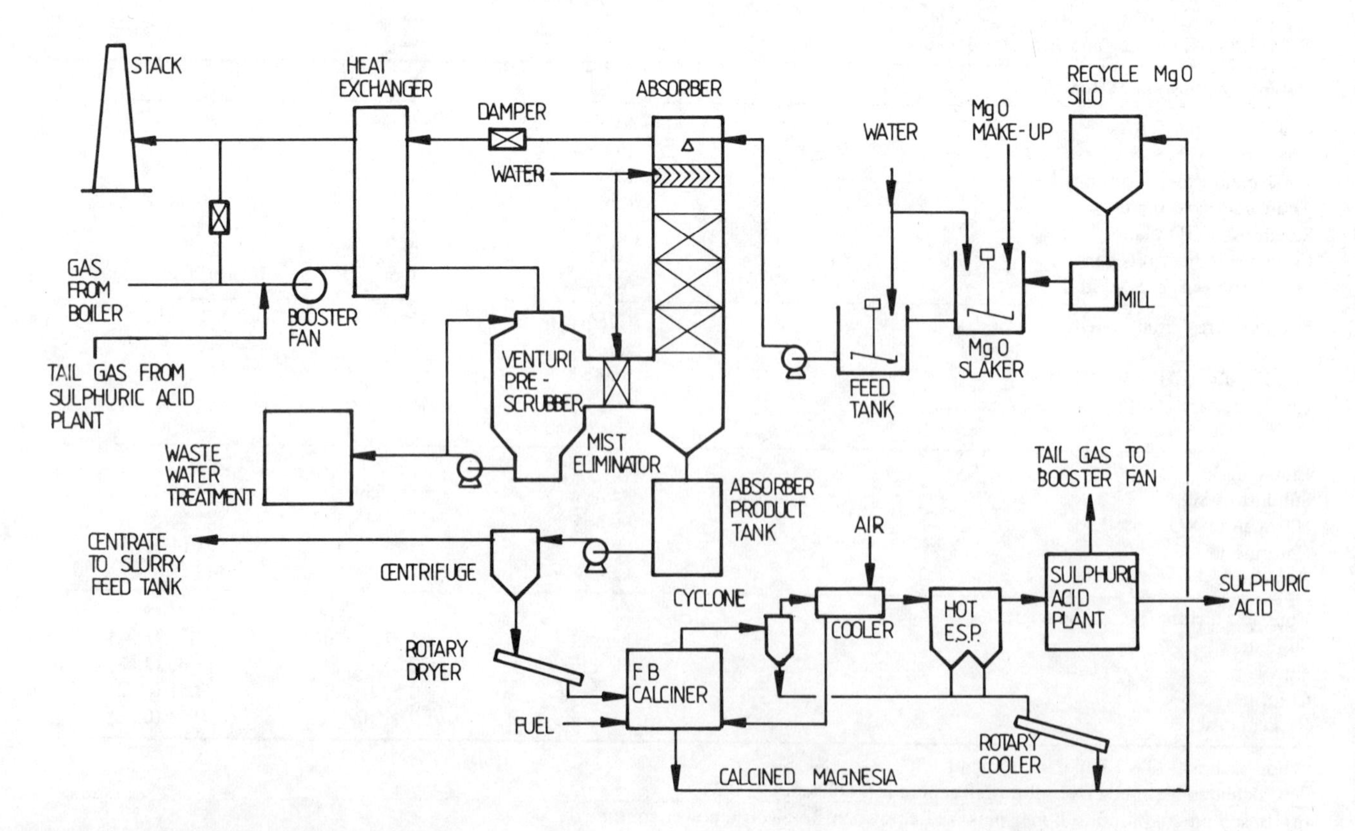

Figure 2.40 Magnesia Slurry Scrubbing FGD Process

furnaces. The two Chemico plants, and the boilers to which they were connected, have been dismantled.

Operating experience shows (see Section 2.3) that availability and reliability are generally high. The principal problems that have been encountered appear to have arisen in the magnesia calcination system: temperature control of the calciner leading to dead-burning of the magnesia, and poor performance of a waste heat boiler. There have also been problems with a booster fan.

Variations and Development Potential: Design and operating variations can include:

- Regeneration by calcination in a rotary calciner (the technique adopted in the early Chemico plants).

- Use of coke to reduce sulphate in the Calciner; this allows a lower calcination temperature to be adopted.

- Alternatives of recovery of: elemental sulphur by a variety of processes (Alliance process, Claus process, Foster-Wheeler 'Resox' process); sulphuric acid (by conventional catalytic oxidation of SO_2); liquid SO_2.

Although complex, the process can achieve high sulphur capture without excessive reagent make-up requirements or production of waste products. The sulphur produced can be marketed as sulphuric acid, liquid SO_2 or as elemental sulphur; or it can be disposed of safely as elemental sulphur. High operating flexibility is obtainable by installing large surge capacity for regenerated and unregenerated reagent.

Potential for process development: ***

Process Requirements for Each Application Considered: These are shown in Table 2.60 for the two applications considered: Datum Combustion Systems 1 and 2.

Table 2.60 Process Requirements

Datum Combustion System		1	2
Inlet Gas at full load			
Volume flow	'000 Nm^3/h	1524*	551
Dry gas	'000 Nm^3/h	1426*	513
Water vapour	'000 Nm^3/h	99*	38
Actual volume flow	'000 m^3/h	2362*	854
Temperature	°C	150	150
Particulates content	mg/Nm^3 (dry)	115	115
SO_2 content	mg/Nm^3 (dry)	4240	3520
NO_x content	mg/Nm^3 (dry)	1025	1440
HCl content	mg/Nm^3 (dry)	340	315
Exit Gas at full load			
Volume flow	'000 Nm^3/h	1654*	595
Dry gas	'000 Nm^3/h	1423*	512
Water vapour	'000 Nm^3/h	231*	83
Actual volume flow	'000 m^3/h	2199*	791
Temperature	°C	90	90
Particulates content	mg/Nm^3 (dry)	15 (25)	15 (25)
SO_2 content	mg/Nm^3 (dry)	400 (784)	650 (937)
NO_x content	mg/Nm^3 (dry)	1025 (1025)	1440 (1440)
HCl content	mg/Nm^3 (dry)	68 (95)	63 (88)
Reaction temperature	°C	53	53
Particulates removal		Separate	Separate
Reagent		$Mg(OH)_2$	$Mg(OH)_2$
Concentration in slurry	wt. %	20	20
Requirements at full load			
Magnesia (97% pure)	tonne/h	0.52	0.14
Process air	tonne/h	8	2
Process water	tonne/h	17	5
No. 6 fuel oil	tonne/h	10	3
Electric Power	MWe	4.3	1.6
Manpower	men/shift	***	***
Average load factor	%	***	***

*Per 500 MWe boiler
Figures in parentheses are annual average emissions for 90% FGD plant availability

It is assumed that the SO_2 content of the gas is to be reduced to the following levels:

- System 1: 400 mg/Nm^3 (dry), equivalent to 90.6% capture of SO_2
- System 2: 650 mg/Nm^3 (dry), equivalent to 81.5% capture of SO_2

It is further assumed that for both systems:

- The particulates content of the gas is reduced in the Prescrubber from 115 to 15 mg/Nm^3, and no further reduction occurs in the Absorber.
- The NO_x content of the gas is unaffected.
- The HCl content of the gas is reduced in the Prescrubber by 80%, and no further reduction occurs in the Absorber.
- The gas to the stack is reheated to 90°C in the Heat Exchanger.

By-products and Effluents: The by-product from the system is sulphuric acid. There is some loss of magnesium sulphite and sulphate to waste, requiring a make-up of fresh magnesia.

The rates of production and composition of the sulphuric acid and waste material produced are summarised in Table 2.61 for the two Datum Combustion Systems considered.

Efficiency and Emission Factors: The efficiency and emission factors for the process are summarised in Table 2.62 for the two applications considered: Datum Combustion Systems 1 and 2.

In calculating the efficiency and emission factors, it is assumed that at full load:

- For Datum System 1, the FGD units for each 500 MWe set consume: 10 tonne/h of fuel oil, equivalent to 17 tonne/h of coal, increasing the effective coal consumption (but not the emissions); and 4.3 MWe of electric power, reducing the power sent out.
- For Datum System 2, the FGD unit consumes: 3 tonne/h fuel oil, equivalent to 4.4 tonne/h coal; and 1.6 MWe of electric power, equivalent to the combustion of a further 0.6 tonne/h of coal at a power station, assuming an overall power generation efficiency of 33%. This additional coal is

Table 2.61 Estimated Properties of Outputs

Datum Combustion System		1	2
Rate of production:			
Wet waste product	tonne/h	9.6*	0.65
Dry waste product	tonne/h	9.0*	0.60
Sulphuric acid (97%)	tonne/h	33.7*	2.3
Composition of wet waste:			
Moisture	wt. %	7	7
$Mg_2SO_3.3\ H_2O$	wt. %	67	67
$Mg_2SO_4.7\ H_2O$	wt. %	25	25
Inerts	wt. %	1	1

*Total from four 500 MWe boiler FGD plants

Table 2.62 Efficiency and Emission Factors

Datum Combustion System		1	2
Coal heat input (gross)	MWt	1292*	469
Coal fired	tonne/h	189*	60.3
FGD plant power consumed	MWe	4.3*	1.6
Equivalent coal input	tonne/h	–	0.9+
Fuel oil	tonne/h	10*	2.9
Coal equivalent to fuel oil	tonne/h	17*	4.4
Useful energy from system	GJe/h	1659	–
	GJt/h	–	1468
Total equiv. coal input	tonne/h	206*(a)	65.3(a)
Efficiency factor	GJe/tonne	8.05(a)	–
	GJt/tonne	–	22.5(a)
Emissions			
Sulphur in SO_2	kg/h	285* (559*)	167 (241)
Nitrogen in NO_x	kg/h	445* (445*)	225 (225)
Chlorine in HCl	kg/h	94* (132*)	31 (44)
Particulates	kg/h	21* (35*)	8 (13)
Emission factors (per tonne coal)		(a)	(a)
Sulphur	kg/tonne	1.38 (2.71)	2.56 (3.69)
Nitrogen	kg/tonne	2.16 (2.16)	3.45 (3.45)
Chlorine	kg/tonne	0.46 (0.64)	0.47 (0.67)
Particulates	kg/tonne	0.10 (0.17)	0.12 (0.20)

* For each 500 MWe boiler FGD plant
+ Calculated assuming overall power generation efficiency = 0.33
(a) Based on coal fired to boiler plus coal equivalent to electric power and fuel oil consumed
Figures in parentheses are annual average emissions for 90% FGD plant availability

arbitrarily assumed to increase the effective coal consumption (but not the emissions).

Effect of plant availability: For illustration purposes, the annual average emissions and emission factors for FGD plant availabilities of 100% and 90% are given in Tables 2.60 and 2.62. Further details are given in Section 1.6.

Effect of load variations: ***

Effect of design variations: ***

Limitations: ***

Costs:

Cost basis: See Section 2.6

Capital costs: See Section 2.6–for retrofit and for new build applications.

Annual running costs and cost factors: See Section 2.6–for retrofit and for new build applications.

Effects of design variations on costs: ***

Effects of annual load patterns on annual running costs: ***

Process Advantages and Drawbacks: Process operated successfully in USA and Japan; high reliability; high sulphur capture efficiency (over 90% capture) can be attained; no requirement for large quantities of reagent; no large quantities of waste products; by-product potentially marketable or can be easily and safely disposed of; de-watered reaction products can be stored in open to await reprocessing (i.e. large surge capacity easily achieved); regeneration by off-site reprocessing possible.

The disadvantages are complexity of process; careful control of temperature and pH needed to avoid bisulphite formation or variable filtration characteristics; careful control of calciner off-gas composition (oxygen concentration) to avoid SO_3 and hence resulphation of regenerated MgO; need for experienced labour.

Evaluation of Bergbau-Forschung Active Carbon FGD Process (Process Code S51.1)

See Section 2.2 for: outline of the basic process; its chemistry; block diagram.

See Section 2.3 for a list of manufacturers offering this type of equipment.

See Section 2.3 for general appraisal of the basic process.

See Section 2.4 for the reason for choosing the Bergbau-Forschung (Code S51.1) basic process.

This basic process type is considered (Section 2.4) to be suitable for application in the UK to utility boilers, and to large industrial boilers having access to reagent reprocessing facilities, and hence it is evaluated here only for Datum Combustion Systems 1 and 2. The process, developed by Bergbau-Forschung GmbH and Foster Wheeler Energy Corporation, is offered by Uhde GmbH (Dortmund, FRG). See Appendix 2 for details of this manufacturer.

Process Description: Figure 2.41 shows a simplified flow diagram for application of the process to a coal fired boiler. It is assumed that the sulphur dioxide produced in the process is treated to recover elemental sulphur.

For dust and sulphur oxides removal, gas at a temperature of 120–150°C from downstream of the electrostatic precipitator or baghouse, boosted by a fan, enters the Adsorber containing activated carbon granules (prepared from coal in the form of extruded granules about 5 mm in length). The carbon, which is contained in louvred channels, with the gas flowing across the bed, adsorbs sulphur dioxide, oxygen and water vapour to form adsorbed sulphuric acid. Nitrogen dioxide (forming 5–10% of the total nitrogen oxides content of the gas) and acid halides are also adsorbed, and the carbon bed filters out much of the particulates content of the gas. The gas temperature rises by 15–20°C across the bed; the cleaned gas passes via a Damper to Stack.

For absorbent regeneration the active carbon is removed continuously from the base of the Adsorber and conveyed to the Regeneration section, which can be either on- or off-site. Fines, including particulates trapped from the gas, are removed in a Classifier, and the oversize material then enters the Regenerator, where it is heated to 400–450°C by mixing with sand heated in a Sand Heater/Lift system. The adsorbed gases, sulphuric acid and nitrogen dioxide, react with the carbon, releasing sulphur dioxide, elemental nitrogen and carbon dioxide, consuming some of the carbon. The carbon remaining is screened on a Vibrating Screen to remove sand and carbon fines, cooled with air in the Carbon Cooler, and returned to the Adsorber; carbon losses are made up by adding fresh carbon. Part of the cooling air from the Carbon Cooler supplies the combustion air for the Sand Heater/Lift system.

For sulphur recovery the SO_2-rich gas from the Adsorber passes through the 'Resox' Reactor, where the SO_2 reacts with carbon to give sulphur vapour. The sulphur is condensed out in the Sulphur Condenser, and collected in the Sulphur Tank. The tail gas is returned to the inlet of the Booster Fan.

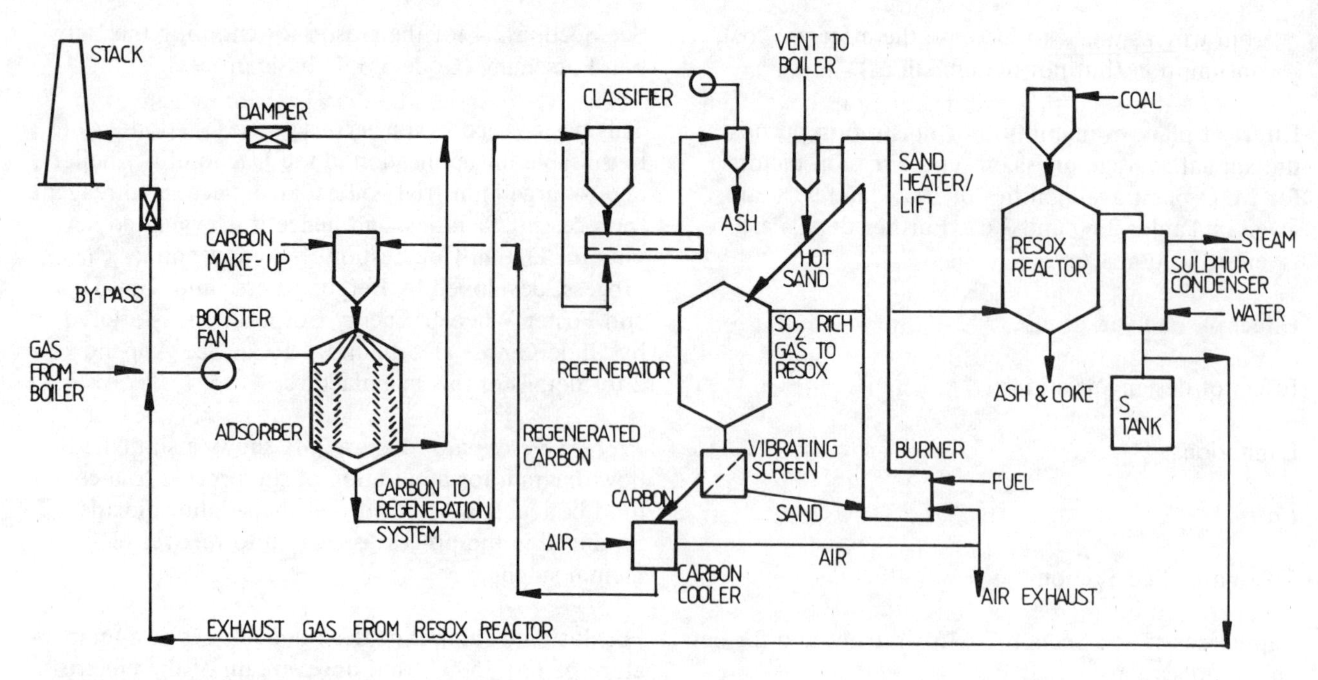

Figure 2.41 Bergbau-Forschung Active Carbon Adsorption FGD Process with Foster Wheeler Resox Sulphur Recovery

Table 2.63 Process Requirements

Datum Combustion System		1	2
Inlet Gas at full load			
Volume flow	'000 Nm^3/h	1524*	551
Dry gas	'000 Nm^3/h	1426*	513
Water vapour	'000 Nm^3/h	99*	38
Actual volume flow	'000 m^3/h	2362*	854
Temperature	°C	150	150
Particulates content	mg/Nm^3 (dry)	115	115
SO_2 content	mg/Nm^3 (dry)	4240	3520
NO_x content	mg/Nm^3 (dry)	1025	1440
HCl content	mg/Nm^3 (dry)	340	315
Exit Gas at full load			
Volume flow	'000 Nm^3/h	1520*	549
Dry gas	'000 Nm^3/h	1423*	512
Water vapour	'000 Nm^3/h	97*	37
Actual volume flow	'000 m^3/h	2439*	881
Temperature	°C	165	165
Particulates content	mg/Nm^3 (dry)	15 (25)	15 (25)
SO_2 content	mg/Nm^3 (dry)	400 (784)	650 (935)
NO_x content	mg/Nm^3 (dry)	945 (953)	1325 (1337)
HCl content	mg/Nm^3 (dry)	68 (95)	63 (88)
Reaction temperature	°C	165	165
Particulates removal		Simultaneous	Simultaneous
Reagent		Active Carbon	Active Carbon
Requirements at full load			
Active carbon	tonne/h	0.95*	0.27
Natural gas	GJt/h	71.5*	19.3
Air	tonne/h	369*	99
Coal for Resox reactor++	tonne/h	1.7	0.5
Electric Power	MWe	4.6*	1.5
Manpower	men/shift	***	***
Average load factor	%	***	***

* Per 500 MWe boiler
++ Regarded as a reagent, not as an energy source
Figures in parentheses are annual average emissions for 90% FGD plant availability

Status and Operating Experience: The status of the Bergbau-Forschung process is indicated in Section 2.3, where it is seen that five installations have been (or are being) erected, in the size range 0.3–370 MWe, totalling about 500 MWe.

Operating experience shows (see Section 2.3) that the principal problems that have been encountered have been: occurrence of hot spots in the Adsorber; poor reliability of the Vibrating Screen for separating carbon from sand; and plugging of the Sulphur Condenser of the Resox system with sulphur and carbon.

Variations and Development Potential: Design and operating variations can include:

- Simultaneous removal of SO_2 and NO_x in a two-stage Adsorber, with ammonia addition to the gas after removal of most of the SO_2 in the first stage; the carbon in the second stage catalyses the reduction of the NO [424, 435].

- Indirect heating of carbon in the Regenerator with hot combustion gases [424].

- Alternatives for sulphur recovery: as elemental sulphur by a variety of processes (Alliance process, Claus process, or by Foster-Wheeler's Resox process); as sulphuric acid (by conventional catalytic oxidation of SO_2); or as liquid SO_2.

Although complex, the process can achieve high sulphur capture without excessive reagent make-up requirements or production of waste products. The sulphur produced can be marketed as sulphuric acid, liquid SO_2 or as elemental sulphur; or it can be disposed of safely as elemental sulphur. High operating flexibility is obtainable by installing large surge capacity for regenerated and unregenerated active carbon.

Potential for process development includes improvements in process control to avoid hot spots in the Adsorber and plugging in the Resox system; improvements in efficiency of sulphur recovery in the Resox system.

Process Requirements for Each Application Considered: These are shown in Table 2.63 for the two applications considered: Datum Combustion Systems 1 and 2.

It is assumed that the SO_2 content of the gas is to be reduced to the following levels:

- System 1: 400 mg/Nm^3 (dry), equivalent to 90.6% capture of SO_2

- System 2: 650 mg/Nm^3 (dry), equivalent to 81.5% capture of SO_2

It is further assumed that for both systems:

- The Resox system has a conversion efficiency of 85%, so that 15% of the sulphur captured leaves the Resox system as various sulphur compounds which are recycled to the boiler.

- The particulates content of the gas is reduced from 115 to 15 mg/Nm^3.

- The NO_x content is reduced by 8% by adsorption of NO_2.

- The HCl content of the gas is reduced by 80%.

- The mechanical breakdown of carbon leads to a carbon make-up rate 50% higher than the theoretical.

- The coal used in the Resox system has a carbon content of 60%; this coal is regarded as a reagent, not as an energy source.

By-products and Effluents: The by-product from the system is elemental sulphur, and there is a carbon waste in addition to the captured flyash. The rates of production of the material produced are summarised in Table 2.64 for the two Datum Combustion Systems considered.

Table 2.64 Estimated Rates of Output Solids Production

Datum Combustion System		1	2
Rate of production:			
Sulphur (99%)	tonne/h	11.07*	0.75
Active carbon fines	tonne/h	1.27*	0.09
Fly-ash	tonne/h	0.57*	0.05

*Total from four 500 MWe boiler FGD plants

Efficiency and Emission Factors: The efficiency and emission factors for the process are summarised in Table 2.65 for the two applications considered: Datum Combustion Systems 1 and 2.

In calculating the efficiency factors, it is assumed that at full load:

- For Datum System 1, the FGD unit for each 500 MWe set consumes: 71.5 GJt/h of natural gas for the Sand Heater/Lift, equivalent to an increase of 2.9 tonne/h of coal; and 4.6 MWe of electric power, reducing the power sent out.

- For Datum System 2, the FGD unit consumes: 19.3 GJt/h of natural gas for the Sand Heater/Lift, equivalent to an increase of 0.7 tonne/h of coal; and 1.5 MWe of electric power, equivalent to the combustion of a further 0.6 tonne/h of coal at a power station, assuming an overall power generation efficiency of 33%. This additional coal is arbitrarily assumed to be included with the coal burned in the boiler for calculating the efficiency factor.

Effect of plant availability: For illustration purposes, the annual average emissions and emission factors for FGD plant availabilities of 100% and 90% are given in Tables 2.63 and 2.65. Further details are given in Section 1.6.

Effect of load variations: ***

Effect of design variations: ***

Limitations: ***

Costs:

Cost basis: ***

Capital costs: ***

Annual running costs and cost factors: ***

Effects of design variations on costs: ***

Effects of annual load patterns on annual running costs: ***

Process Advantages and Drawbacks: Process can be retrofitted; process readily adaptable to combined NO_x-SO_2 abatement; process also removes particulates and other air pollutants; no gas reheat needed; high sulphur capture efficiency (over 90% capture) can be attained; operating flexibility; no requirement for large quantities of reagent; no large quantities of waste products; by-product potentially marketable or can be easily and safely disposed of.

The disadvantages are complexity of process; need for experienced labour.

2.6 Costs for FGD Systems

Costs for six FGD process plants have been included in this section. Details of the procedures and assumptions considered during costing are presented below and in Appendix 3.

Equipment Costs: FGD equipment costs were prepared on the basis of the Equipment Lists presented in Appendix A of EPRI Report No. CS-3342 [95]. The EPRI Report was based on an arbitrary reference coal-fired plant of 1000 MWe (2 × 500 MWe) capacity but, for this study, the Equipment List was modified to represent a plant of 2000 MWe (4 × 500 MWe). Budget prices for items of equipment were obtained, wherever possible, from UK-based vendors and details are given in Appendix 3. Total equipment costs for the six FGD plants are presented in Table 2.66.

As indicated in Appendix 3, some budget costs included for supply and erection, whilst others were exclusively for supply of equipment. It was estimated that total equipment costs compiled in this way included an element for erection which amounted to, on average, 25%. The equipment costs given in Table 2.66 have been adjusted accordingly.

Table 2.65 Efficiency and Emission Factors

Datum Combustion System		1	2
Coal heat input (gross)	MWt	1292*	469
Coal fired	tonne/h	189*	60.2
FGD plant power consumed	MWe	4.6*	1.5
Equivalent coal input	tonne/h	–	0.6+
Natural gas for sand heater	GJt/h	71.5*	19.3
Equivalent coal input	tonne/h	2.9*(a)	0.7(a)
Useful energy from system	GJe/h	1657*	–
	GJt/h	–	1468
Total equiv. coal input	tonne/h	192*(a)	61.5(a)
Efficiency factor	GJe/tonne	8.63(a)	–
	GJt/tonne	–	23.9(a)
Emissions			
Sulphur in SO_2	kg/h	285* (559*)	167 (241)
Nitrogen in NO_x	kg/h	410* (414*)	207 (209)
Chlorine in HCl	kg/h	94* (132)	31 (44)
Particulates	kg/h	21* (35*)	8 (13)
Emission factors (per tonne coal)		(a)	(a)
Sulphur	kg/tonne	1.48 (2.91)	2.72 (3.92)
Nitrogen	kg/tonne	2.14 (2.16)	3.37 (3.40)
Chlorine	kg/tonne	0.49 (0.69)	0.50 (0.72)
Particulates	kg/tonne	0.11 (0.18)	0.13 (0.21)

* For each 500 MWe boiler FGD plant
+ Calculated assuming overall power generation efficiency = 0.33
(a) Based on coal fired to boiler plus coal equivalent to electric power and natural gas consumed
Figures in parentheses are annual average emissions for 90% FGD plant availability

An 'installation factor' for each of the processes was derived from EPRI Report No. CS-3342 [95] by comparing the total equipment costs given in Volume 3 of the Report to the installed equipment costs, referred to as Total Process Capital and given in Volume 1 (non-regenerable processes) or Volume 2 (regenerable processes) of the Report. The installation factor, therefore, provides an indication of the relative complexity of installing each FGD system, and it was assumed that these factors would also be reflected in UK installation costs. It should be noted that equipment and installation costs for an electrostatic precipitator (and associated equipment), referred to as a Particulate Removal System, were deleted in the determination of the installation factors as this component is considered to be part of the boiler system and not of the FGD plant. For the lime spray drying process, however, an electrostatic precipitator or baghouse is an integral part of the plant and an allowance was made for this in estimating the installation factor, based on information from EPRI Report No. CS-3696 [96].

Installation factors and the resultant Total Erection Costs for the six processes are presented in Table 2.66.

Capital Costs: Capital cost estimates for each FGD plant were prepared using the procedures outlined in EPRI Report No. CS-3342 [95] for new-build plants and EPRI Report No. CS-3696 [96] for retrofit plants. In each case, the estimated total erection costs, expressed in terms of pounds sterling per kilowatt of installed boiler capacity (£/kWe), were used as the 'Base Capital Costs' for each process in the estimating procedure.

Details of the EPRI estimating procedure for new-build and retrofit plants appear in Appendix 3.

The following assumptions were made in the estimates for new or retrofit FGD plant fitted to a 2000 MWe coal-fired power station (Datum Combustion System 1):

(a) Scope Adjustments:
- An average factor of 4% of the base capital cost has been selected (retrofit only).

(b) Process Adjustments:
- Unit size = 500 MWe
- Flue gas flowrate = 1.3m^3/s per net MWe
- Sulphur content = 1.6%

(c) Location:
- Seismic and climatic considerations are ignored.
- The soil is of poor bearing capacity and the site is near a river bank.

(d) Retrofit Adjustments:
1) Accessibility and congestion; assumptions:
   - Limited space available, e.g. area around boiler and stack is approximately 2.8 square metres per megawatt,
   - interference with existing structures or equipment which cannot be relocated,
   - special designs are necessary,
   - access for cranes is limited to one side,
   - majority of equipment is on elevated slabs or remotely located.

   (These factors add 25% to the Total Process Retrofit Capital, TPRC. Alternative conditions, as defined in EPRI Report No. CS-3696, would result in an increase of between 2% and 42%).
2) Underground obstructions; assumptions:
   - more than one major obstruction such as circulating water pipe, gas main or ductbank,
   - several minor obstructions, including piping, trenches and plant drainage.

   (These factors add 2% to the TPRC. Alternative conditions, as defined in EPRI Report No. CS-3696, would result in an increase of between 1% to 5%).
3) Ductwork length and distance from scrubber to tie-ins; assumptions:
   - the scrubber is not located symmetrically with the unit and at a distance of over 90 metres from the unit,
   - the tie-in with the existing chimney is from the rear.

   (These factors add 12% to the TPRC. Alternative conditions, as defined in EPRI Report No. CS-3696, would result in an increase of between 2% to 12%).

Table 2.66 Equipment and Erection Costs for Selected FGD Plant (2000 MWe) December 1986

Process	Base Equipment Cost*(£ million)	Installation Factor	Total Erected Cost	
			(£ million)	£/kWe
Limestone with forced oxidation	30.22	3.39	102.6	51.29
Chiyoda	33.96	2.79	94.7	47.35
Saarberg Hölter	32.66	2.85	93.1	46.54
Lime Spray Drying	27.56	2.95	81.3	40.65
Wellman-Lord	52.52	2.91	153.1	76.54
Magnesia (MgO) scrubbing	77.94	2.17	169.0	87.48

*Base equipment costs are given in Part 6 of Appendix 3

(e) Escalation Adjustments:
- Cost estimates are escalated to December 1989 start-up using an escalation rate of 5%.

(f) Project Contingency:
- A single contingency factor of 10% has been selected.

(g) Process Contingency:
- A single contingency factor of 5% has been selected.

(h) General Facilities:
- A factor of 5% has been selected.

(i) Allowance for Funds During Construction:
- The duration of the engineering, procurement and construction phases is assumed to be 3 years. The EPRI allowance (3.7%) has been modified to take account of the revised escalation rate.

(j) Royalty Allowance:
- A royalty of 5% has been selected.

(k) Inventory Capital:
- The inventory Capital Costs indicated in EPRI Report No. CS-3696 [96] have been adjusted to December 1986 Sterling costs.

Capital cost estimates for Datum Combustion System 1 (2000 MWe) FGD plants are outlined in Appendix 3, and summarised in Tables 2.67 (New-Build) and 2.68 (Retrofit).

To estimate capital costs for new or retrofit FGD plant fitted to a 450 t/h of steam coal-fired industrial boiler (Datum Combustion System 2), reference was made to the cost data of Samish [92]. Samish used EPRI cost data to produce capital cost estimates for a number of FGD processes and for the following conditions:

- Two sizes of boiler plant (110.6 MWe and 1106 MWe).
- Two coals of differing sulphur content (0.5% and 4.0%).
- SO_2 emission levels ranging from 0.06 to 2.1 kg SO_2 per GJt fired. (NB 90% sulphur removal for Datum Combustion Systems 1 and 2 gives emission levels of 0.12–0.13 kg/GJt).

These data were used to develop 'cost indices' for estimating capital costs for FGD plants of differing sizes. Datum Case 2 FGD plant capital costs were then derived from Datum Case 1 (2000 MWe) capital costs using the appropriate cost index and the capacity ratio. These capital costs are given in Tables 2.67 (New-build) and 2.68 (Retrofit).

Operating Costs. Operating costs for Datum Combustion System 1 for each of the six processes were estimated using the simplified EPRI procedure in Section 5 of Ref. [96]. An escalation rate of 5% (instead of the EPRI figure of 8.5%) per annum was selected for the estimate. An outline of the procedure is included in Appendix 3, together with cost details for each of the processes. Costs were initially estimated in $/kWe-year (December 1982 prices), converted to sterling at that date and brought up to date (December 1986) using plant cost indices published in *Process Engineering* magazine. First year operating costs are summarised in Table 2.69; it has been assumed that costs for new-build and retrofit plant are identical.

Operating costs for FGD plant of two different sizes were developed by Samish [92] from EPRI cost data. This information was further developed to determine cost indices which could be used to compute operating costs for any size of plant, e.g. Datum Combustion System 2. These cost indices and the operating costs for the smaller plant are given in Table 2.69.

Table 2.67 Capital Costs for FGD Plants (New-build) December 1986

Process	Capital Cost for Datum Combustion System 1 (2000 MWe)* £ Million	Cost Index**	Capital Cost for Datum Combustion System 2 (450 t/h Steam) £ Million
Limestone with forced oxidation	136	0.71	24.7
Chiyoda	126	0.81	18.0
Saarberg-Hölter	124	0.65	26.0
Lime Spray Drying	133	0.65	27.9
Wellman-Lord	164	0.68	32.0
Magnesia (MgO) scrubbing	194	0.62	43.7

* See Appendix 3 (Part 4) for estimates
**Capital Cost of Plant 2 = Capital Cost of Plant 1 × (Capacity Ratio)^y where y = cost index

Table 2.68 Capital Costs for FGD Plant (Retrofit) December 1986

Process	Capital Cost for Datum Combustion System 1 (2000 MWe)* £ Million	Cost Index**	Capital Cost for Datum Combustion System 2 (450 t/h Steam) £ Million
Limestone with forced oxidation	202	0.71	36.7
Chiyoda	186	0.81	26.5
Saarberg-Hölter	183	0.65	38.4
Lime Spray Drying	198	0.65	41.5
Wellman-Lord	242	0.68	47.2
Magnesia (MgO) Scrubbing	286	0.62	64.4

* See Appendix 3 (Part 4) for estimates
**Capital Cost of Plant 2 = Capital Cost of Plant 1 × (Capacity Ratio)^y where y = cost index

Table 2.69 Operating Costs for FGD Plant (New-Build and Retrofit) December 1986

Process	Operating Costs for Datum Combustion System 1 (2000 MWe) £/kWe-year*	Cost Index**	Operating Costs for Datum Combustion System 2 (450 t/h Steam) £/kWe-year
Limestone with forced oxidation	24.02	0.22	40.76
Chiyoda	21.88	0.14	30.63
Saarberg-Hölter	22.90	0.22	38.86
Lime Spray Drying	28.82	0.25	52.56
Wellman-Lord	32.27	0.26	60.28
Magnesia (MgO) scrubbing	25.86	0.22	43.88

* See Appendix 3 (Part 5) for estimate

**Operating costs for Plant 2 = Operating costs for Plant 1/(Capacity Ratio)^y where y = cost index

3. Nitrogen Oxides Abatement Processes (Combustion Techniques)

3. Nitrogen Oxides Abatement Processes (Combustion Techniques)

3.1 Classification of Combustion Techniques

The general classification of NO_x abatement ($deNO_x$) processes adopted in this Manual involves a division into two broad groups according to whether abatement is achieved at the source of combustion, 'Combustion Techniques', or in the flue gas downstream of the boiler, 'Flue Gas Treatment'.

The first group, 'Combustion Techniques', see Figure 3.1, is divided into four distinct basic categories:

- Burner design (Category N10),
- Furnace design or modification (N20),
- Furnace operating methods (N30),
- Fluidised bed combustion

Each category is further sub-divided according to specific design or operation. The first three basic categories are dealt with in this section of the Volume whilst fluidised bed combustion is discussed separately in Section 6.

The second group, 'Flue Gas Treatment', is dealt with in Section 4 of the Volume.

PROCESS/DESIGN CODE NUMBERS

Each major category is assigned a Category Number, e.g. Burner Design–N10. Generalised designs or operating modes within each category are characterised by a change in the last digit of the Category Number, e.g. Externally-Staged Combustion Burners–N12. Specific designs or processes within each sub-division are assigned a Type Number, e.g. Fuel Staging Burner–N12.4.

The basic design or process types have been assigned the Code Numbers shown in Figure 3.1.

SUPPLIERS OF $DeNO_x$ TECHNOLOGY

Full details of all manufacturers quoted in the Manual (names, addresses, telephone and telex numbers, and names of contacts) are listed in Appendix 2.

3.2 Outline Descriptions

PRINCIPLES OF NO_x ABATEMENT FOR PULVERISED COAL BURNING

The term 'NO_x' is used to denote a mixture of the two oxides of nitrogen produced during the combustion of fuels with air; namely nitric oxide, NO, and nitrogen dioxide, NO_2. In flue gas, about 90% of NO_x is in the form of NO, a small part of which will be oxidised to NO_2 in the stack by oxygen from the excess combustion air. However, NO_x concentrations in flue gas are frequently expressed in units of $mg(NO_2)/m^3$.

During the combustion of coal, NO_x is formed by three different processes: 'thermal NO_x' formed by reactions between air nitrogen and oxygen, 'fuel NO_x' formed by oxidation of nitrogen-containing substances originating from coal, and 'prompt NO_x' formed by reactions of air nitrogen with hydrocarbon radicals which take place only at the front of the flame.

For pulverised coal combustion, the reduction of fuel NO_x is of particular importance because coal contains much larger amounts of nitrogen (1–3%) than does oil and gas (0.1–0.5%) and produces a large amount of fuel NO_x accounting for 70–80% of total NO_x.

The fuel NO_x from coal consists of both volatile NO_x (derived from volatile nitrogen compounds in the coal) and char NO_x (derived from residual nitrogen compounds in the char). Usually the former is more predominant than the latter. The volatile NO_x can be abated efficiently by combustion control using a low-NO_x burner (LNB). Char NO_x is not controlled as easily, as oxidation of char nitrogen is relatively insensitive to changes in early oxygen concentration or combustion equipment.

Thermal NO_x can be reduced by decreasing peak temperature, and the local oxygen concentration and residence time at peak temperature. The formation of prompt NO_x can be hindered in the same way as mentioned for thermal NO_x, except for the limitation of the oxygen which is not favourable in this case. Both formations, those of thermal NO_x and prompt

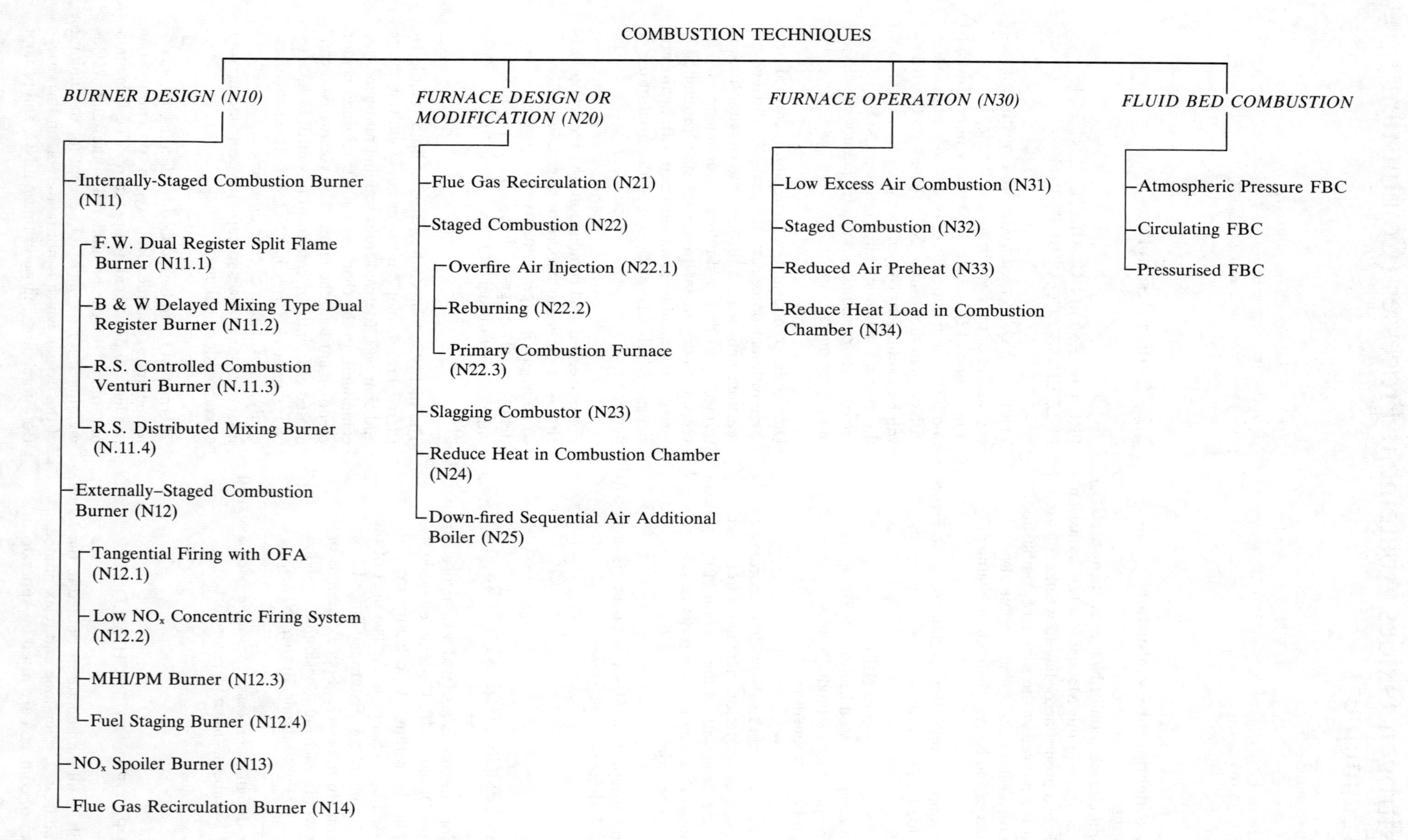

Figure 3.1 Classification of $Deno_x$ Combustion Techniques (Coal)

NO_x, are suppressed by avoiding hot spots in the furnace. The base factor to reduce fuel NO_x is to reduce the oxygen level.

A low-NO_x combustion model of pulverised coal is shown in Figure 3.2. Usually 70–90% of the total air for complete combustion is injected through the burner, with the rest through the overfire air port. In the primary combustion zone (I), volatile matter begins to burn, producing much NO_x due to the combustion of volatile nitrogen compounds. In the secondary zone (II) the volatile matter continues to burn, the char begins to burn, and O_2 concentrations decrease, producing a reducing atmosphere that reduces NO_x to N_2. The reducing atmosphere also exists in the tertiary zone (III), where the NO_x concentrations reach a minimum just before the overfire air port for second stage combustion. The CO concentration remains high.

When the staged air (IV) is injected, the remaining char, CO, and other reducing gases are burned, resulting in a slight increase in NO_x concentration.

To minimise the NO_x level, the gas retention time in the reducing atmosphere should be as long as possible, and the mixing of the secondary air with the flame formed by coal and primary air should be delayed. The resulting long flame with extended reducing zone, however, may promote soot formation and boiler tube wastage. The flame may be unstable, and unburned carbon may increase. Counter-measures for such problems should be taken. Otherwise the NO_x reduction should be limited to where it does not cause an obvious problem.

The fundamental factors in NO_x emission control may be summarised as follows:

(a) Lowering O_2 concentration in combustion zone (for reduction of thermal and fuel NO_x).

(b) Shortening gas residence time in high temperature zone (for reduction of thermal NO_x).

(c) Reduction of fuel nitrogen.

Table 3.1 shows the relationship between the fundamental factors in NO_x emission control and the methods in actual practice. Various combustion modifications are shown schematically in Figure 3.3.

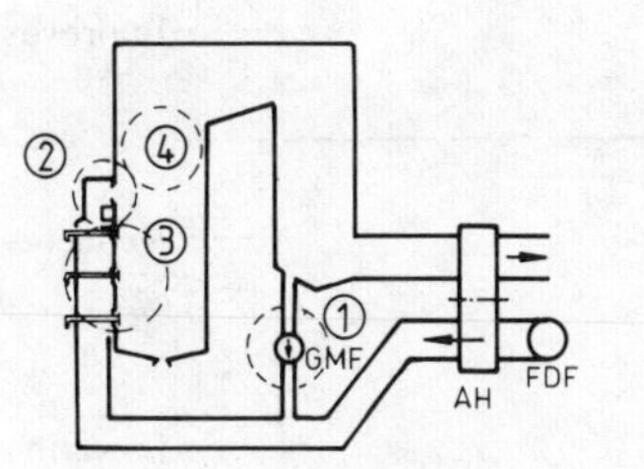

Figure 3.3 Boiler Modifications for NO_x Emission Control [31.10]

TECHNIQUES IN CATEGORY N10–BURNER DESIGN

Burner designers utilize the well-known principles of off-stoichiometric combustion and/or flue gas circulation to reduce NO_x formation. These principles are applied in the flame itself rather than in the total combustion chamber. So burners are designed to control the mixing and stoichiometry of fuel and air in the near-burner region of the furnace to retard conversion of fuel-bound nitrogen to NO_x and formation of thermal NO_x, while still maintaining high combustion efficiency. This is accomplished by controlling the momentum, direction and quantity of fuel and air streams at the burner throat as they are injected into the furnace chamber. Various burner designs have been developed, and they may be categorised as shown in Figure 3.1.

PROCESS CODE N12–EXTERNALLY STAGED COMBUSTION BURNERS

The mechanism for staging in an internally staged combustion burner lies entirely within the burner, using multiple registers (usually dual registers) and/or

Overfire Air
Secondary air (A_2) or A + Flue gas
Coal and primary air
A_2 A_2 I II III IV III

Zone	Burning Matter	NO_x Concentration
I	Most of volatile matter	High
II	Remaining volatile matter Part of char	Lowered (reducing atmosphere)
III	Char	Minimized (reducing atmosphere)
IV	Remaining char Reducing gas (CO, etc.)	Increased (oxidizing atmosphere)

Figure 3.2 Model of Low-NO_x Coal Burning [32.07]

Table 3.1 NO_x Emission Control with Combustion Modification [31.10]

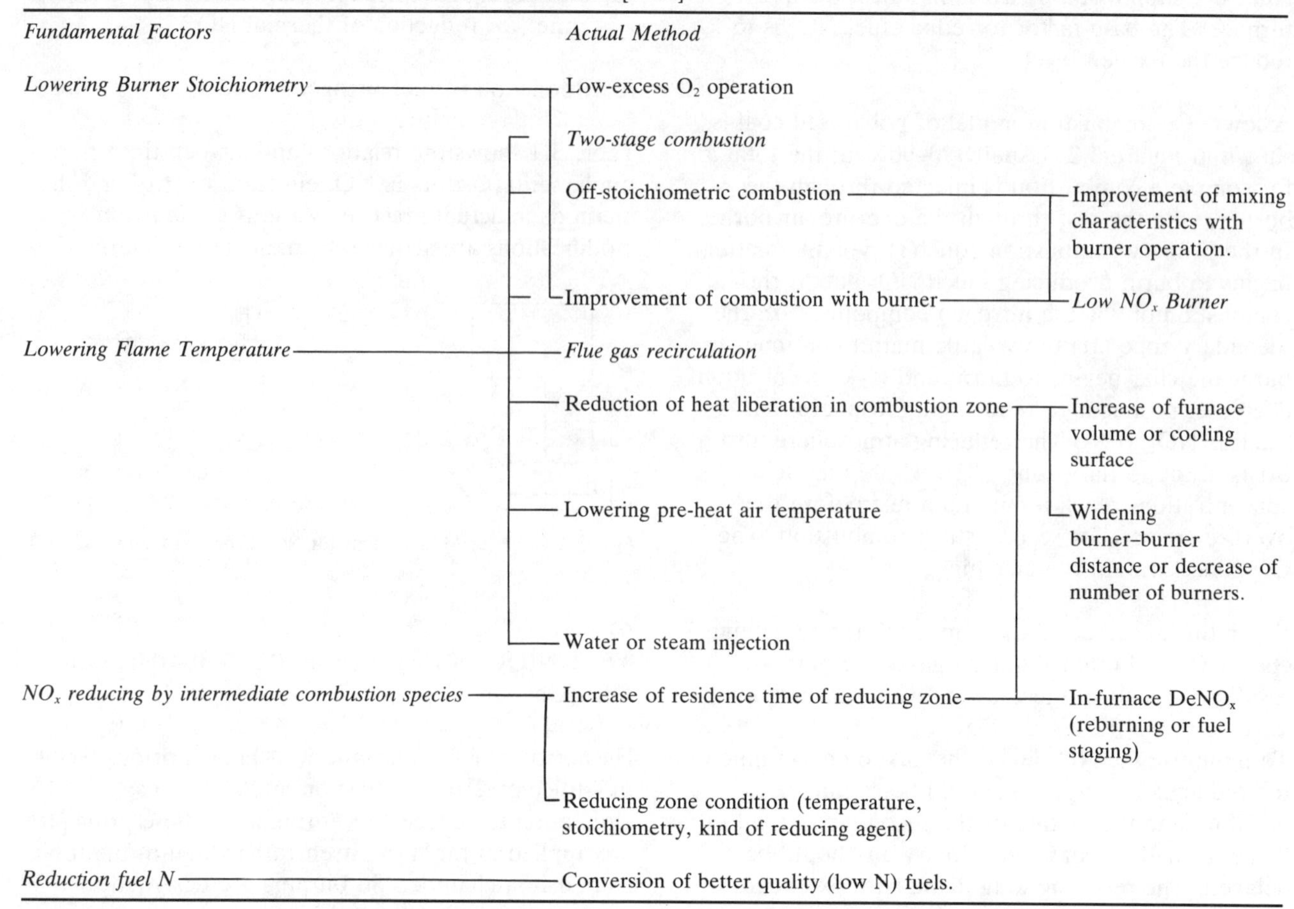

Fundamental Factors	*Actual Method*	
Lowering Burner Stoichiometry	Low-excess O_2 operation	
	Two-stage combustion	
	Off-stoichiometric combustion	Improvement of mixing characteristics with burner operation.
	Improvement of combustion with burner	*Low NO_x Burner*
Lowering Flame Temperature	*Flue gas recirculation*	
	Reduction of heat liberation in combustion zone	Increase of furnace volume or cooling surface
		Widening burner–burner distance or decrease of number of burners.
	Lowering pre-heat air temperature	
	Water or steam injection	
NO_x reducing by intermediate combustion species	Increase of residence time of reducing zone	In-furnace $DeNO_x$ (reburning or fuel staging)
	Reducing zone condition (temperature, stoichiometry, kind of reducing agent)	
Reduction fuel N	Conversion of better quality (low N) fuels.	

flow splitters to create fuel-rich and fuel-lean areas in the (sometimes extended) flame zone and/or delayed mixing between primary and secondary air streams. The variants described here are primarily intended for wall-fired boilers.

Process Code N11.1–Foster Wheeler Dual-Register, Split Flame Burner

The main features of this burner, shown in Figure 3.4 are:

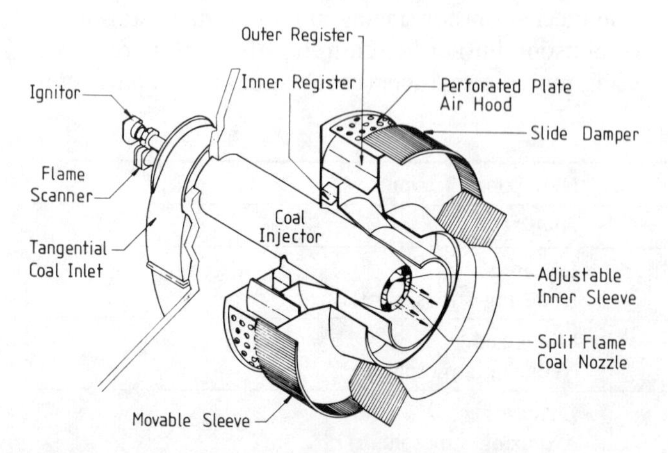

Figure 3.4 Foster Wheeler Controlled Flow/Split Flame Low NO_x Burner [32.08]

(1) The Perforated Plate Air Hood and Movable Sleeve. The pressure drop across the Perforated Plate can be measured to provide an index of relative air flow to the burners and to improve circumferential air distribution to each burner. The Movable Sleeve permits balancing of the secondary air between the burners. Hence, secondary air distribution within the windbox can be controlled both vertically and horizontally.

(2) Two Registers. The series Register arrangement permits independent control of the secondary air swirl in each of the two secondary air annuli. This enables the rate of mixing between secondary air and the primary air/coal stream to be controlled.

(3) An annular Coal Nozzle. The axially movable inner sleeve provides a means of optimising the velocity of the primary air/coal stream at constant airflow. Thus, an optimum velocity ratio between the primary and secondary air streams can be attained.

In this type of burner, the Nozzle was redesigned to split the primary air/coal flow into several distinct streams. Coal particles are concentrated within each stream and, consequently, diffuse more slowly into the secondary air. Using a variable velocity Split Flame Nozzle permits the velocity of the primary air/coal stream to be optimised for minimum NO_x commensurate with flame stability and minimum CO.

Process Code N11.2–Babcock & Wilcox Delayed Mixing Type Dual Register Burner

The design as shown in Figure 3.5 differs from the Foster Wheeler type in that it incorporates a Venturi in the primary air/coal line and an adjustable Plug. This is to provide a homogeneous particle distribution and fine tuning of the primary stream pressure drop.

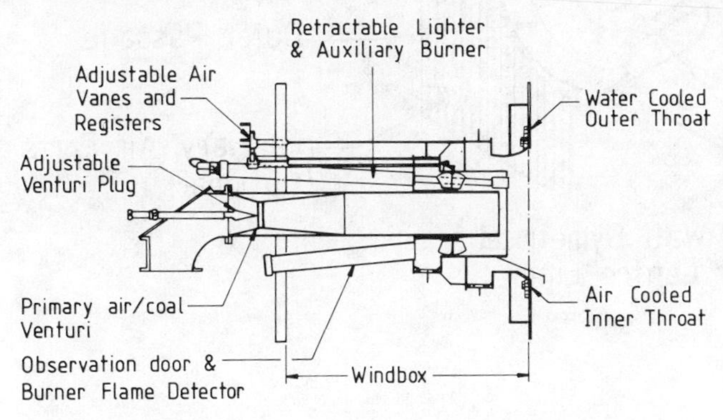

Figure 3.5 Babcock and Wilcox Dual Register Pulverised Coal Burner [31.05, 31.14]

Coal and primary air are introduced in the axis and there are two annular channels accommodating the secondary air. The secondary air system consists of independent registers for complete primary/secondary stream mixing. The inner secondary channel contains swirl vanes to ensure flame stability. The burner is usually employed in combination with a compartmental Windbox which allows the regulation of the airflow between groups of burners.

This delayed type creates a large diffusion type flame, in which the combustion is delayed. This method reduces combustion intensity and acts to reduce peak flame temperatures at each burner, minimising the thermal conversion of the combustion air nitrogen to NO_x. Through controlled fuel/air mixing, the oxygen availability is minimised during the process, thereby reducing fuel nitrogen conversion.

Process Code N11.3–Riley Stoker Controlled Combustion Venturi Burner (C.C.V.)

The C.C.V. burner, as shown in Figure 3.6, is a more recent development of the delayed mixing type burner. It is a circular burner with a single secondary air annulus and Register for switch control. NO_x control is achieved through the Venturi coal Nozzle and Spreader design. This Venturi design serves to concentrate the coal particles in the centre of the Nozzle. The adjustable four-bladed impeller type Spreader imparts swirl to the primary air/coal stream, and also divides the stream into distinct fuel-rich and lean layers before mixing with the secondary air. The Spreader design provides greater flame stability, shorter flame length and improved main flame control under two-staged operation.

Process Code N11.4–Riley Stoker Distributed Mixing Burner

The main feature of the distributed mixing burner, shown schematically in Figure 3.7 and in Figure 3.8,

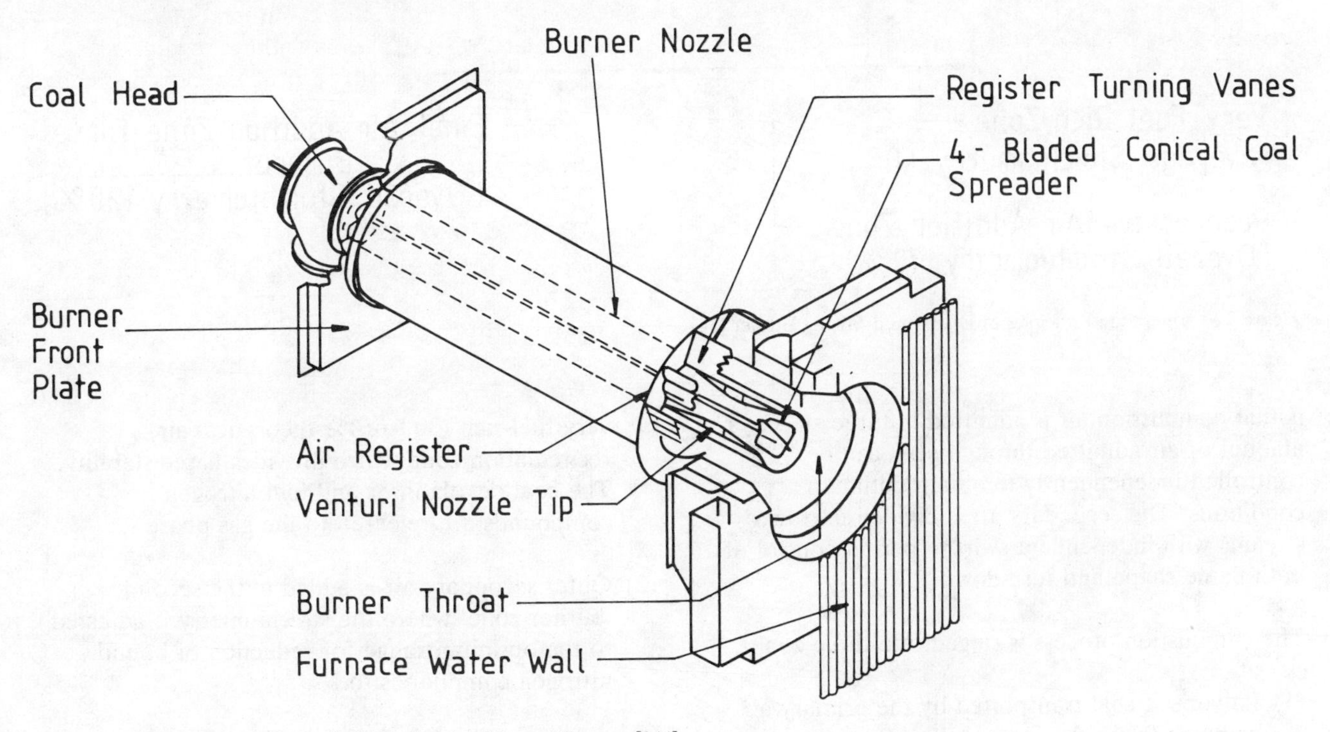

Figure 3.6 Riley Stoker Controlled Combustion Venturi Burner [344]

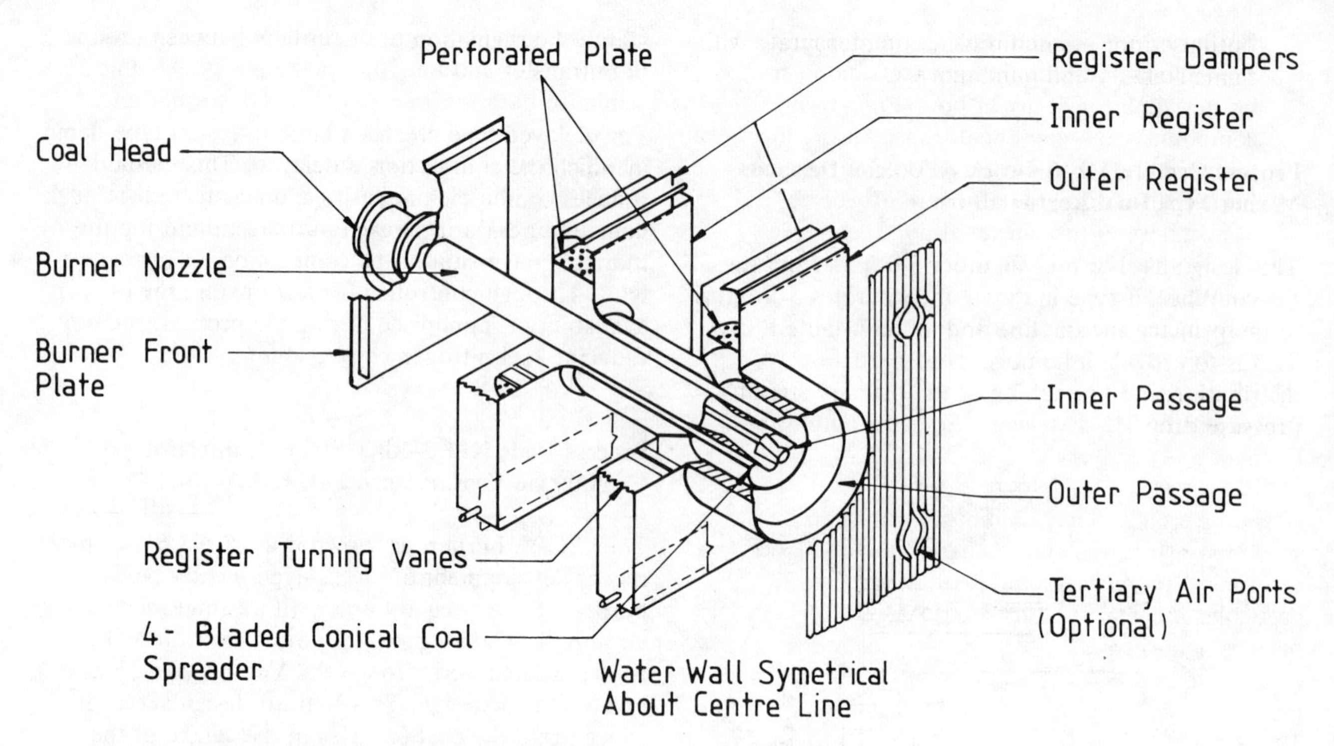

Figure 3.7 Riley Stoker Distributed Mixing Burner [344]

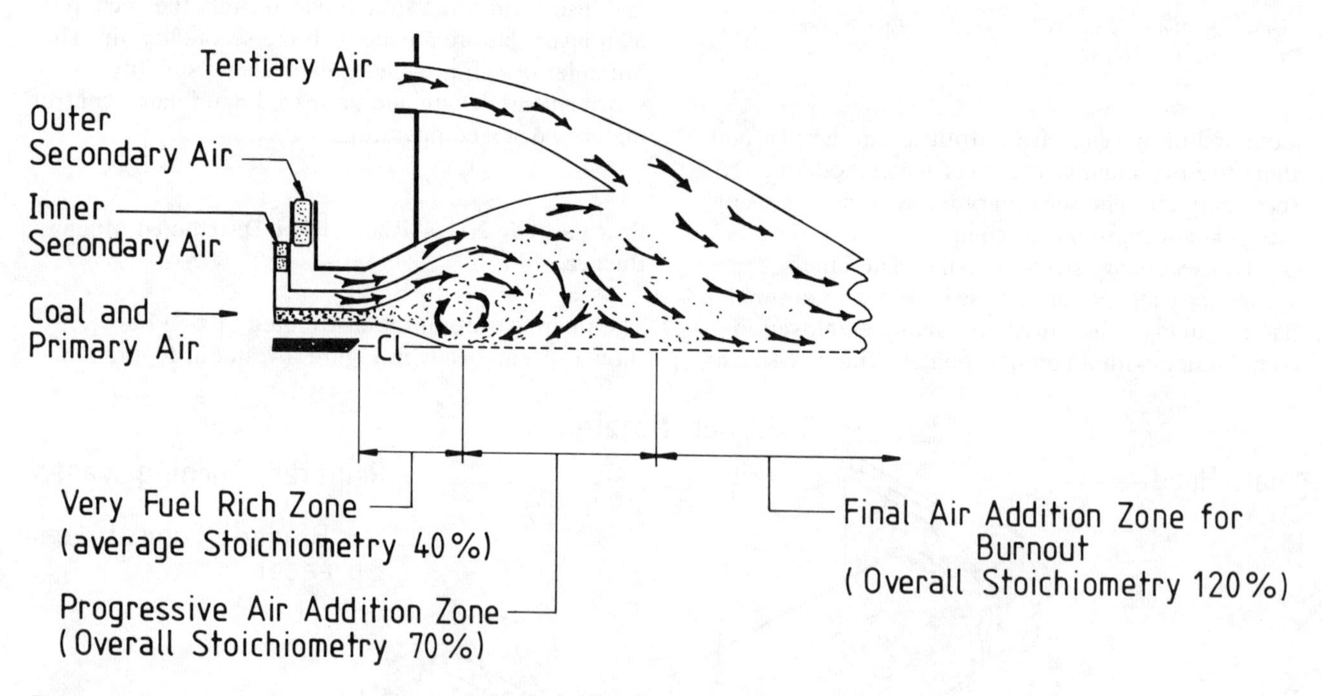

Figure 3.8 Operating Principles of Distributed Mixing Burner [341, 352]

is that combustion air is admitted in three stages. The amount of air admitted through each can be controlled independently to obtain optimum conditions. The secondary air is divided into two streams with independent swirl to permit control of both flame shape and turn-down.

The combustion process is staged into three zones:

(1) Pulverised coal transported by the primary air combined with the inner secondary air to form a very fuel-rich (30 to 50% theoretical air) recirculation zone which provides flame stability. The coal devolatilises and fuel nitrogen compounds are released to the gas phase.

(2) Outer secondary air is added in the second 'burner zone' where the stoichiometry is adjusted to the optimum range for reduction of bound nitrogen compounds to N_2.

(3) For total burn-out, air is supplied through

Tertiary Ports located outside the burner throat. This allows substantial residence time in the burner zone for decay of bound nitrogen compounds to N_2 and heat transfer to reduce peak temperatures. The Tertiary Ports surrounding the burner throat provide an overall oxidising atmosphere in the burner zone. Tests have shown that the fuel:air ratio in the primary combustion zone is the main factor for influencing the formation of NO_x.

Process Code N12–Externally Staged Combustion Burners

In the externally staged combustion burners, secondary (or tertiary) air is added through ports external to the burners and is often directed to particular areas of the furnace. Delayed mixing of primary and secondary/tertiary air occurs; mixing may occur outside the flame zone. Most of the tangential firing systems involve some form of external staging as described below.

Process Code N12.1–Tangential Firing with Overfire Air (OFA)

As shown in Figures 3.9 and 3.10, there are five basic patterns in a typical pulverised fuel (pf) furnace, of which tangentially fired units produce the lowest NO_x emissions. The long diffusion flames occupy a relatively large volume of the furnace, and the mixing rate and combustion intensity are generally less than in other designs. Long gas residence times, good wall heat transfer and long fuel-rich flame zones tend to minimise both thermal and fuel NO_x formation.

The tangential firing system with overfire air (OFA) injection is the oldest, most widely used method of controlling NO_x. This is a technique where a portion

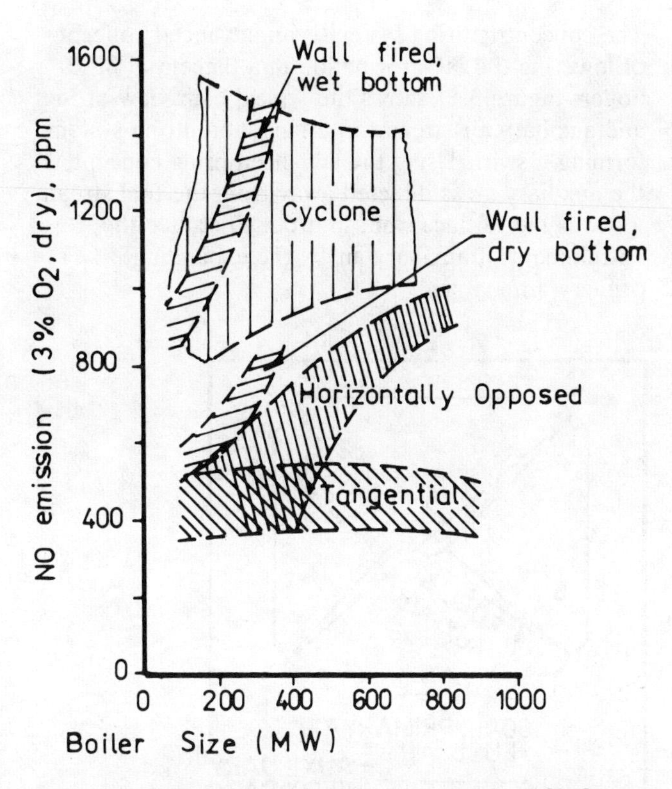

Figure 3.10 Effect of Firing Pattern on NO_x Emissions [343]

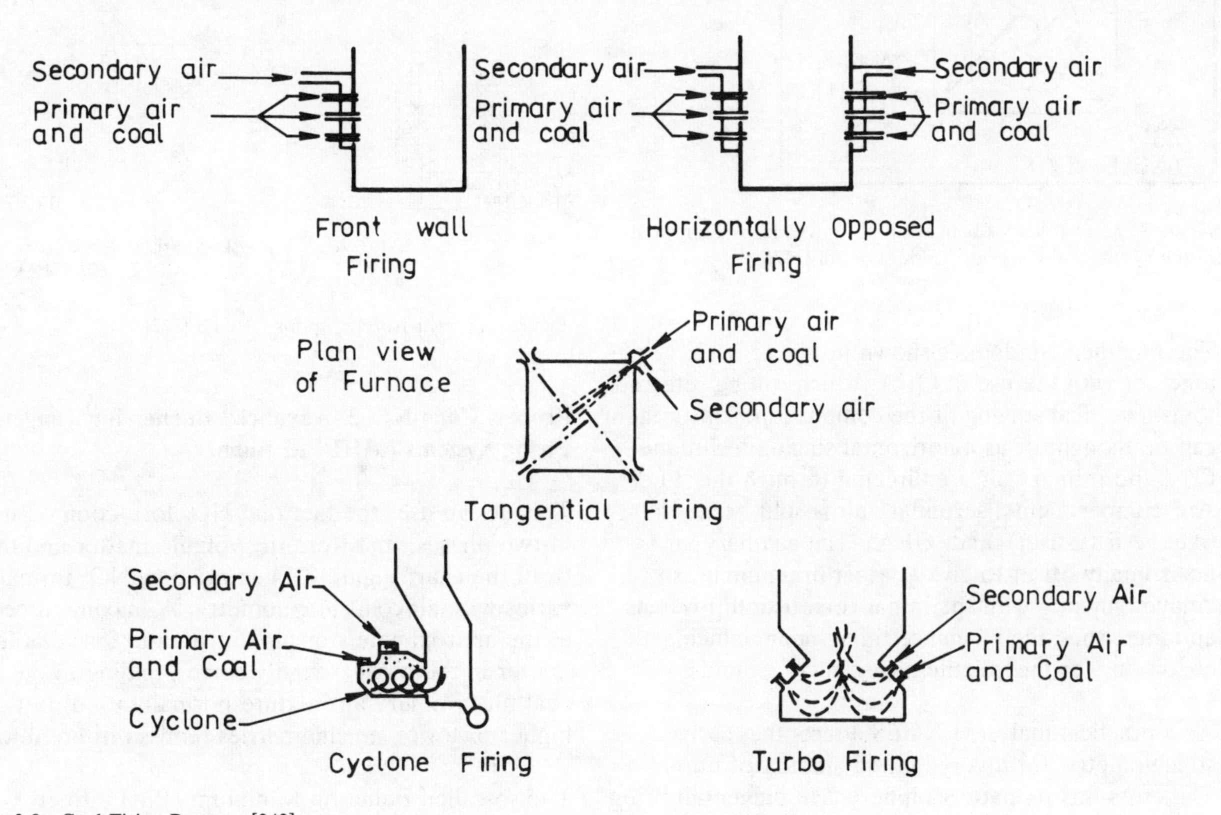

Figure 3.9 Coal Firing Patterns [343]

of the total air required for combustion is diverted away from the main windbox to run with less excess oxygen, which in turn retards the conversion of both fuel and atmospheric nitrogen to NO_x.

Process Code N12.2–Low NO_x Concentric Firing System (LNCFS)

The concentric firing system is an advanced concept of lowering the NO_x formation in a tangential fired boiler. Figure 3.11 shows the typical plan view of fuel and auxiliary air streams in a tangential firing system forming a swirl. Using the advanced firing concept, the auxiliary air is directed away from the fuel stream towards the furnace wall, in order to reduce the entrainment of auxiliary air by the expanding primary air/coal jet.

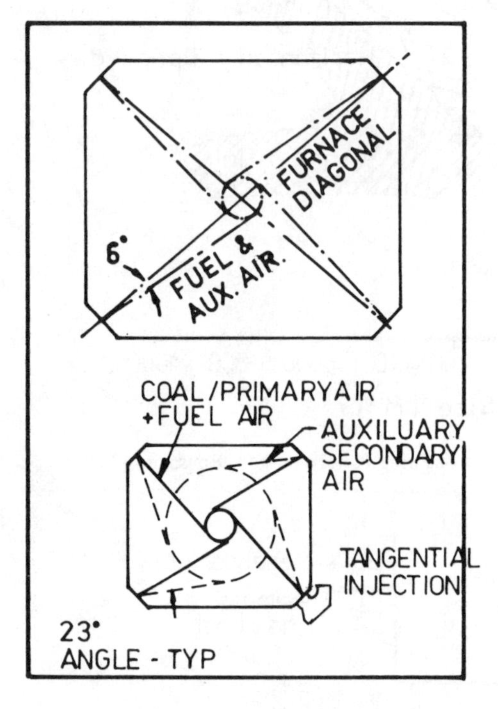

Figure 3.11 Plan View Comparison of Conventional Tangential Firing System and Concentric Firing System [352]

The modified windbox is shown in Figure 3.12, together with the use of OFA, which can be referred to as a vertical staging of the combustion. This system can be thought of as a horizontal staging technique. Coal and primary air are directed through the 'Fuel Air' compartments, secondary air is split between 'Aux. Air (Offset)' and 'OFA'. The auxiliary air is horizontally offset to give a larger firing circle, so delaying mixing with the initial combustion products and increasing the residence time under reducing conditions for the volatile nitrogen compounds.

As a practical matter, LNCFS affects the 'early stoichiometry' for a very limited amount of time. The cross-mixing patterns inherent in tangential firing are massive and the separation of both streams are

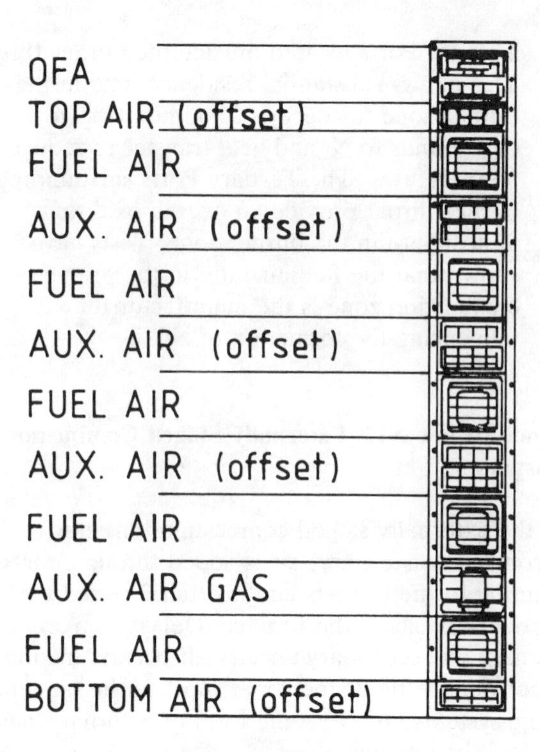

Figure 3.12 Combustion Engineering LNCFS [344]

quickly lost as they penetrate the furnace. Thus, LNCFS, requires the use of 'flame attachment' nozzle tips to accelerate the devolatilisation process. Thus, to retrofit LNCFS, all the air and fuel nozzle tips require replacement, as shown in Figure 3.13. No structural windbox or pressure part changes are normally required.

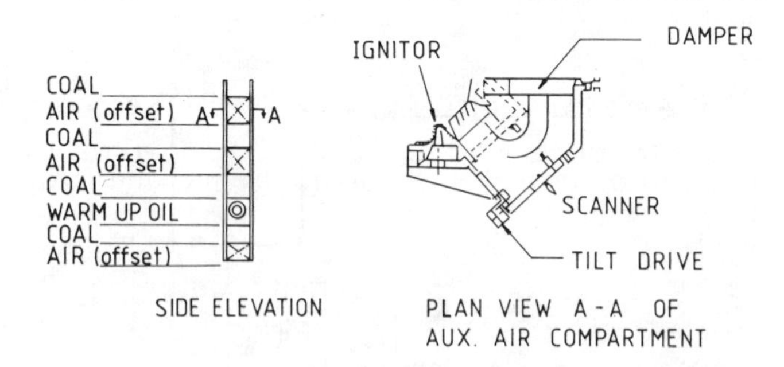

Figure 3.13 Air Nozzle Tip for LNCFS [342]

Process Code N12.3–Advanced Burner for Tangential Firing Systems (MHI/PM Burner)

This design uses the fact that NO_x formation occurs in two phases, first from the volatile matter and then from the char. Figure 3.14 shows how NO_x formation varies with air/coal stoichiometry. A maximum occurs at the approximate stoichiometric ratio for volatile matter in the coals (roughly corresponding to the coal plus primary air mixture normally used) but at higher or lower stoichiometries, emissions are lower.

The so-called Pollution Minimum (PM) burner (developed by Mitsubishi, MHI) takes advantage of

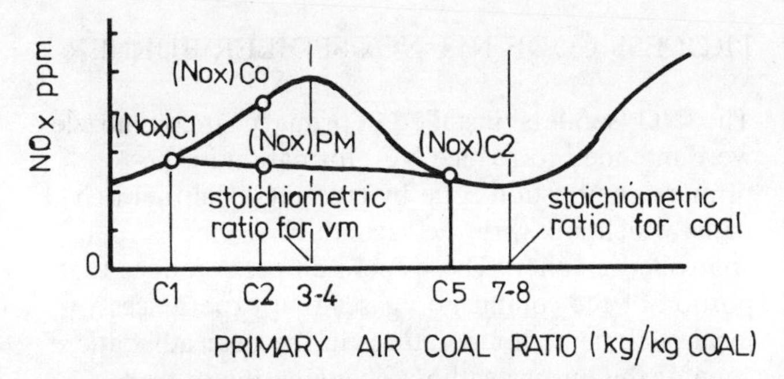

Figure 3.14 Concept of Pulverized Coal Fired Low NO_x Pollution (PM) Burner [367, 376]

this by splitting the primary air plus coal into two streams of different ratios, C_1 (around 2:1) for coal-rich or concentrated-fuel (conc) nozzle and C_2 (around 6:1) for coal-lean or weak nozzle. The NO_x value (NO_x) PM, for the whole burner will be the weighted mean of (NO_x) C_1 and (NO_x) C_2, and will become obviously lower than (NO_x) Co, i.e., the NO_x value obtainable if the whole air/coal mixture with ratio Co burns undivided.

If coal is burned in a conc flame with coal-rich ratio C_1 or a weak flame with coal-lean ratio C_2 only, the NO_x value will be sufficiently low as shown in Figure 3.14 but the combustion is undesirable. The combustion with conc flame only will cause a high percentage of unburned carbon in the fly ash, although the combustion itself is stable. The combustion with weak flame only can essentially maintain low unburned carbon in the fly ash due to the high oxygen concentration, but the effect deteriorates due to poor ignition stability. Thus, only the combined use of the conc flame and weak flame can accomplish the low NO_x combustion with stable ignition and reduced unburned carbon in the fly ash.

The structure of the burner is shown in Figure 3.15. The similarity to BOOS (Burner Out Of Service) operation (i.e. off-stoichiometric combustion) is evident, while the auxiliary air and recirculated flue gas (here termed 'separate gas recirculation' or 'SGR') provide counterparts to OFA.

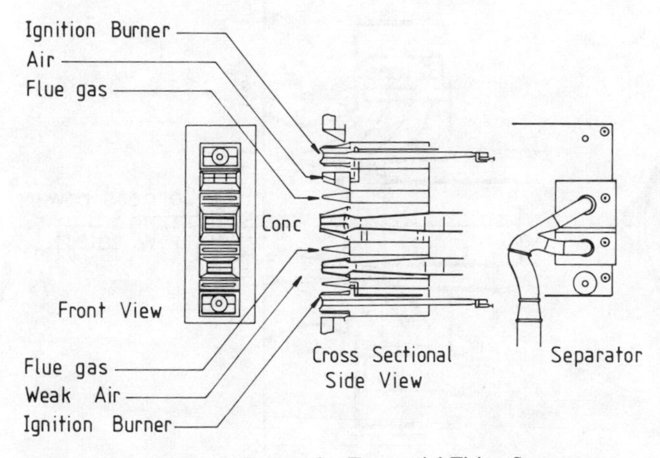

Figure 3.15 Advanced Burner for Tangential Firing Systems MHI/PM Burner

The conc nozzle located at the middle of the burner is provided with a flame holder at its outlet to stabilise the ignition of the rich coal/air mixture injected from the nozzle. SGR nozzles are arranged above and below the conc nozzle, one at each part. An auxiliary air nozzle and an ash nozzle or ignition burner are located below the lower SGR nozzle and a weak (or fuel-lean) nozzle for lean coal/air mixture, and an auxiliary air nozzle and an oil nozzle are located above the upper SGR nozzle.

The upper SGR nozzle is to retard the diffusion of secondary air from the upper auxiliary air nozzle to the conc flame, and the lower SGR nozzle is to retard the mixing of the weak flame from the weak nozzle with the conc flame.

There is a coal feed line incorporated with a separator. The separator divides the coal/air mixture in the line into two streams of different coal/air ratios, and supplies rich coal/air mixture to the conc nozzle and lean coal/air mixture to the weak nozzle.

Process Code N12.4–Fuel Staging Burner

A fuel staging burner does not stage only the total combustion air, but also the injected fuel as shown in Figures 3.16 and 3.17. By staging the fuel a reducing combustion zone is produced within the flame. This method of NO_x reduction is essentially the same as reburning (see below) but is implemented only in the flame.

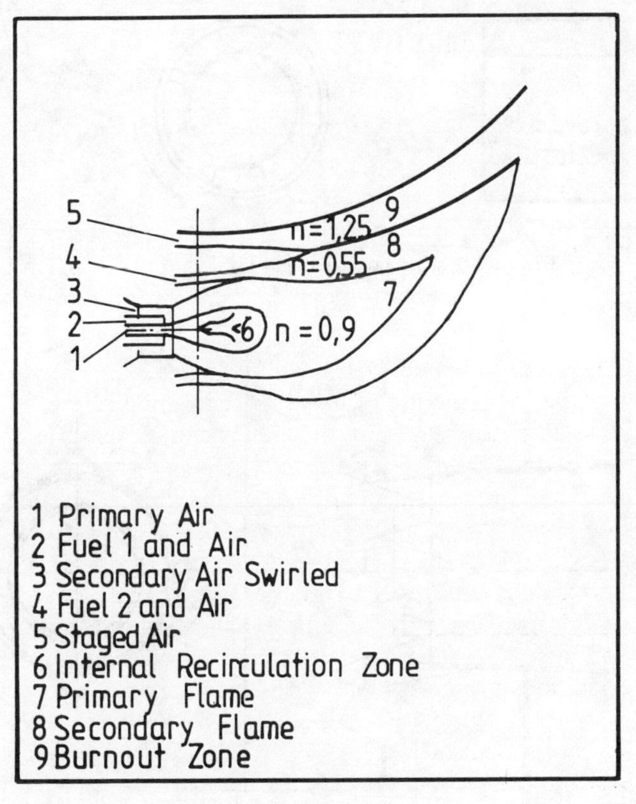

Figure 3.16 Detail of Fuel-Staging Concept [352]

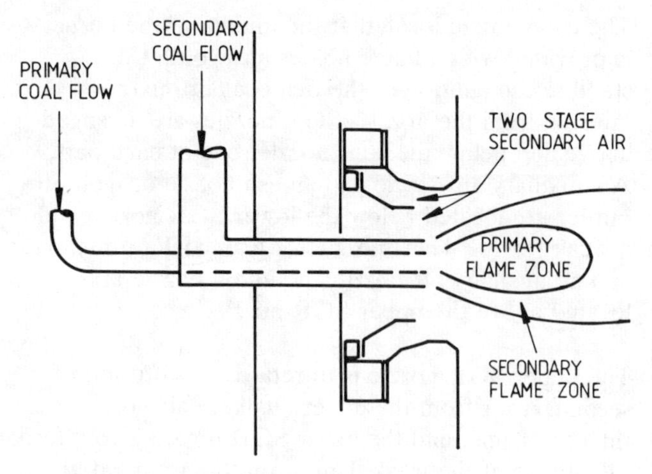

Figure 3.17 Fuel-Staging Low NO_x Burner [368]

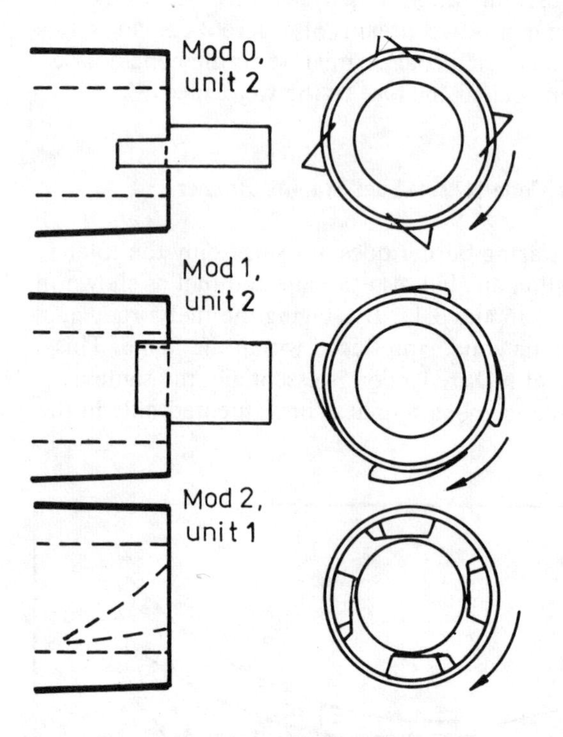

Figure 3.18 KVB NO_x Spoiler Designs [343]

PROCESS CODE N13–NOx SPOILER BURNER

The 'NO_x spoilers' installed in primary air/coal nozzles were intended to reduce NO_x formation in the primary combustion zone by creating local fuel-rich paths which mix with the secondary air in a controlled manner. These fuel-rich paths reduce that portion of the combustion gases that experiences local peak flame temperatures in the near adiabatic zone of the burner without changing the overall burner or furnace stoichiometry.

Thus, the formation of fuel-rich paths is designed to control combustion within the near-burner adiabatic region and then to profile the air/fuel mixing after coal furnace gases are entrained. This staged mixing was accomplished with minor burner modification, and is intended to minimise any air/fuel staging effects which might contribute to reducing atmosphere, wall flame impingement, staging and flame extension into the upper furnace areas.

Figures 3.18 and 3.19 show various configurations of NO_x spoilers that were tested at Four Corners power station in New Mexico, U.S.A.

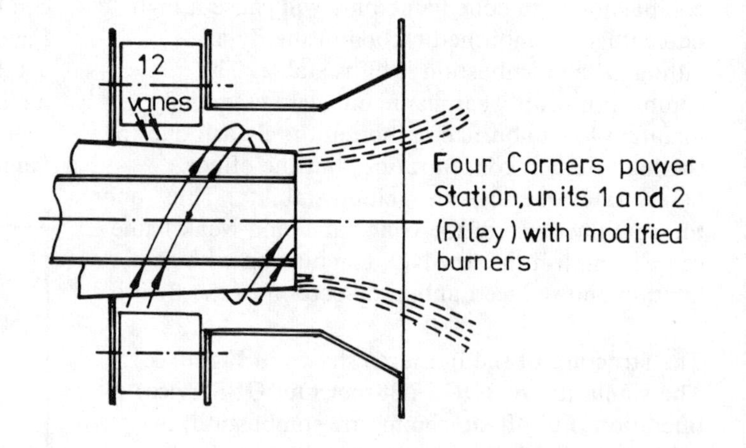

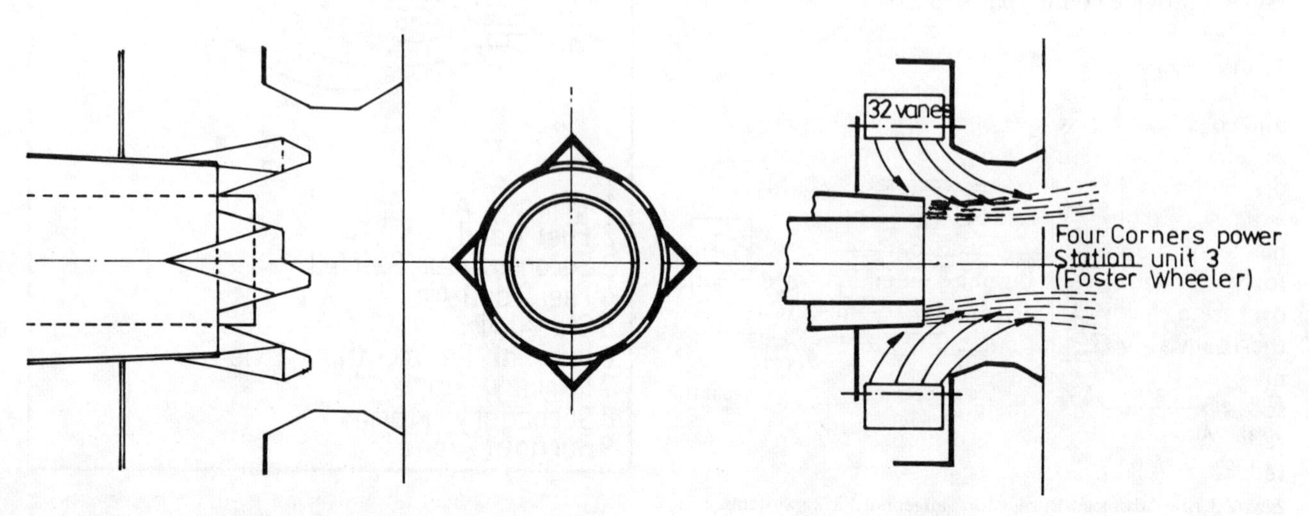

Figure 3.19 KVB Modification 3 NO_x Spoiler Design [343]

PROCESS CODE N14–FLUE GAS RECIRCULATION BURNER

There are two basic types of flue gas recirculation burner. One is the self-recirculation type burner, as shown in Figure 3.20 where the flue gas is directly recirculated at the burner throat without any external piping. The other is as shown in Figure 3.21 where external piping is required.

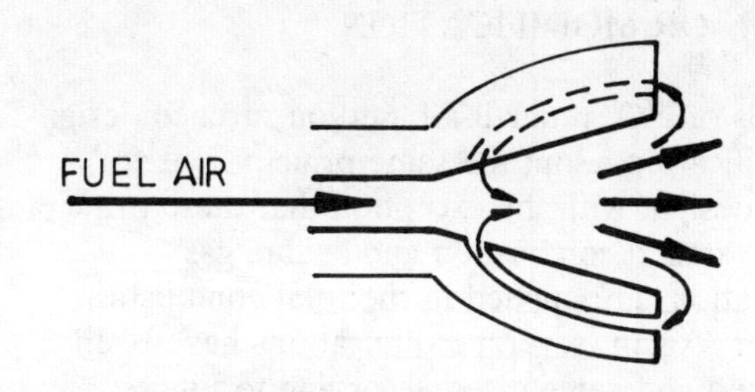

Figure 3.20 Flue Gas Recirculating Burner (Self-Recirculation Type) [350]

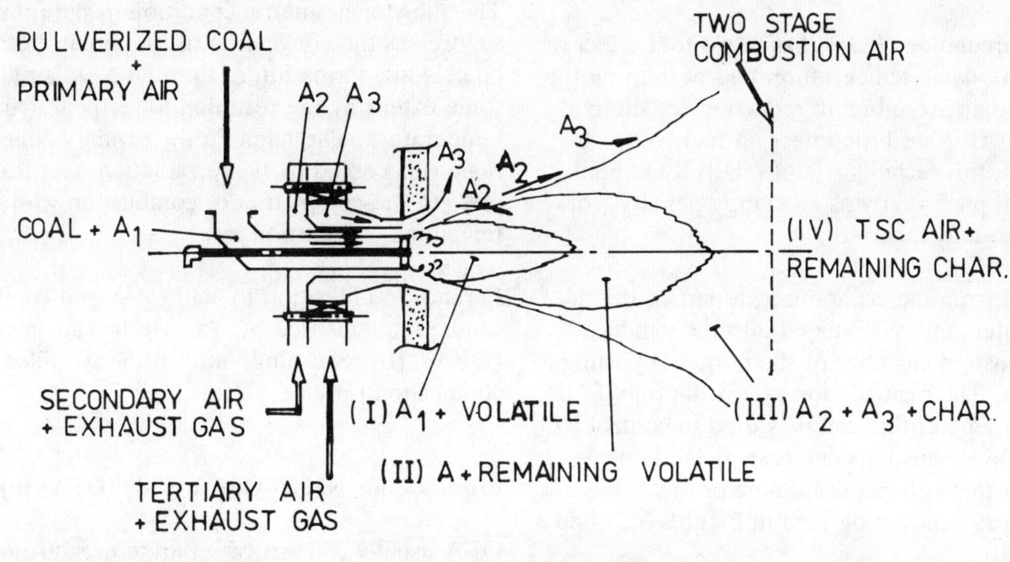

Figure 3.21 shows some of the possible arrangements of such a burner. The combustion exhaust gas is used as inert gas and mixed with combustion air, and NO_x formation is controlled by lowering flame temperature and by lowering oxygen concentration at the burner zone.

Pulverised coal and primary air are supplied through a central coal nozzle equipped with an impeller type coal spreader. Combustion air is supplied through a divided windbox to two air passages where the exhaust gas can be injected. The secondary air passage has to swirl and the tertiary air passage has a register for swirl control. The burner is intended for use with overfire air ports. In the primary combustion zone, the volatiles ignite and most of the fuel-bound nitrogen is liberated to the gas phase. In the secondary zone, the remaining volatiles burn in a reducing atmosphere which promotes NO_x reduction. The process continues through the third zone until staging air is added in stage IV to burn out the remaining combustibles.

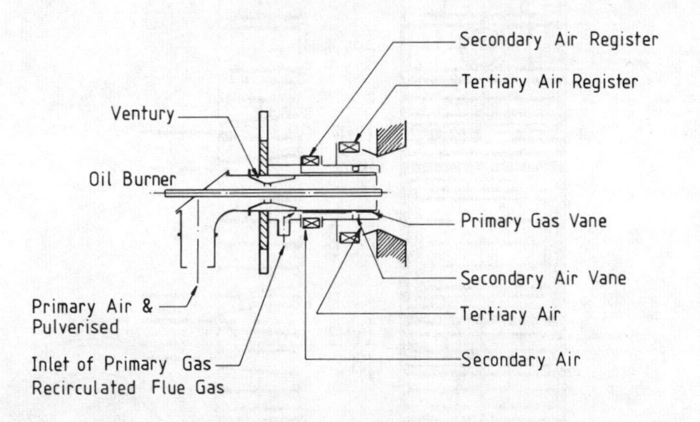

Figure 3.21 Two Examples of Flue Gas Recirculating Burner [352]

Another design is by injecting the flue gas between the fuel and air streams, in order to lower the peak temperature of the burning zone around the burner and restrict oxygen access to the primary combustion zone.

TECHNIQUES IN CATEGORY N20–FURNACE DESIGN OR MODIFICATION

Methods of NO_x reduction based on furnace design or modification adopt the same principles as for burner design, with the exception that these principles (i.e. the staged combustion and/or flue gas recirculation) are applied in the total combustion chamber. Again, several modifications have been developed and categorised according to Figure 3.1.

PROCESS CODE N21–FLUE GAS RECIRCULATION

Flue gas recirculation (FGR) has a two-fold effect of reducing peak flame temperature, and of diluting the oxygen in the air, resulting in reductions of thermal NO_x. However, it has little effect on fuel NO_x. Furthermore, this technique is of relatively limited effectiveness, particularly as an energy penalty (for the fans) is usually incurred.

Downstream from the economiser, a part of the flue gas is separated and recirculated into the windbox or the combustion chamber or the burners by using a hot draft fan. The recirculation rate is the only FGR operating parameter that can be varied to control NO_x reductions. NO_x emissions decrease as the flue gas recirculation rate is increased. Some of the flue gas recirculation system can be seen in Figures 3.21 and 3.22.

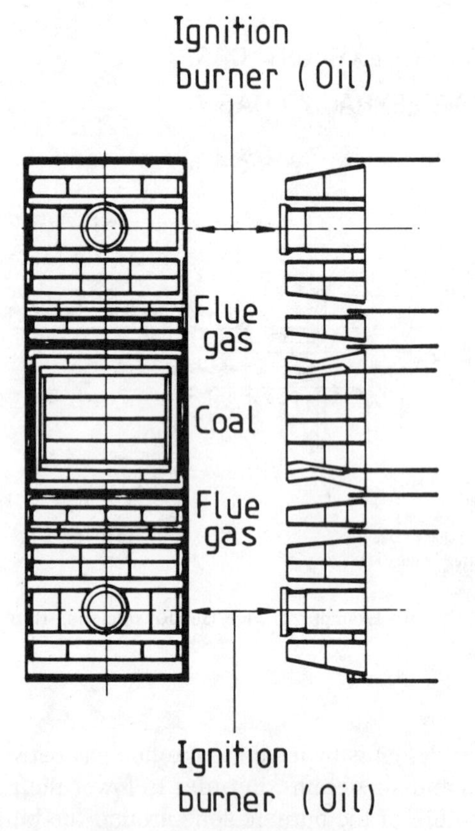

Figure 3.22 Flue Gas Recirculating Burner for Tangential Firing Systems [381]

PROCESS CODE N22–STAGED COMBUSTION

The principle of staged combustion (or off-stoichiometric combustion, OSC) is based on the regulation of oxygen content within the furnace. A similar principle is used in the low NO_x burner concepts described above.

A primary combustion zone is created in the combustion chamber by dividing the total air, where the combustion is performed under a lack of oxygen. To ensure the total burnout at oxygen excess and relatively low temperatures, air is injected into an additional secondary combustion zone.

The sub-stoichiometric condition in the primary zone suppresses the conversion of fuel bound nitrogen to NO_x. Also, formation of thermal NO_x is reduced to some extent by the resulting lower peak flame temperature. The flame in the primary zone (fuel-rich region) is cooled by flame radiation heat transfer prior to the completion of combustion with the remaining combustion air.

Furnace modification to achieve staged combustion can be implemented by: (a) overfire air injection (OFA), (b) reburning, and (c) the so-called primary combustion furnace.

Process Code N22.1–Overfire Air (OFA) Injection

OFA usually consists of separate pressure port openings in the furnace above the main windboxes as shown in Figure 3.23. Within the opening a tilting nozzle tip is installed which has the ability to direct the air towards or away from the main windbox. The tilting feature provides the operator with a means of adjusting the residence time between the main fuel and the OFA. In practice, it is a means of fine tuning NO_x. The fuel delivery system is unaffected.

The burners operate sub-stoichiometrically so that in the centre of the boiler a strongly turbulent fuel-rich flame is formed. The remaining air is added above through ports which are designed or modified for this purpose.

Process Code N22.2–Reburning (Fuel Staging)

The 'reburning' process or 'fuel staging' is also known as in-furnace NO_x reduction (IFNR). As shown schematically in Figure 3.24 the combustion furnace is divided into three regions. In the first stage of the furnace, the main supply of fuel undergoes fuel-lean combustion. This is followed by a reburning stage where additional fuel is injected. In this region, NO_x formed in the fuel-lean first stage is reduced significantly after undergoing interactions with the flame radical species in a fuel-rich environment.

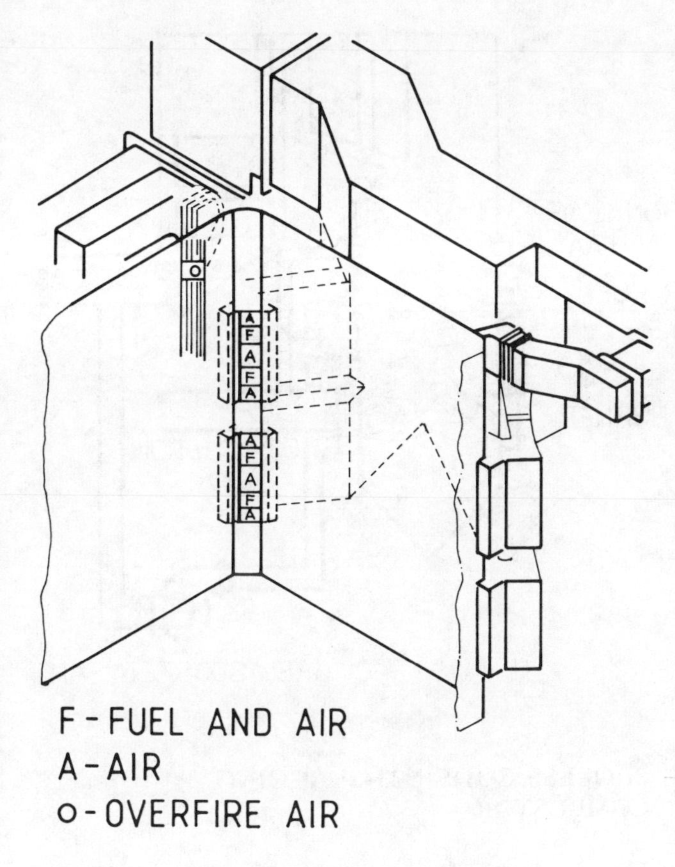

Figure 3.23 Typical Retrofitted OFA System [342]

Downstream of the reburning region, additional air is injected to complete burnout.

A number of fuel-staged firing configurations can be created with alternate fuel-rich/air-rich zones in the furnace, as shown in Figure 3.25. From a process view point, their main differences depend on whether the chemical energy which fuels the second stage flame is supplied externally (through reburn fuel injection) or internally (through fuel-rich first stage combustion produced gases).

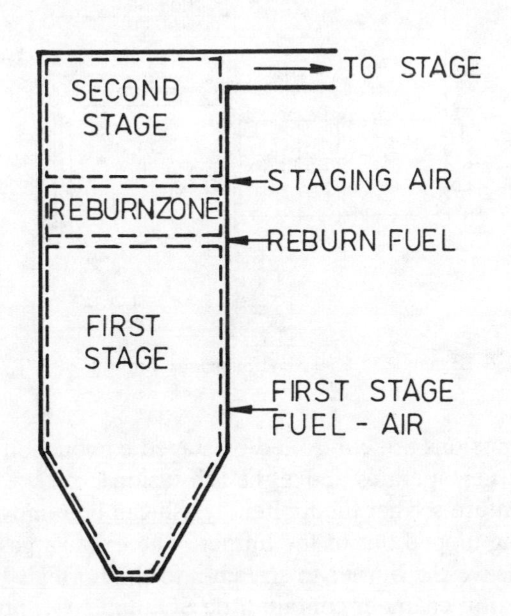

Figure 3.24 Basic Configuration for the Process of IFNR [387]

One such technique has been developed by MHI and is called MACT (Mitsubishi Advanced Combustion Technology) as shown in Figure 3.26. (A similar technique is also used by the Babcock Hitachi IFNR and KAWASAKI KHI KVC system). It is designed to take advantage of the high temperature kinetic characteristics of the nitrogen oxide molecule. Nitrogen oxide is relatively unstable at the elevated temperatures commonly found in the mid-furnace section of most modern boilers. Above 1200°C, the NO molecule can be reduced by reacting with certain radicals at high temperatures. The high temperature chemistry is not in itself new, but what is unique to the MHI process is the use of the primary boiler fuel as the radical (reducing agent).

MACT diverts a small percentage of fuel from the main burner combustion zone and injects the fuel through UFI (upper fuel injection) ports with an inert propellent, usually flue gas. The fuel injected into the 'De-NOxing zone' has insufficient oxygen to burn, and upon reaching sufficient temperature generates, for a brief period of time, high temperature radicals which have such a high affinity for oxygen that they literally strip the oxygen from the NO molecule. This high temperature thermal decomposition of NO is called the 'DN' process.

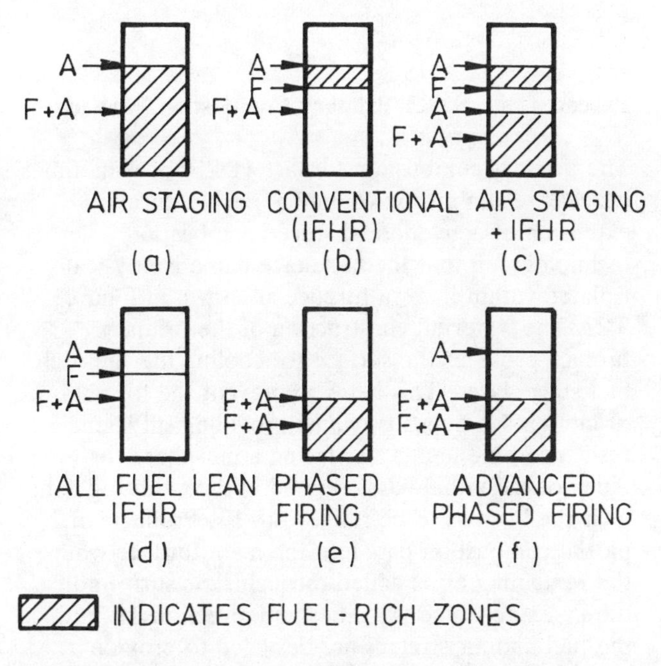

Figure 3.25 Examples of Possible IFNR Firing Arrangements [387]

Since the fuel injected into the mid-furnace area in the DN process must be burned to completion to salvage the heating value of the fuel and to prevent hydrocarbon emission, the balance of the required combustion air is injected at the 'AA' elevation (where 'AA' denotes additional air supply: see Figure 3.26). The 'AA' process is the straightforward oxidation of the products of incomplete combustion

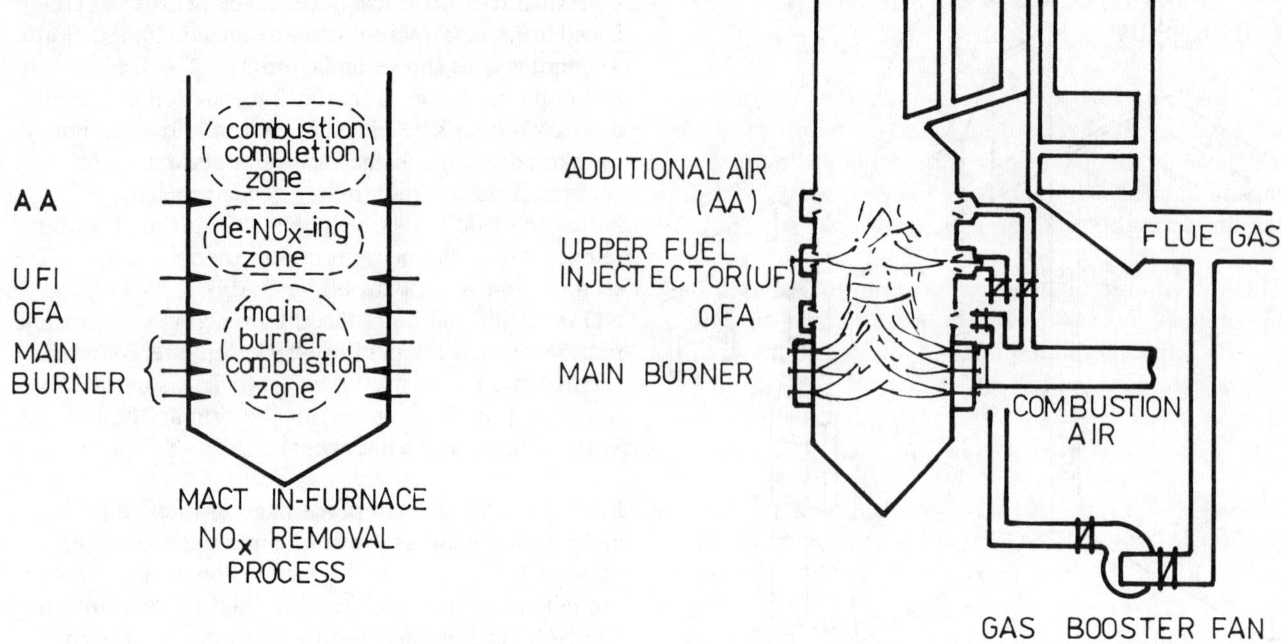

Figure 3.26 Schematic Diagram of MACT for Steam Generator [342]

leaving the DN process. The MACT system does a very effective job of destroying most of the NO in the DN process, but regenerates some of the NO in the 'AA' process. The result is a net reduction of approximately 50% of all the NO entering the MACT process.

Process Code N22.3–Primary Combustion Furnace

The primary combustion furnace (PCF) or sometimes called 'precombustion chambers', is a natural extension of conventional staged combustion technology, in that the first stage flame is physically isolated within its own furnace as shown in Figure 3.27. The waterfall construction of the primary furnace provides the surface for cooling the fuel-rich, first stage gases. This helps to prevent the formation of molten ash deposits, whilst providing sufficient residence time under a reducing atmosphere for much of the initial NO_x, formed in the oxidising zone near the burner, to be reduced to N_2. Products of partial combustion pass into the main furnace where the remaining air is added through slots surrounding the first stage. The velocities and directions of both the fuel and air streams are designed to provide a long diffusion flame in the second stage. The NO_x emission can be diminished by limiting the second stage flame-zone temperature.

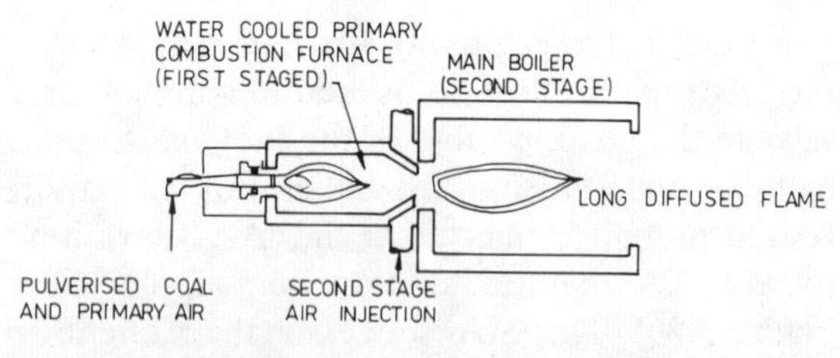

Figure 3.27 The B & W Primary Combustion Furnace (PCF) [355]

PROCESS CODE N23–SLAGGING COMBUSTOR

The so-called high temperature pre-burner or slagging combustor is a burner which allows the conversion of a gas-and oil-fired boiler to pulverised coal firing with minimal derating. It uses a burner concept which is able to control both SO_2 and NO_x simultaneously and remove fly ash (see Section 5).

The principle construction concept is a burner chamber outside the combustion chamber of the boiler, as shown in Figure 3.28. The combustion process is mainly finished when the exhaust gas streams into the boiler.

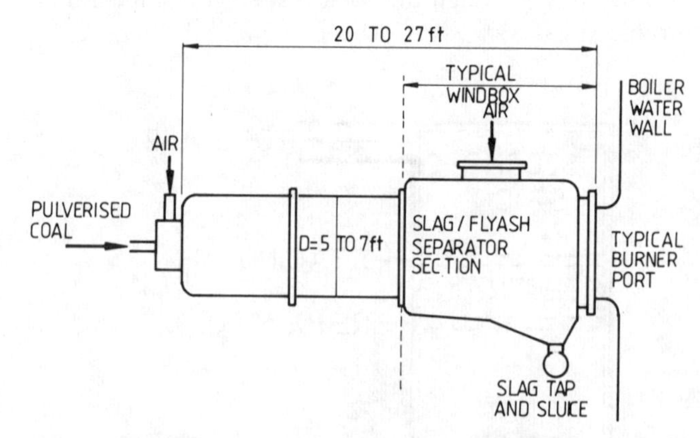

Figure 3.28 Example of a 30 MWt Slagging Combustor [427]

NO_x emissions are controlled by staged combustion. The burner operates above the ash fusion temperature so that the molten fly ash can be removed with slag tapped out of the burner. The exhaust gases which leave the burner to stream into the normal combustion chamber contain little SO_x and NO_x, no slag and little fly ash.

PROCESS CODE N24–REDUCE HEAT IN COMBUSTION CHAMBER

Older designs of boiler were frequently very compact, to maximise the heat output for a given unit of space. This usually involved using very turbulent, rapidly mixing flames, with high peak temperatures and therefore resulted in high NO_x emissions.

The low volumetric heat release criterion is based on the philosophy that a lower volumetric heat output, e.g. by reducing the intensity of mixing, could lead to a corresponding fall in NO_x emissions.

PROCESS CODE N25–DOWN-FIRED SEQUENTIAL AIR ADDITION DSAA BOILER (OR ARCH FIRING)

A typical arrangement of the DSAA boiler is shown in Figure 3.29. Pulverised coal and primary air are injected vertically downwards into the furnace from an arch, the remainder of the combustion air (secondary and tertiary) being added through the front wall in a sequential manner.

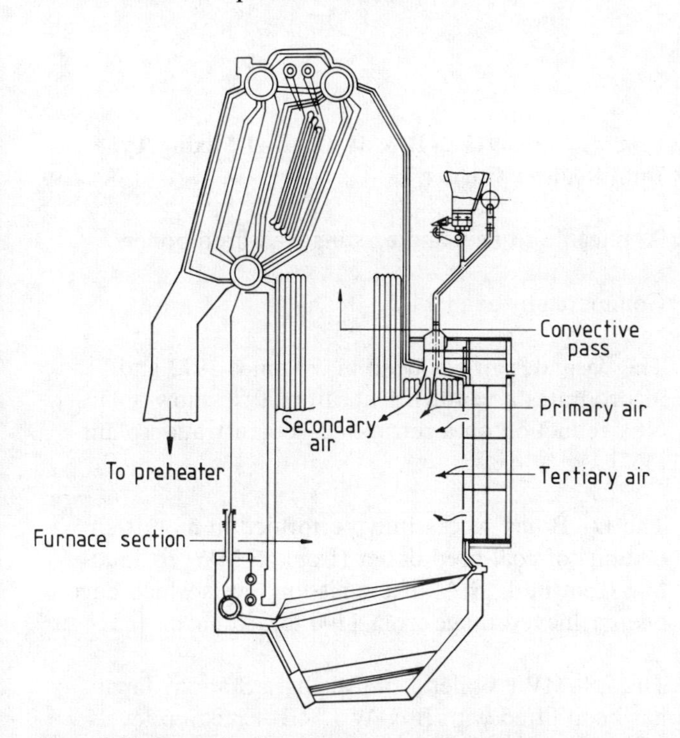

Figure 3.29 Example of an 80 MW Down-Fired Sequential Air Addition Furnace [343]

The possibility of having low NO_x emissions is suggested by the combustion characteristics of the arch fired boilers as follows:

(1) Flame stability. The injected coal receives a relatively high heat flux (thus rapid volatilisation of the fuel), the magnitude of which depends on the design of the arch.

(2) Extended heat release zone. The sequential air addition delays the mixing of the coal/primary air mixture and the secondary air.

(3) Variable radiant heat flux. A variable flame shape affords a controllable radiant heat flux to the furnace walls that is relatively independent of load.

TECHNIQUES IN CATEGORY N30–FURNACE OPERATION

This is a technique where no modification is required to the existing unit, since only the operating condition of the furnace is changed and it can be implemented in several ways, as indicated in Figure 3.1.

PROCESS CODE N31–LOW EXCESS AIR COMBUSTION

Excess air is normally provided in a furnace to ensure complete combustion of the coal. As the excess air is increased, the oxygen concentration in the flame zone increases and the temperature decreases.

Low excess air (LEA) firing causes a reduction of the local flame zone concentration of oxygen by which both the thermal and the fuel NO_x formation is reduced.

LEA can also be used in combination with other combustion modifications, such as staged combustion and operating at reduced load. However, requirements on combustion control increase if a combination of such measures is installed.

PROCESS CODE N32–STAGED COMBUSTION

The principle of off-stoichiometric combustion (OSC) or staged combustion has been described above. This principle can be applied on existing furnaces to reduce NO_x emission by making the burners operate under Biased Burner Firing (BBF) or under Burners Out Of Service (BOOS). It is mainly applicable to boilers which operate with an array of burners (e.g. utility boilers).

BBF means that by appropriate measures, a fuel-rich combustion zone is produced at the lower rows of burners, and a fuel-lean combustion zone at the upper rows of burners.

BOOS means that by operating most of the burners (of a multiple burner array) under fuel-rich conditions, the remaining burners (normally the upper rows) are out of service injecting only air. This is to create a fuel-rich region, in which the coal is devolatilised, and allowing as long a residence time within that zone as possible before the admixing of the secondary combustion air.

PROCESS CODE N33–REDUCED AIR PREHEAT [352]

Thermal NO_x production can sometimes be decreased by reducing the amount of combustion air preheating, which again reduces peak temperature in the primary combustion zone. The reduced air preheat operation appears relatively inefficient in suppressing fuel nitrogen conversion.

PROCESS CODE N34–REDUCE HEAT LOAD IN COMBUSTION CHAMBER

A reduced load operation decreases the heat release per volume and per surface area and thus lowers peak flame temperatures, resulting in lower thermal NO_x formation. Fuel/air mixing rates are also decreased and this may lower NO_x emissions from fuel-bound nitrogen.

3.3 General Appraisal of Each Technique

PROCESS CODE N11–INTERNALLY STAGED COMBUSTION BURNERS

Process Code N11.1–F.W. Dual Register Burner

This burner is only applicable to wall-fired units and can be used for new units and retrofitting (with minimal modifications).

This type of burner is commercially available.

Operating experience is mainly from two large utility boilers in the U.S.A. retrofitted with the controlled flow/split flame (CF/SF) burner. One is the 350 MWe San Juan No. 1 single wall-fired unit burning a high ash sub-bituminous coal; the other is a 525 MWe opposed fired unit burning a high moisture, western, sub-bituminous coal. NO_x emission levels below 490 mg/m^3 have been achieved. Further NO_x reduction (down to below 250 mg/m^3) are obtainable with moderate overfire air. Owing to the potential for slagging and high unburned carbon levels when large amounts of overfire air are used, 20% overfire air (minimum burner stoichiometry of 96%) seems to be the upper limit. [348, 369]

In the summer of 1986, a 500 MWe single wall-fired utility boiler at Eggborough Power Station in the U.K. was equipped with this type of burner [352].

A modified version, developed with IHI has been retrofitted to the pf horizontally opposed firing boiler (called dual flow convergent fuel nozzle (DF-CN)) 265 MWe Isogo plant in Japan. Emissions are around 145 ng/J (150–200ppm) compared with 220–295 ng/J with overfire air only. The NO_x reduction on the 175 MWt pf horizontally front firing boiler was 30–40%. [348]

In early 1986, Allegheny Power Service Corp. (APS) installed FW controlled flow/split flame burners along with a set of boundary air ports on one of the 626 MWe opposed–wall firing boiler. The boundary air technique helps maintain oxidising conditions on the sidewalls. With the overfire air ports closed, NO_x emissions decreased from 1.0 lb/million Btu to less than 0.5 lb with the new burners. When the overfire air ports were opened, the decrease was from 0.7 lb/million Btu to 0.4 lb. In both cases, with and without the new burners, unburned carbon in the fly ash remained below 2.5% while the CO content remained below 50 ppm. [401]

Process Code N11.2–B & W Delayed Mixing Type Dual Register Burner

Applicable to new and existing wall-fired boilers.

Commercially available.

The basic design has been in use since 1972 and incorporates a venturi. Data from 1976 show a 45% NO_x reduction on a retrofitted demonstration plant [367].

The DRB has successfully performed in more than 60 units of coal-fired boiler (from 65 MWt to 1300 MWt) in the U.S.A. The emission levels which have been achieved range from 1300 to 380 mg/m^3 [352].

The 250 MWe boiler 1 of Takehara Station, Japan, has been fitted with B & W DRB. Emission levels were 200 ppm with OFA and 300–360 without [367].

The newly built 700 MWe boiler 3 of Takehara Station, Japan has been fitted with B & W primary gas DRB. Emission levels were about 250 ppm. (150 ppm with selective catalytic reduction) [367].

Full-scale tests of an industrial boiler (80 t/h) using the advanced type of DRB indicate an NO_x emission reduction compared to the pre-retrofit level by 30–40% down to 300 to 410 mg/m^3 (6% O_2) and no change in unburnt carbon loss [352].

Process Code N11.3–Riley Stoker Controlled Combustion Venturi Burner

Applicable to industrial and utility boilers. It was developed for retrofitting and is equally applicable to new units.

Commercially available.

Operating experience shows [385, 383, 344] that retrofitted to a 400 MWe front-fired and a 360 MWe opposed-fired utility boiler in the U.S.A. 50–60% NO_x reductions, down to 200–250 ng/J, have been achieved.

Results from retrofitting to a 700 MWe boiler show that emissions of 170–210 ng/J, can be achieved without adverse effects on the boiler system.

Process Code N11.4–Distributed Mixing Burner

Applicable for all new wall-fired industrial and utility boilers.

Retrofitting may require significant modification to the firing wall to accommodate the large number of tertiary air ports in the burner zone.

This burner is under development.

In October 1979 the 700 MWe boiler of Saarbergwerke, FRG was modified to DMB (24 burners, each of 70 MWt). The NO_x emission was reduced by 50% (from 1540 mg/m^3 to 690 mg/m^3). A larger NO_x reduction was found with increasing air staging. The flame length did not increase while using a special air regulation system (staging of the air according to the load) [352].

The Western Illinois Power Corporation (WIPCO) Pearl Station's 20 MWe front wall-fired boiler has been retrofitted with DMB. Under normal operation a 50% reduction in NO_x emission has been achieved. 70% reduction down to 260 ppm was achieved under controlled condition. 130 ng/J (180 ppm) is achievable, but at the expense of stability problems. This DMB incorporates FW's proprietary controlled flow burner exit geometry to improve flame stability. The long term test at this facility resulted in a little capacity derating due to increased carbon loss [352, 384].

The Riley Stoker type DMB has been tested on the 29 MWt simulator furnace, California. An emission level of 220 ppm can be achieved [383, 385].

PROCESS CODE N12–EXTERNALLY STAGED COMBUSTION BURNER

Process Code N12.1–Tangential Firing with Overfire Air (OFA)

Suitable for new build and retrofit applications.

Widely in commercial use. A proven, well established technology.

In operating experience, 10%–35% NO_x emission reductions have been indicated, with usually no reduction in boiler efficiency, but the effectiveness is very site-specific and depends on the design and the placement of the ports for proper mixing, the furnace firing configuration and the coal type. OFA is usually incorporated with other combustion modification techniques [352].

Possible operation problems of incomplete burnout, increased smoke and particulate matter emissions and the risk of slagging and corrosion (due to the long flame produced, with its extended reducing zone) can be minimised by proper maintenance of excess air levels and proper distribution of air between burner OFA ports. This side-effect can limit its applicability and will incur additional operating costs.

Process Code N12.2–Low NO_x Concentric Firing System (LNCFS)

Suitable for new build and retrofit applications. (In retrofitting, all the air and fuel nozzle tips require replacement).

The system is commercially available.

Test results from Utah Power & Light's 400 MWe Hunter No. 2 Station indicated a 20% NO_x reduction from the standard firing mode for LNCFS with OFA without significant side-effects on unburnt carbon loss [352].

This firing system has been tested on a 61 MWe utility boiler. NO_x emission reductions of 24% to 40% were obtained [352].

The U.K., the Netherlands and Canada have planned to retrofit a utility boiler with this unit.

Process Code N12.3–Mitsubishi PM Burners

Suitable for new build and retrofit applications. (As a retrofit, the PM burner is not a 'plug in' replacement of the original windbox. The windbox height must be increased which can present interference problems).

Test show excellent results but not yet fully implemented.

Operating experience shows [344] that in a test furnace, the PM system using 20% OFA has achieved emissions of 50–70 ng/J compared with 130–190 ng/J for a conventional burner with 20% OFA.

At a Japanese new power plant's boiler (350 MWe) firing Australian Coal using the PM system, a NO_x concentration of 310 mg/m^3 has been achieved at full load. The carbon residue in the fly ash was not higher than 5% [352].

In 1982, tests on an industrial boiler of the Daishowa Paper Co. showed that by increasing overfire air from 5 to 17%, the NO_x concentration decreased from 210 to 150 ppm with little increase of unburned carbon from 4 to 4.5% [342].

Test results showed that NO_x concentration may be reduced to 60–80 ppm for low N coal or 100–150ppm for high N coal [342].

Process Code N12.4–Fuel Staging Burners

Mainly applicable to new units. Retrofitting is possible.

The burners are under development.

Operating experience shows [352] that using this burner concept it is anticipated that the NO_x concentration from a coal-fired dry bottom boiler may be reduced from 750 mg/m^3 to 250 mg/m^3.

Results from a 15 MWt test furnace show that NO_x emissions between 200–300 mg/m^3 with 20% secondary fuel can be achieved.

See also below.

PROCESS CODE N13–NO_x SPOILER BURNER

This burner is suitable for new build and retrofit applications.

This technology was adopted in the 1970's, but inconsistency in performance appears to have led to its being abandoned.

Operating experience [343, 36–38] is as follows: In 1976/77, units 1 and 2 (each 175 MWe) and the 225 MWe unit 3 of Four Corners Power Plant, New Mexico were retrofitted with NO_x spoiler burners. Various configurations of NO_x spoilers were tried (described below). NO_x emission levels below 520 ppm have been achieved but appear to be non-repeatable.

The NO_x level was reduced from 910 to 750 ppm after installation of the unmodified spoiler (see Figures 3.19 and 3.20).

With the first modified design, the NO_x level could be reduced below 600 ppm. Following the completion of this design, it appeared that the difficulty in producing the desired fuel rich streams resulted from a 4 to 1 ratio between the secondary and primary stream velocities. The low velocity stream would rapidly lose its identity as the result of shear forces due to high velocity gradients at the boundaries of the low streams. Therefore, the second modification was designed to increase blockage of the primary air/coal nozzle to 40. This lowered the ratio between secondary and primary air stream velocities to 2.5 to 1. With the second modification, the NO_x level was below 520 ppm but its performance appears to be inconsistent.

The third modification of the spoiler was specially designed for the unit 3 where the air register is closer to the burner face and the end of the coal nozzle is in the register area. The secondary air is not confined between the coal nozzle and the burner throat and produces a different primary and secondary mixing pattern as compared with unit 1 and 2. Test results showed that the flame was positioned at the burner throat and did not allow the four coal streams per burner to maintain their identity for controlled combustion and reduced NO_x formation.

PROCESS CODE N14–FLUE GAS RECIRCULATION BURNER

This burner is suitable for new build and retrofit applications. (As a retrofit, it can be installed at wall-fired units equipped with separate FGR, unless it is of the self recirculating type).

Information on the self recirculating type burner is limited, and status is therefore uncertain. The basic type is commercially available.

Operating experience shows [367, 352] that the dual air register burners (with overfire air) installed in the 250 MWe Takehara Thermal No. 1 boiler in Japan were retrofitted with circulated air registers. A NO_x reduction rate of 31% was achieved by conducting two stage combustion, 39% by adoption of the dual register alone, 63% by combining the two, and 75% by using two-stage combustion and additional flue gas mixing; the NO_x concentration when using only conventional circular burners was assumed to be 100%.

See also below.

PROCESS CODE N21–FLUE GAS RECIRCULATION

This system is suitable for new build and retrofit situations, particularly for utility slag tap boilers. When retrofitted, the increase in gas mass flow for a given heat output can cause boiler operating problems.

It is especially applicable to utility boilers of the slag tap type to reduce peak combustion temperature and available oxygen.

In combination with other control methods such as off stoichiometric combustion, FGR has been effective with a limited recirculation rate.

There are indications that on special types of grate firings, NO_x reductions up to 50% can be achieved by a combination of LEA and FGR, although some other authors report that NO_x emissions from spreader stoker boilers equipped with FGR were not affected. Boiler efficiency improvements can be achieved with FGR for spreader stoker boilers with high or intermediate oxygen levels.

The new 70 MW tangentially coal-fired Austrian Vortsberg III boiler has been equipped with FGR. A NO_x concentration of 600 mg/m^3 has been achieved at a FGR ratio of 15%.

FGR has been installed at an 150 MWt wet bottom boiler in Germany where a NO_x concentration of 1000 mg/m^3 base load was achieved using combination of FGR (30%) and low NO_x burner.

The FGR system requires a greater investment at existing and new facilities than other combustion modifications like low NO_x burners.

There is about 1% efficiency penalty on the recirculation fan load.

PROCESS CODE N22–STAGED COMBUSTION

Process Code N22.1–Overfire Air Injection

This system is suitable for new build and retrofit applications. (Retrofitting to an existing boiler with OFA requires additional duct work and furnace wall penetration).

It is better applied to larger units than to small units, because total burn-out of the fuel needs a longer residence time using OFA injection. The potential gas residence time tends to be shorter in small units. This effect may cause incomplete combustion; to avoid this, an increased excess air rate through the OFA ports may be required, which leads to a decreased boiler efficiency.

A NO_x emission reduction of about 30% has been reported from Western Siberian Power Plant in the USSR when OFA has been used.

Data from the Gardanne power station, France show that NO_x reduction efficiency of 30% can be achieved by injecting 20% of the total excess air through overfire air ports.

Process Code N22.2–Reburning

Mainly applicable to new units.

With existing boilers in most cases the retrofit of a reburning system may be difficult because of burnout problems.

Test results from the MACT system with low NO_x (OFA) combustion on the 1150 kg/h boiler in Japan show that a NO_x emission level of 50 ppm can be achieved [342].

The MACT concept has been implemented in a 600 MWe boiler which has been in operation since September 1983. It utilises fuel oil as the 'MACT fuel' and pulverised coal as the main fuel [342].

The % reduction across the MACT system appears to be independent of the inlet NO_x concentration. This MACT can be utilised in conjunction with low NO_x burner systems such as the PM burner [342].

The MHI IFNR system has been applied commercially to a few oil-fired boilers (e.g. the 85 MWt oil fired industrial boiler of Mitsubishi Petrochemical at Yokkaichi). NO_x emission levels were reduced from 200–250 to 110–130 ppm; tests have also been conducted with coal-firing at pilot plants [342].

There is also a KVB, Inc. version of the IFNR system which is still under development [347].

The quantity of unburnt carbon in the ash may be of concern for future developments, as it may be high enough for corrosion problems to arise in the long term. Further substantiation of the extent of carbon burnout is required [352].

Process Code N22.3–Primary Combustion Furnace

Mainly applicable to new boilers.

Test results [343, 355] from the 1.2 MWt test furnace at B & W's Alliance Research Centre, Ohio show that NO_x emissions from the PCF depend on low second stage flame front temperatures and gradual mixing of fuel and air in the second stage flame. NO_x emissions lower than 100ppm have been recorded.

A conceptual design for a 650 MWt furnace incorporating PCF has been prepared by B & W, who expect it to meet the 120 ppm (85g.NO_2/GJ) emission goal.

PROCESS CODE N23–SLAGGING COMBUSTOR

This combustor is suitable for conversion of gas or oil-fired boilers to pulverised coal firing.

Operating experience [352, 427] is limited. The goal of research is to achieve NO_x levels of 100 to 200 mg/m^3 and simultaneously reduce SO_x emission by up to 90%.

Slagging combustors are under development by several companies and research organisations in the USA.

See Section 5 for further details of this design.

PROCESS CODE N24–REDUCE HEAT IN COMBUSTION CHAMBER

This method is suitable for new build and retrofit applications.

Operating experience is various.

PROCESS CODE N25–DOWN-FIRED SEQUENTIAL AIR ADDITION BOILER

Mainly applicable to new boilers.

Operating experience shows [343] that although not commonly used today, the Foster Wheeler Energy Corporation has been marketing arch-fired furnaces with capacities as large as 350 MWt for anthracite firing in Europe.

Test results on 70, 125 and 275 MWt arch firing boilers show NO_x emissions level of about 350, 250 and 200 ppm respectively which varied by ± 10% with excess air levels. Their differences were probably due to a different firing pattern and/or conservative boiler design [343].

PROCESS CODE N31–LOW EXCESS AIR COMBUSTION

LEA can be used in combination with other combustion modifications such as staged combustion and operating at reduced load. However, requirements on combustion control increase if a combination of such measures is installed.

Operating Experience shows [352] that LEA is easy to implement, there is no need to change the construction.

The boiler efficiency may increase.

The control of the combustion process has to ensure the safe operation of the boiler with LEA. This control has also to be effective at load swings and variations of fuel quality.

The use of LEA at firing installations is limited by possible negative side-effects such as increased fouling and corrosion by the reducing atmosphere, increased CO emissions and incomplete burnout.

LEA is not in widespread use as a NO_x control technique for industrial boilers.

NO_x reductions of up to 44% and efficiency increases of up to 2.5% have been reported using LEA firing.

PROCESS CODE N32–STAGED COMBUSTION (BBF, BOOS)

Suitable for adoption on existing boilers; no equipment changes are needed.

With some installations, the use of staged combustion results in the boiler being derated, i.e. when the lower burners cannot be used to supply coal to obtain the full boiler output. In these instances, reduction of NO_x emission by the use of OSC may not be a tenable solution.

A 30–40% reduction in NO_x emission can be obtained by the use of BOOS.

PROCESS CODE N33–REDUCED AIR PREHEAT (RAP)

Suitable only for new small units designed for low feed-water inlet temperatures, in which safe ignition conditions can be provided.

RAP appears relatively inefficient in suppressing fuel nitrogen conversion.

It can cause significant reductions in thermal efficiency, if no counter measures (i.e. enlarging the economiser) are applied.

PROCESS CODE N34–REDUCE HEAT LOAD IN COMBUSTION CHAMBER

Suitable for new build and retrofit applications.

Operating experience is various.

3.4 Comparison and Selection of Techniques for Detailed Study

COMPARISON OF INTERNALLY STAGED COMBUSTION BURNERS

For the internally-staged combustion burner, a variety of mechanical configurations is possible for controlling the mixing of fuel and air to reduce NO_x formation in a pulverised coal-fired boiler. Burner manufacturers employ different approaches in hardware and design philosophy for different types of boilers. They are all using the same principles of dual register to control the secondary air flow to the flame.

A comparison of the performance of each burner is shown in Table 3.2. The Riley Stoker Controlled Combustion Venturi Burner employs a four-bladed coal spreader to create an inner fuel-rich and outer fuel-lean zone in the flame. The Foster Wheeler Controlled Flow-Split Flame combines a split flame annular coal nozzle with dual air registers. The Babcock & Wilcox Dual Register relies on dual air registers and the U.S. EPA's Distributed Mixing Burner couples a low-NO_x burner with outboard air ports in the furnace wall.

There are variations of FW and B&W type burners, each of which has operating experience in large utility boilers. Thus they are chosen for further evaluation study in Section 3.5. The Riley Stoker CCV is a relatively more recently developed burner compared to FW DF/SF and BW D-R burners. From the experiences of the 2 utility boilers retrofitted with the CCV burner, the reduction in NO_x emission level is comparable with, if not greater than, the FW DF/SF or BW-DR burners.

The DMB was initially developed by the International Flame Research Foundation (IFRF) and Steinmuller. It was subsequently incorporated into the FW CF or Riley Stoker CCV burner designs. Air flows are adjusted so that primary air, coal and inner secondary air create a very fuel rich zone (40% of theoretical stoichiometry) encouraging fuel nitrogen

Table 3.2 Comparison of Internally-Staged Burners

Burner Type	Foster Wheeler Corp. Controlled Flow–Split Flame	Babcock & Wilcox Dual Register	Riley Stoker Corp. Controled Combustion Venturi	US EPA Distributed Mixing Burner
Process Code	N11.1	N11.2	N11.3	N11.4
Techniques (all use dual air registers principle).	Combine a split-flame annular coal nozzle with dual air register.	Relies on dual air Registers.	Employs a four-bladed coal spreader.	Couples a low NO_x burner with out-board air ports on the furnace wall.
NO_x level	150–200 ppm	200–300 ppm	250 ppm	Predicted 120 ppm is possible 200–300 ppm (tested).
Used in commercial plants	a) 350 & 525 MWe San Juan utility boiler U.S.A. b) 500 MWe Eggborough utility boiler U.K. c) 2 × 265 MWe Isogo plant in Japan*. d) 626 MWe Alegheny Power Service Corp, U.S.A.*	a) 700 MWe boiler 3 Takehara, Japan* b) 250 MWe boiler 1, Takehara, Japan c) 60 units of various size (65 MWt to 1300 MWt) in U.S.A.	a) 360 & 400 MWe utility boilers in the U.S.A.	a) 700 MWe boiler in Germany. b) 20 MWe WIPCO boiler in U.S.A.* c) 29 MWt simulator furnace California.+
Other versions	*Dual flow–Convergent fuel nozzle (DF–CN) developed by FW/IHI	*Primary Gas (PG-DRB)		*Incorporate FW's proprietary controlled flow +Riley Stoker type
Applicability (all applicable to new unit)	Flame length is not affected. Thus suitable for retrofitting	Elongated flame due to its delayed mixing combustion. Thus might cause flame impringement on furnace wall on existing furnace design	Applicable to retrofitting	Retrofitting may be a problem due to space requirement for tertiary air ports

* Alternative versions of burner installed
\+ (see above)

to enter the gas phase. Addition of the outer secondary air raises the stoichiometry to 70% of theoretical to maximise conversion of bound nitrogen to N_2. Tertiary air raises the overall stoichiometry to 120% (an overall oxidising environment is maintained in the furnace to minimise slagging and corrosion). Initial pilot tests showed that emissions were strongly dependent on the overall stoichiometry of primary and secondary air, though flame instability resulted if the stoichiometry was too low.

Field evaluation of the DMB showed NO_x emission levels of 200–300 ppm (though, it is thought that a 120ppm NO_x emission level with this burner is achievable). Various modifications are necessary in an existing boiler to accommodate this burner, so the preferred DMB application is in a new boiler. Since this DMB is still under development and literature is limited, it will not be further evaluated at the present time.

Finally, uncertainties remain over some aspects of low NO_x burner design. Retrofitting of a low NO_x burner to a tightly designed boiler, with a high heat release rate, may lead to problems of flame impingement on the furnace walls and a consequent risk of increased corrosion by sulphide formation. Increased corrosion is again the main worry if the coal has a high chloride content.

COMPARISON OF EXTERNALLY STAGED COMBUSTION BURNERS

The performance of various externally staged combustion techniques (mainly for tangential firing systems) is compared in Table 3.3. Each system is believed to have a maximum percent NO_x reduction which represents the maximum technical and economic limits for that technology.

Overfire Air (OFA) is regarded as the first step for units that do not already have it as part of the original design. For most fuels and with existing furnace configurations, OFA can achieve an 18% to 30% reduction in NO_x emission.

OFA does a poor job of controlling NO_x formed early in the combustion process, because the 15% to 20% of the total air that goes to the OFA is a very small part of the O_2 available to the volatile matter being burned and fuel nitrogen being released.

Thus, to control the all-important fuel conversion, large quantities of air must be withheld from the fuel for the duration of the devolatilisation and small char particle combustion process. By doing so, the system is simultaneously deprived of a sufficient quantity of O_2 to oxidise nitrogen to NO. One way of accomplishing this is through a concept called the Low NO_x Concentric Firing System (LNCFS). To avoid the two jets (fuel and fuel air and secondary air) from mixing, they are pointed in diverging directions, forming two concentric circles. Like OFA, air is effectively withheld from the fuel; but unlike OFA, LNCFS affects the very early stoichiometry of the fuel. A 20% reduction in NO_x from the standard tangential firing mode with OFA has been demonstrated in a 400 MWe utility boiler, and the effect of LNCFS on carbon loss is insignificant.

The goal of the Pollution Minimum (PM) burner is the same as the LNCFS which is to reduce NO_x by reducing the local stoichiometry (available O_2) around the coal particle early in the combustion process. However, it utilises a novel concept of dividing the coal stream into fuel-lean and fuel-rich jets in-between a flue gas recirculation jet. Effectively it is an optimisation of both the OFA and BOOS operation within one design (as described in Section 3.2). The PM burner is regarded as part of a family of ultra-low-NO_x burners where laboratory and field test results have shown superior performances to LNCFS. However, there may be doubts about retrofitability due to space requirements.

Table 3.3 Comparison of Four Externally-Staged Burners

Burner Type	Tangential Firing with OFA	LNCFS (C.E.)	PM Burner (MHI)	Fuel Staging or IFNR
Process Code	N12.1	N12.2	N12.3	N12.4/22.2
Techniques (all use tangential firing system).	Use of stoichiometry or staged combustion.	Diverts some combustion air along the furnace walls away from the flame.	Creates alternating fuel-rich and fuel-lean flames.	Both air and fuel are staged with final reburning inside furnace.
NO_x level achievable.	250–300 ppm	250–150 ppm	200–120 ppm	120 ppm
Used in commercial plants.	Various	400 MWe Utah utility Boiler 61 MWe utility boiler	350 MWe Boiler in Japan Industrial Boiler of Daishowa Paper Co.	15 MWt Test furnace Test in 600 MWe utility boiler in Japan
Applicability (all applicable to new unit)	Applicable to retrofitting	Applicable to retrofitting	More suitable to new unit	More suitable to new unit

In fact, no matter how closely the combustion process is controlled, the prerequisites for completing the combustion process (temperature, time and excess oxygen) will be sufficient to generate at least some NO_x. Therefore, there is always a minimal value below which further reduction in NO_x is impractical. Thus, for very low NO_x requirements, techniques have been developed actually to destroy NO_x already formed by the combustion process. One such IFNR (In Furnace NO_x Reduction) or reburning technique is the MHI MACT (Mitsubishi Advanced Combustion Technology). This technique consists of a second combustion zone in the boiler. Test results showed a net reduction of approximately 50% of all the NO_x entering the MACT process.

The IFNR technique seems very encouraging, but, further development is required especially for coal burning units. This system is probably not completely applicable to retrofits and cannot yet be regarded as proven technology in EEC countries. Nevertheless, due to its very promising results, it is chosen for further studies in this Manual.

COMPARISON OF FURNACE DESIGN OR MODIFICATION TECHNIQUES

Table 3.4 shows a comparison of furnace design or modification techniques. The B & W's Primary Combustion Furnace (PCF) is an extension of staged combustion technology. The substoichiometric gaseous mixture is isolated physically within its own furnace, limiting potential corrosion problems to a relatively small area of the boiler which may be constructed of corrosion resistant materials or as a replaceable item. NO_x emissions lower than 100 ppm have been recorded from a 1.2 MWt pilot plant [343]. This technology is still under development.

Similarly, the slagging combustor is still under development. One aim of the research is to investigate a burner which allows the conversion of gas- and oil-fired boilers to pulverised coal firing with minimal derating, using a burner concept able to control SO_2 and NO_x simultaneously, and permit the removal of fly ash (see Section 5). It is expected that NO_x levels of 50–100 ppm can be achieved.

The Flue Gas Recirculation (FGR) technique is of relatively limited effectiveness, particularly as an energy penalty (for the fans) is usually incurred. Partial FGR may be incorporated in other forms of combustion technique such as in the LNB and in the Mitsubishi PM system.

The OFA technique is very effective for NO_x reduction and may be used with most fuels and most combustion systems. It is particularly suitable for tangentially-fired boilers, and is usually incorporated with LNB or other new tangential firing systems (e.g. LNCFS). The following operational problems may appear:

Table 3.4 Comparison of Furnace Design or Modification Techniques

Furnace Design or Modification	FGR	OFA	Reburning	Primary Combustion Furnace	Slagging Combustor	Reduce Heat in Furnace	Arch Firing
Process Code	N21	N22.1	N22.2	N22.3	N23	N24	N25
Techniques	Low combustion temp. Low O_2 concentration.	2 stages combustion.	3 stages combustion.	Water cooled primary combustion chamber.	High temp. primary slagging combustion.	Low volumetric heating rate.	Down-fired sequential air addition.
NO_x level achievable (or % reduction).	up to 15%	10–35%	120 ppm	100 ppm	50–100 ppm	(20–25%)	200 ppm
Problem(s)	– Cost up. – Expansion of flame. – Energy penalty (for the fan)	– Incomplete combustion – Extension of flame. – Tube erosion			High furnace exit gas temperature may be unacceptable	Possible derating of boiler	Very limited application (only applicable to DSAA boiler)
Application to existing furnace	Possible	Possible	Not easy	Not easy	Convert oil gas fired to pulverised coal fired	More applicable to new boiler	No
Operating experience	Various	Various	15 MWt test furnace	1.2 MWt pilot plant	30 MWt plant	Various	275 MWt DSAA boiler
Status	In operation	In operation	Under development	Under development	Under development	In operation	Under review

increased potential for boiler tube erosion due to local reducing conditions.

tendency for slag accumulation in the furnace.

decrease in efficiency of combustion.

Most of the techniques in this category are either under development or are applicable only to boiler designs which are not in common use today. Therefore, with the exception of reburning, these techniques will not be further evaluated. Both FGR and OFA are usually incorporated with other forms of combustion technique, since there are limitations on the percentage NO_x reduction that can be achieved by FGR or OFA alone.

Both Biased Burner Firing (BBF) and Burners Out Of Service (BOOS) techniques use the principle of OSC or staged combustion. They are usually applicable to boilers with multiple rows of burners. The top rows of burners may be operated under fuel-lean (BBF) or air-alone (BOOS) conditions. The advantage is that no modifications are necessary to the furnace water wall.

Most of these techniques (if being applied alone) are usually a temporary measure to reduce the NO_x emission levels on an existing furnace. They can be used in combination with other combustion modifications to provide a further reduction in NO_x emission levels, although the requirements for combustion control would also increase.

COMPARISON OF FURNACE OPERATION TECHNIQUES

Table 3.5 shows a comparison of furnace operation techniques. Low excess air firing causes a reduction of the local flame zone oxygen concentration by which NO_x formation is reduced. The use of the LEA firing is a simple, feasible and effective technique. There may be negative side effects such as increased fouling and corrosion by the reducing atmosphere, increased CO emissions and incomplete burnout.

Reduced air preheat operations can cause significant reductions in thermal efficiency and fuel penalties, if no counter measures (e.g. enlarging the economiser) are applied, and its applicability is limited with respect to ensuring safe ignition conditions.

REVIEW OF COMBINATION OF COMBUSTION TECHNIQUES

Viewing the $DeNO_x$ combustion techniques as a whole, they can be classified into LNB, OSC, FGR, LVHR and reburning processes as shown in Table 3.6. Various combinations of the processes are usually employed to achieve a lower NO_x emission level. The Electric Power Development Company (EPDC) of Japan has conducted extensive tests of combustion modifications for NO_x abatement, jointly with MHI, BHK, IHI, and KHI. Test results shown in Figure 3.30 indicate that, by using a combination of LNB with staged combustion (OFA) and FGR, NO_x can be reduced by 60–80% (to 100–150 ppm with low nitrogen coal and to 150–200 ppm with nitrogen-rich coal).

Table 3.5 Comparison of Furnace Operation Techniques

Furnace Operation	LEA	Biased Firing	BOOS	RAP	Derating
Process Code	N31	N32.1	N32.2	N33	N34
Technique	Low excess air combustion. This reduces O_2 concentration.	Upper row operates fuel-lean and lower row operates fuel-rich.	Upper row burners are out of service injecting only air.	Lowering of air pre-heating temp. Thus lowering combustion temp.	Reduction of heat load in furnace. Thus lowering combustion temp.
NO_x level achievable (or % reduction)	40%	30–40%	30–40%	20–30%	20–30%
Reduction efficiency on Thermal NO_x	*	o	o	*	*
Fuel NO_x	o	o	o	x	x
Problem	Unburnt carbon CO emissions fouling & corrosion		Derating of unit	Reduction in thermal efficiency and fuel penalties	
Application to existing furnace.	Easy	Applicable to burners with multiple array.	Applicable to burners with multiple array.	Easy	Possible

* Effective o Some effect x Little or no effect

Table 3.6 Comparison of Performances of Combustion Modifications

Measure	Reduction in NO_x Emission (%)	Applicability		Side-Effects/Restrictions
		New	Retrofit	
Low NO_x Burner (LNB)	20–50	YES	YES	– Possible increase of CO in flue gas and unburnt carbon in fly ash.
Reburning	30–50	YES	Site specific	– Not enough full scale experience using coal as secondary fuel for reburning. – Evaporation process of the boiler may be influenced. – Limited by the potential of corrosion by reducing atmosphere and increase of CO in flue gas and unburnt carbon in fly ash.
Off-Stoichiometric Combustion (OSC)	10–40	YES	Site specific	– Limited by the potential of corrosion by reducing atmosphere and increase in CO in flue gas and unburnt carbon in fly ash. – Extension of flame.
Low Volume Heating Rate (LVHR)	20–25	YES	NO	– Less slagging and fouling.
Flue Gas Recirculation (FGR)	up to 15	YES	Site specific	– Evaporation process of the boiler may be influenced. – Stability of ignition is a limiting factor. – Possible increase of CO in flue gas and near boiler walls causing higher corrosion potential.
LVHR+LNB	36–62			
LVHR+LNB+FGR	36–67			
LVHR+LNB+OSC	42–77			
LVHR+LNB+REBURNING	55–81			

3.5 Evaluation of Selected Techniques

FW DR Burners

The first FW LNB was the controlled flow (CF) burner as shown in Figure 3.31. Emissions of NO_x from this burner are typically 40 to 50% lower than those from the original high turbulent intervane burner. The split flame (CF/SF) burner was then developed and a further 10 to 20% NO_x reduction was achieved. A modified version called Dual Flow Convergent Fuel (DF/CN), Figure 3.33, was developed with IHI. It was fitted to an 265 MWe Isogo plant and has achieved a NO_x level of 170 ppm. The concept of a fuel staging burner may be incorporated into the CF/SF burner (see Figure 3.36), although its prime purpose is to switch more from oil to coal burning during low load and start-up. A further reduction on NO_x emission levels may be achieved.

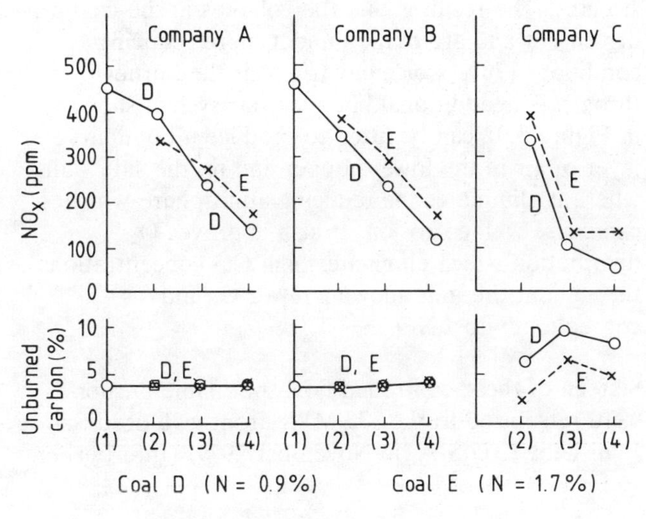

Figure 3.30 Examples of Combustion Modification Tests Conducted by EPDC and Boiler Manufacturers (MHI, BHK, IHI, KHI) [367]

The CF burner and the CF/SF burner are similar in design and operation except for the split flame nozzle. NO_x emission is reduced by controlling the fuel/air distributions within the flame while the throat stoichiometry is maintained at normal levels. Delayed combustion NO_x burners, in contrast, produce a long flame as do externally-staged burners.

There are two series secondary air flow registers for flame control. The outer register is electrically driven and controls the overall flame shape. The manually adjustable inner register controls the position of the throat and the air/fuel mixture in the sub-stoichiometric region of the flame. The movable sleeve allows controlling and optimising of the air flow without changing the registers. The perforated plate performs two functions: by measuring the pressure drop, the air flow can be monitored; it also distributes the air equally around the perimeter of the

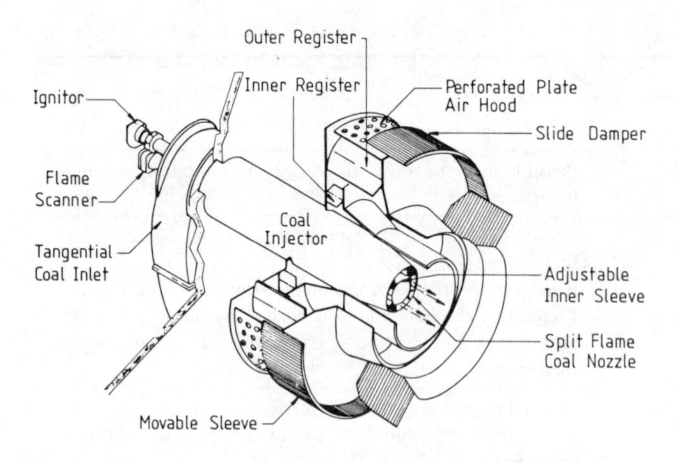

Figure 3.31 F.W. Controlled flow Low NO_x Burner [368]

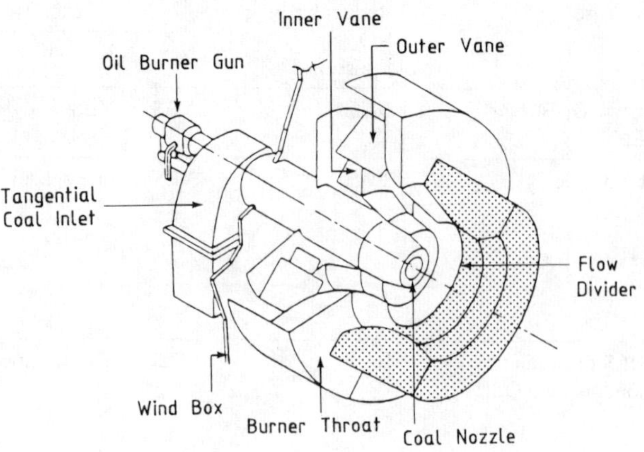

Figure 3.33 IHI–FW DF Burner [348]

air register. The burner is equipped with a variable velocity coal nozzle. The velocity of primary air and coal stream can be optimised with respect to the velocity of secondary air stream. This, in combination with the inner registers, forms the mechanism which allows adjustment of the fuel/air distribution within a burner.

The additional component to the CF/SF burner is the split flame nozzle which forms four distinct coal streams. The result is that the volatiles in the coal are driven out and are burned under more reducing conditions (70% stoichiometric near the burner throat). The additional boundary air system shown in Figure 3.32 can be fitted to produce a boundary layer of air in the lower hopper and up the side walls where it eliminates the reducing atmosphere which can cause wall corrosion. It also improves O_2 distribution which eliminates high CO concentrations throughout the unit allowing lower O_2 and consequently low NO_x.

Sixteen of these controlled flow-split flame burners were retrofitted to the 375 MWe front-wall-fired San Juan Unit 1, U.S.A. in November 1979. (Interburner spacings were 3.3m vertically and 3.0m horizontally. Each burner has a capacity of 83.5 MWt max.) With these burners NO_x levels have been held to about 170g/GJ. Full operating experience of this burner has been outlined in Section 3.3 and is summarised in Figures 3.35 and 3.36.

Another fuel staging burner similar to Figure 3.38 is under development. The concept (see Figure 3.37) requires a small primary coal flame within the main coal flame that is operated sub-stoichiometrically. The devolatilised carbon from the primary flame passes out through the main flame providing additional NO_x reduction sites in the main flame. Owing to the high rise in oil prices during the 1970's, the fuel staging burner was modified to the so-called low load burner system. It is intended to eliminate 95% of the oil used during the start-up and cycling of a coal-fired unit. The low load burner allows operation down to 10% of fuel load without a

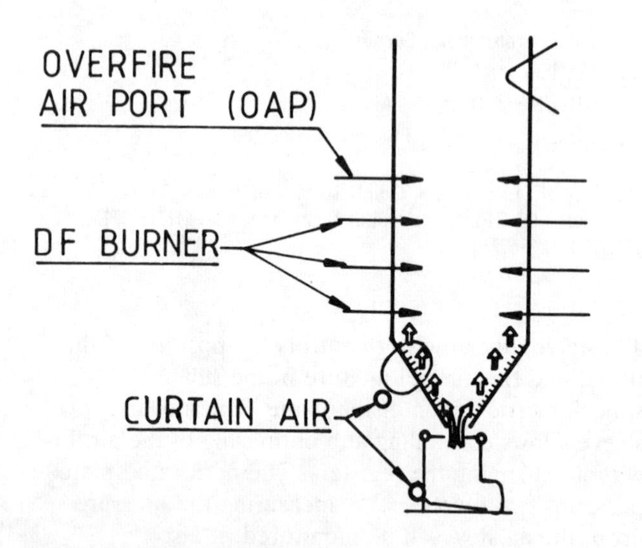

Figure 3.32 IHI's Boundary Air System [367]

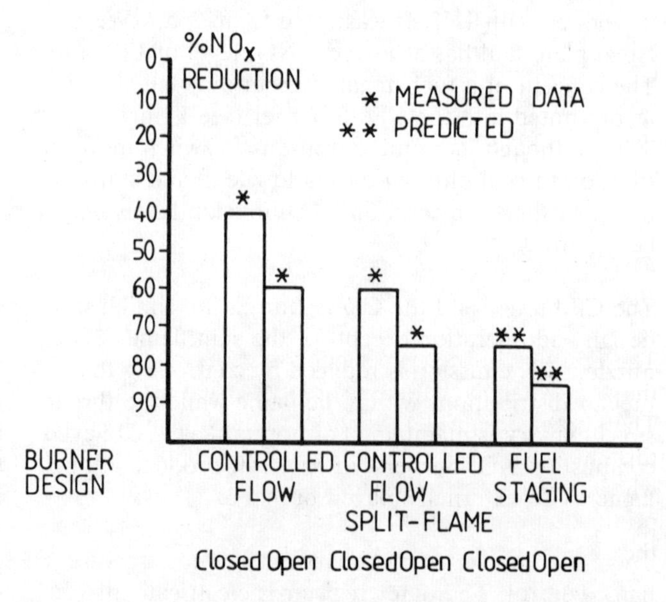

Figure 3.34 NO_x Reduction Comparison of FW Burners [368]

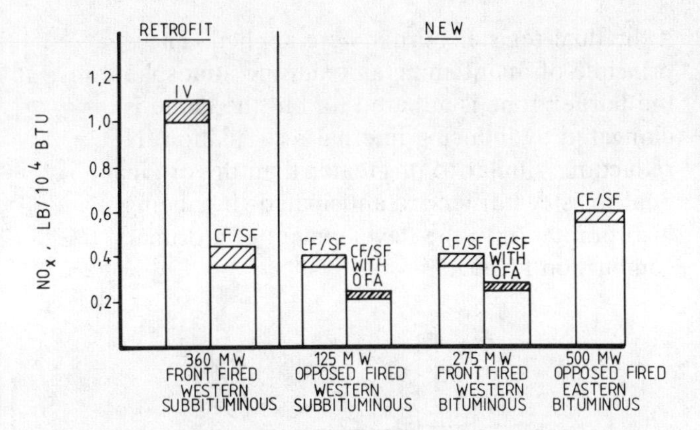

Figure 3.35 Utility Boiler NO_x Reduction [368]

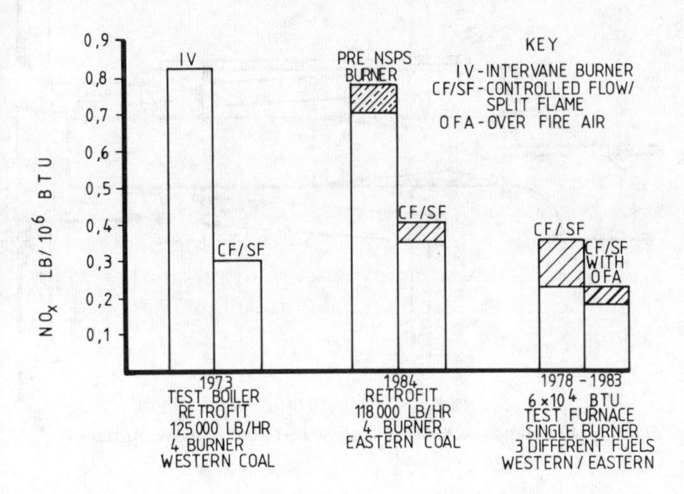

Figure 3.36 Industrial Scale Low NO_x Burner Retrofits [368]

stabilising oil flame. The low load burner can be integrated into the CF/SF burner for low NO_x operation. The low load burner with its swirler is located entirely inside the split flame coal nozzle. Centralised in the swirler is the low load coal nozzle, the entire arrangement produces the flame within the flame concept of the fuel staging burner. The burner (without split flame coal nozzle) has been used at Gibson Station in the U.S.A. The combination of CF/SF low NO_x burner and the low load burner must be optimised for low NO_x operation. Figure 3.34 shows the potential NO_x reduction for the combination with respect to existing burners.

FW/IHI DF CN (Dual Flow Convergent Fuel Nozzle) Burner

The DF type burner is designed to reduce NO_x, using the aerodynamic characteristics of swirled air flow. The burner design developed permits sufficient turbulence in the throat to maintain stable ignition which controls the mixing of fuel and air while limiting production of unburned matter. Figure 3.33 shows the general configuration of the burner.

The secondary air providing most of the total combustion air is divided into outer and inner flow by an annular Flow Divider located between the Coal Nozzle and throat ring. The inner swirled air is supplied mainly to the burner axis zone, where fuel rich combustion takes place, and flame stabilisation is achieved. The outer swirled air is supplied along the periphery of the Burner Throat and controls the char burnout rate, while strengthening the external recirculation flow of combustion products.

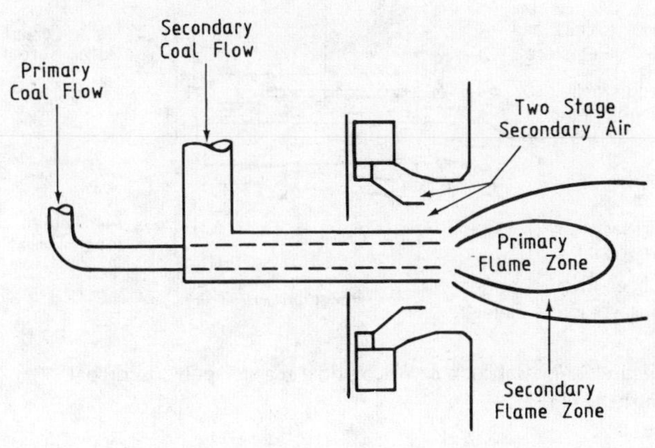

Figure 3.37 Fuel-staging Low NO_x Burner Concept [368]

The NO_x reduction principles with the DF type burner are:

(1) Minimum primary/secondary air mixing prior to completion of the combustion of the volatile fraction (fuel-rich combustion at the flame centre zone).

(2) Ignition stability at the injector.

(3) Dilution of the secondary air with self-recirculating combustion products. Thus the DF burner can be effective for both thermal NO_x and fuel NO_x.

The DF-CN burner and a boundary air system (Figure 3.32) has been used at the Isogo Station of EPDC, Japan for two 265 MWe boilers (built in 1967 and 1969) using low nitrogen coal (about 1.0%N) rich in volatile matter (about 40%). The boundary air creates a layer of an oxidising atmosphere close to the lower furnace walls to prevent slag formation or accumulation and possible corrosion.

Operating experience on this plant is as follows:

Date		*Achieved NO_x Level (ppm)*	*Modification*
Before	1973	570	None
After	1973	380–510	OFA
	1977	240–250	DF-CN + OFA
	1979	170–190	DF-CN + more OFA (20%)

B & W Delayed Mixing Type Dual-Register Burner

The B & W dual-register burner (Figure 3.38) was developed as a replacement for the cell burner. The burner design was first demonstrated as a burner retrofit at the Gaston Steam Plant in the U.S.A. where a reduction in NO_x emissions of 45% was attained.

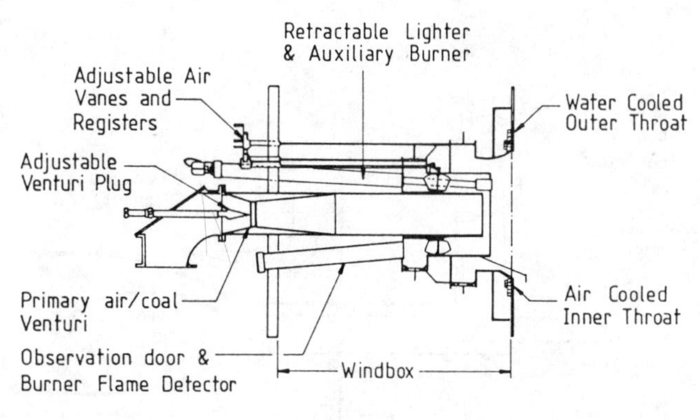

Figure 3.38 Babcock & Wilcox dual register pulverised coal burner [343, 352]

The B & W PG (Primary Gas) DRB (Figure 3.39) uses a combination of B & W dual-register LNB, two-stage combustion and flue gas recirculation. It has a primary gas port for injection of recirculated flue gas between the secondary air and the flame formed by coal and the primary air. The radially swirled combustion air is added in stages around a central primary combustion zone. Primary combustion occurs sub-stoichiometrically with the addition of primary gas (PG), and peak temperatures are reduced and the high temperature zone shortened. It has been fitted in the 700 MWe Takehara plant in Japan. Emission levels of approximately 150 ppm have been achieved.

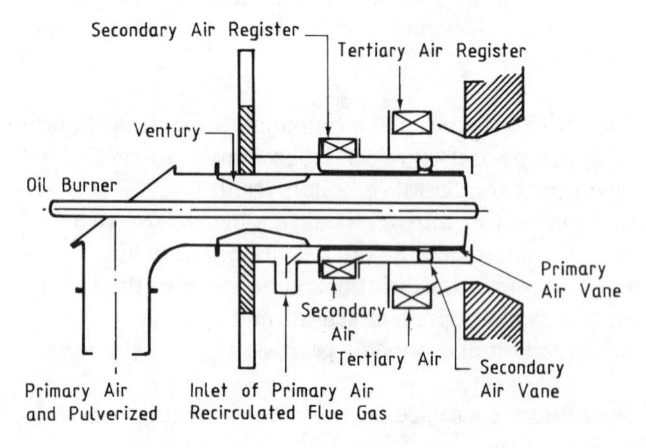

Figure 3.39 Babcock & Wilcox PG-Dual air register burner [352]

An advanced B & W dual-register burner (Figure 3.40) was then developed. Its major advantage is that modifications to pressure parts necessary to increase the burner throat size to reduce secondary air velocities, are not as extensive as would be required if the dual-register burner were applied. The principle of maintaining an oxidising atmosphere in the burner zone is adhered to, but the flame is elongated by inducing internal recirculation. NO_x reductions similar to or greater than the original dual-register burner are anticipated. It is being incorporated into the development of a primary combustion furnace.

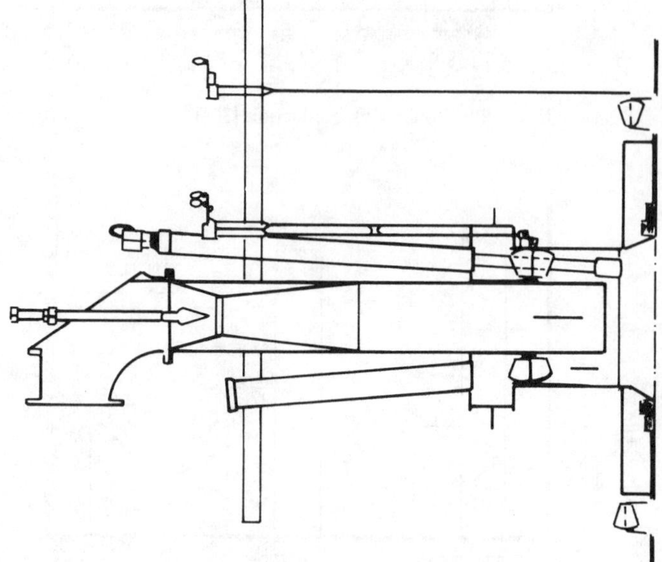

Figure 3.40 Babcock & Wilcox advanced dual register burner [343]

Babcock-Hitachi H.K. in Japan have modified the dual-register burner to amplify its staging effect (Figure 3.41). Greater internal staging is achieved with an air separation plate that extends the secondary air sleeve into the burner throat. Outer tertiary air is directed away from the core of the flame. This sharply reduces the stoichiometry of the flame base during the coal devolatilisation stage, inhibiting NO_x formation. This new burner is also called the high temperature NO_x reduction (HT-NR) burner since one of the features of this burner is to reduce NO_x efficiently by keeping a high fuel-rich flame. It was developed in conjunction with the IFNR technology (see below).

It was claimed that it will reduce NO_x by 50% beyond what is achievable with conventional low NO_x burners, and reduce slagging in the furnace as well. The burners have yet to be demonstrated in full-scale utility applications.

In-Furnace NO_x Reduction (IFNR) [347, 368, 349, 378, 342, 352]

IFNR is one of the new low NO_x combustion technologies with multi-burners. Full demonstrations on actual boilers (both gas-fired and oil-fired) have been performed satisfactorily, but very little test data on full-scale coal-fired boilers are available at the

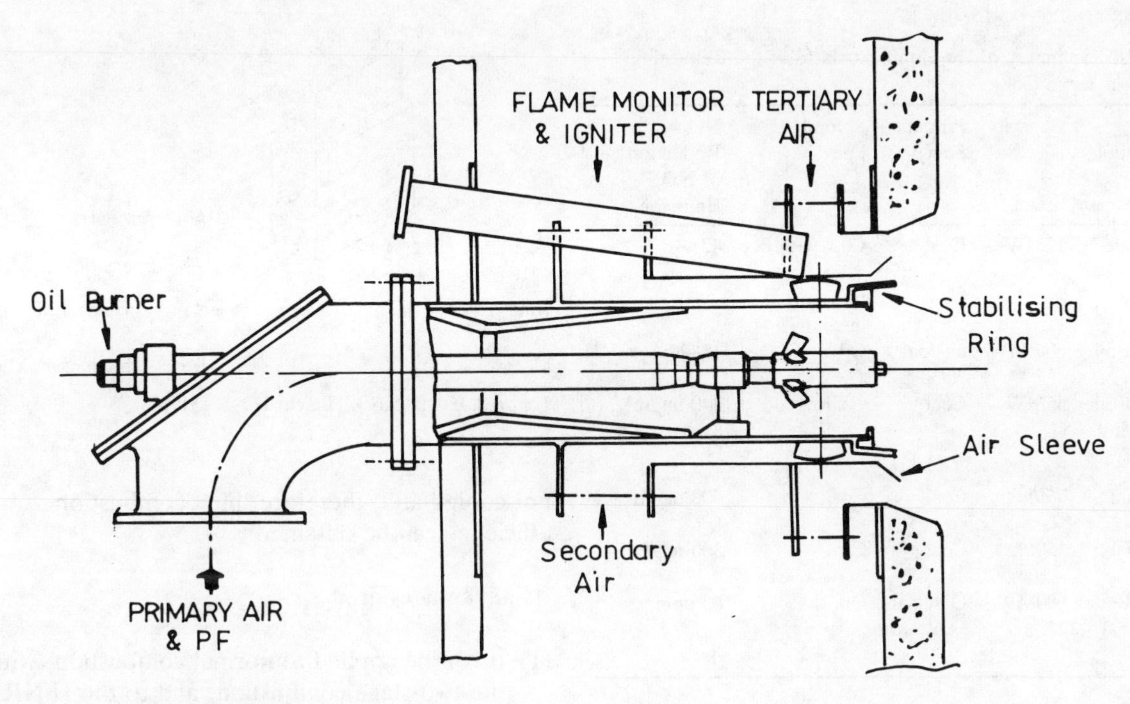

Figure 3.41 Babcock Hitachi HT-NR Burner

present moment. This system is the three-stage combustion with a NO_x decomposition process. (It has been known since the 1970's that the decomposition reaction proceeds in the presence of hydrocarbon in its flame. In particular, this reaction is remarkable in fuel-rich flames) [352].

The expected NO_x emissions in Japan using the reburning process plus LNB on a coal-fired boiler are 120–300 mg/m^3, compared to 200–400 mg/m^3 with LNB alone. The results of field tests conducted on two Japanese boilers suggest that: reburning is effective over a wide load range of 30% to 100% of the nominal; no damage to boilers or to subsequent equipment occurs; boiler efficiency is kept as high as in ordinary operation; and steam temperature and flue gas temperature remain unaffected.

Reburning tests were carried out at a 55 MWt pf front-firing boiler. As shown in Figure 3.42, the

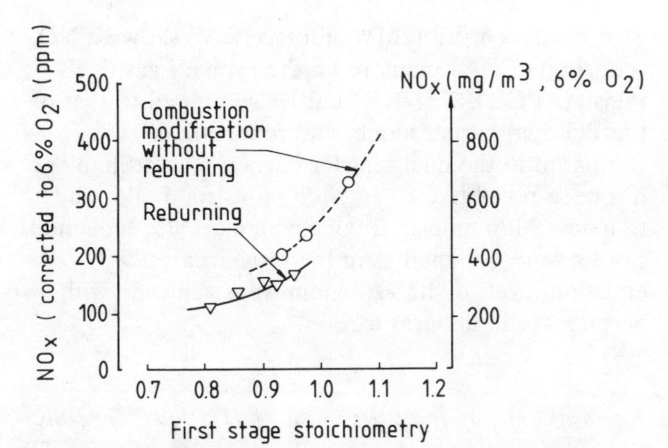

Figure 3.42 Relation between NO_x and first stage stoichiometry with 10% reburning [352]

reburning process results in NO_x emission levels that are lower than with conventional firing, falling with decreasing first stage stoichiometry. Reburning also appeared effective in increasing burnout, as shown in Figure 3.43.

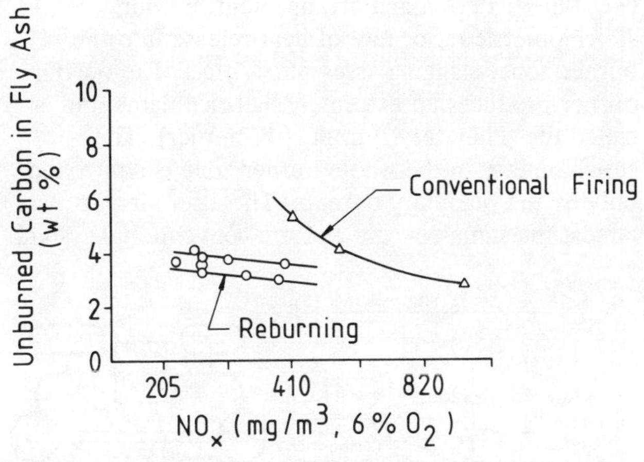

Figure 3.43 Relation between NO_x and unburnt carbon in fly ash with and without reburning [352]

Two boilers (75t/h and 150t/h), which use the reburning concept firing coal as primary and secondary fuel, were built in 1984 in Japan. The guaranteed NO_x emission level of 400 mg/m^3 (at the 75t/h boiler) is achieved at a primary air ratio of 1.0. At the 150t/h boiler the guaranteed NO_x emission level of 380 mg/m^3 is achieved at a primary air ratio of 1.15 and 300 mg/m^3 at 1.1.

The development status of the reburning process can be seen in Table 3.7.

Table 3.7 Summary of Development status of the Reburning Process in Japan [352]

Unit Description	Capacity	Primary Fuel	Secondary Fuel	NO_x Reduction or NO_x Emission
Pilot scale test (1 burner)	12MW	Coal	Coal with flue gas as transport medium	43%
Utility Boiler	600 MW	oil/coal 70/30%	Oil	15–20%
Utility Boiler front firing system	55MW	Coal	Coal	240 mg/m^3
Industrial Boiler	75t/h	Coal	Coal	400 mg/m^3
Industrial Boiler	150t/h	Coal	Coal	380 mg/m^3
Utility Boiler Tangential Firing	600MW	Oil/Coal	Oil	60 ppm

There are various IFNR techniques developed by various companies using the same fundamental concept.

Babcock & Hitachi IFNR System

Figures 3.44 and 3.45 show a schematic view of the Babcock Hitachi (BHK) IFNR technique. The main burner design is the same as for conventional systems (i.e. burner design and arrangement, burner stoichiometric ratio, rate of heat release into the burner zone, slagging assessment, etc.). The deNO_x burner produces an extremely fuel-rich flame and is called the 'Planetary Burner' ('P' burner). The stoichiometry in the whole burner zone is controlled only by the planetary burners. The after air port uses almost the same concept as for a conventional system.

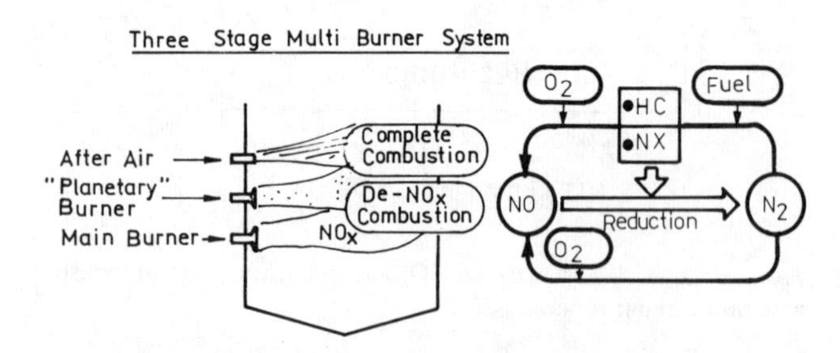

Figure 3.44 BHI In-Furnace NO_x Reduction Technique [349]

At present, BHK is developing a new extremely low NO_x coal fired burner. The concept of IFNR with multi-burners as mentioned before, is packed into the new extremely low NO_x pulverised coal burner itself. It is called the High Temperature NO_x Reductions (HT-NR) burner. Figure 3.46 shows the flame structure of the HT-NR burner. The feature of this burner is to reduce NO_x efficiently by keeping a high temperature fuel-rich flame (zone D). The flame is

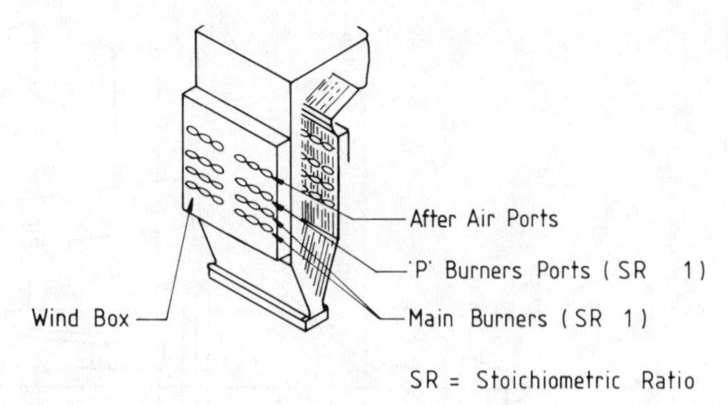

Figure 3.45 Babcock Hitachi IFNR [349]

not cooled and, therefore, high combustion efficiency can be obtained.

Its advantages are:

(1) It can be applied to normal combustion, without the two-stage combustion, and to the IFNR.

(2) It can reduce the corrosion of the exterior surface of furnace waterwalls, because the atmosphere on the wall near the burners is kept in an oxidised condition.

(3) The flue gas recirculation through the burners can be reduced.

(4) It can be interchanged with conventional burners.

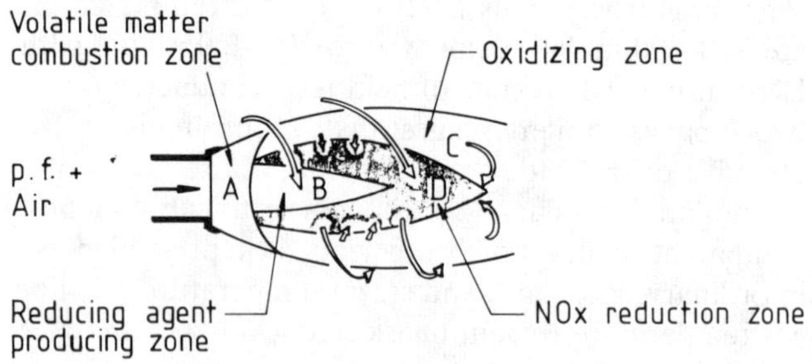

Figure 3.46 New Extremely Low NO_x pf Burner concept [349, 378]

Test results on the 12MWt furnace have shown a NO_x reduction of 50% or more on the primary gas dual register (PG-DR/HT-NR) burner compared to that of the PG-dual burner alone, and reduction of 80% compared to the dual register burner. This technology has been retrofitted to an 80t/h industrial boiler and to a new 90t/h industrial boiler. Similar NO_x reduction results were obtained as in the test furnace. NO_x emission levels of 100–200 ppm were achieved with no increase in unburnt carbon.

Kawasaki Heavy Industries Ltd. (KHI) KVC Systems

The KHI's IFNR version is called KVC (Kawasaki Volume Combustion System). The conceptual

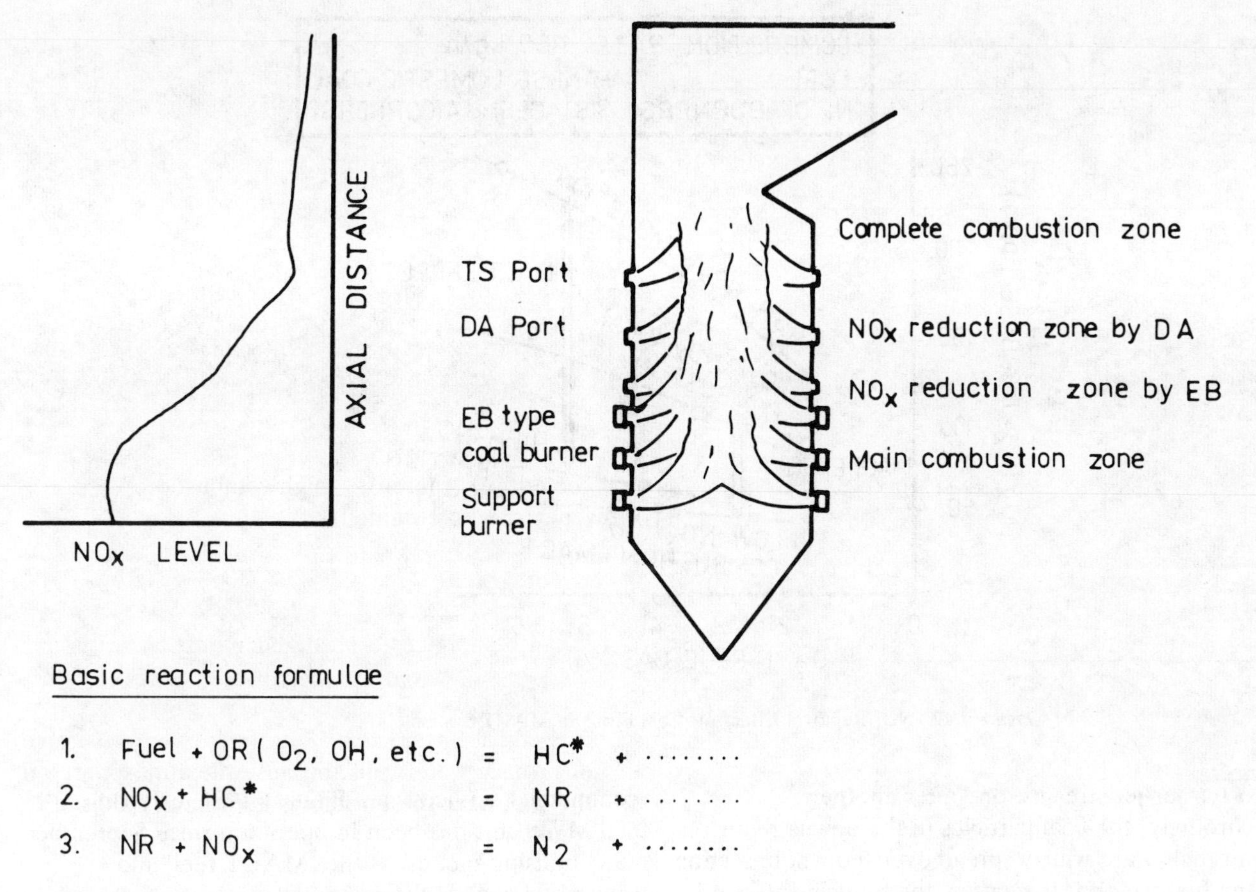

Figure 3.47 Concept of KVC System [347]

combustion process model is presented together with the NO_x generating mode, as is illustrated in Figure 3.47. This combustion zone is divided into three stages:

(1) Lowest support burner region in which a specific amount of pulverised coal is burning, as a heat shield, with a low-NO_x burner to maintain the downstream gas temperature at an appropriate level for the stable combustion.

(2) Intermediate NO_x decomposition region in which EB (Each-Air-Bias) burners in multiple rows are properly arranged at every interval between them and burner jets, generating the reducing flames which repeatedly decompose the NO_x, once formed in the upstream region, to molecular nitrogen gas due to reducing intermediate products.

(3) Upper burnout region in which the remaining combustibles are finally burned out with TS-air injected from TS-air ports.

The EB burner is a pair of a pulverised coal-fired burners with spinning vanes to impart swirl for controlling the coal-air mixing, and an EB air nozzle to bias oxygen gas distribution along the flame front, as shown in Figure 3.48. Thus, the coal particles are fired under the partial combustion condition that

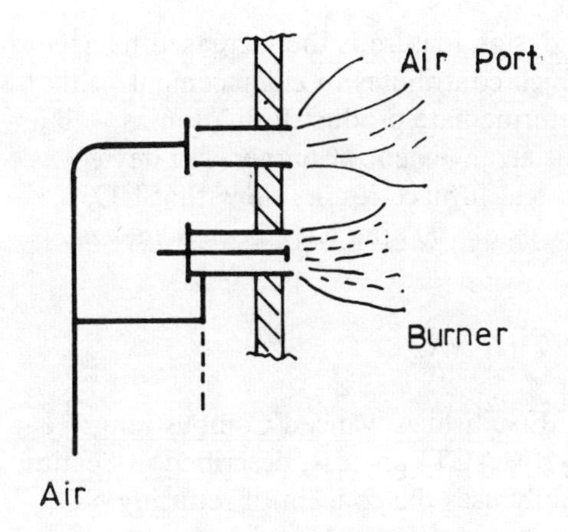

Figure 3.48 E.B. Burner and Each-Air-Bias System [347]

generates the reducing flame in this region, which is termed the Each-Air-Bias (EB) system.
Furthermore, the stoichiometric ratio is altered as a whole along the axial distance of the furnace in which the lower volatile phase combustion is in the oxidising condition; the reducing flame zone is fuel-rich and the burnout-zone is air-rich by a controlling means of Total-Air-Bias (TB) system.

In this process, the NO_x (once formed) passes through the next burning flame region, where the NO_x is mixed with the reducing intermediate products of activated hydrocarbons in the flame fronts of the

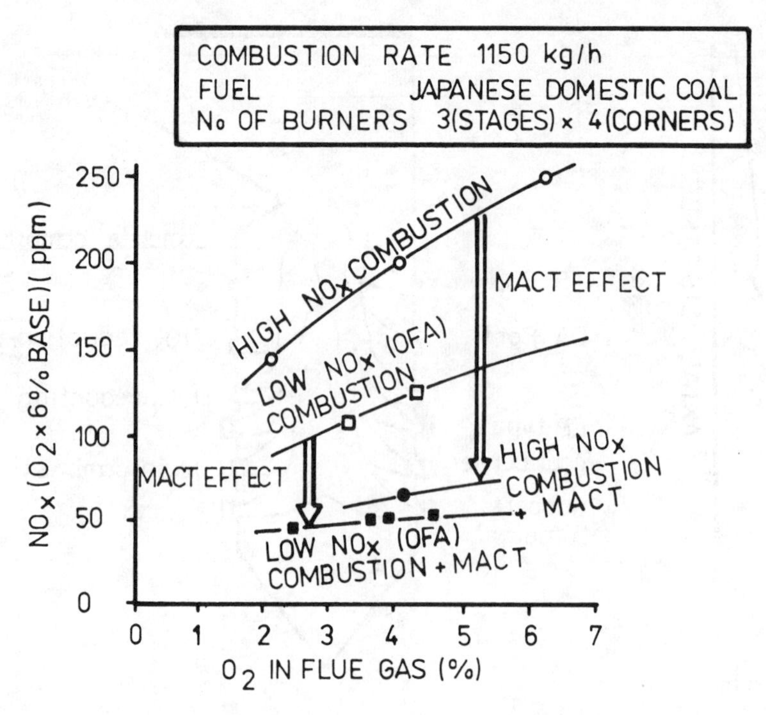

Figure 3.49 NO_x Removal Effect for Coal Firing in MACT System [342]

next burner jet streams, one after another. Accordingly, the coal particles in the downstream flame region are widely spread over the whole section of the furnace and, thereafter, the reducing region is filled with the flame where the NO_x decomposition reaction occurs entirely in the furnace volume.

The boiler design feature is the increased furnace volume, which contributes to enhancement of the reducing intermediate product formation as well as the inherent arrangement of burner and devices. Large scale test furnace results show that NO_x emission levels of 100 ppm have been achieved.

MHI MACT (IFNR)

The MHI Mitsubishi Advanced Combustion Technology (MACT) process, described in Section 3.2, essentially uses the concept of reducing NO_x already formed by the initial combustion process.

Test results on a 1150 kg/h test furnace showed NO_x levels as low as 50 ppm. They also showed a net reduction of approximately 50% of all the NO_x entering the MACT process (see Figure 3.49). The percent reduction across the MACT system also appears to be independent of the inlet NO_x concentration. This is significant because it means that MACT can be utilised in conjunction with low NO_x burner systems, such as the MHI PM (Pollution Minimum) burner (which in itself is capable of achieving very low NO_x levels) to produce NO_x concentrations on new boilers as low as 70 ppm to 115 ppm.

MHI has had an oil-fired PM/MACT combination in commercial service since January 1981 in Japan. A 600 MWe unit has been in operation since September 1983 utilising fuel oil as the 'MACT fuel' and pulverised coal as the main fuel.

3.6 Costs

The recent paper by Leggett [102] summarising progress on the O.E.C.D. Project on Control of Major Pollutants (MAP) was used to provide the limited cost data shown in the following Tables 3.8 and 3.9. Within the MAP study, differences in data were assessed and presented as ranges rather than specific values due to substantial uncertainties regarding the starting emission level, the abatement efficiency and other similar factors.

The prices and ranges in the cost effectiveness summary tables are not intended to be indicative of the cost, or cost effectiveness, for any individual facility–rather, they attempt to represent a typical facility in that source category so that, on average, the estimated costs should be representative.

A questionnaire issued to UK burner vendors did not return any usable costing information.

Cost-effectiveness is defined in terms of US 1984 dollars per tonne ($/t) of NO_x abated (converted to £/t). The costs were derived by annualising the total capital investment required for the pollution abatement technique. Assumptions used to annualise the capital costs were made as consistent as possible, using a real discount rate of 5 per cent and equipment lifetimes that vary with the type of facility (eg. 30 years for combustion modifications on new boilers or 15 years for retrofits; 3–5 years for catalysts for

Table 3.8 Utility Boilers–Coal Fired (NO_x)

Source Level of control	% Abated	Annual K Cost (£/kWt)	Annual O&M (p/kWh)	£/t Abated
Utility Boilers Existing				
Low NO_x Burners (LNB)	40–50	1.8–10.0	0.0007–0.006	43–528
2nd Generation LNB	50–70			132
Combustion Modifications				
Selective Catalytic Reduction (SCR)	60–90	37.8–62.9	2.6–8.6	150–614
Utility Boilers New				
Low NO_x Burners	45–50	0.4	Negligible	14
Combustion Modifications				
LVHR & LNB	40	1.9	Negligible	29
LVHR & LNB & IFNR	70	−7.4		678–1264
Selective catalytic Reduction	80–90	18–53	0.06–0.22	557–1900
SCR & LVHR	88	9.3	0.13	943
SCR & IFNR & LNB	94	−14.8	0.2–0.3	1330–1770

LVHR = low volume heating rate
IFNR = in-furnace NO_x reduction

selective catalytic reduction etc). The annual charge for operation and maintenance (O & M) including by-product sales and changes in fuel consumption, is added to the annualised capital charge. The resulting net annual cost is divided by the annual tons of NO_x abated.

For further details the reader should refer to the paper by Leggett [102].

Table 3.9 Industrial Boilers–Coal Fired (NO_x)

Source level of control	% Abated	Annual K cost (£/kWt)	£/t Abated
Industrial Boilers Existing			
Low NO_x Burners (LNB)	50	2.8	136
Combustion Modifications			
LNB & OFA (Japan)	35–65	4.1–5	643–2020
SCR (USA)	90	47	1280

OFA = over fire air injection
SCR = selective catalytic reduction

4. Nitrogen Oxides Abatement Processes (Flue Gas Treatment)

4.1 Classification of the Processes

4.2 Outline Descriptions

4.3 General Appraisal of Processes

4.4 Processes for Detailed Study

4.5 Evaluation of Selected $DeNO_x$ FGT Processes

4.6 Costs

4. Nitrogen Oxides Abatement Processes (Flue Gas Treatment)

4.1 Classification of the Processes

The general classification of NO_x abatement (deNO_x) processes adopted in this Volume involves a division into two broad groups according to whether abatement is achieved at the source of combustion, 'Combustion Techniques', or in the flue gas downstream of the boiler, 'Flue Gas Treatment'.

The first group, 'Combustion Techniques', is dealt with in Section 3 of this Volume.

The second group, 'Flue Gas Treatment' (FGT), is covered in this section. Figure 4.1 shows a classification of these FGT processes into two basic types:

- Dry processes (Category N40)
- Wet processes (Category N50)

Dry processes may be either catalytic or non-catalytic, whilst wet processes involve either direct absorption or gas-phase oxidation prior to absorption. One further sub-division occurs in the case of dry, non-catalytic processes, i.e. reduction or adsorption.

Only those flue gas treatment processes which selectively remove NO_x are dealt with in this section; those processes which are non-selective or have been developed specifically for simultaneous removal of NO_x and SO_2 are dealt with in Section 5.

PROCESS/DESIGN CODE NUMBERS

Each major category is assigned a Category Number, e.g. Dry Processes–N40. Generalised processes within each category are characterised by a change in the last digit of the Category Number, e.g. Selective Non-Catalytic Reduction processes–N42.

The basic design or process types have been assigned the Code Numbers shown in Figure 4.1.

SUPPLIERS OF DeNO_x TECHNOLOGY

Full details of all manufacturers quoted in the Volume (names, addresses, telephone and telex numbers, and names of contacts) are listed in Appendix 2.

4.2 Outline Descriptions

DEFINITION OF NO_x

The term 'NO_x' is used to denote a mixture of the two oxides of nitrogen produced during the combustion of fuels with air; namely nitric oxide, NO, and nitrogen dioxide, NO_2. In flue gas, about 90% of NO_x is in the form of NO, a small part of which will be oxidised to NO_2 in the stack by oxygen from the excess combustion air. However, NO_x concentrations in flue gas are frequently expressed in units of $mg(NO_2)/m^3$.

CATEGORY N40

CATEGORY N41–SELECTIVE CATALYTIC REDUCTION

Outline of Processes [352]: A simplified block diagram of the process is presented in Figure 4.2. Several alternative locations for the Catalytic Reactor are possible depending on the operating temperature of the Electrostatic Precipitator and the type of FGD process installed (if any). These alternative arrangements are considered in Section 4.3. Flue gas from the Boiler enters the Catalytic Reactor at a temperature of 300°–400°C. Ammonia is injected into the flue gas immediately upstream of the Reactor and at near stoichiometric conditions (relative to the nitric oxide, NO, in the gas). The nitrogen oxides in the flue gas are converted to nitrogen and water in the presence of the catalyst, and no deleterious waste products are formed. The treated gas is discharged to the Stack via an Air Preheater and Electrostatic Precipitator. This particular arrangement is referred to as the 'High-Dust System' because dust is removed from the flue gas after denitrification.

Chemistry of Process [352, 419]: Ammonia reduces the nitrogen oxides in the flue gas to nitrogen and water according to the following equations:

$$6NO + 4NH_3 = 5N_2 + 6H_2O$$
$$6NO_2 + 8NH_3 = 7N_2 + 12H_2O$$

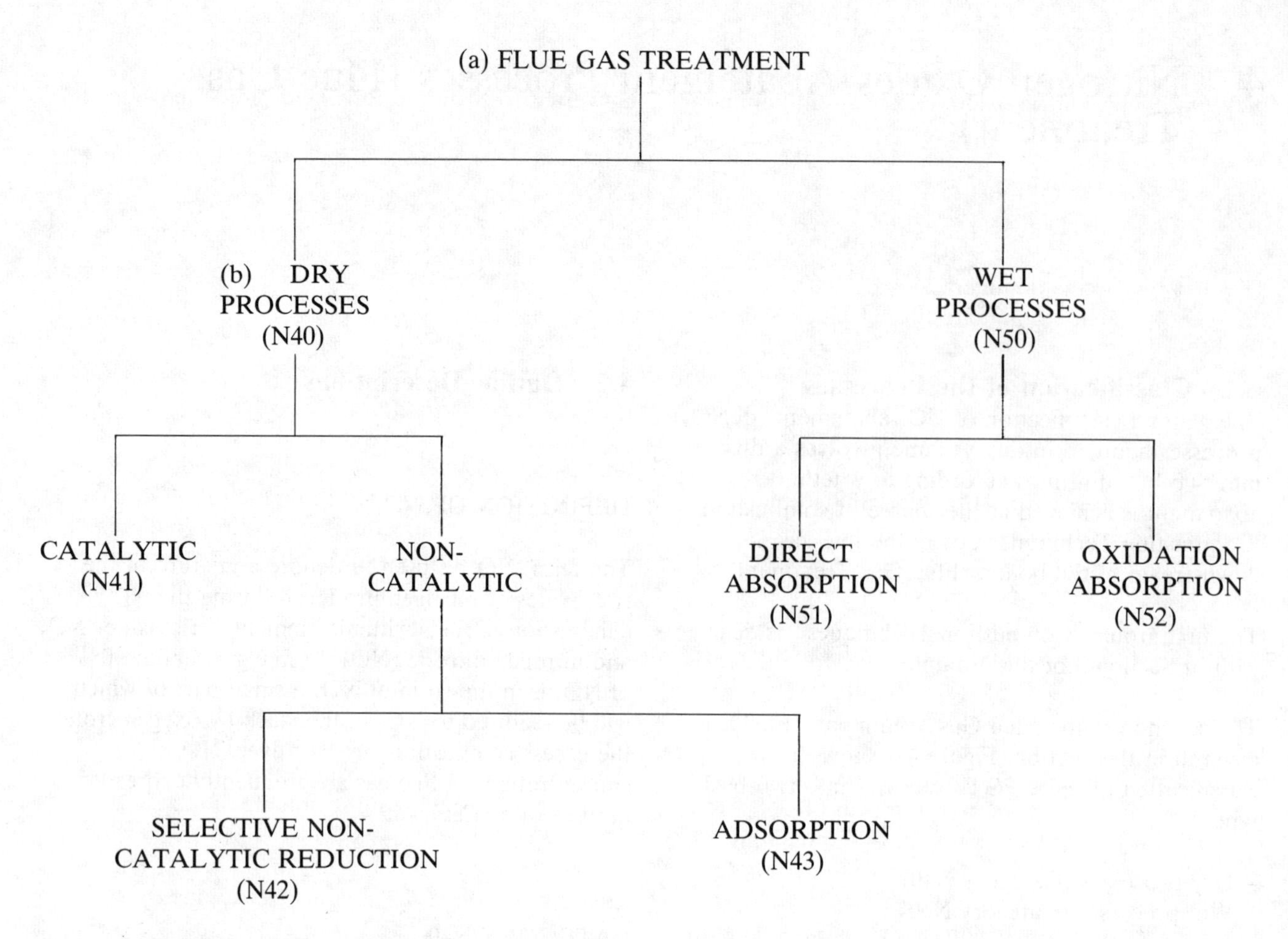

Figure 4.1 Classification of $Deno_x$ Flue Gas Treatment Processes

The reducing reaction also proceeds in the presence of oxygen, but with higher ammonia consumption:

$$4NO + O_2 + 4NH_3 = 4N_2 + 6H_2O$$
$$2NO_2 + O_2 + 4NH_3 = 3N_2 + 6H_2O$$

The oxidation of flue gas SO_2 to sulphur trioxide (SO_3) is also promoted by the catalyst. The SO_3 subsequently reacts with ammonia to form ammonium salts:

$$SO_3 + H_2O + NH_3 = NH_4HSO_4$$
$$SO_3 + H_2O + 2NH_3 = (NH_4)_2SO_4$$

The presence of SO_3 is undesirable not only for its reaction with ammonia but also because it increases the acid dewpoint of the flue gas. For this reason, catalysts used in SCR $deNO_x$ processes tend to have poor SO_2 oxidation capability.

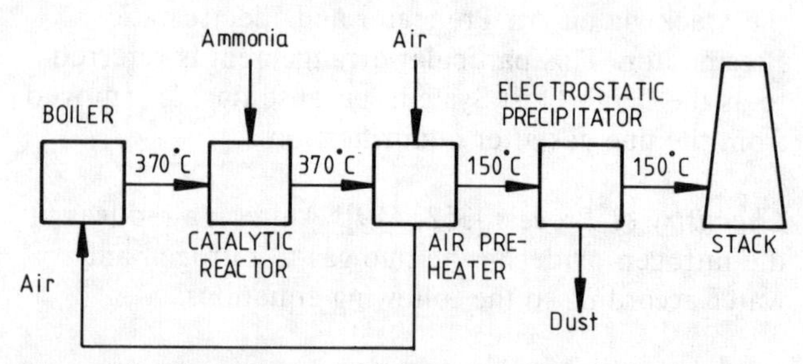

Figure 4.2 Block Diagram of Selective Catalytic Reduction Denitrification Process

CATEGORY N42–SELECTIVE NON-CATALYTIC REDUCTION

Outline of Process [352, 419]: A simplified block diagram of the process is presented in Figure 4.3. Denitrification of the flue gas is achieved by reaction with ammonia. As no catalyst is used in this process, much higher temperatures are required than with selective catalytic reduction processes, typically 930° to 1030°C. To achieve these temperatures, the ammonia is injected directly into the upper section of the Boiler. Optimum operating temperatures can be reduced to as low as 700°C by injecting hydrogen or natural gas with the ammonia. Exit gases from the Boiler pass to an Air Preheater and Electrostatic Precipitator prior to discharge to the Stack. No additional downstream equipment is required for this process.

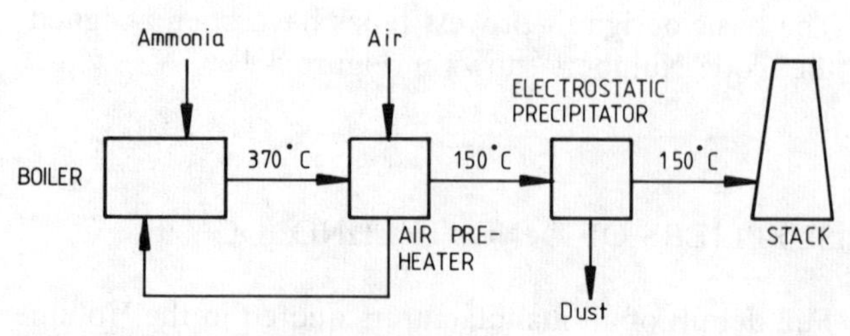

Figure 4.3 Block Diagram of Selective Non-Catalytic Reduction Denitrification Process

Chemistry of Process [352]: The overall reactions between flue gas nitric oxide, excess air and ammonia are as follows:

$$5O_2 + 4NH_3 = 4NO + 6H_2O$$
$$4NO + O_2 + 4NH_3 = 4N_2 + 6H_2O$$

The first reaction is dominant above 1100°C, whilst the second reaction dominates in the range 800° to 1000°C.

CATEGORY N43–DRY ADSORPTION

Outline of Process [352, 419]: A simplified block diagram of the process is presented in Figure 4.4. Adsorption processes using activated carbon were initially developed for SO_2 removal from flue gases, and later extended to simultaneous SO_2/NO_x removal (see Section 5). The optimum operating temperature for the extended process is 130°–150°C, but if selective and increased NO_x removal is possible at temperatures as low as 80°C, then an activated carbon reactor can be installed downstream of a separate FGD unit. Low levels of NO_x removal (40–60%) are reported [419] for this process, but the possibility exists to extend the use of the activated carbon bed to catalyse the reducing reaction between NO_x and added NH_3 (see above). The adsorbent is regenerated by blending with hot sand at a temperature of 650°C, whereupon nitrogen is evolved. Lower regeneration temperatures, e.g. 400°C, have been reported for simultaneous FGD/deNO_x applications.

Chemistry of Process [419]: Adsorbed NO_x is converted to N_2 in the Regenerator as follows:

$$2NO + C = N_2 + CO_2$$
$$2NO_2 + 2C = N_2 + 2CO_2$$

CATEGORY N50

CATEGORY N51–ABSORPTION OXIDATION (DIRECT ABSORPTION)

Outline of Process [419]: A simplified block diagram of the process is presented in Figure 4.5. Nitric oxide is absorbed by an aqueous salt solution containing a liquid-phase oxidising agent. Although Absorption of NO is slow, due to its low solubility, the absorbed gas is rapidly converted to nitrate salts and subsequently removed from the waste water stream. The oxidising agent readily reacts with absorbed SO_2, as well as NO, so it is necessary to install a separate SO_2 Absorber upstream of the NO_x Absorber to minimise consumption of the expensive oxidising agent by the more soluble SO_2.

Flue gas from the Boiler and Air Preheater enters a Prescrubber where halogen acid gases and most particulates are removed. The pre-scrubbed gases pass through an SO_2 Absorber and into the large NO_x Absorber where NO is gradually absorbed by a recirculated scrubbing agent. Gas leaving the NO_x Scrubber is reheated prior to discharge to the Stack.

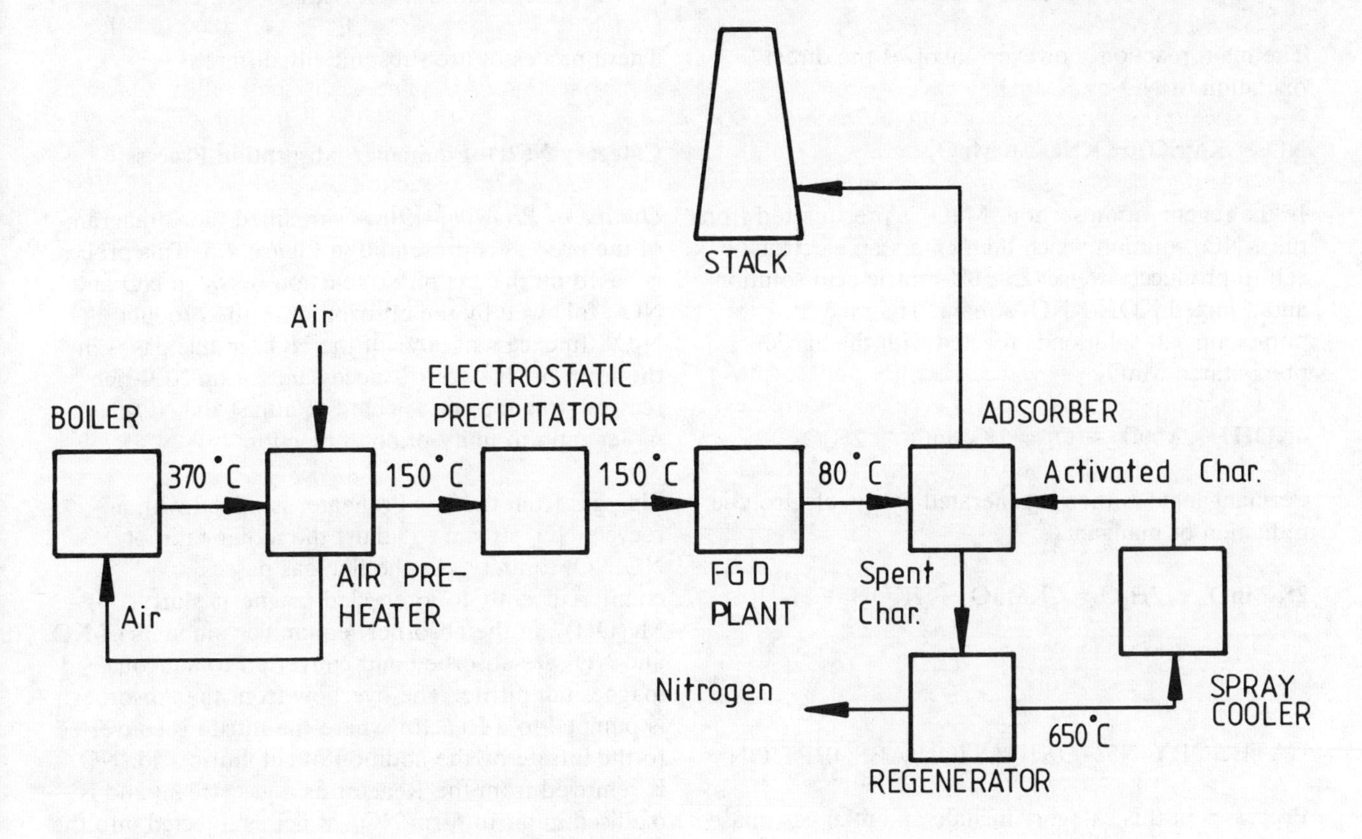

Figure 4.4 Block Diagram of Dry Adsorption Denitrification Process

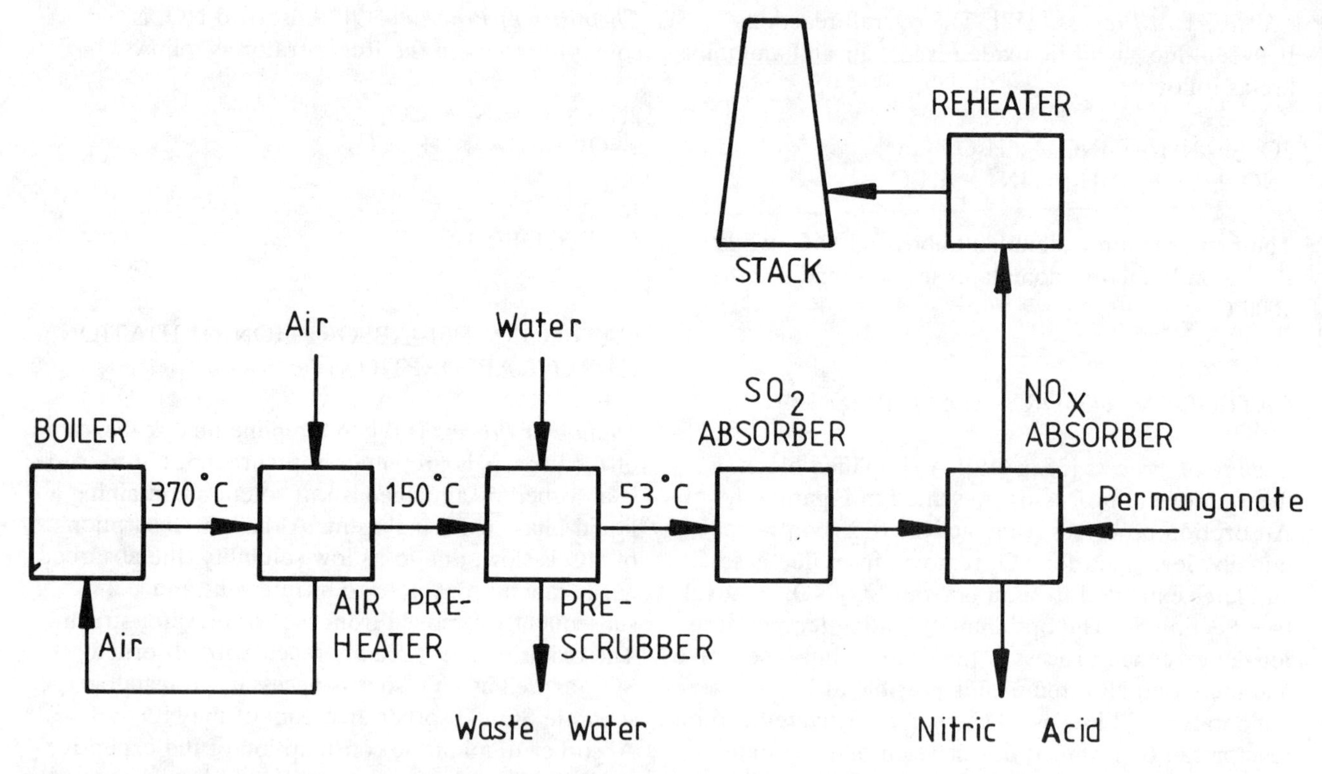

Figure 4.5 Block Diagram of Absorption-Oxidation Denitrification Process

Chemistry of Process [419]: In the absorber, NO and NO_2 react with a potassium hydroxide scrubbing solution containing potassium permanganate:

$$NO + 2KOH + KMnO_4 = K_2MnO_4 + KNO_2 + H_2O$$
$$3NO + 2KOH + KMnO_4 = MnO_2 + 3KNO_2 + H_2O$$
$$NO_2 + K_2MnO_4 = KMnO_4 + KNO_2$$

The main reaction, however, involves the direct oxidation of NO by $KMnO_4$:

$$NO + KMnO_4 = KNO_3 + MnO_2$$

In the regeneration section, MnO_2 is precipitated from the KNO_3 solution which then enters an electrolytic cell to produce a weak (25–30%) nitric acid solution and a mixed KOH/KNO_3 stream. The mixed potassium salt solution is reacted with the earlier precipitated MnO_2:

$$4KOH + 2MnO_2 + O_2 = 2K_2MnO_4 + 2H_2O$$

Permanganate is then regenerated by the electrolytic reduction of manganate:

$$2K_2MnO_4 + 2H_2O = 2KMnO_4 + 2KOH + H_2$$

CATEGORY N52–OXIDATION ABSORPTION

Processes in this category include an initial gas phase oxidation stage followed by absorption in a scrubbing solution. Oxidation of the absorbed NO_x occurs to produce a by-product nitrate solution. In general, there are presently two types of process in this category:

- Equimolar NO-NO_2 absorption (category N52.1)
- NO_2 or N_2O_5 absorption (category N52.2)

These processes are substantially different.

Category N52.1–Equimolar Absorption Process

Outline of Process [419]: A simplified block diagram of the process is presented in Figure 4.6. This process is based on the gas-phase reaction between NO and NO_2, followed by the absorption of the product N_2O_3. In excess of 90% of the NOx in flue gas is in the form of NO, so it is necessary for an NO_2-rich recycle stream to be injected to adjust the NO_2:NO molar ratio to unity prior to reaction.

Flue gas from the Air Preheater is mixed with a recycle NO_2 stream to adjust the molar ratio of NO:NO_2 to unity. As the flue gas passes countercurrently to a recycled magnesia slurry, $Mg(OH)_2$, in the absorber, equimolar amounts of NO and NO_2 are absorbed and converted to aqueous magnesium nitrite. The overflow from the Absorber is pumped to a Reactor where the nitrite is converted to the nitrate by the addition of sulphuric acid. NO is removed from the Reactor as a gas stream and is oxidised in air to form NO_2, which is injected into the flue gas prior to the NO_x Absorber.

Chemistry of Process [419]: In the NO_x absorber, a recycle NO_2 stream is mixed with the flue gas to adjust the molar ratio of $NO:NO_2$ to unity. An $Mg(OH)_2$ slurry absorbs equimolar amounts of NO and NO_2 as follows:

$$NO + NO_2 = N_2O_3 \quad (1)$$
$$Mg(OH)_2 + N_2O_3 = Mg(NO_2)_2 + H_2O \quad (2)$$

As the NO_x concentration in the flue gas falls below 200ppm, the reaction decreases to a negligible rate and NO_2 is absorbed:

$$2NO_2 = N_2O_4 \quad (3)$$
$$2Mg(OH)_2 + 2N_2O_4 = Mg(NO_3)_2 + Mg(NO_2)_2 + 2H_2O \quad (4)$$

If NO_x concentrations of less than 150–200ppm are required, a second denitrification stage (using ozone) is necessary:

$$NO + O_3 = NO_2 + O_2 \quad (5)$$

with the NO_2 being absorbed as shown in equations 3 and 4 above.

In the reactor, sulphuric acid is added to convert the nitrite salts to nitrate:

$$3Mg(NO_2)_2 + 2H_2SO_4 = 2MgSO_4 + Mg(NO_3)_2 + 4NO + 2H_2O \quad (6)$$

NO is insoluble in this solution and passes to the oxidation tower, where air converts it to NO_2 prior to re-injection into the NO_x absorber:

$$2NO + O_2 = 2NO_2 \quad (7)$$

Category N52.2–N_2O_5 Absorption Process

Outline of Process [419]: A simplified block diagram of this process is presented in Figure 4.7. This process is based on gas-phase oxidation of NO_x to N_2O_5, using excess ozone, followed by absorption in a weak (8–10%) aqueous nitric acid solution. The flue gas is adiabatically cooled and humidified prior to absorption. A small purge stream from the Absorber is concentrated to 60% acid in a Steam Evaporator. The cleaned flue gas flows to a second Absorber where it is contacted with a calcium sulphite slurry to remove the remaining ozone, prior to reheating and discharge to Stack. Make-up $CaSO_3.{}^1/_2H_2O$ is added to the Ozone Absorber and a small purge stream is removed as a waste sludge. The absence of a nitrate waste water treatment system limits the process to clean flue gas streams i.e. free of dust or SO_x.

Chemistry of Process [419]: Flue gas NO_x rapidly and selectively reacts with ozone as follows:

$$NO + O_3 = NO_2 + O_2$$
$$2NO_2 + O_3 = N_2O_5 + O_2$$

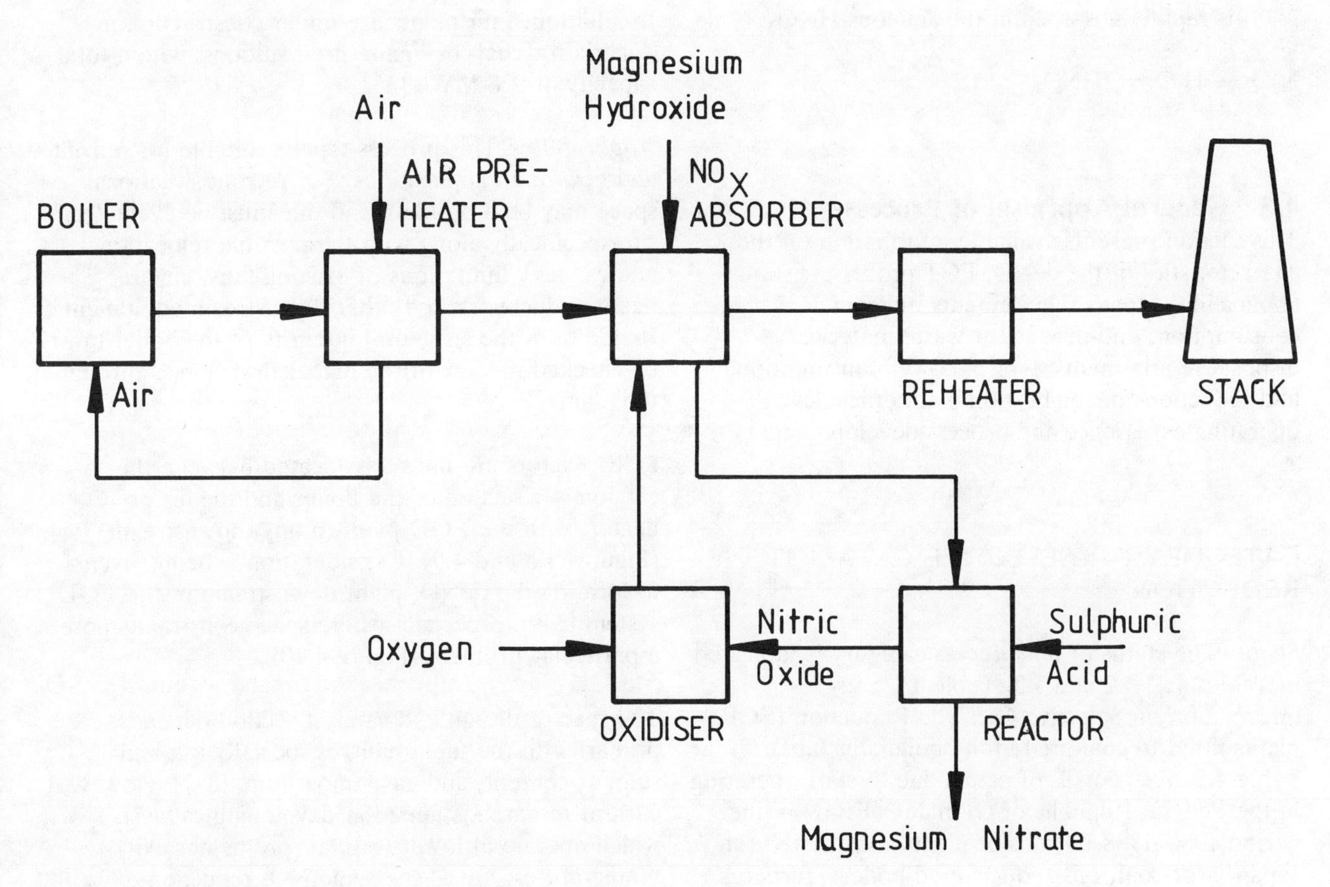

Figure 4.6 Block Diagram of Oxidation-Absorption Denitrification Process (1)

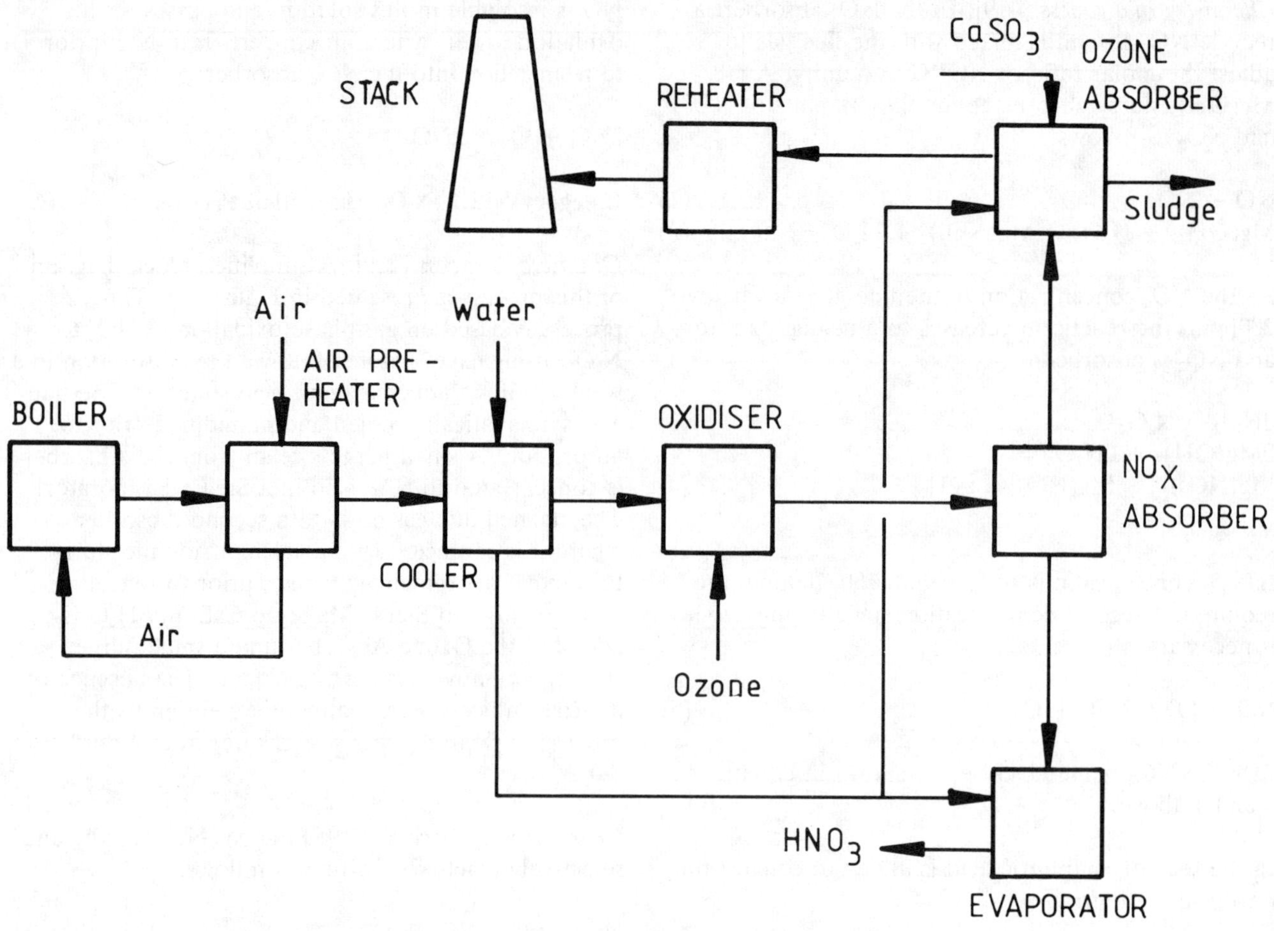

Figure 4.7 Block Diagram of Oxidation-Absorption Denitrification Process (2)

N_2O_5 is rapidly absorbed in the aqueous HNO_3:

$$N_2O_5 + H_2O = 2HNO_3$$

4.3 General Appraisal of Processes

This section presents available information on the characteristics of the deNO$_x$ FGT processes: status, applicability, space requirements, reagent consumption, end-product or waste materials disposal requirements, typical power consumption and reductions in combustion plant efficiency, operating experience and process developments.

CATEGORY N41–SELECTIVE CATALYTIC REDUCTION

Status: The status of this process category is indicated in Tables 4.1, 4.2 and 4.3. Table 4.1 presents information on Selective Catalytic Reduction (SCR) plants fitted to coal-fired utility boilers in Japan. Table 4.2 gives details of plants due to start operation in the Federal Republic of Germany (FRG) in the period 1986–1988. Details of plants constructed (in Japan?) for coal, oil- and gas-fired boilers, furnaces etc., are given in Table 4.3.

In addition, four plants are under construction in Austria for coal- or lignite-fired stations, with a total capacity of 928 MWe [352].

Applicability: This process type is suitable for retrofit and new build applications. For retrofit situations, space may be a problem and this must be evaluated site-specifically along with the possible relocation and capacity limitations of existing fans, air pre-heaters, ductwork and other downstream equipment. In addition, the structural integrity of the boiler must be checked in view of the higher draft losses through the plant.

SCR reactors are normally located between the economiser section of the boiler and the air pre-heater, with the FGD plant, if any, downstream (Figures 4.8 and 4.9). Consideration is being given to locating the deNO$_x$ plant downstream of the FGD system [349], especially where space constraints pose a particular problem (Figure 4.10).

The reactor design and catalyst section depends primarily on the fuel quality, especially ash and sulphur content, and gas temperature [347]. However, current research is aimed at developing catalysts which operate at lower (e.g. air pre-heater exit) temperatures, but at the same NO$_x$ reduction efficiencies [429], in order to simplify retrofitting.

Table 4.1 Status of Selective Catalytic Reduction Plants in Japan. Coal-Fired Utility Boilers Only

Vendor	No of Plants	Plant Size Range (MWe)	Total MWe	Dates	% Denox	Reference
Babcock-Hitachi	10	88–700	2208	1980–85	53–80	349, 350
IHI	4	250–600	1850	1983–87		348, 350
Kawasaki	1	–	125	1980	75	350, 347
Mitsubishi	11	75–700	2973	1980–86	up to 90	342
Total	26		7156			

Table 4.2 Status of Selective Catalytic Reduction Plants in the Federal Republic of Germany. Coal-Fired Boilers [28]

Vendor	No. of Plants(a)	Approx. Total MWe
Babcock-Hitachi	4	1,000
IHI	5	900
Kawasaki	6	2,200
Mitsubishi	5	1,500
Norton	3	105
Total	23	5,705

(a) for initial start-up during 1986–1988

Table 4.3 Status of Selective Catalytic Reduction Plants (All Sources) [350]

Vendor	Utility Boilers		Other Plant		Additional References
	No. of Plants	Total (a) Flue Gas Volumes	No. of Plants	Total (a) Flue Gas Volumes	
Asahi Glass			1	70	
Babcock Hitachi & Hitachi Limited	21 (b)	19,540 (c)	8	1630	
Hitachi Zosen			9	1867	
IHI	15 (c)	18,505 (b)	1	200	
JGC			5	393	
Kawasaki HI	2	750	1	25	347, 417
Kobe Steel			1	101	
Kurabo Engineering			1	30	419
Mitsubishi HI	32	25,950	7	713	
Mitsubishi KK			10	244	
Mitsui Toatsu			13	1297	
Mitsui Engineering			12	773	
Nippon Kokan			1	1320	
Sumitomo Chemical			6	999	
Sumitomo Chem. Eng.			6	130	
Total (end 1983)	70	64,754	82	9792	

(a) Units of thousands Nm^3/h
(b) Total of 27 in 1984 [349]
(c) Total of 'more than' 20 in 1984 [348]

SCR plants are optimally designed for NO_x reduction efficiencies of 80% [342]. However, higher efficiencies of up to 90% are feasible, but require a significant increase (30%) in catalyst volume and cost [371]. This is an important consideration, since it has been reported that catalyst costs amount to 40–60% of annual levelised operating costs [410]. Capital costs for SCR plant are dependent on unit size, fuel type, NO_x reduction, location, reagent and catalyst costs, etc, but are significantly higher than for combustion techniques [345]. It has been reported that flue gas treatment on a new coal-fired utility boiler in the USA would add 4–5% to the total plant cost, with significantly higher percentages expected for smaller boilers and retrofits [341, 342].

It is believed that a combination of SCR and combustion modifications is essential for coal-fired applications where high NO_x reductions are required [347], and that for lower reduction levels only combustion modifications may be necessary [341].

Space Requirements: SCR plant have only three major components: the reactor, catalyst and ammonia storage/injection system. Very little information on space requirements is available in the literature but

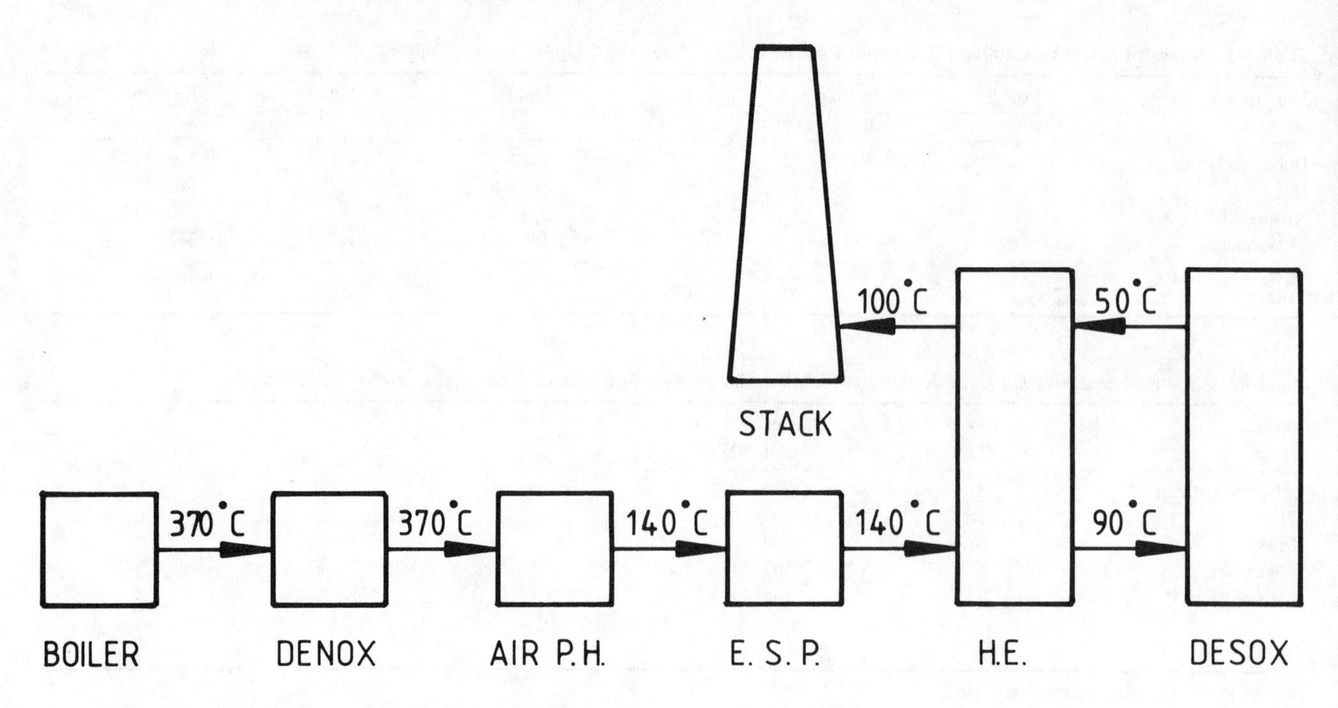

Figure 4.8 Denitrification Plant Locations–High Dust System

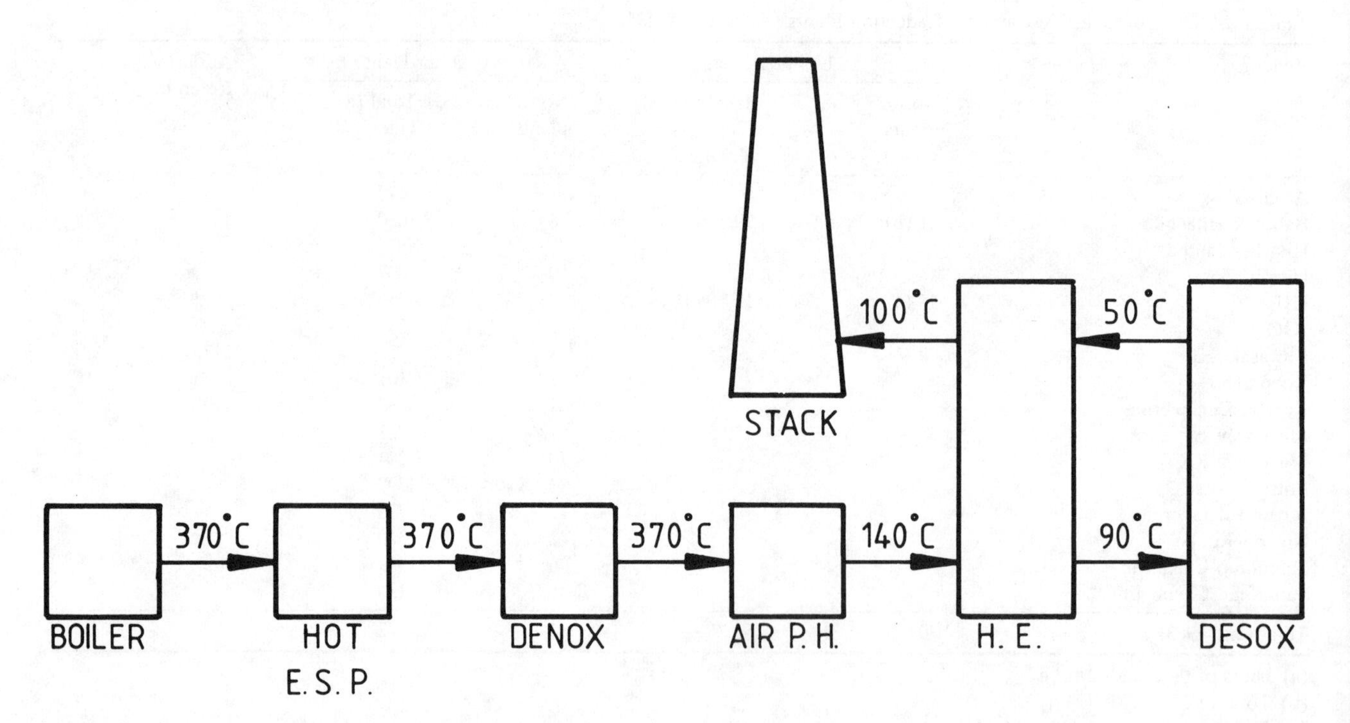

Figure 4.9 Denitrification Plant Locations–Low Dust System

Pruce [371] quotes a typical SCR reactor envelope of 6.1 m × 7.6 m × 18.3 m for an 80 MWe system. Kawasaki present outline specifications for a 700 MWe pulverised-coal boiler with two reactors, each of 13.4 m × 14.5 m cross-sectional area, and a total catalyst volume of 700 m³ [347].

Other catalyst volumes are quoted in Table 4.4 for 175 MWe, 500 MWe and 700 MWe plants, being 300–407 m³, 310 m³ and 804–1043 m³, respectively [352]. It can be noted that these catalyst volumes vary with NO_x removal efficiency and initial flue gas NO_x concentrations.

As mentioned above, the catalyst is a major cost factor and the required catalyst volume is dependent not only on the two factors previously mentioned, but also on the allowable ammonia slippage (leakage) and any redundancy necessary for coal quality variation [347].

Consumables: In a SCR plant, there are three main consumables: ammonia, catalyst and electricity. Electricity consumption, and its effect on power plant operating efficiency, is dealt with later in this section.

Typical ammonia consumption figures have been

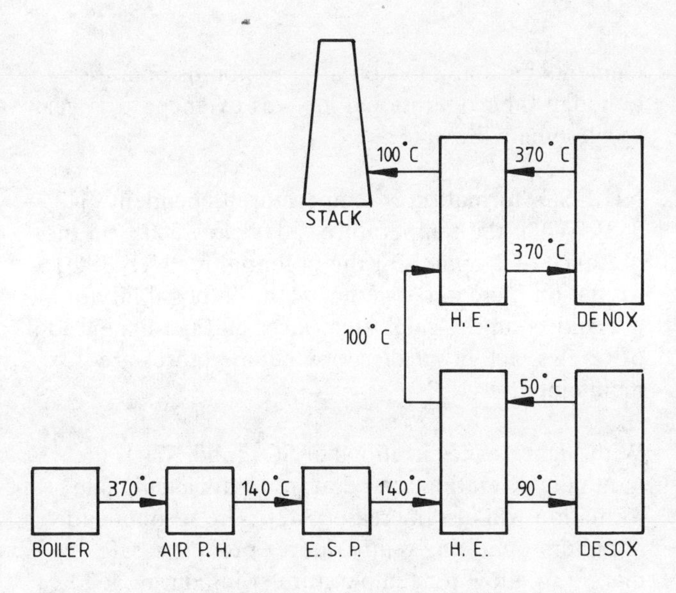

Figure 4.10 Denitrification Plant Locations–Tail Gas System

Table 4.4 Ammonia Consumption of SCR Units [352]

A. 100 MWt plant:

Initial NO_x Concentratioin (mg NO_2/m^3)	Final NO_x Concentration (mg NO_2/m^3)	NO_x Reduction (%)	Ammonia Consumption (kg/h)
1800	200	89	215
1500	200	87	170
1300	200	85	145
800	200	75	80
650	200	69	60
500	200	60	40

(Consumption is 0.130–0.134 kg/h per mg NO_2/m^3 reduction)

B. 500 MWe utility boiler:

Flue gas flowrate = 1,500,000 m^3/h (4% vol. O_2)
Flue gas temperature = 400°C
Inlet NO_x concentration (as NO_2) = 800 mg/m³
Inlet NO_2 concentration (maximum) = 40 mg/m³
Outlet NO_x concentration (as NO_2) = 200 mg/m³
Outlet NH_3 concentration = 3.8 mg/m³

Ammonia consumption = 360 kg/h
Electricity consumption (for NH_3 injection) = 24 kW
Hot water consumption (for NH_3 evaporation) = 4000 kg/h
Hot water inlet/outlet temperatures = 120°/90°C
Pressure loss (NH_3 injection + reactor) = 12 mbar
Catalyst volume = c.310 m³ (honeycomb/molecular sieve)

C. 700 MWe conventional boiler:

Flue gas flowrate (m_3/h)	2.4×10^6	2.25×10^6
Inlet NO_x concentration (mg/m³)	1000	500
NO_x removal efficiency (%)	80	80
Outlet NH_3 (ppm)	3	3
NH_3/NO_x mole ratio	0.82	0.84
Catalyst volume (m³)	1,043	804
Pressure drop (mbar)	9	8
Power consumption (% generated)	0.22	0.22

D. 175 MWe wet bottom boiler:

Flue gas flowrate (m³/h)	0.57×10^6	0.57×10^6
Inlet NO_x concentration (mg/m³)	2000	2000
NO_x removal efficiency (%)	80	90
Outlet NH_3 (ppm)	3	3
NH_3/NO_x mole ratio	0.81	0.91
Catalyst volume (m³)	300	407
Pressure drop (mbar)	10	11
Power consumption (% generated)	0.23	0.25

quoted as 0.7–0.8 kg/h per MWe of installed capacity for 75% NO_x reduction to an output concentration of 200 mg/m³; see Table 4.4 [352]. Alternative sources express ammonia consumption in terms of the mole ratio of NH_3 to NO_x reacted. Typically, 60 to 85 percent NO_x reduction is achieved with 0.61 to 0.90 mol. NH_3 per mol. of NO_x reacted, with 1–5 ppm slippage [350]. NO_x reductions of 80%, with 3 ppm leakage and an outlet NO_x concentration of 200 mg/m³, require a NH_3/NO_x ratio of 0.82. This equates to 0.89 kg/h NH_3 per MWe of installed capacity (assuming 90% of NO_x is in the form of NO).

In 'high dust' systems (Figure 4.8), ammonia leakage may be absorbed by the fly ash and removed in the electrostatic precipitator [347]. The ammonia is released if the fly ash is subsequently heated and can be recycled to the reactor to reduce ammonia consumption.

The catalyst accounts for 40–60% of the annual levelised costs [410]. This is due to the short lifetime guarantees, normally 1–2 years, being offered for the catalyst [343]. Kawasaki claim operational periods of more than 3 years with coal-fired units [347] with the possibility of prolonging the life of the catalyst by soot-blowing prior to shutdown and monitoring the operating performance. Catalyst replacement can be carried out in one of two ways: full volume replacement every 2 to 3 years or partial replacement at yearly intervals [347]. The latter method is normally preferred for reasons of plant economy.

For treating flue gases from coal-fired boilers, the catalyst should possess the following properties [342]:

(i) Resistance to toxic materials,
(ii) Resistance to abrasion,
(iii) Mechanical strength (especially where soot blowing occurs),
(iv) Resistance to thermal cycling,
(v) Resistance to oxidation of SO_2 to SO_3,
(vi) Resistance to plugging.

Catalyst technology is considered proprietary by most vendors and, therefore, little information is available. Generally a base metal of iron, vanadium, chromium, manganese, cobalt, nickel, copper, barium, platinum, carbon, molybdenum or tungsten, or oxides, sulphates or combinations of these metals, is deposited on a carrier of alumina (Al_2O_3), titania (TiO_2) or silica (SiO_2) [345, 346]. Many of these materials are damaged by SO_2 and SO_3 in the flue gas, and under these conditions the most stable and most widely adopted catalyst contains vanadium compounds on a titania carrier [345].

Abrasion tests by Hitachi have shown a weight loss of less than one percent over 20,000 hours (2.3 years) with dust loadings of 50g/Nm³ [2,429]. It was claimed that this weight loss did not affect catalyst performance. Kawasaki and Mitsubishi, however, do acknowledge losses of performance. Kawasaki reports a 4–6% loss of catalyst activity after 10,800 hours of operation [408], whilst Mitsubishi noted an increase in ammonia leakage to 10 ppm after 25,000 hours operation at a 175 MWe plant and with constant 50% NO_x reduction (500 to 250 ppm NO_x in flue gas) [413]. Erosion is normally minimised by operating with low flue gas velocities through the catalyst reactor [342]. In view of the high cost of the catalyst, regular inspection is carried out during normal boiler outages [342].

Pressure Losses: Honeycomb and plate catalyst reactors have low pressure drops due to their high voidage. Table 4.4 indicates typical pressure drops of 8–12 mbar across the reactor and ammonia injection nozzles [352, 347].

Power Consumption: Only limited power consumption figures are available. Table 4.4 estimates overall consumption to be of the order of 0.20–0.25% of the power generated for boilers of 175 MWe and 700 MWe capacity [352]. The same reference cites 360 kWe consumption for a SCR plant located after the economiser per 100 MWe of boiler capacity, and 600 kWe per 100 MWe if installed downstream of a wet FGD plant.

Operating Experience: In Japan, all SCR reactors are installed between the boiler economiser and the air preheater, irrespective of whether hot or cold electrostatic precipitators are installed. In Europe, there is some interest in the installation of deNO_x reactors downstream of the FGD plant, due to space limitations in retrofit situations [349]. There is little experience of the latter configuration at present, although six plants of this type are under construction in the FRG [28].

There is a range of catalysts for applications with all types of fuel, from those with low SO_2 oxidation rates to those with high activity [347]. With coal-fired boilers, the choice of catalyst will be a compromise between NO_x reduction and SO_2 oxidation [348]. The oxidation of SO_2 to SO_3 and ammonia leakage is detrimental to downstream equipment due to the subsequent formation of sulphuric acid, ammonium sulphate and bisulphate (NH_4HSO_4) [347]. NH_4HSO_4 may stick to the catalyst surface causing reduced catalyst activity [348], cause fouling, erosion and reduced heat transfer in the air preheater, lead to baghouse blinding, modifications in scrubber chemistry in downstream FGD plant, and increase the disposal problems for fly ash and scrubber byproducts [343, 26]. Ammonium sulphate is not removed completely by electrostatic precipitators and may result in an unacceptable plume formation. In Japan, the EPDC has monitored sulphate deposition in a 250 MWe coal-fired power station; air preheater elements became blocked after 6 months operation, although their operational life was extended to a year by washing with water.

NH_4HSO_4 formation is temperature dependent [342, 343]. When the temperature falls below 300°C in the presence of 1 ppm SO_3, the potential for NH_4HSO_3 formation increases together with the possibility of deposition on the catalyst surfaces and permanent loss of catalyst activity if these low temperatures are maintained.

With higher concentrations of SO_3 (and NH_3), the minimum operating temperature to avoid sulphate formation will also increase [342]. The recommended minimum operating temperatures provide a safe margin to allow for temperature swing during load changes, the NH_4HSO_4 formation temperature being reported as 210°C in the presence of 10 ppm (36 mg/Nm^3) SO_3 and 10 ppm (8 mg/Nm^3) NH_3 in the flue gas [343]. In practice, the formation and effects of NH_4HSO_3 are minimised by the provision of an economiser flue gas bypass (to increase temperature), the reduction of SO_3 and NH_3 concentrations in the reactor outlet gases or by modifying the air preheater design; all these options increase plant costs or reduce the efficiency of power generation [343].

A maximum operating temperature is imposed by the possible dissociation of ammonia above 400°C; if temperatures of 480°C–540°C are exceeded, sintering of the catalyst may occur thereby permanently impairing its activity [342].

At 80% NO_x reduction, NH_3 leakage is negligible. At 90% reduction, NH_3 leakage is 10–20 ppm or higher [371].

An outlet flue gas concentration of 5 ppm is regarded as the acceptable limit [344] and NH_4HSO_3 deposition on the air preheater is avoided altogether if NH_3 leakage is 3 ppm (2.3 mg/Nm^3) [350]. The NH_3 concentration can be minimised by uniform distribution within the reactor inlet duct by means of a multi-nozzle arrangement [347].

About 70–80% of the ammonia leakage is captured by the fly ash, but this does not present any problems in the air preheater provided both the SO_3 concentration and NH_3 leakage do not exceed 5 ppm [348]. The fly ash does, however, have a deleterious effect upon the NO_x removal efficiency if it contains significant quantities of alkali metal (Na,K) oxides, although alkali earth metal (Ca,Mg) oxides do not have the same effect [348].

It has been reported that there is no noticeable deterioration of titania-based catalysts caused by the presence of halogen compounds [348]. Granular and pelletised catalysts have largely been superseded by plate and honeycomb arrangements. Metal plate type catalysts with a thickness of 1 mm remain relatively dust free and usually do not need soot blowers to maintain low pressure drops through the catalyst bed [349].

Process Development: Spherical catalysts in a fixed bed are unsuitable for flue gases from coal-fired units due to dust deposition. This led to the development of moving catalyst beds which introduced construction complexities and abrasion of the catalyst [347]. The openness of the pipe, honeycomb and grid plate catalysts has now superseded bed catalysts for $deNO_x$ applications.

Some technical difficulties have been experienced in the formation of these configurations with some catalyst materials. Catalysts based on Al_2O_3 and ferric oxide (Fe_2O_3) are poisoned by SO_x in the flue gas [350], so TiO_2 with a small amount of vanadium pentoxide (V_2O_5) is overwhelmingly in use for SO_x-containing gases.

It has been reported that V_2O_5/TiO_2 catalysts can be used in flue gases containing up to 5700mg/m^3 SO_2 and 360mg/m^3 SO_3 [352]. This catalyst combination is highly active but the V_2O_5 does promote some SO_2 oxidation; a low oxidation catalyst can be produced by replacing the V_2O_5 partly or fully with oxides of tungsten or molybdenum [350].

SO_x poisoning is not a problem in situations where the SCR reactor is located downstream of an FGD plant [347]. The layout of the catalytic reactor incorporates an inactive stage at the upstream end of the active catalyst [342] which often takes the form of end-face hardening, of 12–50 mm depth, to cope with fly ash abrasion [347].

It is reported that improved catalysts have erosion rates of about 0.00015%wt per hour [347]. For coal applications, vertical downward gas flow in the catalyst chamber is preferred to minimise ash deposition on the catalyst surfaces [342]. Catalyst life expectancy is often quoted as 1–2 years [343], or as high as 3 years [347].

It is generally expected that the performance of a catalyst in a coal-fired unit should match that of an oil-fired system, where operational lifetimes in excess of 6 years can be achieved [348].

Extensive field trials are taking place in the FRG, on a pilot-plant scale, to investigate the effect of variable load patterns on the catalyst [28]. German boilers are considered to have more frequent load swings than those in Japan, so thermal shock on the catalyst structure is likely to be much greater. Activated coke is being considered as the catalyst in reactors located downstream of FGD plant, because it is effective at a relatively low temperature, e.g. 90°–120°C, and so less reheating of the flue gas stream is required prior to entering the $deNO_x$ reactor [28].

Another interesting development involves pilot-scale test programmes and field trials in the FRG aimed at coating the plates of a rotary regenerative-type of air preheater with catalytic material, so eliminating a separate SCR reactor [28]. In this arrangement, catalyst temperatures would typically range from 150°–350°C.

Appraisal:

1.	Information available	2
2.	Process simplicity	1
3.	Operating experience	2
4.	Operating difficulty	1
5.	Loss of power	2
6.	Reagent requirements	1
7.	Ease of end-product disposal	2
8.	Process applicability	2
	Total	13

CATEGORY N42–SELECTIVE NON-CATALYTIC REDUCTION

Status: A Selective Non-Catalytic Reduction (SNR) process was first developed by Exxon Research Engineering Corporation (ERE) in 1972, and marketed as the Thermal Denox Process [414]. A similar kind of process was independently developed by Mitsubishi (MHI) in Japan and tested at a 375 MWe oil-fired power station [24]. To date, this is the largest plant on which full-scale tests have been carried out [352].

The process has been commercially demonstrated in oil- and gas-fired utility boilers, steam boilers and process furnaces, and tests have been conducted on a municipal incinerator, oil-field steam generator and glass melting furnace, see Table 4.5 [414]. At present, there is no commercial experience with coal-fired boilers, other than test work [409]. The largest coal-fired unit at which limited testing has taken place is 265 MWe [352].

The largest boiler equipped with SNR has a capacity of 180 MWe (equivalent) [352].

Applicability: This process type is suitable for retrofit and new build applications.

The principles of SNR are similar to those of SCR except that, in the absence of a catalyst, a much higher operating temperature is required, typically 900°–1000°C, for the NH_3–NO_x reactions to take place. It is, therefore, necessary to inject the ammonia directly into the upper section of the boiler or process firebox. Operating temperatures are critical because of the narrow temperature window over which the reactions are effective and this is one of the major drawbacks of the method.

If temperatures are too high, i.e. higher than 980°C, side reactions produce NO with NO_x concentrations in the flue gas actually increasing at temperatures in excess of 1090°C [414]. If the temperature is too low, reaction rates are reduced and higher concentrations of NH_3 appear downstream.

The effective operating temperature range can be extended by adding hydrogen (H_2) or natural gas (CH_4). The optimum reaction temperature decreases with increasing H_2 (or CH_4): NH_3 ratio, with the result that at a ratio of 2.0 (and NH_3:NO_x = 1.7), the effective temperature range is 700°C ± 50°C [352].

In full-scale commercial operations, ensuring that NH_3 is injected at the correct temperature is difficult, especially during load swings. This is largely overcome by the inclusion of multiple NH_3 injection points and the controlled use of H_2, as indicated above.

SNR systems installed in early commercial boilers in Japan achieved 50–60% NO_x reduction. The recent development of wall injectors (see later) has extended the system's capabilities to about 80% [352], although higher efficiencies (90%) have been claimed to be possible [409].

While this method is harder to control and generally less efficient than SCR techniques, it is much less expensive to install, especially on existing boilers and heaters [345]. In addition it is free from the problems associated with catalysts, namely blinding/plugging and SO_2 oxidation to SO_3 [345].

Higher NH_3 consumption and the use of H_2 (or CH_4) have an effect on operating costs, although these must be compared to the comparatively high cost of the catalyst in an SCR system, e.g. 40–60% of total operating costs [410].

No commercial experience, only test work, has been obtained with this system in coal-fired boilers, and problems with locating early designs of the NH_3 injection system, especially in retrofit situations, have been noted [409]. It is claimed, however, that the new wall injector design can be easily retrofitted to existing boilers with minimal impact on the boiler structure, and without the need to relocate ducting, air preheaters, stacks, etc [409]. Proper design and location require the use of a detailed kinetic model of the system [409].

To date, most SNR applications have been on flue gas containing virtually no SO_3 [409]. Unreacted NH_3 will combine with SO_3 to produce ammonium sulphate and bisulphate, with the possibility that the bisulphate will cause fouling of downstream equipment. In addition, unreacted ammonia will be absorbed by fly-ash causing possible problems in the electrostatic precipitators [24] and possibly influencing the quality of the fly-ash [4].

The EPDC in Japan have concluded that SNR is not practical for coal-fired boilers [24], although this is partly based on the problems associated with the design and performance of the older NH_3 injection system.

Ando [350] limits the suitability of SNR to SO_x-lean gases from small boilers or furnaces, and to gases which contain impurities that are likely to reduce the activity of SCR catalysts, e.g. from refuse incinerators. However, Exxon [414, 416] believe the process to be suitable for coal-fired boilers as a result of tests carried out on flue gases from coal combustion in the late 1970's and the development of methods

Table 4.5 Status of Selective Non-Catalytic Reduction Plants in Japan & USA Oil & Gas Fired Units

Vendor	No. of Units	Type of Plants	Dates	Range (Nm^3/h)	% Denox (Typical)	Reference
EXXON	6	Utility Boilers	to 1985		60–80(b)	414, 417
EXXON	7	Industrial Boilers	to 1985		50–80(b)	414, 417
EXXON	33	Petroleum Heaters	to 1985		75–85(b)	414, 417
EXXON	2	Incinerators	to 1985		70–80(b)	414, 417
EXXON	4(a)	Oil Field Steamers	to 1985			414, 417
EXXON	1(e)	Glass Melting Furnace	to 1985		50–80(b)	414, 417
MKK(c)	c.20 (f)	Small gas-fired boilers & heaters	1975–78	5,000–20,000	40–60	419
MHI(d)	10 min (f)	Refuse incinerators	to 1985	30,000–100,000	40–60	419

(a) 3 commercial, 1 demonstration. In USA only
(b) Improved technology, i.e. wall injection. Industrial boilers and furnaces in Japan with early technology: 20–60% [350]
(c) Mitsubishi Kakoki
(d) Mitsubishi Heavy Industries
(e) In USA
(f) In Japan only

for minimising the effects of ammonium salts on downstream equipment.

Space Requirements: No details of space requirements are reported in the literature, but this is expected to be insignificant, i.e. the space required for the ammonia storage, mixing and injection system. Compared to the SCR system, major space savings are to be expected because of the absence of the catalytic reactor [352].

Consumables: Typical NH_3 consumption figures have been presented by Exxon [416] for a sample 83 t/h of steam oil- or gas-fired industrial boiler; see Table 4.6.

Ammonia consumption is also expressed in terms of an NH_3:NO_x molar ratio. Quoted ratios are variable (Table 4.7), but are generally of the order of 2.0 for NO_x reductions of about 60%. These compare with a molar ratio of 0.61 for 60% deNO_x with SCR [350].

Hydrogen (or natural gas) consumption is dependent upon temperature variations in the boiler. For example, with a NH_3/NO_x ratio of 1.7 and no hydrogen, optimum flue gas temperatures are in the region of 970°C [352]. As the H_2/NO_x molar ratio increases to 0.5, 1.3 or 2.4, the optimum flue gas temperature falls to 825°C, 750°C and 700°C, respectively. It is claimed that with the new wall injector design, the injection of hydrogen may be unnecessary [414].

Pressure Losses: No information on typical pressure losses is quoted in the literature. Pressure drops through SCR systems have been reported as 8–12 mbar across the ammonia injection nozzles and catalyst reactor, so SNR systems are expected to sustain significantly lower losses than these. In addition, the newer wall injector systems will reduce the pressure losses still further.

Power Consumption: Limited information on power requirements in SNR plant is available. Reference to Table 4.6 suggests that the annual consumption of 55 MWh for a 83 t/h industrial boiler approximates to about 0.05% of the equivalent power output [416].

Operating Experience: The main factors affecting the injection of ammonia into an SNR process are [352]:

- Temperature,
- Residence time at temperature,
- Temperature profile,
- initial NO_x concentration,
- NH_3/NO_x molar ratio,
- H_2/NH_3 molar ratio,
- Mixing conditions.

Early commercial or test facilities involved the location of one or more injection nozzles in the flue gas stream at the appropriate temperature, so that a mixture of NH_3 and its carrier gas (steam or air) could be injected [352]. The carrier gas is usually required to prevent overheating of the grid [409]. In more recent applications, simpler and less expensive wall injectors have largely superseded the injection grids. The wall injectors can be easily retrofitted to an existing boiler with minimum impact on the structure of the boiler and without the need to relocate ducting, air preheaters, stacks etc [409].

Table 4.6 Ammonia Consumption in SNR plant

BOILER DESIGN CONDITION:	83 t/h 42 bar 370°C
Fuel:	Oil or Gas
Initial No_x Concentration	200 ppm(vol)
NH_3:NO_x Ratio:	2.0:1
NH_3 Vaporiser:	Electric Element; direct contact
NH_3 Storage Capacity:	38 m^3 (30 days)
Annual NH_3 Consumption (a):	164 tonnes
Annual Electric Consumption (a):	55 MWh

(a) 65% load factor

Table 4.7 Typical NH_3: NO_x Molar Ratios for SNR

Mol. NH_3 / Mol.No_x	% Deno$_x$	Other Plant Details	Ref (Year)
0.3–0.5	10–20	Oil Refinery/Petrochemical Plant Inlet NO_x = 400 mg/m^3	352 (1986)
0.8–2.0	30	Wall Injectors (new design)	350 (1985)
1.0	30–40	Unreacted NH_3 = 10–15 mg/m^3	350 (1985)
1.0–2.0	45–55	Oil Refinery/Petrochemical Plant	352 (1986)
1.0–2.0	48–63	235 MWe Utility Boiler Injection Grid (old design)	414 (1985)
1.5	35–45	375 MWe Plant, 0.2%S oil. Unreacted NH_3 = 13–23 mg/m^3	352 (1986)
1.5–2.0	50–60	Commercial Boilers	345 (1981)
1.5–2.5 (Optimum)	c.90	Oil Fired Steam Generator	416 (1982)
3.0	70		343 (1980)
4.0	60		352 (1986)

The ammonia must be injected within a specific and narrow temperature range which varies according to boiler load. To accommodate these requirements the wall injectors are arranged in two or more zones, and the amount of NH_3 is controlled. It is necessary to inject varying quantities of hydrogen into the system with the NH_3 in order to obtain good NO_x removal rates at low boiler loads. This adds to the cost and complexity of the operation [409]. Variations in temperature (and flue gas composition) within the large-size ducts, pose additional operating difficulties, thereby limiting the extent to which NO_x reduction may be achieved with SNR techniques [371].

Other difficulties associated with SNR may be:

(a) The leakage and subsequent emission of ammonia (generally below 45 mg/m^3) and by-products. In this aspect, SNR differs little from SCR although, with higher operating NH_3/NO_x molar ratios, leakage rates are expected to be higher.

(b) The potential fouling of downstream equipment, especially the air preheater, with ammonium bisulphate. In the high temperature flue gas, sulphur oxides from the fuel do not interfere with the SNR reaction chemistry [414, 417]. In addition, the injected NH_3 does not promote the oxidation of SO_2 to SO_3: this occurs in SCR due to the presence of the catalyst. However, unreacted ammonia reacts with SO_3 and H_2O in the flue gases, as the temperature falls, to form ammonium bisulphate and, at still lower temperatures, ammonium sulphate. The sulphate is a dry solid which creates neither corrosion nor unacceptable fouling problems in the air preheater [417], whereas the bisulphate is a sticky, corrosive liquid at preheater temperatures. Exxon research [414, 417] has predicted that ammonium bisulphate formation can be minimised by one or more methods:

- Maintaining an NH_3:SO_3 molar ratio above 2.0 and providing sufficient reaction time for the formation of sulphate in preference to the bisulphate,

- Limiting NH_3 leakage to 5 ppm (4 mg/Nm^3) or less,

- Maintaining the air preheater outlet flue gas temperature above 204°C,

- Maintaining the temperature at the NH_3 injection point lower than the sulphate formation temperature, which is dependent on the concentrations of NH_3 and SO_3.

The first method is considered the most practical for SNR processes since NH_3 leakage rates (c.50 ppm or 38 mg/Nm^3) are generally twice those of SO_3 [417]. In addition, extensive testing in Japanese oil-fired boilers indicate that sulphate/bisulphate deposits can easily be removed by water-washing at intervals [417]. Ammonium sulphate creates a 'blue haze' problem at the stack [409].

(c) Thermal damage to the NH_3 injection nozzles and the difficulties associated with keeping them cool. Corrosion and erosion of the injection grid is a further potential problem with coal-fired boilers [409]. These problems are avoided with the improved wall injectors.

The SNR process has been demonstrated commercially in oil- and gas-fired boilers, but there has been no commercial experience with coal-fired boilers. Operation at a 375 MWe oil-fired (0.2%S) utility boiler in Japan has been virtually trouble free, except for problems associated with ammonium sulphate and bisulphate deposits in the air preheater and plume formation at the stack [352].

Process Development: The development of wall injectors has eliminated many of the disadvantages associated with injection grids and reportedly improved NO_x reduction levels [409, 414]. The proprietary designed injectors are located at or near the boundary walls of the injection zone, and can be readily retrofitted to existing boilers. Optimum design and location of the simple, large jets does require the use of a detailed kinetic model of the unit to be fitted.

Wall injectors can be located at the optimum flue gas temperature, even within the combustion zone of the boiler [414]. Two sets of injectors may be used to accommodate load variations with the result that hydrogen injection, to promote low flue gas temperature performance, may be unnecessary.

The first commercial unit installed with wall injectors started operation in late 1984, although NO_x reduction levels were kept relatively low (30%) [350].

Exxon have developed a fundamental kinetic model for SNR which has broadened the understanding of the process, allowing them to predict accurately the performance and NH_3 leakage rate for any type of fired unit [414].

In the late 1970's, tests were carried out in Japan and the USA to increase $deNO_x$ efficiency by including a small amount of SCR catalyst with an SNR system in oil-fired utility boilers [350]. NO_x removal efficiencies of 50–60% and NH_3 leakage rates below 10 ppm (8 mg/Nm^3) were targeted. The tests were not successful and the SNR facilities were removed.

Appraisal:

1.	Information available	1
2.	Process simplicity	2
3.	Operating experience	1
4.	Operating difficulty	0
5.	Loss of power	2
6.	Reagent requirements	1
7.	Ease of end-product disposal	2
8.	Process applicability	2
	Total	11

CATEGORY N43–DRY ADSORPTION

Activated carbon (coke or char) is capable of catalysing the reaction between NO_x and NH_3 at significantly lower temperatures than for SCR [352, 4]. The ability of activated carbon to adsorb SO_2 from the flue gases has led to the development of this process for simultaneous removal of NO_x and SO_2 (see Section 5). However, there is considerable interest in the FRG in the use of activated coke as a catalyst in cold-side SCR applications, i.e. downstream of the FGD plant [28]. Operating temperatures would be in the region of 90° to 120°C, thereby requiring significantly less reheat of the saturated flue gases than would be the case with conventional cold-side SCR catalysts. In addition, with reduced SO_3 concentrations in the flue gas at the NH_3 injection point, the deposition of ammonium bisulphate and consumption (and leakage) of NH_3 would be minimised [352].

Disadvantages of the process include high carbon losses and a low NO_x removal efficiency of 40–60% [343]. Higher deNO_x efficiencies (80%) have been reported at a 90 MWt demonstration plant for simultaneous NO_x/SO_2 removal in Japan, with normal flue gas temperatures of 120°–155°C [24], whilst it is recognised that for commercial applications a low-cost activated coke is required to counter the considerable coke consumption rate [350].

Appraisal (for NO_x reduction only):

1.	Information available	0
2.	Process simplicity	1
3.	Operating experience	0
4.	Operating difficulty	0
5.	Loss of power	2
6.	Reagent requirements	1
7.	Ease of end product disposal	2
8.	Process applicability	2
	Total	8

CATEGORY N50–FLUE GAS TREATMENT-WET PROCESSES

Although several wet processes have been developed or advocated for the removal of NO_x from flue gases, they are all limited by the low solubility of NO in water. The general approach is either to oxidise the NO to the more soluble NO_2 in the gas phase prior to absorption or absorb the NO_x directly in the absorption solution. In addition, the absorbed NO_x can either be reduced or oxidised in solution. Therefore, four major categories of wet NO_x-removal process are recognised:

(i) Gas-phase oxidation, absorption and liquid-phase reduction (oxidation-absorption-reduction)
(ii) Gas-phase oxidation, absorption and liquid-phase oxidation (oxidation-absorption)
(iii) Direct absorption and liquid-phase reduction (absorption-reduction)
(iv) Direct absorption and liquid-phase oxidation (absorption-oxidation)

Most scrubbing solutions for wet NO_x removal also remove SO_x, so current developments are aimed at simultaneous NO_x-SO_x processes, which are particularly suitable for coal or high-sulphur oil applications [345].

Oxidation-absorption-reduction processes are, in most cases, developments of commercially-available FGD techniques [345], and absorption-reduction processes are being developed specifically for the simutaneous removal of NO_x and SO_x to avoid the use of the expensive gas-phase oxidant [419]. Of the remaining two wet deNO_x categories, absorption-oxidation processes were originally developed to treat nitric acid plant tail gases and have a major problem with the high consumption of the expensive liquid-phase oxidant [343]. This problem is caused by the oxidant reacting with absorbed SO_2 as well as NO, and so a separate (and more conventional) SO_2 absorber is necessary. This factor, plus the need for an extremely large NO_x absorber to ensure adequate NO absorption rates, makes the absorption-oxidation processes unattractive and expensive when compared to other NO_x and simultaneous NO_x/SO_x processes [343].

The oxidation-absorption category encompasses a number of processes that have no common mechanism other than a gas phase oxidation stage followed by an absorption stage. The two main processes in this category involve liquid phase (oxidation) reactions which convert the NO_x to nitrate salts or nitric acid, respectively. Of all the wet processes, (ozone) oxidation processes are epected to exhibit the highest NO_x removal efficiencies [347]. The main disadvantages are a very high oxidant consumption rate and the need to scrub the flue gas of SO_2 and particulates prior to the NO_x removal stage [343]. The latter problem makes the process more complicated than similar wet processes where SO_2 and NO_x are simultaneously removed in a single absorber.

Small units have been built in Japan to test wet deNO_x technologies, but it has been reported that these processes will be of little importance for some time [4]. This is primarily due to the waste water problems, caused by the presence of nitrite, nitrate and nitrogen-sulphur compounds, which make wet deNO_x processes of little importance compared to dry techniques [350, 352].

Kawasaki Heavy Industries in Japan have developed wet and dry deNO_x processes since 1970 [347]. Owing to economic and technical difficulties, their effort since 1977 has been concentrated solely on the development of the dry SCR process.

Appraisal (for NO_x reduction only):

1.	Information available	0
2.	Process simplicity	0
3.	Operating experience	0
4.	Operating difficulty	0
5.	Loss of power	1*
6.	Reagent requirements	0
7.	Ease of end product disposal	0
8.	Process applicability	2
	Total	3

*Assumed in absence of data

4.4 Processes for Detailed Study

The selection of processes for detailed study in this Volume has been based upon their suitabiliy for application in the U.K. for the three datum combustion systems (Section 1.4) considered:

- 2000 MWe (4 × 500 MWe) power station (Datum system 1).

- Large (450 tonne steam/h) industrial boiler (Datum system 2).

- Small (13 tonne steam/h) factory boiler (Datum system 3).

In principle, all of the processes listed in section 4.1 can be applied to all combustion plant, but the attraction of many processes diminishes with factors such as decrease in plant operating scale, and increases in deNO_x process complexity, reagent costs, and end-product disposal difficulty.

Appraisal of Processes

To evaluate some of these factors, a rough appraisal of each process type has been made in Section 4.3 by assigning 'merit points' for a number of features; merit points have been awarded according to the scale:

0 Below average merit

1 Average merit

2 Above average merit

3 Outstandingly above averge merit

The features to which these points have been assigned are described in Section 1.5; they are briefly:

1 Information available

2 Process simplicity

3 Operating experience–extent and difficulties encountered

4 Operating difficulty–availability, reliability

5 Loss of power sent out–by installation of the FGD process

6 Reagent requirements–quantities

7 Ease of end-product disposal

8 Process applicability–e.g. for retrofit

All of the processes listed in Section 4.1 and outlined in Section 4.2 are appraised in Section 4.3. The merit points assigned to the process types for each of the above features are summarised in Table 4.8. It should be noted that the number of points in the merit point system adopted in other Sections of the Manual are not strictly comparable with those considered here.

Processes suitable for the U.K.

The principal purpose of assigning merit points to each of the processes was to aid in the selection of processes that could be considered suitable for applicaton in the U.K. It was arbitrarily assumed that suitable deNO_x processes would be those having a total of 10 or more merit points. It is worthy of note that one process, Dry Adsorption, was assigned 8 points, differing from SCR and SNR processes primarily in the lack of information (for deNO_x alone) and operating experience. Although the process has recognised potential, it is being developed primarily as a simultaneous NO_x-SO_2 abatement process and is discussed further in Section 5.

Selection of processes for detailed study

No flue gas treatment processes are considered to be suitable for the smallest operating scale dealt with in this Volume (Datum System 3), primarily because the lower NO_x reduction efficiencies required to meet expected NO_x emission standards can be readily, and less expensively, attained by combustion modifications alone.

For utilities and large industrial boilers, two processes for consideration in the U.K. are:

- Selective Catalytic Reduction (SCR) process (Process Code N41; merit rating 13 points). SCR is a well-established process with commercial experience in coal-fired units dating back to 1980, and much further for oil- and gas-fired applications. It is a relatively simple process and can achieve NO_x reduction efficiencies of over 90%.

- Selective Non-Catalytic Reduction (SNR) process (Process Code N42; merit rating 11 points). SNR has been commercially applied to oil- and gas-fired units but, to date, not to coal-fired systems; full-scale testing has, however, been carried out at coal-fired plants. It is a very simple and relatively inexpensive process, capable of achieving NO_x reduction efficiencies of 70–80% for utility and industrial boiler applications.

Table 4.8 Summary of $DeNO_x$ Process Appraisals

Code No.	Name	Merit Points for Feature No:								Total Points
		1	2	3	4	5	6	7	8	
N41	Selective Catalytic Reduction	2	1	2	1	2	1	2	2	13
N42	Selective Non-Catalytic Reduction	1	2	1	0	2	1	2	2	11
N43	Dry Adsorption	0	1	0	0	2	1	2	2	8
N50	Wet Processes	0	0	0	0	1	0	0	2	3

Features: 1. Information available
2. Process simplicity
3. Operating experince
4. Operating difficulty
5. Loss of power
6. Reagent requirements
7. Ease of end-product disposal
8. Process applicability

4.5 Evaluation of Selected Denox FGT Processes

EVALUATION OF SELECTIVE CATALYTIC REDUCTION PROCESS (CODE N41)

See Section 4.2 for: outline of basic process, chemistry and schematic diagram.

See Section 4.3 for a list of manuacturers offering this type of equipment and for a general appraisal of the basic process.

See Section 4.4 for the reason for choosing the selective Catalytic Reduction $DeNO_x$ process.

This basic process type is considered to be suitable for application in the U.K. only to utility boilers and large industrial boilers. The prohibitive cost of the catalyst and the scale of operation make SCR unsuitable for application with small factory boilers, where combustion techniques would be more cost effective. Hence, the SCR process is evaluated here only for Datum Combustion Systems 1 and 2.

Process Description: Figure 4.11 shows a simplified flow diagram for application of the process to a pulverised fuel fired water tube boiler.

A mixture of ammonia gas and air is injected into the flue gas upstream of a catalytic reaction chamber which, in a typical arrangement, is located between the outlet of the economiser section of the boiler and the flue gas inlet of the air preheater. The flue gas mixture flows vertically downwards through the reactor where the ammonia reacts with NO_x to form nitrogen gas and water vapour. The Reactor consists of several levels of catalyst modules which, in coal fired applications, are typically preceded by an inactive stage in order to protect the active catalyst sections against erosion. The vertical flow path through the reactor minimises the amount of fly-ash settling on the surface of the catalyst. Some fly-ash is collected in Hoppers at the bottom of the Reactor chamber.

Flue gas leaving the Reactor is cooled in a Heat Exchanger prior to entering an Electrostatic Precipitator at a temperature of about 150°C. The gas is discharged to the Stack via an induced-draught Fan.

There are several possible locations for an SCR reactor. The system described above is termed a 'High Dust' system (Figure 4.8) because the flue gas contains a high fly-ash loading when it is introduced to the SCR reactor. If an FGD plant is also installed, it would be located downstream of the Electrostatic Precipitator, posibly after being further cooled in a Heat Exchanger. An alternative arrangement is for the SCR reactor to be located downstream of a high temperature electrostatic precipitator in order to minimise erosion of the catalyst and deposition in the reactor. This is termed a 'Low Dust' system (Figure 4.9).

A third arrangement, which is favoured by some German companies, locates the SCR reactor downstream of an FGD plant, the so-called 'Tail Gas' system (Figure 4.10). This alleviates many of the problems associated with retrofitting $deNO_x$ technology to boiler plant already ftted with an FGD process. In addition to the benefits of the low dust system, this arrangement significantly reduces the SO_2/SO_3 concentrations in the SCR reactor, thereby limiting the problems caused by ammonium bisulphate on downstream equipment, such as the Air Preheater and minimising wastage of amonia. The major disadvantage of this arrangement is the need to reheat the saturated flue gasses from the FGD plant (typically 50°C) to a temperature that is suitablefor SCR, e.g. 370°C. The use of low temperature catalysts such as activated carbon may ultimately lead to low temperature reheat requirements.

Status and Operating Experience: The status of the SCR process is indicated in Section 4.3 and Tables 4.1, 4.2 and 4.3 where it can be seen that in the region of 200 SCR plants have been, or are being, installed in Japan, the FRG and USA. Of these, 53 are known to be fitted to coal-fired plant with a total capacity of about 13.8 GWe.

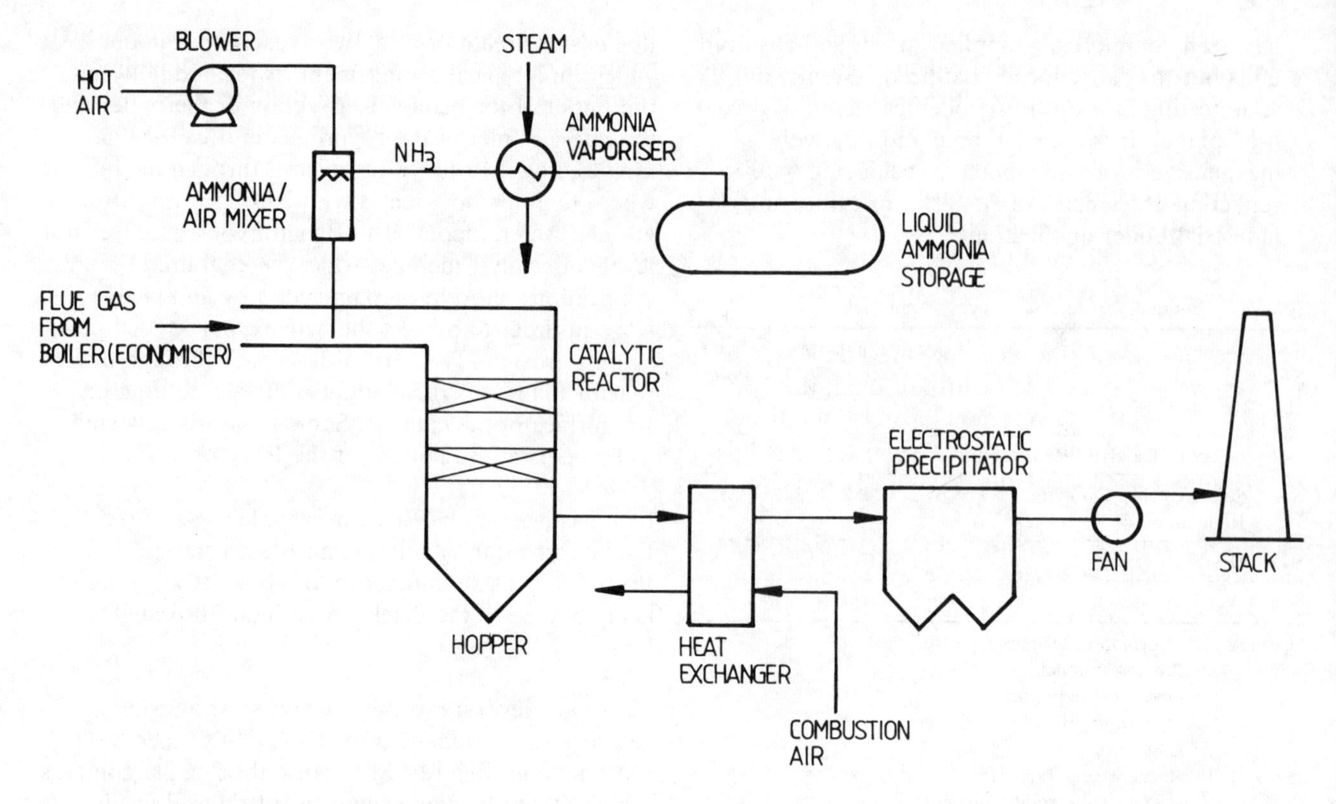

Figure 4.11 Selective Catalytic Reduction Process

Despite some potential problems (see later) SCR is the most widely utilised and developed process for NO_x reduction. It is finding widespread application in the FRG because the stringent deNO_x regulations call for reductions beyond the capabilities of combustion modifications, i.e. at least 50–70%, and because of the early deadlines (late 1980's) for compliance.

In its early stages of development, a number of operational problems were identified and solved. These were:

(a) SO_x poisoning of the catalyst. TiO_2-based catalysts have largely replaced those based on Al_2O_3 and Fe_2O_3 to minimise damage caused by SO_2 and SO_3. The 'Tail Gas' arrangement will also minimise the presence of sulphur oxides in the flue gases entering the SCR reactor.

(b) Catalyst promotion of SO_2 oxidation to SO_3. Low-oxidation catalysts are now used to minimise this reaction. V_2O_5 is added to TiO_2 in small amounts to reduce SO_x poisoning, but it does promote some SO_2 oxidation. Other oxides, such as those of tungsten or molybdenum, can be used to replace all or part of the V_2O_5 [350].

(c) Catayst dust plugging. Parallel flow catalysts, e.g. honeycomb, plate or tube designs, are used to minimise dust capture. (N.B. The honeycomb design is the most popular among vendors bcause of its strength and ease of handling). Flue gases with high dust loadings are normally directed downwards through the reactor to reduce settling on the catalyst. Soot blowing can also be employed to remove the dust periodically from the reactor.

(d) Fly-ash erosion of catalysts. Moderate gas velocities in the reactor, the selection of a harder catalyst, and the use of dummy catalyst sections upstream of the active catalyst all serve to minimise erosion.

(e) Ammonium bisulphate deposition in the catalyst. Maintaining flue gas temperatures above 330°C in the reactor inhibits the reactions which produce the bisulphate. This can be achieved by installing a hot gas bypass upstream of the economiser. Temperature control is especially important during perods of load swing.

(f) Ammonium bisulphate deposition in the air preheater. The potential for bisulphate formation increases, with decreasing temperature, making the air preheater (and induced-draft fan) vulnerable to this problem. The main counter-measure is to minimise the concentration of the main reactants, NH_3 and SO_3, in the gas stream leaving the reactor. NH_3 leakage from the reactor should, typically, be less than 5 ppm (4 mg/Nm^3). Alternative measures are the installation of soot blowers or design changes to the air preheater plates. Solutions for minimising the effects of bisulphates either increase costs or reduce the efficiency of power generation. It should be noted, however, that EPDC field tests in Japan have shown that without these measures, air preheater elements could become blocked in only 6 months of operation; with water washing 12 months operation could be achieved [343].

(g) Ammonia contamination of fly-ash and FGD waste-water. In Japan, NO_x removal efficiencies are normally limited to 80% to avoid significant NH_3 leakage to downstream equipment.

(h) maintenance and accuray of NO_x and NH_3 instrumentation. The control of the NH_3-injection rate was reported [408] to be a major operating difficulty at the 215 MWe gas/oil-fired Huntington Beach Unit 2 Station in the U.S.A.

In Japan it has been reported that, as a result of the developments listed above, SCR has become highly reliable and less costly [350]. All plants installed since 1979 have been operated automatically and with virtually no trouble.

The first full-scale SCR system for a coal-fired boiler was intalled at Chugoku Electrics 175 MWe Shimonoski No. 2 Station in 1980, and was of the High-Dust design [413]. The plant was designed for 50% NO_x removal (500 ppm to 250 ppm). Since start-up in May 1980, there had been little change in deNO_x efficiency, although the NH_3 outlet concentration (leakage) had increased gradually to about 10 ppm after 26,000 hours (3 years) of operation. This was attributed to performance loss of the catalyst which had operated continuously without change. As the increase in NH_3 leakage had been accompanied by an increased draught loss in the air preheater, the catalyst was partially replaced before continuing operation.

Hokkaido Electrics 350 MWe coal-fired boiler at Tomato-Atsuma was the first SCR with a 'hot' electrostatic precipitator arrangement. The fuel has a low sulphur (0.3% content, and 1.2% nitrogen, with one-quarter of the flue gas being treated wth SCR and one-half being treated with a limestone-gypsum FGD process. The reported [352] smooth operation of the plant since October 1980 demonstrates the feasibility of the 'Low-Dust' system. Furthermore, MHI research has indicated that there is no difference between High-Dust and Low-Dust systems with regard to the decrease in catalyst activity due to ageing [412].

The first full-scale demonstration SCR at a dry-bottom coal-fired boiler in Europe started operation at the end of 1985 at the 420 MWe Altbach No. 5 station in the FRG, whilst the first unit at a wet-bottom coal-fired boiler is under construction at the 345 MWe Knepper power station in the FRG.

In an extensive programme of field trials in the FRG starting in early 1985, it has been verified [28] that installations based on Tail-Gas (post FGD) SCR systems can expect to show favourable reliability. Furthermore, it is anticipated that catalyst life in such an arrangement can expect to be better than for 'Hot-Side' SCR layouts.

When an SCR system is installed at a coal-fired station, its influence on the other equipment should be considered. Typical operational considerations are [413]:

(a) Boiler–control of SCR inlet temperature (if necessary).

(b) Air preheater/fan–increase of draught loss due to SCR, and prevention of air preheater plugging and corrosion.

(c) Electrostatic precipitator–influence of leakage NH_3 on fly-ash (low temperature E.P. only).

(d) FGD plant–prevention of plugging and corrosion of a gas–gas heater (if installed), and influence of NH_3 in waste-water.

Variations and Development Potential: NO_x is reduced by NH_3 at a temperature of 150°–450°C in the presence of a catalyst. The optimum temperature is in the range 300°–400°C; at lower temperatures the reaction is too slow, whilst at higher temperatures the activity of the catalyst tends to be lowered by thermal effects [350]. These temperatures are readily achieved by locating the SCR reactor between the economiser section of the boiler and the air preheater, so that no flue gas reheat is required. Low temperature catalysts for use at 150°–250°C and high temperature catalysts for use at 400°–550°C have been developed for special gases [350] and there is much interest in the FRG in activated coke as a low-temperature (90°–120°C) catalyst in Tail Gas SCR systems [28].

Ammonia is a suitable reducing agent for SCR applications because it selectively reacts with NO_x. Other reducing agents such as hydrogen, carbon monoxide and methane could be used as alternatives, but consumption is generally high due to reaction with the oxygen present in the flue gas.

As seen previously, apart from the Tail-Gas arrangement, where the SCR reactor is located downstream of an FGD plant, there are two basic 'hot-side' arrangements, e.g. High-Dust and Low-Dust systems. A comparison of the features of these two latter systems is given in Table 4.9.

The most popular catalyst material at present is V_2O_5 on a TiO_2 base. V_2O_5 is highly active with a comparatively long life, whilst TiO_2 is affected very little by SO_3. Other catalyst materials that can be considered are the oxides of iron (ferric),

Table 4.9 Features of High-Dust & Low-Dust SCR Systems [24]

	HIGH-DUST SYSTEM	*LOW-DUST SYSTEM*
SCR REACTOR	1. Countermeasures against catalyst erosion required. Install rectifying plates upstream of catalyst layer, and use erosion-resistant catalyst. 2. Catalyst should have open shape to prevent plugging by coarse dust.	No special measures against erosion are necessary; erosion of catalyst is insignificant.
AIR PREHEATER	Ammonium bisulphate deposition tends to be small because of washing effect of coarse dust. To prevent plugging, it is necessary to select suitable element shape, operate suitable soot blower and control NH_3 leakage to 5 ppm (4 mg/Nm^3).	Ammonium bisulphate deposition tends to be large because of less washing effect of dust. To prevent plugging, it is necessary to select suitable element shape, operate soot blower and control NH_3 leakage to 5ppm (4 mg/Nm^3).
ELECTROSTATIC PRECIPITATOR	1. Performance is low in cases of low sulphur coal combustion. Otherwise, performance is stable. 2. Lower cost.	1. Performance variation due to coal composition is small. 2. Radiation losses from the casing affects the boiler efficiency. 3. Higher flue gas volumes may be handled if gas is recirculated back to the boiler from downstream of ESP. 4. Higher cost.
FGD PLANT	SCR leakage NH_3 has no effect on FGD; NH_3 reacts with SO_3 and the products adhere to fly-ash and are collected in ESP. NH_3 concentration in waste-water from FGD is kept low.	SCR leakage NH_3 reacts with SO_3 and products are collected in FGD plant. The NH_3 concentration in waste-water from FGD is higher.
ASH HANDLING	Ammonia is contained in the fly-ash because the ESP is installed downstream of the SCR reactor.	The ash collected in the ESP contains no NH_3.

Table 4.10 Advantages & Disadvantages of Parallel-Flow Catalysts [352]

MAIN ADVANTAGES	*MAIN DISADVANTAGES*
1. No plugging dust, even at high dust loadings.	1. Shutdown required to replace catalyst (compared to moving-bed designs).
2. Smaller frontal area; granular catalysts must have thin beds and large frontal areas to avoid high pressure drop.	2. Higher cost per unit surface area.
3. Higher gas velocity, thus improving transfer to the catalyst surface.	3. Larger reactor volume because of space velocity.
4. Minimum attrition compared to moving-bed design.	4. Higher cost for installing the catalyst, compared to dumping the granular type.
5. Less pore blinding.	

Table 4.11 Advantages of Molecular Sieve Catalyst [352]

1. Low catalyst volume because of high inner surfaces. Only small reactors required.
2. No oxidation of SO_2 to SO_3 at temperatures below 450°C.
3. NH_3 is stored in the catalyst causing highly flexible behaviour at load variations.
4. Low NH_3 leakage, e.g. below 5 ppm, even at high removal efficiencies.
5. No poisoning of the catalyst can occur, so lifetime is limited only by mechanical erosion.
6. The used catalyst can be used as a raw material in the ceramic industry.

molybdenum, tungsten and chromium. An alternative base material is silica.

In addition to the selected catalyst material, the NO_x removal performance will be affected by such criteria as the molar ratio of NH_3 to NO_x, and the volume, temperature, NO_x concentration and O_2 concentration of the flue gas entering the reactor.

The design of the catalyst bed is determined by the dust and SO_3 concentration in the flue gas and the need to minimise pressure drop. Granular or pellet type catalysts have been largely superseded by the parallel-flow catalysts which take the shape of a honeycomb, plate or ring arrangement. The honeycomb shape appears to give the best combination of specific surface (area per volume) and pressure drop [352]. The advantages and disadvantages of the parallel-flow catalysts compared to granular types are listed in Table 4.10. A special catalyst that is totally based on ceramics is the molecular sieve catalyst. The reported advantages of this type of catalyst are given in Table 4.11. A zeolite type catalyst has been licensed for use in three small SCR plants in the FRG using the 'Tail-Gas' arrangement [28].

This is a well-established and proven design for gas- and oil-fired systems which has found widespread application to coal-fired units in the last six years. It is capable of achieving high NO_x reduction efficiencies (90% or more), which are beyond individual combustion techniques and, possibly, the selective non-catalytic reduction method.

Potential for Process Development: In Europe, guarantees of catalyst life are generally up to 2 years. This is partly due to the lack of operational experience of SCR plant in Europe, but also because Japanese experience has indicated actual catalyst life to be little more than 3 years for coal-fired boilers [352]. It is anticipated that the future will see much effort being made by catalyst manufacturers to extend the lifetime of the catalyst to at least 5 years, as is presently the case for oil-fired systems. It is further expected that low SO_2-oxidation catalysts will be developed for application with flue gases having high SO_2 concentrations [2].

Low-temperature SCR systems, with activated coke as the catalyst, are being tested in the FRG. Such a system is of particular interest in 'Tail-Gas' arrangements because it needs little or no preheating, experiences no catalyst poisoning, and the catalyst can either be regenerated or disposed of as a fuel [361].

One of the most interesting developments in the field of SCR is also taking place in the FRG. This is the installation of catalyst-coated plates on the surfaces of a rotary regenerative air preheater [28]. The catalyst operates over a temperature range of 150°C to 350°C, and the NH_3 is injected into the flue gas behind the economiser. The idea is attractive insofar as the catalyst surface area required for a moderate deNO$_x$ efficiency is comparable to that available in typically-installed regenerative heat exchangers. In this way, a separate SCR reactor would not be required, saving space and cost. NO_x removal efficiencies in excess of 60% have been reported in small-scale field tests, but an early test on larger scale (150 MWe) in late 1985 produced lower efficiencies [28]. An alternative source of information on this development indicates that deNO$_x$ efficiencies of 90% with low-sulphur (0.2%) coals decreased rapidly when higher sulphur coals were tested [352]. The potential for this arrangement seems to be as a low-cost deNO$_x$ method, used in addition to combustion modifications.

A great deal of research and field testing has been carried out in Japan and the FRG to establish the design of air preheaters for conventional SCR applications and gas–gas heaters in a 'Tail-Gas' SCR arrangement. Important design criteria are as follows [28]:

(a) Heating surfaces must have a configuration that can be easily cleaned by soot blowing.

(b) The cross-over of heating-element layers should be avoided in areas most likely to be affected by ammonium bisulphate deposition.

Table 4.12 Process Requirements

Datum Combustion System		1	2
Gas at full load:			
Volume flow	'000 Nm³/h	1524*	551
Dry gas	'000 Nm³/h	1426*	513
Water Vapour	'000 Nm³/h	99*	38
Actual volume flow	'000 Nm³/h	3589*	1298
Temperature	°C	370	370
Particulates content	mg/Nm³ (dry)	115	115
SO_2 content	mg/Nm³ (dry)	4240	3520
HCl content	mg/Nm³ (dry)	340	315
Inlet gas at full load:			
NO_x content (as NO_2)	mg/Nm³ (dry)	1025	1440
Outlet gas at full load:			
NO_x content (as NO_2)	mg/Nm³ (dry)	200 (282)	200 (324)
NH_3 content (a)	ppm	3	5
Reagent		Ammonia	Ammonia
NH_3:NO_x Molar Ratio		0.82	0.91
Requirements at full load:			
Ammonia	kg/h	2219	334
Hot water (120°–90°C)	tonnes/h	24.7	3.7
Electric power	MWe	4.1	0.34
Manpower	men/shift	0	0
Average load factor	%	***	***

*Per 500 MWe boiler

***Data not available

(a) 1 ppm NH_3 = 0.76 mg/Nm³

Figures in parentheses are annual average omissions for 90% deNO$_x$ plant availability

(c) Enamelled heating elements should be used, especially where SO_3 concentrations exceed 2 ppm (7 mg/Nm³).

(d) highly efficient soot-blowing devices should be used to limit outages.

(e) Provision for frequent off-line water washing of the air preheater should be considered in cases where SO_3 concentrations in excess of 2 ppm are encountered.

Process Requirements for Each Application Considered: Process requirements are shown in Table 4.12 for the two applications considered: Datum Combustion Systems 1 and 2.

It is assumed that for System 1:

- The NO_x content of the gas is to be reduced to 200 mg/Nm³, corresponding to a NO_x reduction of about 80%.

- NH_3:NO_x molar ratio is 0.82.

It is assumed that for System 2:

- The NO_x content of the gas is to be reduced to 200 mg/Nm³, corresponding to a NO_x reduction of about 86%.

- NH_3:NO_x molar ratio is 0.91.

In both cases, it is assumed that:

- The SCR reactor is located immediately downstream of the boiler.

- NO_x consists of 90% NO and 10% NO_2 by volume.

- Particulate settling in the reactor is negligible.

- Reactions between NH_3, SO_2 and HCl can be ignored.

Byproducts and Effluents: The reaction between NO_x and ammonia produces nitrogen gas and water vapour, so essentially no problem waste material is formed. Side-reactions between NH_3 and SO_3 in the flue gas result in the formation of ammonium bisulphate/sulphate which cause fouling problems on downstream equipment; these reactions are inhibited by the use of low SO_2-oxidation catalysts and by limiting NH_3 leakage to less than 5 ppm (4 mg/Nm³).

Where an FGD plant is located downstream of the SCR reactor, leakage NH_3 and ammonium salts will be flushed from the gas stream by the FGD scrubbing liquor and discharged in the waste water stream. The presence of these gas-borne wastes will affect the FGD reagent consumption and the nature of the waste water. Typically, an NH_3 leakage rate of 5 ppm (4 mg/Nm³) from an SCR unit fitted to a 2000 MWe utility boiler is equivalent to 23 kg/h of NH_3 to be neutralised in the FGD plant.

Efficiency and Emission Factors: The efficiency and emission factors for the process are summarised in Table 4.13 for the two applications considered: Datum Combustion Systems 1 and 2.

In calculating the efficiency factors, it is assumed that at full load:

- The performance without the incorporation of the SCR plant would be as shown in Table 1.4.

- For Datum System 1, the SCR plant overall power consumption is 0.22% of the energy generated (including hot water for ammonia evaporation), i.e. 4.1 MWe.

- For Datum System 2, the SCR plant overall power consumption is 0.23% of the equivalent energy generated (148 MWe), i.e. 0.34 MWe. This is taken to be equivalent to the combustion of 0.13 tonne/h of coal at a power station, assuming an overall power generation efficiency of 33%. This additional coal is arbitrarily assumed to be included with the coal burned in the boiler for calculating the efficiency and emission factors.

The efficiency of the SCR process is primarily determined by the activity of the catalyst, although it is influenced by reaction temperature, NH_3:NO_x molar ratio and the flue gas space velocity through the catalyst [352].

Effect of plant availability: For illustration purposes, the annual average emissions and emission factors for deNO_x plant availabilities of 100% and 90% are given in Tables 4.12, 4.13 and 4.14.

Effect of load variations: Reduced boiler loads are accompanied by drops in the temperature of the flue gas leaving the boiler economiser and, hence, entering the SCR reactor. Lower operating temperatures in the reactor promote the formation of ammonium bisulphate (and sulphate) which cause fouling problems in downstream equipment. This factor, and the potential damage caused by thermal shock to the catalyst by frequent cycling (as is common in German power stations and is being rigorously investigated there), can be countered by installing a by-pass around the economiser to maintain acceptable temperature levels. This leads to a reduced plant efficiency factor, although no data are available on the precise effort.

Effect of design variations: Two major design variations have an effect on NO_x emission factors: 'Tail Gas' SCR arrangements with conventional and

Table 4.13 Efficiency and Emission Factors

Datum Combustion Systems		1	2
Coal heat input (gross)	MWt	1292*	468
Coal fired	tonnes/h	189*	60.3
SCR plant power consumed	MWe	4.1	0.34
Equivalent coal input	tonne/h	–	0.13(b)
Useful energy from system	GJe/h	6681	–
	GJt/h	–	1468
Total equiv. coal input	tonnes/h	756	60.43
Efficiency factor	GJe/tonne	8.84	–
	GJt/tonne	–	24.3
Nitrogen emission (in NO_x)	kg/h	347 (489)	31 (51)
NO_x emission factor (a)	kg/tonne	0.46 (0.65)	0.51 (0.83)

*For each 500 MWe boiler
(a) Emission factor expressed as kg nitrogen per tonne of coal
(b) Calculated assuming overall power generation efficiency = 0.33
Figures in parentheses are annual average emission for 90% deNO_x plant availability

activated carbon catalysts, respectively. The latter arrangement is in the development stage and will not be considered further. Estimated emission factors for the former arrangement, located downstream of a limestone-gypsum process, are presented in Table 4.14. It is assumed that the process conditions leaving the FGD plant are as given in Table 2.41, and that the gases leaving the FGD absorption tower are heated to the SCR reactor operating temperature by regenerative-type heat exchangers (see Figure 4.8). Emission factors without the SCR plant would be as indicated in Table 2.43 and efficiency factors for this and the more conventional (hot-side) arrangement are expected to be as presented in Table 4.13.

Limitations: Selective Catalytic Reduction is limited in its application by its low cost effectiveness in comparison to combustion techniques. It is most likely to be applied where high NO_x reduction efficiencies (80–90%) are required, although at the higher end of the range there remains the possibility that increased leakage rates of NH_3 will result in fouling and waste water problems downstream. These higher efficiencies can be attained by a combination of combustion techniques and by the more recent developments in the field of selective non-catalytic redution, although the latter still has problems with temperature control.

Costs:

Capital Costs: See Section 4.6

Annual running costs and cost factors: ***

Effect of design variations on cost: ***

Effect of annual load patterns on annual running costs: ***

Process Advantages and Drawbacks: The advantages are:

1. NO_x removal efficiencies in excess of 90% can be achieved.
2. The process has been commercially applied to flue gas from coal-fired boilers since 1980.
3. Low leakage NH_3 rates, i.e. less than 5 ppm (4 mg/Nm^3), are experienced for NO_x removal efficiencies of up to 80%.

Table 4.14 Efficiency and Emission Factors for Tail Gas System

Datum Combustion System		1	2
Gas at full load:			
Volume flow (dry)	'000 Nm^3/h	1423*	512
Temperature	°C	370	370
NO_x concentrations:			
At inlet	mg/Nm^3	945	1325
At exit	mg/Nm^3	200 (274)	200 (312)
NO_x removal efficiency	%	79	85
Nitrogen emission (in NO_x)	kg/h	346 (476)	31 (49)
NO_x emission factor (a)	kg/tonne	0.46 (0.63)	0.52 (0.81)

*For each 500 MWe boiler
(a) Emission factor expressed as kg of nitrogen per tonne of coal
Figures in parentheses are annual average emissions for 90% deNO_x plant availability

The disadvantage is that catalyst lifetime is limited to 2–3 years, accounting for a high proportion (40–60%) of the annualised operating costs.

EVALUATION OF SELECTIVE NON-CATALYTIC REDUCTION PROCESS (PROCESS CODE N42)

See Section 4.2 for outline of basic process, chemistry and schematic diagram.

See Section 4.3 for a list of manufacturers offering this type of equipment and for a general appraisal of the basic process.

See Section 4.4 for the reason for choosing the selective non-catalytic reduction $DeNO_x$ process.

This basic process type is considered to be suitable for application in the U.K. only to utility boilers and large industrial boilers. The comparatively lower NO_x reduction efficiencies required for small factory coal-fired boilers, to attain the same NO_x emission concentrations as for larger boilers, can be achieved more cost effectively using combustion techniques. Hence, the SNR process is evaluated here only for Datum Combustion Systems 1 and 2.

Process Description: The selective non-catalytic reduction process (SNR), sometimes referred to as Thermal $DeNO_x$, is based on the reaction between flue gas NO_x and injected ammonia gas to produce nitrogen and water vapour. The reaction is very sensitive to temperature with peak NO_x reduction rates occurring over a limited temperature range of about 900° to 1000°C. Generally, the NH_3 is injected into the flue gas stream by means of an air stream carrier gas at a location (or locations) specifically selected to provide an optimum reaction temperature and residence time. Hydrogen (or natural gas) can be injected with the ammonia to extend the effective NO_x reduction reactions down to temperatures of about 700°C (for H_2:NH_3 ratio of 2:1) with the result that lower optimum reaction temperatures are achieved with higher H_2 injection rates.

The major factors affecting the process are:

- Temperature,
- Residence time at temperature,
- Temperature profile,
- Initial NO_x concentration,
- NH_3:NO_x molar ratio,
- H_2:NH_3 molar ratio,
- Mixing condition,
- Interaction of flue gas constituents, especially O_2, H_2O and free radicals.

To achieve the temperatures required for the reaction, the NH_3 (and H_2) is normally injected into the upper section of the boiler, either upstream or downstream of the superheater, between two adjacent superheater sections, or at two or more of these locations. Each boiler needs to be considered separately for optimum location of the NH_3 injection system:

A schematic diagram of a typical system is depicted in Figure 4.12.

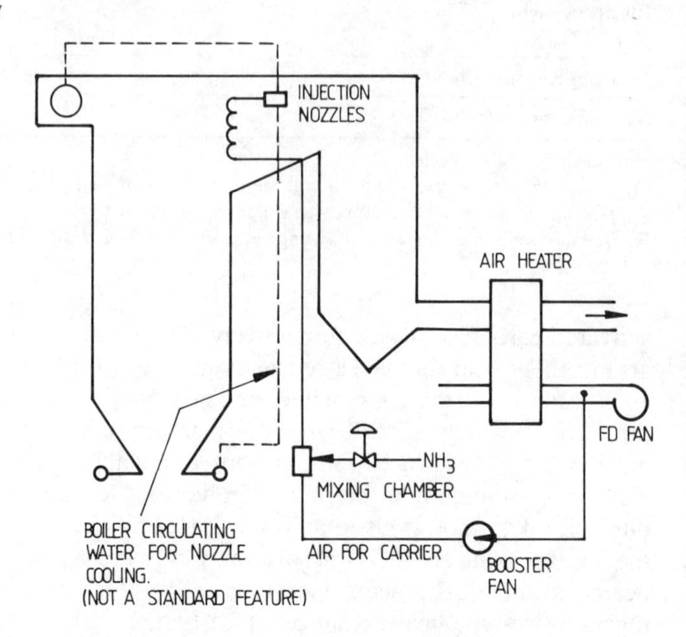

Figure 4.12 Schematic Diagram of NH_3 Injection System

Status and Operating Experience: The status of the SNR process is indicated in Section 4.3 and Table 4.5, where it can be seen that more than 80 commercial SNR plants have been installed in Japan and the U.S.A. All are gas- or oil-fired units, of which thirteen are known to have been fitted to utility and industrial boilers. At present, there is no commercial experience with coal-fired boilers, other than test work.

The largest boiler equipped with SNR has an equivalent capacity of 180 MWe, whilst the largest oil and coal-fired plants on which test work has been carried out are 375 MWe and 265 MWe, respectively.

Early applications of this $DeNO_x$ method involved the location of an injection grid within the flue gas stream, at the appropriate temperature, for the injection of NH_3 and carrier gas (air or steam). Sometimes, multiple grids were used (at several locations) or hydrogen injected with the NH_3 to account for changes in temperature, especially during load swings. The injection grid was subject to high temperature, corrosive, erosive and fouling conditions, whilst multiple grids and H_2 injection increased the cost and complexity of the system.

The injection grids have now been largely superseded by wall injectors, which consist of large jets located at or near the boundary walls of the injection zone. The advantages of wall injectors in an SCR application have been listed [414] as:

- Higher performance,
- Lower investment cost,
- Better load following (without hydrogen),
- Lower fouling tendency,
- No cooling requirement,
- Simple installation,
- Easily retrofitted,
- Greater cost effectiveness.

Despite their simplicity and lower cost, wall injectors require careful design and location to achieve the optimum performance. It is now possible to locate wall injectors at the optimum flue gas temperature, and even within the combustion zone of the boiler. Two or more sets of injectors can be installed, at little extra cost, to allow for temperature fluctuations due to load changes. As such, the injection of H_2 will be minimised, thereby reducing operating costs.

Wall injectors can be retrofitted to existing boilers without the need to relocate ducting or other downstream equipment, and with minimal impact of the boiler structure. If actual and anticipated flue gas temperatures in the region of the wall injectors differ significantly, it is relatively easy to relocate the injectors.

The injection grid system was first applied in Japan in early 1974 where NO_x reduction efficiencies of up to 65% were achieved. Wall injector technology has been applied since Autumn 1984 to a 200 t/h boiler generating steam and power in a Japanese chemical plant. NO_x reduction efficiencies of 70% were achieved at a cost estimated to be 20% of that of a comparable SCR facility [414]. Efficiencies of up to 80% for industrial and utility boilers (and up to 90% for other applications) have been achieved with this technology [416].

SNR, like SCR, has experienced downstream ammonium bisulphate fouling problems caused by the low temperature reaction between leakage NH_3 and the SO_3 present in the flue gas stream. SNR NH_3 leakage rates are somewhat higher (c.50 ppm or 38 mg/Nm^3) than for SCR, but flue gas SO_2 oxidation to SO_3 is reduced due to the absence of a promoting catalyst. It is claimed that the problems of ammonium bisulphate fouling can be minimised by satisfying one or more of the following criteria [414]:

1. Maintaining an NH_3:SO_3 molar ratio above 2.0, and providing sufficient reaction time for the formation of ammonium sulphate rather than the bisulphate.

2. Limiting NH_3 leakage to 5 ppm (4 mg/Nm^3) or less.

3. Maintaining the air preheater flue gas outlet temperature above 200°C.

4. Maintaining the temperature at the NH_3 injection point lower than the sulphate formation temperature.

Method '1' is the most practical method for SNR processes because typical NH_3 leakage rates are generally twice those of SO_3 [417].

NH_3 leakage in SNR systems is extremely complicated because it is dependent upon the interaction of numerous factors, including the concentration of flue gas components, time-temperature relationships, NH_3 mixing efficiency, and the design of downstream equipment. The amount of NH_3 leakage can be determined for particular fired systems, permitting Exxon to guarantee leakage rates as low as 5 ppm (4 mg/Nm^3) where favourable conditions exist [414].

Operation at a 375 MWe utility boiler in Japan, firing on low-sulphur (0.2%) coal, has been virtually trouble-free except for problems associated with ammonium bisulphate/sulphate deposits in the air preheater, and plume formation at the stack [352].

Results from testing at a 265 MWe coal-fired unit in Japan in 1977 led to the conclusion that SNR was not a practical deNO_x process for coal-fired boilers [24]. However, more recent tests on flue gases generated from coal combustion and using the improved wall injector design, have demonstrated that the process is applicable to coal-fired utility boilers [414].

Variations and Development Potential: Although the older injection grid technology is still installed in a number of oil- and gas-fired units, it has been largely superseded by the newer wall injection system. These developments, plus the use of a fundamental kinetic model of the process chemistry and a three-dimensional flow model, form the current state of the art for SNR.

The use of multiple injection nozzles allows for fluctuations in flue gas temperature away from the optimum NH_3/NO_x reaction temperature range of 900°–1000°C. The injection of a readily oxidisable gas, such as H_2 or CH_4, with the NH_3 lowers the optimum temperature range progressively to about 700°C ± 50°C for a H_2:NH_3 ratio of about 2:1 [352].

SNR is a flue gas deNO_x treatment process that has, in the last few years, achieved high NO_x reduction efficiencies comparable to those of SCR. It is claimed that the process is up to five times as cost effective as SCR [414] primarily due to the relative ease of installation and lack of catalyst replacement costs. Its potential appears to be in the middle ground between the low cost, lower efficiency combustion techniques and the high cost, high efficiency SCR methods.

Whilst the new wall injector nozzles appear to have eliminated many of the disadvantages associated with early grid designs, their proper design and location requires the use of a detailed kinetic model of the system to be fitted. This recently developed kinetic model is based on the fundamental chemistry of the process, and it is claimed to predict accurately the performance and NH_3 leakage rates for any type of fired unit [414].

The use of wall injectors has also promoted the development of three-dimensional, turbulent flow modelling techniques to ensure adequate mixing of NH_3 and flue gas NO_x, especially in larger boilers. This technique also plays a part in predicting $DeNO_x$ performance (and NH_3 leakage) for specific applications.

Process Requirements for Each Application Considered: Process requirements are shown in Table 4.15 for the two applications considered: Datum Combustion Systems 1 and 2.

It is assumed that for System 1:

- The NO_x content of the gas is to be reduced to 300 mg/Nm³, corresponding to a NO_x reduction of about 71%.

- NH_3:NO_x molar ratio is 2.5.

It is assumed that for System 2:

- The NO_x content of the gas is to be reduced to 400 mg/Nm³, corresponding to a NO_x reduction of about 72%.

- NH_3:NO_x molar ratio is 2.5.

In both cases, it is assumed that:

- H_2 is not required for performance control.

- NO_x consists of 90% NO and 10% NO_2 by volume.

- Reactions between NH_3, SO_3 and HCl can be ignored.

NO_x reduction efficiencies of the order of 70% have been selected, being an average over the range of efficiencies to be expected with utility and industrial boilers fitted with the latest technology [417].

Table 4.15 Process Requirements

Datum Combustion System		1	2
Gas at full load:			
volume flow	'000 Nm³/h	1524*	551
Dry gas	'000 Nm³/h	1426*	513
Water vapour	'000 Nm³/h	99*	38
Actual volume flow	'000 Nm³/h	6827*	2468
Temperature	°C	950	950
Particulates content	mg/Nm³ (dry)	115	115
SO_2 content	mg/Nm³ (dry)	4240	3520
HCl content	mg/Nm³ (dry)	340	315
Inlet gas at full load:			
NO_x content (as NO_2)	mg/Nm³ (dry)	1025	1440
Outlet gas at full load:			
NO_x content (as NO_2)	mg/Nm³ (dry)	300 (372)	400 (504)
NH_3 content (a)	ppm	50	50
Reagent		Ammonia	Ammonia
NH_3:NO_x molar ratio		2.5	2.5
Requirements at full load:			
Ammonia	kg/h	5944	771
Hot water (120°C–90°C)	tonnes/h	66.0	8.6
Electric power	MW	0.93	0.07
Manpower	men/shift	NIL	NIL
Average load factor	%	***	***

***Data not available
* per 500 MWe boiler
(a) 1 ppm NH_3 = 0.76 mg/Nm³
Figures in parentheses are annual average emissions for 90% $deNO_x$ plant availability

By-Products and Effluents: The reaction between NO_x and ammonia produces nitrogen gas and water vapour, so essentially no waste material is formed. Side reactions between NH_3 and SO_3 in the flue gas result in the formation of ammonium bisulphate which causes fouling problems on downstream equipment; these reactions are normally inhibited in SNR systems by maintaining NH_3:SO_3 molar ratios above 2.0, so that the less troublesome ammonium sulphate is formed in preference to the bisulphate.

Where an FGD plant is located downstream of the SNR reactor, leakage NH_3 and ammonium salts will be flushed from the gas stream by the FGD scrubbing liquor and discharged in the waste water stream. The presence of these gas-borne wastes will affect FGD reagent consumption and the nature of the waste water. Typically, an NH_3 leakage rate of 50 ppm from an SNR unit fitted to a 2000 MWe utility boiler is equivalent to 230 kg/h of NH_3 to be neutralised in the FGD plant.

Efficiency and Emission Factors: The efficiency and emission factors for the process are summarised in Table 4.16 for the two applications considered: Datum Combustion Systems 1 and 2.

In calculating the efficiency factors, it is assumed that at full load:

- The performance without the incorporation of the SNR plant would be as shown in Table 1.4.
- For Datum System 1, the SNR plant overall power consumption is 0.05% of the energy generated (including hot water for NH_3 evaporation) i.e. 940 kWe.
- For Datum System 2, the SNR plant overall power consumption is 0.05% of the equivalent energy generated (148 MWe) i.e. 74 kWe. This is taken to be equivalent to the combustion of 29 kg/h of coal at a power station, assuming an overall power generation efficiency of 33%. This additional coal is arbitrarily assumed to be included with the coal burned in the boiler for calculating the efficiency and emission factors.

Exxon have indicated ranges of expected NO_x reduction efficiencies for different types of fired equipment, but stress that the SNR process is plant specific and would need to be evaluated for each application [417].

Effect of plant availability: For illustration purposes, the annual average emissions and emission factors for DeNOx plant availabilities of 100% and 90% are given in Tables 4.15 and 4.16. Further details are given in Section 1.6.

Effect of Load Variations: Reduced boiler loads are accompanied by flue gas temperature fluctuations in the upper sections of the boiler. This would be accompanied by a loss of NO_x reduction efficiency if it were not for the multiple location of NH_3 injection nozzles and the facility to inject H_2 into the flue gas (with NH_3) to reduce the optimum reaction temperature. It is anticipated, therefore, that emission factors would change very little at reduced loads, but possibly at the expense of increased H_2 consumption.

Effect of Design Variations: Systems fitted with the older injection grid layouts would be expected to exhibit lower emission factors than for the newer wall injection arrangement. However, it is expected that only the latest technology will be available commercially.

Limitations: Selective non-catalytic reduction is limited in its application by its lower cost effectiveness in comparison to combustion techniques, and by its lower NO_x reduction efficiency in comparison to selective catalytic reduction techniques

Table 4.16 Efficiency and Emission Factors

Datum Combustion System		1	2
Coal heat input (gross)	MWt	1292*	468
Coal fired	tonnes/h	189*	60.3
SNR plant power consumed	MWe	0.93	0.07
Equivalent coal input	tonne/h	–	0.03(b)
Useful energy from system	GJe/h	6693	–
	GJt/h	–	1468
Total equiv. coal input	tonnes/h	756	60.33
Efficiency factor	GJe/tonne	8.85	–
	GJt/tonne	–	24.3
Nitrogen emission (NO_x)	kg/h	520 (646)	62 (79)
NO_x emission factor (a)	kg/tonne	0.69 (0.86)	1.04 (1.30)

*For each 500 MWe boiler
(a) Emission factor expressed as kg nitrogen per tonne of coal
(b) Calculated assuming overall power generating efficiency = 0.33
Figures in parentheses are annual average emissions for 90% deNO_x plant availability

[416]. In both cases, however, it is beginning to approach the required performance. It may find applications in combined low-NO_x burners and SNR arrangements where the high deNO_x levels attainable by SCR are required, but at a substantially lower investment cost. SNR is more difficult to control than SCR, and the comparatively high leakage rates of NH_3 present potential downstream fouling and waste water problems.

Costs

Capital Costs: See Section 4.6

Annual Running Costs and Cost Factor: ***

Effect of Design Variations on Cost: ***

Effects of Annual Load Patterns on Annual Running Costs: ***

Process Advantages and Drawbacks

The advantages are:

1. Catalysts and specific reactors are not required, thereby avoiding plugging problems and SO_2 oxidation.

2. Preheating is unnecessary because the reaction takes place in a high temperature zone.

3. Space for a catalytic reactor is unnecessary.

4. Much less expensive to install than SCR, especially on existing boilers.

5. Capable of dealing with particulate-laden flue gas streams.

The disadvantages are:

1. The technique has not been installed commercially in a coal-fired boiler.

2. Considerable amounts of NH_3 leakage occur, which could result in the formation of ammonium bisulphate and subsequent fouling of downstream equipment.

3. NH_3 consumption is high, and H_2 is sometimes required to maintain deNO_x efficiencies.

4. Only a narrow temperature range exists over which optimum deNO_x efficiencies can be achieved. Hence, control can sometimes be difficult.

5. Lower NO_x reduction efficiencies are possible, compared to SCR.

6. Variations in flue gas temperature and composition within the duct can present operational problems, especially where duct sizes are large.

4.6 Costs

The recent paper by Leggett [102] summarising progress on the O.E.C.D. Project on Control of Major Pollutants (MAP) was used to provide the limited cost data shown in the following Tables 4.17 and 4.18. Within the MAP study, differences in data were assessed and presented as ranges rather than specific values due to substantial uncertainties regarding the starting emission level, the abatement efficiency and other similar factors.

Table 4.17 Utility Boilers–Coal Fired (NOx)

Source Level of control	% Abated	Annual K Cost (£/kWt)	Annual O&M (p/kWh)	£/t Abated
Utility Boilers Existing				
Low NO_x Burners (LNB)	40–50	1.8–10.0	0.0007–0.006	43–528
2nd Generation LNB	50–70			132
Combustion Modifications				
Selective Catalytic Reduction (SCR)	60–90	37.8–62.9	2.6–8.6	150–614
Utility Boilers New				
Low NO_x Burners	45–50	0.4	Negligible	14
Combustion Modifications				
LVHR & LNB	40	1.9	Negligible	29
LVHR & LNB & IFNR	70	−7.4		678–1264
Selective catalytic Reduction	80–90	18–53	0.06–0.22	557–1900
SCR & LVHR	88	9.3	0.13	943
SCR & IFNR & LNB	94	−14.8	0.2–0.3	1330–1770

LVHR = low volume heating rate
IFNR = in-furnace NO_x reduction

The prices and ranges in the cost effectiveness summary tables are not intended to be indicative of the cost, or cost effectiveness, for any individual facility–rather, they attempt to represent a typical facility in that source category so that, on average, the estimated costs should be representative.

A questionnaire issued to UK burner vendors did not return any usable costing information.

Cost-effectiveness is defined in terms of US 1984 dollars per tonne ($/t) of NO_x abated (converted to £/t). The costs were derived by annualising the total capital investment required for the pollution abatement technique. Assumptions used to annualise the capital costs were made as consistent as possible, using a real discount rate of 5 per cent and equipment lifetimes that vary with the type of facility (e.g. 30 years for combustion modifications on new boilers or 15 years for retrofits; 3–5 years for catalysts for selective catalytic reduction etc.). The annual charge for operation and maintenance (O & M) including by-product sales and changes in fuel consumption, is added to the annualised capital charge. The resulting net annual cost is divided by the annual tons of NO_x abated.

For further details the reader should refer to the paper by Leggett [102].

Table 4.18 Industrial Boilers–Coal Fired (NO_x)

Source level of control	% Abated	Annual K cost (£/kWt)	£/t Abated
Industrial Boilers			
Existing			
Low NO_x Burners (LNB)	50	2.8	136
Combustion Modifications			
LNB & OFA (Japan)	35–65	4.1–5	643–2020
SCR (USA)	90	47	1280

OFA = over fire air injection
SCR = selective catalytic reduction

5. Combined SO_2–NO_x Abatement Processes

5. Combined SO_2–NO_x Abatement Processes

5.1 Classification of the Processes

The general classification of processes for the simultaneous abatement of SO_2 and NO_x emission adopted in this Manual is presented in Figure 5.1. The processes are divided into four categories (NS10–NS40) according to the scheme:

- First division: distinction between dry processes (NS10 and NS20) and wet processes (NS30 and NS40).

- Second division: for dry processes, distinction between catalytic (NS10) and non-catalytic (NS20); for wet processes, distinction between direct absorption (NS30) and oxidation-absorption (NS40).

- Third division: for dry catalytic processes, distinction between regenerable (NS11) and non-regenerable (NS12); for dry non-catalytic processes, distinction between those using burner modifications (NS21), radiation (NS22) and absorption (NS23); for wet processes (both direct absorption and oxidation absorption) distinction between liquid-phase NO_x oxidation (NS31 and NS41) and liquid-phase NO_x reduction (NS32 and NS42). Note that the burner modifications (NS21) are not processes but are essentially modified combustion techniques.

Some of the processes yielding a dry end-product involve spray-drying of solutions or slurries as an integral feature of the process. Such processes are frequently described in the literature as 'dry', 'semi-dry' or 'wet-dry' processes. These terms are regarded as misleading for such processes, which are referred to in this Manual as solution-based or slurry-based (as appropriate) absorption processes with dry end products. In the Manual, the terms 'wet' and 'dry' are used to describe the state of the reagent contacted with the gas; the terms 'semi-dry' and 'wet-dry' are not used.

Process Code Numbers: A number of basic process types occur within each process category. Basic process types are assigned a Code Number comprising the relevant Category Number followed by a Type Number; e.g. the electron beam radiation processes fall in Category NS22 and have been assigned the Code No. NS22.1.

The Code Numbers assigned to the basic process types are shown in Table 5.1.

Table 5.1 Classification of Combined SO_2/NO_x Abatement Processes

Category NS10: Dry processes–Catalytic	
Category NS11: Regenerable catalyst	
NS11.1	Active carbon adsorption/selective catalytic reduction (SCR)
NS11.2	Copper oxide
Category NS12: Non-regenerable catalyst	
NS12.1	Non-selective catalytic reduction of NO_x and SO_2
NS12.2	Catalytic reduction of NO_x and catalytic oxidation of SO_2
Category NS20: Dry processes–Non-catalytic	
Category NS21: Burner modifications	
NS21.1	Limestone Injection into Multistage Burner (LIMB)
NS21.2	Slagging combustor
Category NS22: Radiation	
NS22.1	Electron beam
Category NS23: Adsorption	
NS23.1	Sodium carbonate process
Category NS30: Wet processes–Direct absorption	
Category NS31: Liquid phase NO_x oxidation	
NS31.1	Lime spray dryer
Category NS32: Liquid-phase NO_x reduction	
NS32.1	Asahi chemical process
NS32.2	Limestone scrubbing with chelating compound
NS32.3	Sulf-X process
Category NS40: Wet processes–Oxidation absorption	
Category NS41: Liquid phase NO_x oxidation	
NS41.1	Oxidation plus ammonia scrubbing
NS41.2	Kawasaki process
Category NS42: Liquid phase NO_x reduction	
NS42.1	Oxidation plus limestone slurry scrubbing–IHI Process
NS42.2	Oxidation plus limestone slurry scrubbing–Moretana Process

5.2 Outline Descriptions of Combined SO_2–NO_x Abatement Processes

Process Code NS11.1–Active Carbon Adsorption/SCR

Outline of Process [352, 424]: A simplified block diagram of the process is presented in Figure 5.2.

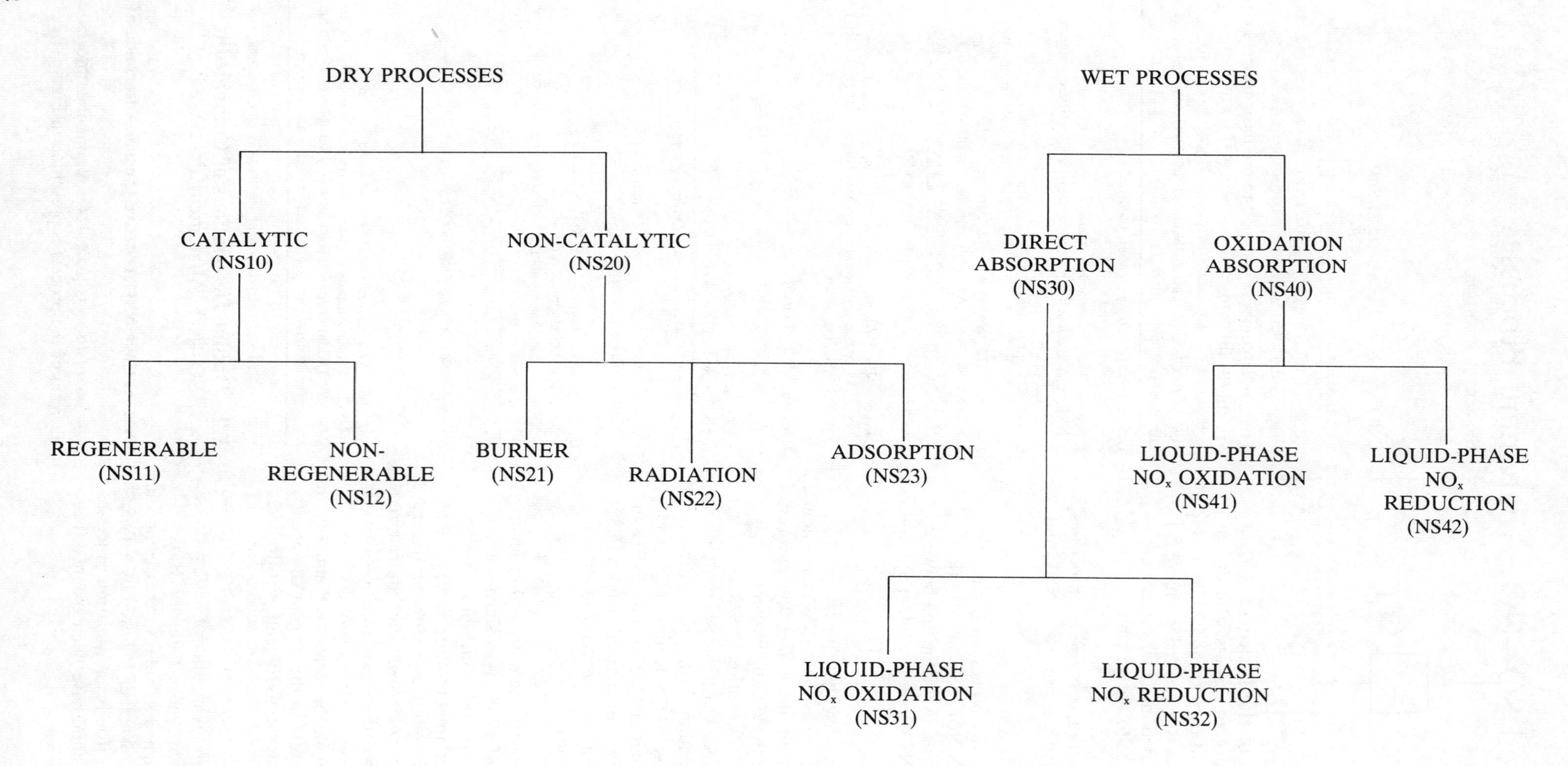

Figure 5.1 Simultaneous SO_2/NO_x Abatement Processes

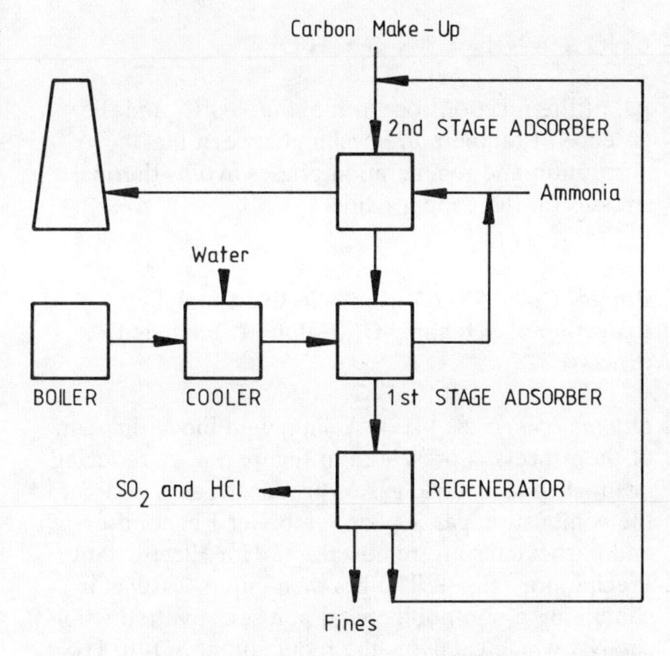

Figure 5.2 Block Diagram of Carbon Adsorption/SCR Combined Abatement Process

Gas from the Boiler at up to 150°C is cooled, if required, to about 120°C by injection of water. It then flows through the First Stage Adsorber containing a moving bed of activated carbon pellets sized about 10 mm, or of coke, from the Second Stage Adsorber. The carbon adsorbs SO_2, together with NO_2 which forms 5–10% of the NO_x content of the gas. The carbon catalyses the oxidation of SO_2 to SO_3, which reacts with water vapour also adsorbed from the gas, forming sulphuric acid. Ammonia is added to the gas leaving the First Stage Adsorber, and the gas then enters the Second Stage Adsorber, where the carbon catalyses the reaction of ammonia with NO to form nitrogen and water; it behaves as a selective catalytic reduction (SCR) catalyst. Some reaction also occurs between adsorbed ammonia and SO_2 adsorbed in the Second Stage to form ammonium sulphate and bisulphate. There is a temperature rise of 15–20°C in the Adsorbers, and the cleaned gas is exhausted to stack. Carbon leaving the base of the First Stage Adsorber is transferred to the Regenerator, where it is heated in tubes to 400–450°C by hot combustion gases outside the tubes. The sulphuric acid and ammonium salts are desorbed and decomposed by carbon to produce SO_2, H_2O, CO_2, oxygen and nitrogen. The SO_2 is treated: by liquefaction for sale as liquid SO_2; by oxidation to SO_3 for the manufacture of sulphuric acid; or by reduction to elemental sulphur for sale or safe disposal. The regenerated carbon is cooled and returned to the Second Stage Adsorber, together with fresh carbon to replace that lost in regeneration. The carbon bed also acts as a panel bed filter, capturing particulates which are screened out before regeneration. Further, it adsorbs halogen acid gases, which are released during regeneration; they are separated from the regenerator tail gas before treatment of the SO_2.

Chemistry of the Process [352, 424]: In the First Stage Adsorber, SO_2 and water vapour are adsorbed by the activated carbon, and they react on the carbon surface with oxygen in the gas. The reaction can be represented by:

$$2\,SO_2 + O_2 + 2\,H_2O = 2\,H_2SO_4$$

The sulphuric acid formed remains adsorbed in the pores of the carbon. The reaction is exothermic, and there is a 15–20°C rise in gas temperature.

At the higher temperature (400–450°C) in the Regenerator, the sulphuric acid decomposes to sulphur trioxide and water; the trioxide is reduced by the carbon to form the dioxide:

Decomposition: $H_2SO_4 = SO_3 + H_2O$

Reduction: $2\,SO_3 + C = SO_2 + CO_2$

The reduction involves transient formation of a surface carbon-oxygen species, C. . .O:

$$2\,H_2SO_4 + 2\,C = 2\,SO_2 + 2\,H_2O + 2\,C...O$$
$$2\,C...O = C + CO_2$$

In the Second Stage Adsorber, the carbon catalyses the reduction of NO by NH_3:

NO Reduction: $6\,NO + 4\,NH_3 = 5\,N_2 + 6\,H_2O$
$4\,NO + 4\,NH_3 + O_2 = 4\,N_2 + 6\,H_2O$

Reaction also occurs between ammonia and residual SO_2 adsorbed in the Second Stage Adsorber to form ammonium sulphate and bisulphate:

$2\,SO_2 + O_2 + 2\,H_2O = 2\,H_2SO_4$
Sulphate $H_2SO_4 + 2\,NH_3 = (NH_4)_2SO_4$
Bisulphate $H_2SO_4 + NH_3 = NH_4HSO_4$

In regeneration, these reactions are reversed, and the ammonia released is decomposed by reaction with the surface carbon-oxygen species, C. . .O; this involves no loss of carbon:

NH_3 Decomposition:
$2\,NH_3 + 3\,C...O = N_2 + 3\,H_2O + 3\,C$

A carbon make-up is required to replace that consumed in the reduction of H_2SO_4 formed in the First Stage Adsorber, which can account for 70–80% of the total sulphur dioxide captured in the process (0.09 kg/kg SO_2 removed in the First Stage). The loss of carbon during reduction increases porosity, and hence the internal surface of the carbon remaining, which is therefore further activated. However, the reduction weakens the carbon particles, so that there are carbon break-down losses which also have to be replaced, resulting in a carbon make-up rate (typically

0.12–0.18 kg/kg SO_2 removed) that is higher than the theoretical.

Process Code NS11.2–Copper Oxide Absorption/SCR

Outline of Process [145, 352]: A simplified block diagram of the process is presented in Figure 5.3. The process operates at a temperature of about 400°C, and the Reactors are therefore located after the Boiler Convection Passes, and upstream of the Air Heater and Electrostatic Precipitator (ESP). There are two Reactors, containing copper oxide supported on an alumina base, and operated cyclically: one Reactor is on stream, with the gas flowing over (not through) the copper oxide, whilst the other is being regenerated by passing hydrogen over the sulphated copper oxide, reducing it to metallic copper and giving an SO_2-rich off-gas. When the Reactor is put back on stream, the copper is oxidised by oxygen in the gas, reforming copper oxide for further reaction.

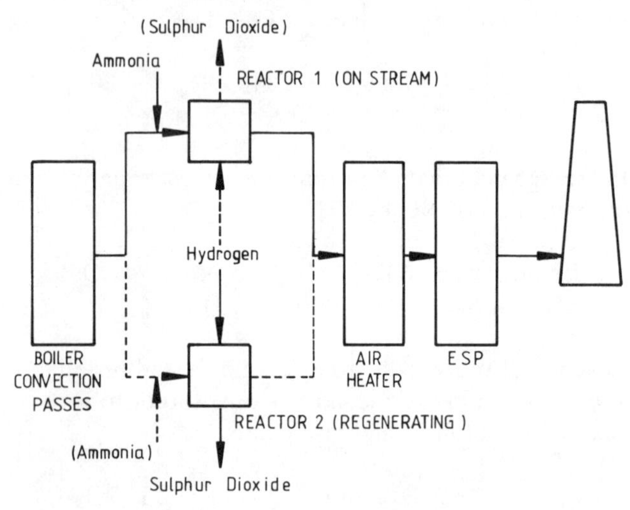

Figure 5.3 Block Diagram of Copper Oxide Combined Abatement Process

Chemistry of Process [145, 352]: The SO_2 absorption reaction is:

Absorption $2\ CuO + 2\ SO_2 + O_2 = 2\ CuSO_4$

In the regeneration phase, hydrogen reduces the copper sulphate to metallic copper. The copper is re-oxidised when the absorption phase is resumed. The reactions are:

Reduction $CuSO_4 + 2\ H_2 = Cu + SO_2 + 2\ H_2O$

Oxidation $2\ Cu + O_2 = 2\ CuO$

Ammonia reacts with nitrogen oxides in the presence of copper oxide and sulphate which act as catalysts for the selective reduction of NO_x; the copper sulphate has the greater catalytic activity. The reduction reaction is:

$6\ NO + 4\ NH_3 = 5\ N_2 + 6\ H_2O$

All of the reactions occur at about 400°C, and the absence of temperature cycling between the absorption and regeneration phases avoids thermal stresses on the copper oxide.

Process Code NS12.1–Non-Selective Catalytic Reduction of SO_2 and NO_x (Ralph M. Parsons Co. Process)

Outline of Process [419]: A simplified block diagram of the process is presented in Figure 5.4. A reducing agent such as natural gas or producer gas is added to the combustion gas leaving the boiler Economiser, and particulates are removed in a Hot Electrostatic Precipitator (ESP). The gas then enters a Reactor containing a non-noble metal non-selective reduction catalyst which catalyses the reduction of SO_2 to H_2S and of NO_x to elemental N_2. The gas is then cooled by passage through the Air Heater, and the H_2S is converted to elemental sulphur in a Stretford Absorber. The gas is reheated in the Reheater before being exhausted to stack.

Chemistry of Process [419]: The reduction agents are hydrogen and carbon monoxide; when natural gas is used, H_2 and CO can result from reforming of methane, e.g. by the reaction:

Reforming $CH_4 + H_2O = 3\ H_2 + CO$

The reduction reactions, yielding H_2S and N_2, can be represented as follows:

Reduction of SO_2 $SO_2 + 3\ H_2 = H_2S + 2\ H_2O$

Reduction of NO $2\ NO + 2\ CO = 2\ CO_2 + N_2$

Reduction of NO_2 $2\ NO_2 + 4\ CO = 4\ CO_2 + N_2$

In the Stretford Absorber, H_2S is absorbed by sodium carbonate solution; this can be simplified as:

H_2S Absorption
$H_2S + Na_2CO_3 = NaHS + NaHCO_3$

The Stretford absorber solution, which also contains soluble vanadium compounds, is then oxidised by air to precipitate sulphur and regenerate the sodium carbonate; simplified as:

$2\ NaHS + 2\ NaHCO_3 + O_2 = S_2 + 2\ Na_2CO_3 + 2\ H_2O$

The vanadium acts as the oxygen carrier, being oxidised to the pentavalent state by the oxidising air, and reduced to the tetravalent state in the reaction with NaHS.

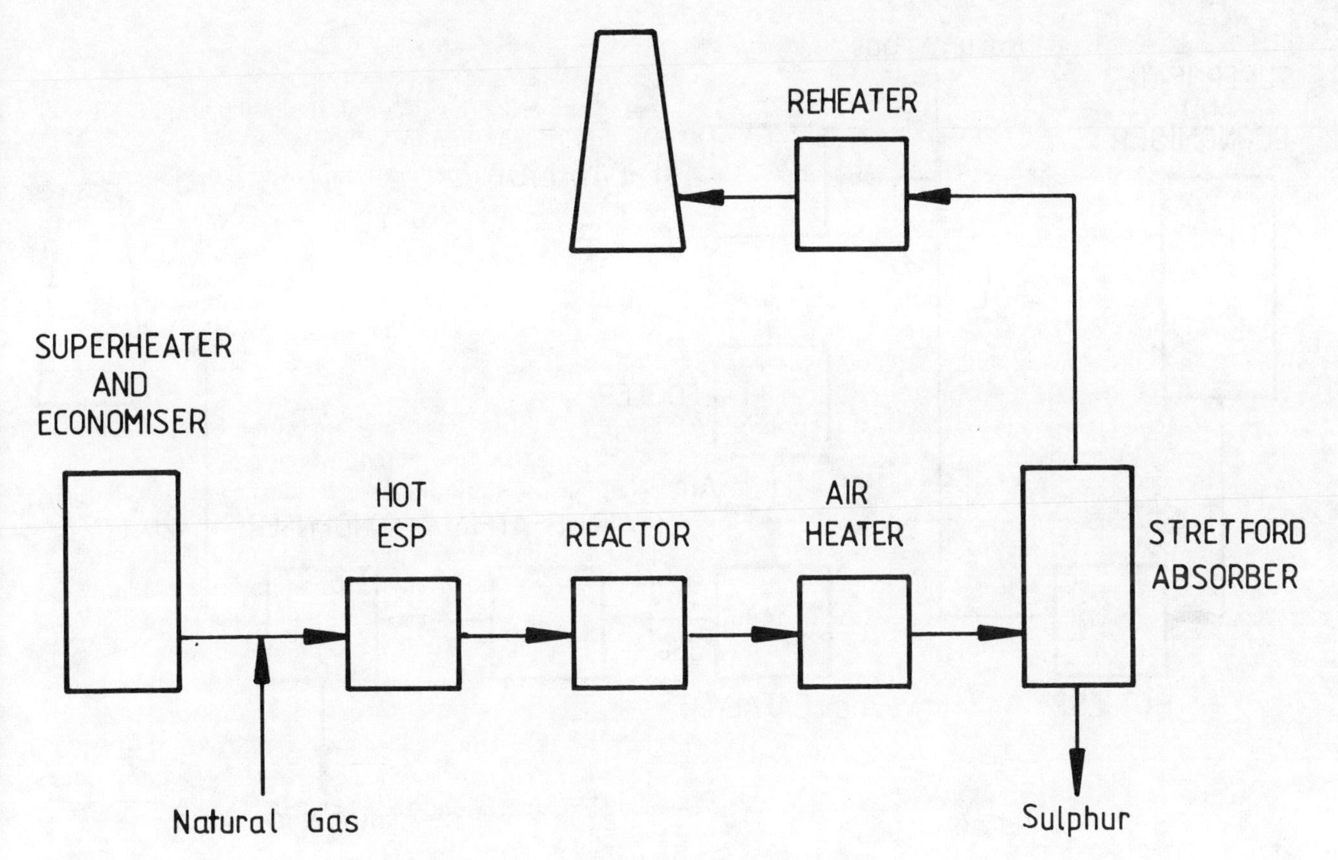

Figure 5.4 Block Diagram of Combined Abatement Process–Non-Selective Catalytic Reduction of NO_x and SO_2

Process Code NS12.2–Catalytic Reduction of NO_x and Catalytic Oxidation of SO_2

Outline of Process [352]: A simplified block diagram of the process is presented in Figure 5.5. Gas leaving the Superheater and Economiser section of the boiler at a temperature of 450°C passes through a Hot Electrostatic Precipitator (ESP) to remove particulates. Natural gas is injected into the gas stream to reduce the oxygen content before it is contacted with the First Catalyst, and to reduce nitrogen oxides to nitrogen in contact with the First Catalyst. The gas is then cooled to about 360°C by passing over further economiser heat transfer surface in the Cooler. Oxidising conditions are restored by the addition of air, and the gas is then contacted with the Second Catalyst to oxidise sulphur dioxide to trioxide. The cleaned gas passes through the boiler Air Heater and the air-cooled Condenser before being exhausted to stack. The condensate by-product is 93% sulphuric acid.

Chemistry of the Process [352]: The addition of natural gas results in establishing reducing conditions for reaction with the First Catalyst, and nitrogen oxides are reduced:

$$4\ NO + CH_4 = CO_2 + 2\ H_2O + 2\ N_2$$

After restoring oxidising conditions by the addition of air, the oxidation of SO_2 to SO_3, catalysed by e.g. vanadium pentoxide, and subsequent formation of sulphuric acid, occurs as in the Cat-Ox FGD process (see Section 2.2).

Oxidation $$2\ SO_2 + O_2 = 2\ SO_3$$

Acid formation $$SO_3 + H_2O = H_2SO_4$$

Process Code N21.1–Limestone Injection into Multi-Stage Burner (LIMB)

Outline of Process [431]: Figure 5.6 is a schematic diagram of a Limestone Injection/Multi-staged Burner (LIMB).

The technology is a combination of low-NO_x combustion techniques (see Section 3.2) and the injection of dry limestone into the furnace for SO_x control (see Section 2.2). In the LIMB design, however, the finely-ground limestone is injected into the furnace either through the burner itself or in close proximity to it. The reaction products and any unreacted limestone leaving the furnace are removed with the fly-ash in an electrostatic precipitator or baghouse.

Chemistry of Process [297]: When finely-ground limestone is injected into a flame, it undergoes a calcination reaction:

$$CaCO_3 = CaO + CO_2$$

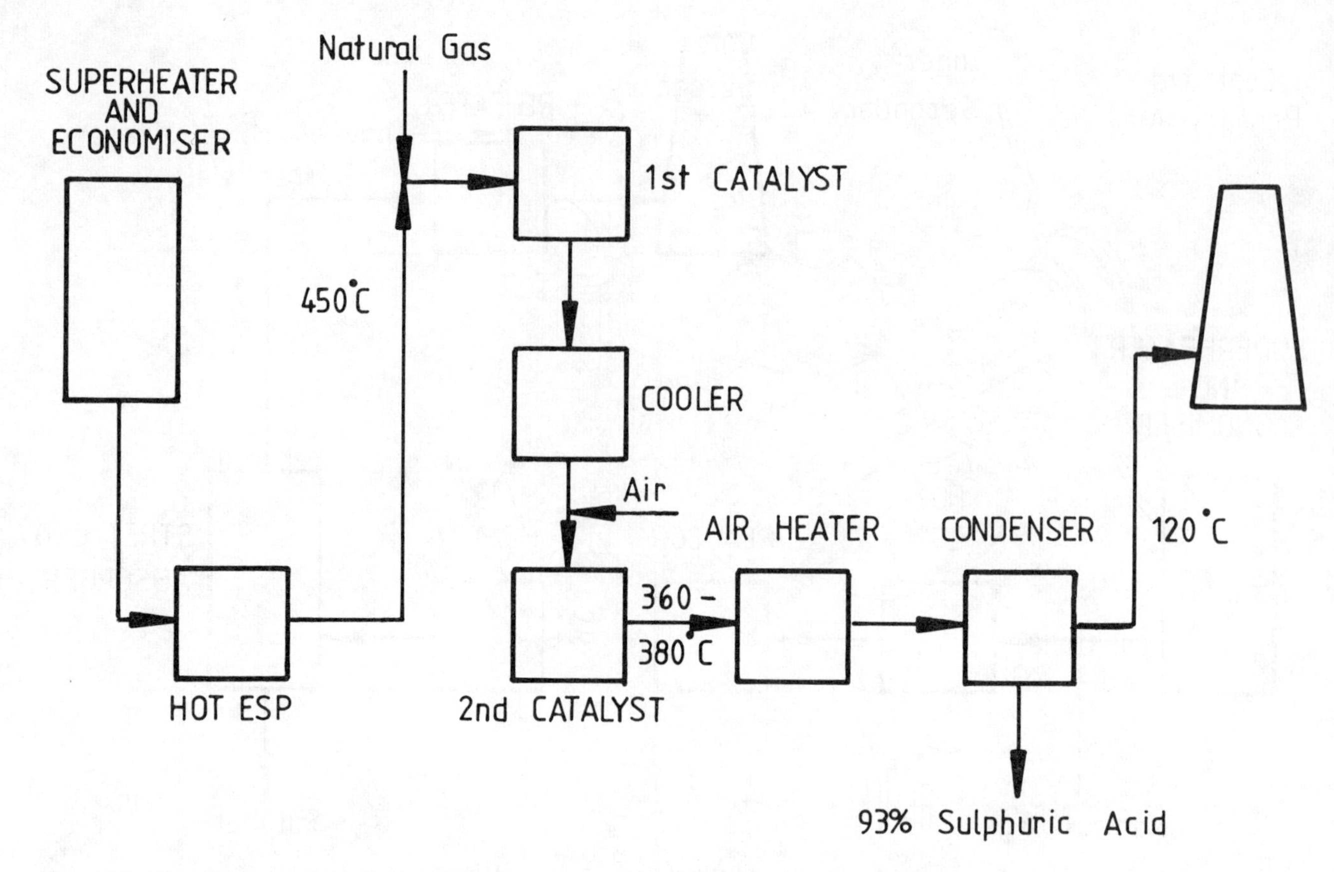

Figure 5.5 Block Diagram of Catalytic NO_x Reduction/Catalytic SO_2 Oxidation Combined Abatement Process

Experimentally-confirmed thermodynamic considerations indicate that calcination starts at about 800°C. With pulverised coal flame temperatures being significantly higher than this, rapid calcination is expected to take place thus allowing reaction with SO_2:

$$2\ CaO + 2\ SO_2 + O_2 = 2\ CaSO_4$$

Intermediate stages in this overall reaction involve the formation of calcium sulphite ($CaSO_3$) or the direct reaction between CaO and SO_3.

At high temperatures, $CaSO_4$ becomes thermally unstable and the above reaction may become reversible.

Sulphur released as H_2S in the reducing zone of the burner is captured by the calcined limestone but not by direct reaction with the limestone.

Process Code NS21.2–Slagging Combustor

Outline of Process [427]: Rockwell International are developing a unique design of multi-stage slagging burner which is intended to control NO_x and SO_x entirely within the burner itself. Combustion is initiated within the burner, external to the boiler, using pulverised coal and typical preheated (350°C) air.

SO_x and NO_x control is accomplished by control of reducing combustion conditions within the burner. Fuel sulphur is captured as a solid calcium-sulphur compound which is either removed in the separator section of the burner or in a conventional electrostatic precipitator or baghouse downstream. NO_x is converted to molecular nitrogen.

The slag and a major part of the (molten) fly-ash is removed in a simple, inertia-type separator which is located in a part of the burner operating at a temperature well in excess of the ash-fusion temperature. The slag is tapped from the burner external to the boiler. If the burners are installed in boilers designed for coal firing, the slag/fly-ash separator is optional, and the slag can be run into the boiler for removal from the furnace floor.

Process Code NS22.1–Electron Beam Radiation Process

Outline of Process [352, 419]: A simplified block diagram of the process is presented in Figure 5.7. Gas from the Boiler Electrostatic Precipitator (ESP) or Baghouse is cooled and humidified in the Cooler by injection of water; ammonia is also added, and the gas passes through a Reactor in which it is subjected to an intense field from an electron beam. The field brings about reactions between ammonia, sulphur dioxide and nitrogen oxides to form solid ammonium nitrate and ammonium nitrate sulphate. The solids

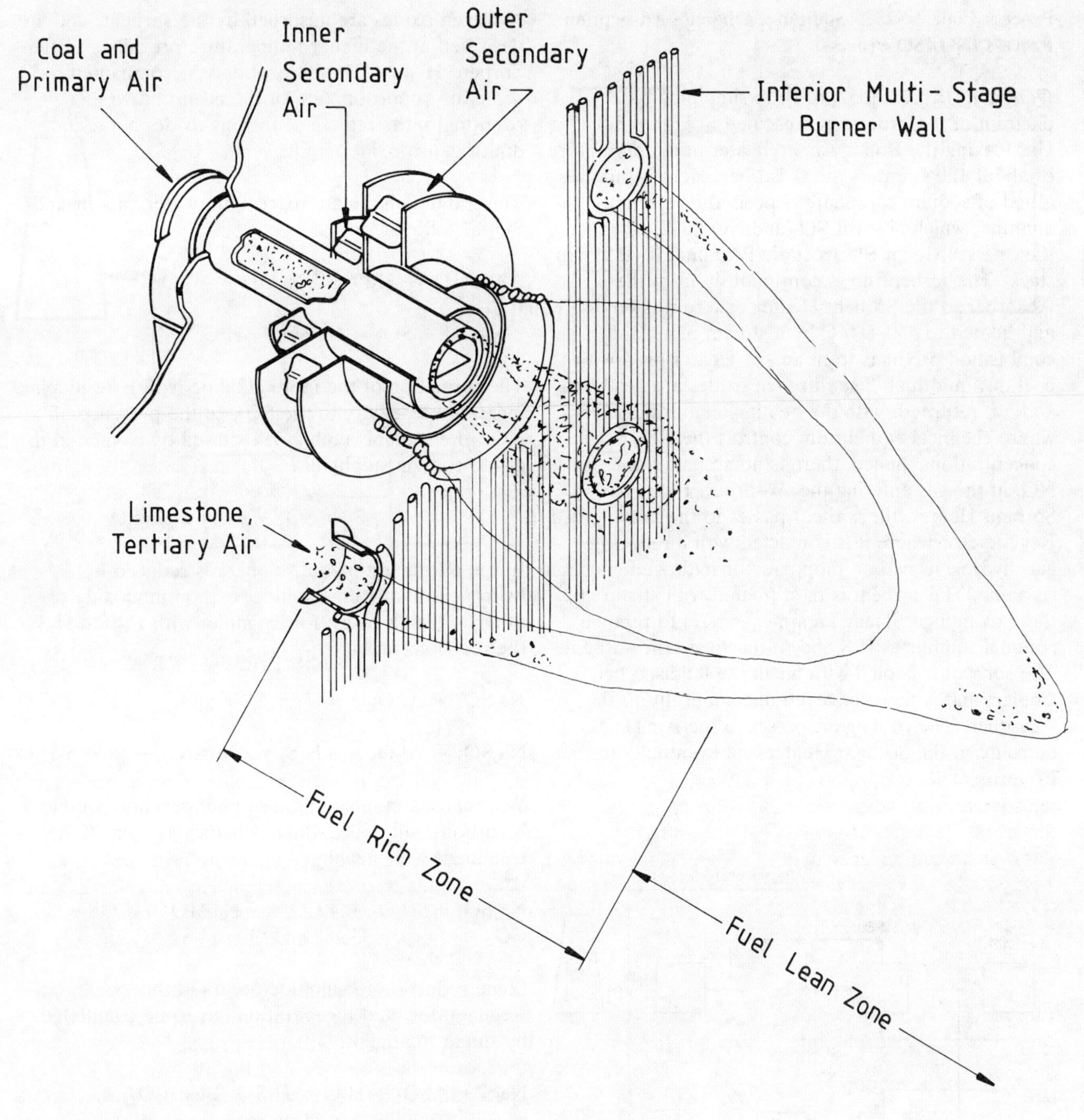

Figure 5.6 Limestone Injection Multi-Stage Burner for Simultaneous NO_x and SO_x Control

are removed in a second ESP or Baghouse, and are potentially saleable as a fertiliser by-product. The gas is exhausted to stack.

Chemistry of the Process [352, 419]: Under the influence of the electron beam radiation, SO_2 and NO_x react with ammonia in the presence of water vapour to form ammonium sulphate and ammonium nitrate-sulphate:

Sulphate formation $2\ SO_2 + O_2 + 2\ H_2O = 2\ H_2SO_4$
$H_2SO_4 + 2\ NH_3 = (NH_4)_2SO_4$

Nitrate formation $4\ NO + 3\ O_2 + 2\ H_2O = 4\ HNO_3$
$NH_3 + HNO_3 = NH_4NO_3$

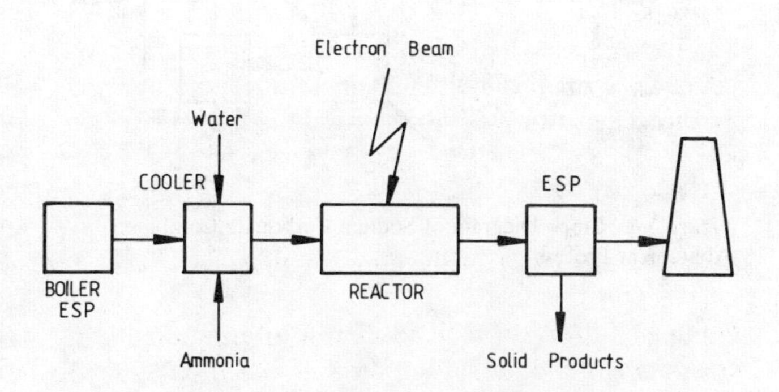

Figure 5.7 Block Diagram of Electron Beam Radiation Combined Abatement Process

Process Code NS23.1–Sodium Carbonate Adsorption Process (NOXSO Process)

Outline of Process [352, 434]: A simplified block diagram of the process is presented in Figure 5.8. Gas leaving the Boiler Air Preheater at a temperature of about 120°C enters the Adsorber. Here it fluidises a bed of sodium carbonate deposited on gamma-alumina, which absorbs SO_2 and NO_x. The clean gas then passes via an Electrostatic Precipitator (ESP) to stack. The sorbent flows continuously from the Absorber to the Sorbent Heater, where it is fluidised and heated to 550–600°C by a stream of air and combustion products from an Air Heater fired with coal or liquid fuel. The nitrogen oxides are desorbed and are returned, with the heating air, to the boiler where chemical equilibrium controls the NO_x concentrations; hence, there is no accumulation of NO_x in the gas entering the Adsorber. From the Sorbent Heater, the sorbent passes to the moving-bed Regenerator where it is contacted with a reducing gas–hydrogen, carbon monoxide or hydrogen sulphide. The sorbent is then treated with steam in the moving-bed Steam Treatment vessel to remove residual sulphur as H_2S and to reactivate the sorbent. The sorbent is cooled with air in the fluidised bed Cooler, and is then conveyed pneumatically to the Adsorber. The cooling air passes to the Air Heater en route to the Sorbent Heater and eventually to the Boiler.

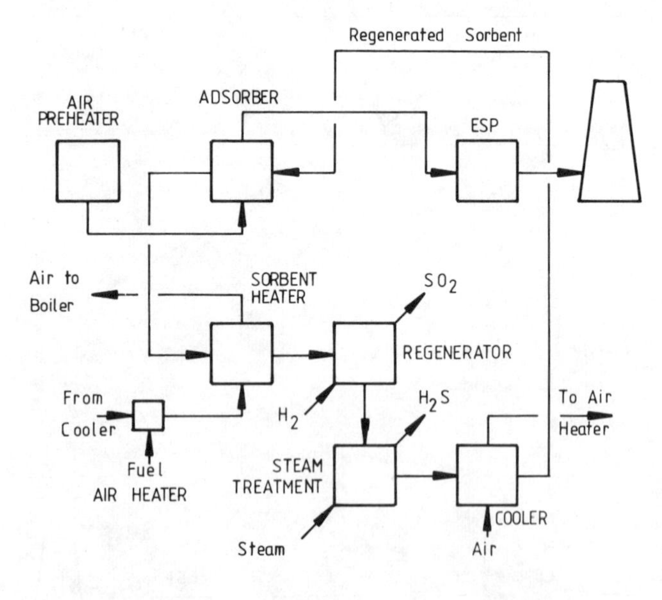

Figure 5.8 Block Diagram of Sodium Carbonate Combined Abatement Process

Chemistry of the Process [352, 434]: The sodium carbonate and the gamma-alumina substrate react to form sodium aluminate, which is the actual sorbent:

$$Na_2CO_3 + Al_2O_3 = 2\ NaAlO_2 + CO_2$$

Nitrogen oxides are adsorbed by the sorbent, and are desorbed at the higher temperature prevailing in the Sorbent Heater. The desorption is accompanied by some disproportionation of the adsorbed NO_x, resulting in the release of nitrous oxide, N_2O, and dinitrogen trioxide, N_2O_3.

The sodium aluminate reacts with water vapour and SO_2 as follows:

$$2\ NaAlO_2 + H_2O = 2\ NaOH + Al_2O_3$$

$$2\ NaOH + SO_2 = Na_2SO_3 + H_2O$$

The formation of the more alkaline NaOH results in increased reactivity to the SO_2. In the presence of NO_x, the sodium sulphite is oxidised by oxygen in the gas to sodium sulphate:

$$2\ Na_2SO_3 + O_2 = 2\ Na_2SO_4$$

In the Regenerator, the sulphate is reduced by hydrogen, hydrogen sulphide, carbon monoxide or other reducing gases; for example, with H_2 and H_2S the reactions are:

$$Na_2SO_4 + Al_2O_3 + 4\ H_2 = 2\ NaAlO_2 + H_2S + 3\ H_2O$$

$$Na_2SO_4 + Al_2O_3 + 3\ H_2S = 2\ NaAlO_2 + 4S + 3\ H_2O$$

With carbon monoxide the sulphur reaction product is carbonyl sulphide, which is hydrolysed on further treatment with steam:

$$Na_2SO_4 + Al_2O_3 + 4\ CO = 2\ NaAlO_2 + COS + 3\ CO_2 \qquad COS + H_2O = H_2S + CO_2$$

Some reduction to sulphide occurs in the Regenerator, and regeneration has to be completed by steam treatment:

$$Na_2S + Al_2O_3 + H_2O = H_2S + 2\ NaAlO_2$$

Process Code NS31.1–Lime Spray Dryer Process

Outline of Process [352, 432]: A simplified block diagram of the process is presented in Figure 5.9. Gas from the Boiler, at a temperature 120–160°C, enters the top of a Spray Dryer into which a slurry of lime containing an additive, e.g. sodium hydroxide, is sprayed. The lime, which is hydrated and slurried in the Slaker, is pumped to the slurry atomiser nozzles via a Feed Tank, where it is mixed with the additive. The gas passes down the Spray Dryer co-current with the slurry spray droplets; the sulphur oxides, nitrogen oxides and acid halides in the gas react with the lime and alkali, and water is evaporated from the slurry droplets to produce a dry product which also contains the particulates present in the in-going gas. Coarse particles of product are collected at the base of the Spray Dryer, and fine particles are separated from the gas in a Baghouse or Electrostatic Precipitator (ESP). The gas leaves the system at 90–100°C, and is exhausted to the stack. Part of the dry end product is recycled to the Feed Tank to increase the conversion of lime.

The dry end product contains calcium and sodium sulphite, sulphate, nitrite, nitrate and halides, together with some carbonate, unreacted lime and coal ash. It is usual to recycle part of the dry end product to the Slurry Preparation stage to increase the lime utilisation.

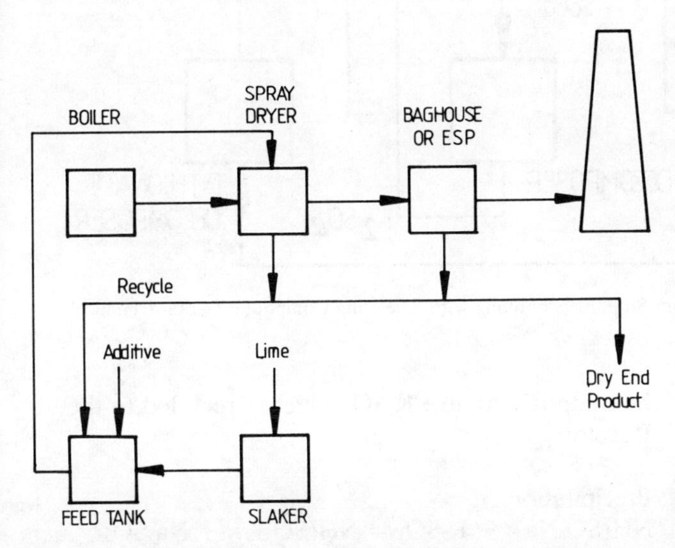

Figure 5.9 Block Diagram of Lime Spray Dryer Combined Abatement Process

Chemistry of Process [122]: The sulphur oxides, SO_2 and SO_3, acid halides, HF, HCl and HBr (represented below as 'HHa') and nitrogen oxides react in the Spray Dryer with the lime and alkali in the slurry as by reactions including the following

$$Ca(OH)_2 + SO_2 = CASO_3 + H_2O$$

$$Ca(OH)_2 + SO_3 = CaSO_4 + H_2O$$

$$Ca(OH)_2 + 2HHa = CaHa_2 + 2H_2O$$
$$2\ NaOH + 2\ NO + O_2 = NaNO_2 + NaNO_3 + H_2O$$
$$2\ NaOH + 2\ NO_2 = NaNO_2 + NaNO_3 + H_2O$$

The absorption of NO_x does not occur in the absence of SO_2, and the efficiency of NO_x abatement increases with increase in the molar ratio of SO_2/NO_x. For example [352] the abatement is negligible at a zero molar ratio, about 30% at a molar ratio of 0.3 and about 70% at a molar ratio of 1. However, it has been shown [444] that NOx absorption at low SO_2/NO_x molar ratios is enhanced by the use of an ionising electron beam; see also above.

The reaction temperature is higher than would be desirable for sulphur capture alone (see above), but this is offset by the presence of NaOH, which enhances the sulphur capture activity of the slurry.

Process Code NS32.1–Sodium Sulphite Scrubbing Process with Chelating Compound (Asahi Chemical Process)

Outline of Process [419]: A simplified block diagram of the process is presented in Figure 5.10. Gas from the Boiler passes through the Prescrubber to remove particulates and acid halides and to cool the gas, and then enters the Absorber where SO_2 and NO are absorbed by a solution of sodium sulphite containing the ferrous salt of ethylene diamine tetracetic acid (Fe^{++}.EDTA). The clean gas is heated in the Reheater and exhausted to the Stack. The solution leaving the Absorber is recirculated (together with sodium carbonate make-up) via a Reducer, where the nitric oxide chelate complex of the FE^{++}.EDTA reacts with sodium sulphite to form sodium sulphate, imidodisulphonate and dithionate, and to release elemental nitrogen. The sulphate and dithionate are separated from a side stream of the circulating solution in the Crystalliser system and sent to the Dithionate Decomposer; the mother liquor is treated with potassium sulphate in the Reactor to precipitate potassium imidodisulphonate which is separated from the solution and heated in the Decomposer to yield potassium sulphate (recycled to the Reactor) and sulphite together with SO_2 and N_2. The dithionate is thermally decomposed in the Dithionate Decomposer to sulphate, with release of SO_2. The sulphate, together with SO_2 from the Decomposer and Dithionate Decomposer, water and limestone, are fed to the Converter system where sodium sulphite absorbent is recovered and recirculated to the Absorber, and gypsum is produced.

Chemistry of the Process [419]: The absorbent is a solution of sodium sulphite and the ferrous salt of ethylene diamine tetracetic acid (Fe^{++}.EDTA) at a pH of 6.3. The principal reactions in the Absorber form sodium bisulphite and dithionate (with SO_2), and a ferrous chelate complex (with NO):

Bisulphite $SO_2 + Na_2SO_3 + H_2O = 2\ NaHSO_3$

Dithionate $4\ NaHSO_3 + O_2 = 2\ Na_2S_2O_6 + 2\ H_2O$

Chelate $NO + Fe^{++}.EDTA = Fe^{++}.EDTA.NO$

Oxygen in the gas causes some oxidation of the ferrous EDTA to the ferric EDTA, Fe^{+++}.EDTA, and of sulphite to sulphate:

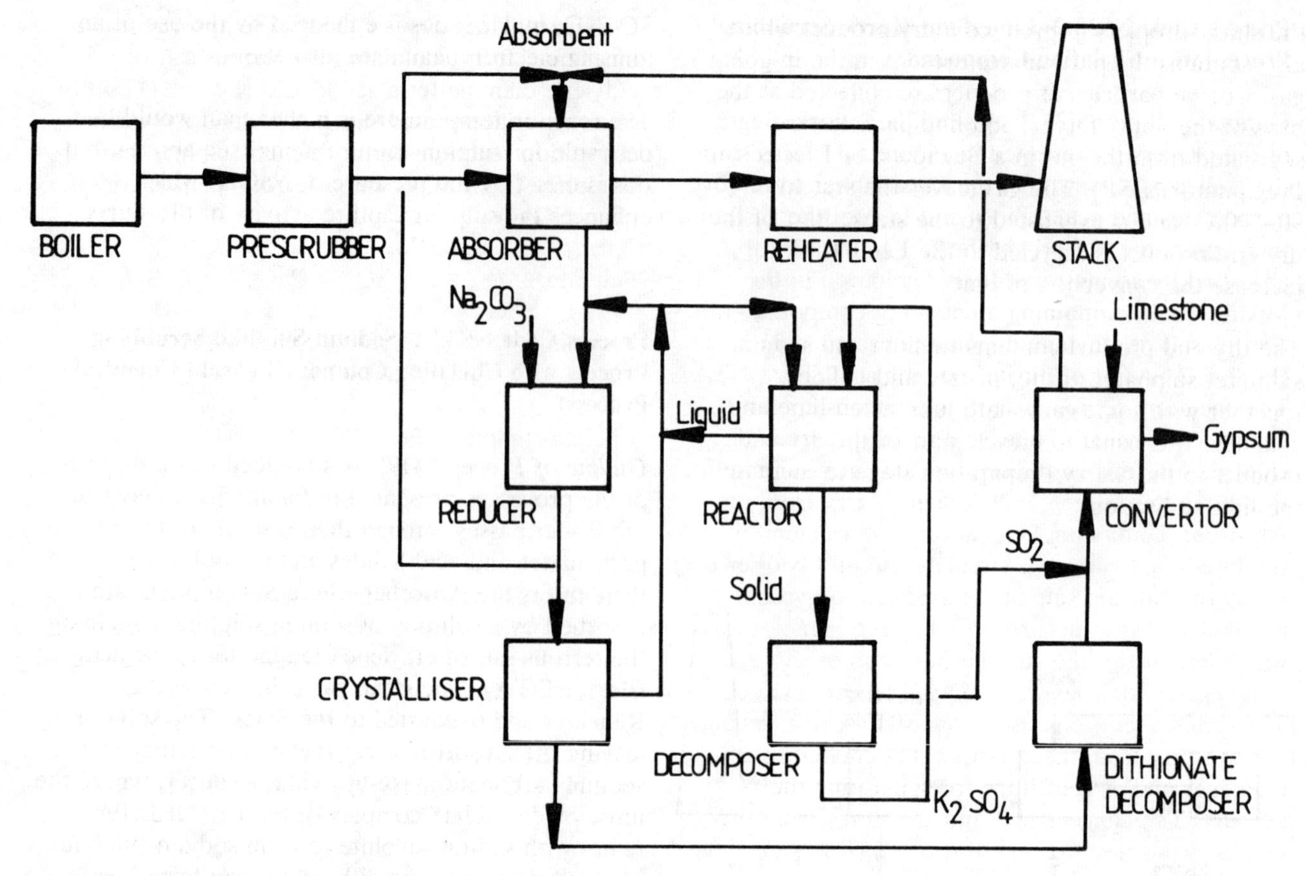

Figure 5.10 Block Diagram of Combined Abatement Process Sodium Sulphite Scrubbing with Chelating Compound (Asahi Chemical)

In the Reducer the main reactions restore the absorbent for recirculation to the Absorber:

Chelate reduction
$2\ Fe^{++}.EDTA.NO + 2\ Na_2SO_3 = 2\ Fe^{++}.EDTA + 2\ Na_2SO_4 + N_2$

Bisulphite reaction
$Na_2CO_3 + 2\ NaHSO_3 = 2\ Na_2SO_3 + H_2O + CO_2$

The Fe^{+++}.EDTA is reduced to Fe^{++}.EDTA by the sulphite ion, and some sodium imidodisulphonate, $NH(SO_3Na)_2$ is formed.

Sodium dithionate dihydrate, $Na_2S_2O_6.2H_2O$, and sodium sulphate decahydrate, $Na_2SO_4.10H_2O$, are crystallised out from a side-stream in the Crystalliser system, dehydrated and treated in the Dithionate Decomposer where the dithionate is thermally decomposed to sulphate:

Dithionate decomposition $Na_2S_2O_6 = Na_2SO_4 + SO_2$

The mother liquor, containing sodium imidodisulphonate, is treated with potassium sulphate solution in the reactor to precipitate potassium imidodisulphonate, which is centrifuged out. The concentrate is returned to the Reducer, and the solid is thermally decomposed in the Decomposer to give K_2SO_4 which is recycled to the Reactor:

Precipitation
$NH(SO_3Na)_2 + K_2SO_4 = NH(SO_3K)_2 + Na_2SO_4$

Decomposition
$2\ NH(SO_3K)_2 = 2\ SO_2 + K_2SO_4 + K_2SO_3 + H_2O + N_2$

The SO_2 released by decomposition of dithionate and imidodisulphonate are sent to the Converter system, where sodium sulphate from the Dithionate Decomposer, together with a feed of water and limestone, are used to produce sodium sulphite for recycle to the Absorber, and gypsum, $CaSO_4.2H_2O$. A complex series of operations occurs in the Converter system; the reactions can be simplified as the formation of calcium sulphite hemihydrate, and its conversion to gypsum:

Sulphite
$2\ CaCO_3 + 2\ SO_2 + H_2O = 2\ CaSO_3.(1/2)H_2O + 2\ CO_2$

Gypsum
$2\ CaSO_3.(1/2)H_2O + 2\ Na_2SO_4 + 3\ H_2O = 2\ Na_2SO_3 + 2\ CaSO_4.2H_2O$

Process Code NS32.2–Limestone Slurry Scrubbing Process with Chelating Compound

Outline of Process [352]: A simplified block diagram of the process is presented in Figure 5.11. Gas from the Boiler passes through the Prescrubber to remove particulates and acid halides and to cool the gas, and then enters the Absorber where SO_2 and NO are absorbed by a limestone slurry (prepared in the Mixing Tank) containing a chelating compound–the ferrous salt of ethylene diamine tetracetic acid (Fe^{++}.EDTA) or of nitrilo triacetic acid (Fe^{++}.NTA). The chelating compound forms a ferrous chelate complex with nitric oxide, and in the presence of chloride ions this is reduced by calcium bisulphite in the slurry to nitrogen and oxygen. Slurry from the Absorber, containing SO_2 captured as calcium sulphite hemihydrate, is circulated to an Oxidiser (in which the calcium sulphite is oxidised to gypsum by air bubbled into the slurry) and to a Regenerator (in which any of the chelate that has been oxidised to the ferric condition is reduced to the ferrous state by a reducing agent, e.g. ascorbic acid). A side stream of the slurry from the Oxidiser is circulated to a thickener and/or centrifuge for separating the gypsum. The concentrate, and the slurry from the Oxidiser and the Regenerator, are recirculated to the Absorber.

Chemistry of the Process [352]: The absorbent is a limestone slurry containing the ferrous salt of ethylene diamine tetra-acetic acid (Fe^{++}.EDTA) or of nitrilo triacetic acid (Fe^{++}.NTA). The principal absorption reactions form: calcium sulphite and bisulphite (with SO_2); and a ferrous chelate complex (with NO):

Sulphite $SO_2 + CaCO_3 = CaSO_3 + CO_2$

Bisulphite $CaSO_3 + SO_2 + H_2O = Ca(HSO_3)_2$

Chelate complex $Fe^{++}.EDTA + NO = Fe^{++}.EDTA.NO$

$Fe^{++}.NTA + NO = Fe^{++}.NTA.NO$

The ferrous chelate complex is reduced by bisulphite present in the slurry in the presence of chloride ions but the ferrous EDTA and NTA become oxidised to the ferric state:

$$4\ Fe^{++}.EDTA.NO + Ca(HSO_3)_2 = 2\ N_2 + O_2 + Ca(HSO_4)_2 + 4\ Fe^{+++}.EDTA$$

$$4\ Fe^{++}.NTA.NO + Ca(HSO_3)_2 = 2\ N_2 + O_2 + Ca(HSO_4)_2 + 4\ Fe^{+++}.NTA$$

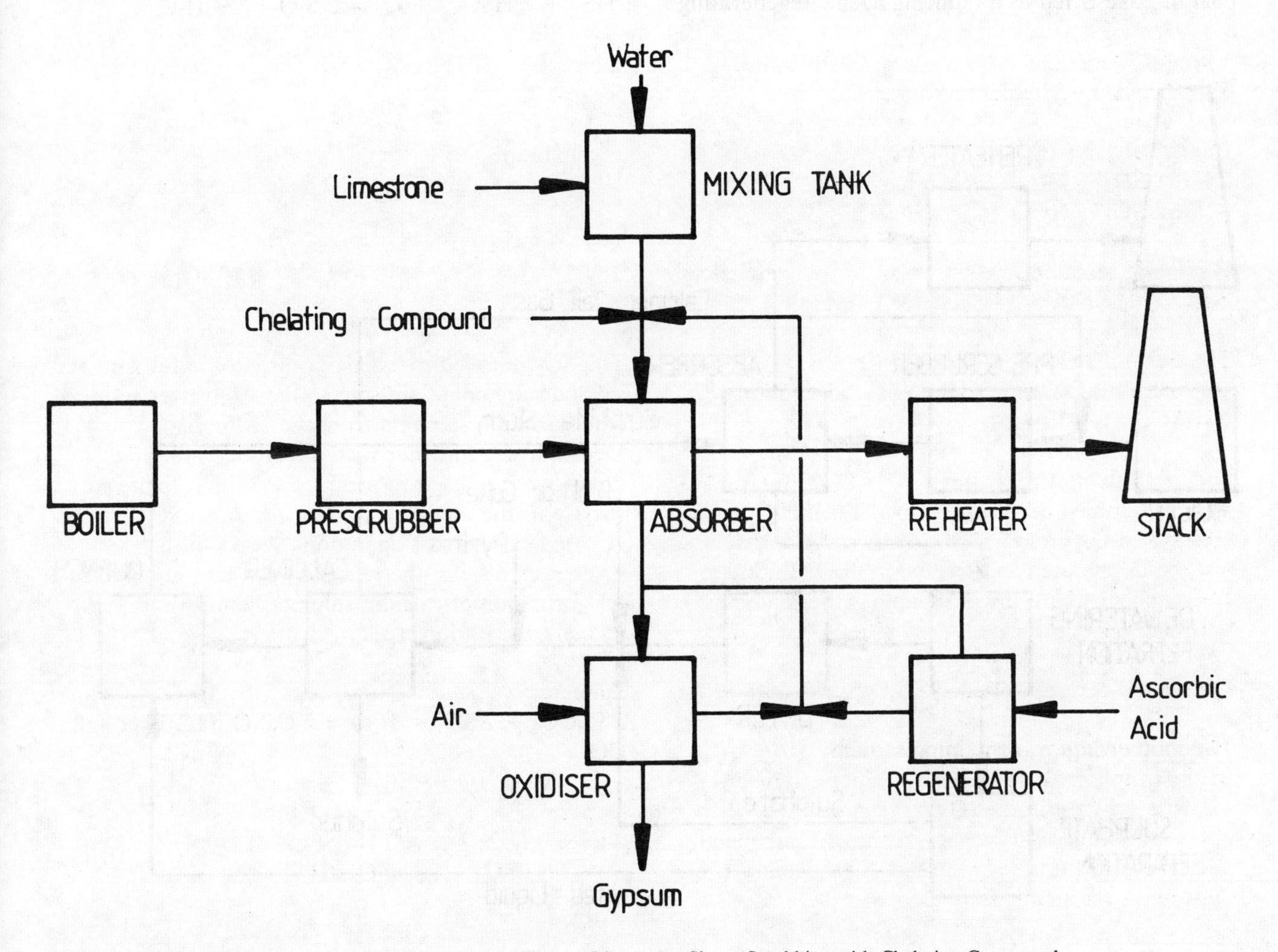

Figure 5.11 Block Diagram of Combined Abatement Process Limestone Slurry Scrubbing with Chelating Compound

They are restored to the ferrous state by reduction with e.g. ascorbic acid in the Regenerator.

In the combined abatement modification of the Saarberg-Holter-Lurgi FGD process (Section 2.5) the oxidation of sulphite to gypsum occurs in the Absorber, and formic acid is added to the absorbent slurry to accelerate SO_2 absorption and gypsum formation.

Process Code NS32.3–Sulf-X Process

Outline of Process [145, 175] (see Section 2.2): A simplified block diagram of the process is presented in Figure 5.12. Gas from the Boiler first passes through a Prescrubber to remove acid halides and particulates, and then into the Absorber where it is scrubbed with a slurry containing a mixture of iron sulphides in an aqueous sodium sulphide solution. The reactions with sulphur dioxide produce a range of insoluble iron/sulphur compounds, some soluble iron sulphate, and trace amounts of elemental sulphur. Reduction of NO_x to elemental nitrogen also occurs. The solid reaction products are separated from the solution in the Dewatering and Filtration system, dried in a steam-heated Dryer, and roasted at 650–750°C in an indirectly-fired Calciner to which coal or coke is fed as a reducing agent, regenerating the iron sulphides and releasing sulphur. Part of the solution from the Filtration stage is treated in a Sulphate Separation system to crystallise out sodium sulphate which is transferred to the Calciner. The solids from the Calciner pass to the Quench, where they are mixed with liquid from the Filtration and Sulphate Separation stages to reform the sulphide slurry sent to the Absorber.

Sulphur vapour in the tail gas from the Calciner is condensed for marketing or safe disposal, and the gas is returned to the Prescrubber inlet.

Chemistry of Process [95]: The sorbent slurry is a complex mixture of iron compounds in an aqueous sodium sulphide solution, with a pH controlled at 6.0–6.4 by the buffering action of $Fe(OH)_2$. The main absorbing compound is FeS. The sulphur capture reactions occurring in the Absorber include:

Ionisation of FeS:

$$FeS + H_2O = Fe^{++} + HS^- + OH^-$$

Ionisation of SO_2:

$$SO_2 + H_2O = H^+ + HSO_3^-$$

Thiosulphate formation:

$$2\,HS^- + 2\,HSO_3^- + O_2 = 2\,S_2O_3^{--} + H_2O$$

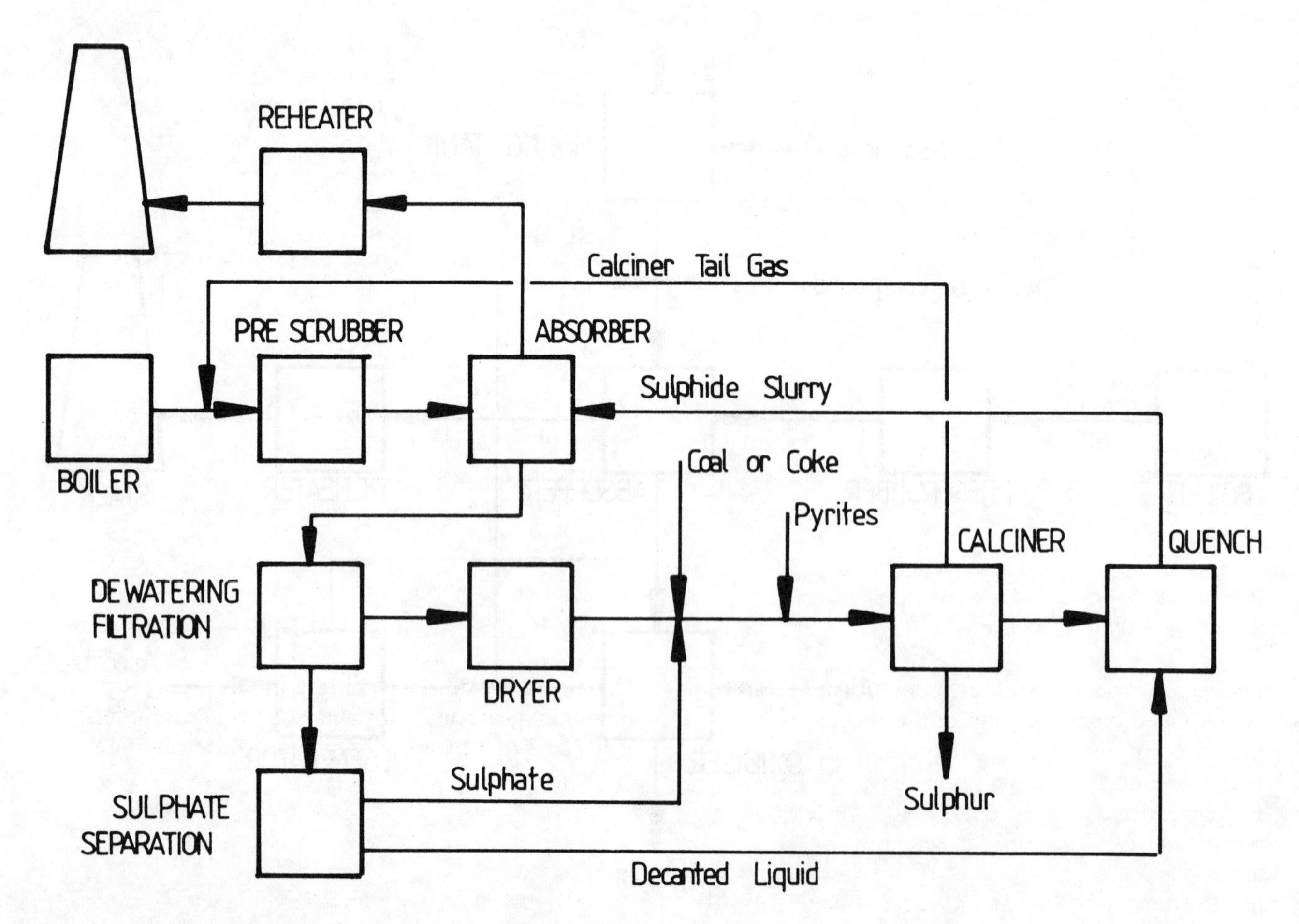

Figure 5.12 Block Diagram of Sulf-X Combined Abatement Process

Sulphur release: $S_2O_3^{--} + H^+ = HSO_3^- + S$

Sulphate formation:
$2\ HSO_3^- + O_2 = 2\ SO_4 + 2\ H^+$

The overall reaction can be expressed by the following simplification (note that the equation does not balance):

Overall reaction:
$FeS + SO_2 + O_2 = Fe^{++} + SO_4^{--} + H^+$

Nitrogen oxides are reduced by ferrous sulphide, releasing elemental nitrogen and forming ferrous sulphate:

Nitric oxide reduction:
$4\ NO + FeS = 2\ N_2 + FeSO_4$

$2\ NO_2 + FeS = N_2 + FeSO_4$

These reactions result in the formation of complex mixtures of iron sulphides, Fe_xS_y, together with $FeSO_4$, $Fe(OH)_2$ and Na_2SO_4. The key controlling factors are maintenance of the correct pH, the correct Na_2S concentration, and the buffering action of the $Fe(OH)_2$.

The main regeneration reaction is the thermal decomposition of the iron sulphides to give the sorbent FeS and sulphur; e.g. for FeS_2:

$FeS_2 = FeS + S$

Sodium sulphate is crystallised out from the solution, and the solid is sent to the Calciner where it is reduced by carbon to the sulphide. The ferrous sulphate is subsequently reduced in solution:

$Na_2SO_4 + 2\ C = Na_2S + 2\ CO_2$

$FeSO_4 + Na_2S = FeS + Na_2SO_4$

There are some losses of iron from the system, and these are made up by feeding pyrites, FeS_2, to the Calciner.

Process Code NS41.1–Oxidation plus Ammonia Scrubbing (Walther Process)

Outline of Process [Walther promotional literature]: A simplified block diagram of the process is presented in Figure 5.13. Ammonia is added to gas from the Boiler, and the gas is cooled in a Heat Exchanger (H.E.) before passing through three scrubbers (S1, S2 and S3) in series. In S1, the ammonia is absorbed, and sulphur oxides and acid halides removed, with ammoniacal effluent from S3. Ozone is then added to the gas to oxidise NO to NO_2, and more ammonia is also added before the gas enters S2, where the ammonia is absorbed by water and NO_2 is removed. Ammonia remaining in the gas leaving S2 is absorbed by water in S3, and residual ozone and NO_2 are reduced by a bleed of effluent from S1 containing sulphite and bisulphite. The effluent from S3 is pumped to S1, and the clean gas is reheated in the H.E. and exhausted to the stack. The effluents from S1 and S2 pass respectively to Oxidisers OX1, where ammonium sulphite and bisulphite are oxidised to sulphate by air and ammonia, and OX2, where ammonium nitrite is oxidised to nitrate by air, and carbonate is decomposed by addition of acid. The solutions from OX1 and OX2 are dried in the Spray Dryer to give a solid product from the base of the Spray Dryer and the Electrostatic Precipitator (ESP). Gases from OX1, OX2 and the ESP are returned to the inlet to the H.E. The solid product is potentially marketable as a fertiliser.

Chemistry of the Process [Walther promotional literature]: Sulphur oxides and acid halides ('HHa' = HCl and HF) are absorbed by ammonium hydroxide solution in Scrubber S1, where the pH is 6:

Sulphite $SO_2 + 2\ NH_4OH = (NH_4)_2SO_3 + H_2O$

Bisulphite $(NH_4)_2SO_3 + SO_2 + H_2O = 2\ NH_4HSO_3$

Sulphate $SO_3 + 2\ NH_4OH = (NH_4)_2SO_4 + H_2O$

Halides $HHa + NH_4OH = NH_4Ha + H_2O$

In Oxidiser OX1 at a pH of 4.0, sulphite and bisulphite are oxidised to sulphate:

Bisulphite oxidation $NH_4HSO_3 + NH_3 = (NH_4)_2SO_3$

Sulphite oxidation
$2\ (NH_4)_2SO_3 + O_2 = 2\ (NH_4)_2SO_4$

Ozone oxidises NO to NO_2, and this is absorbed by ammonium hydroxide in S2 at pH 8 to 9 to form ammonium nitrite and nitrate:

NO Oxidation $NO + O_3 = NO_2 + O_2$

Absorption
$2\ NO_2 + 2\ NH_3 + H_2O = NH_4NO_2 + NH_4NO_3$

The nitrite is oxidised to nitrate in OX2, and ammonium carbonate formed in S2 is decomposed at pH = 3 to 4 by nitric or sulphuric acid addition:

Nitrite oxidation $2\ NH_4NO_2 + O_2 = 2NH_4NO_3$

Acid treatment
$(NH_4)_2CO_3 + 2\ HNO_3 = 2NH_4NO_3 + CO_2 + H_2O$
$(NH_4)_2CO_3 + H_2SO_4 = (NH_4)_2SO_4 + CO_2 + H_2O$

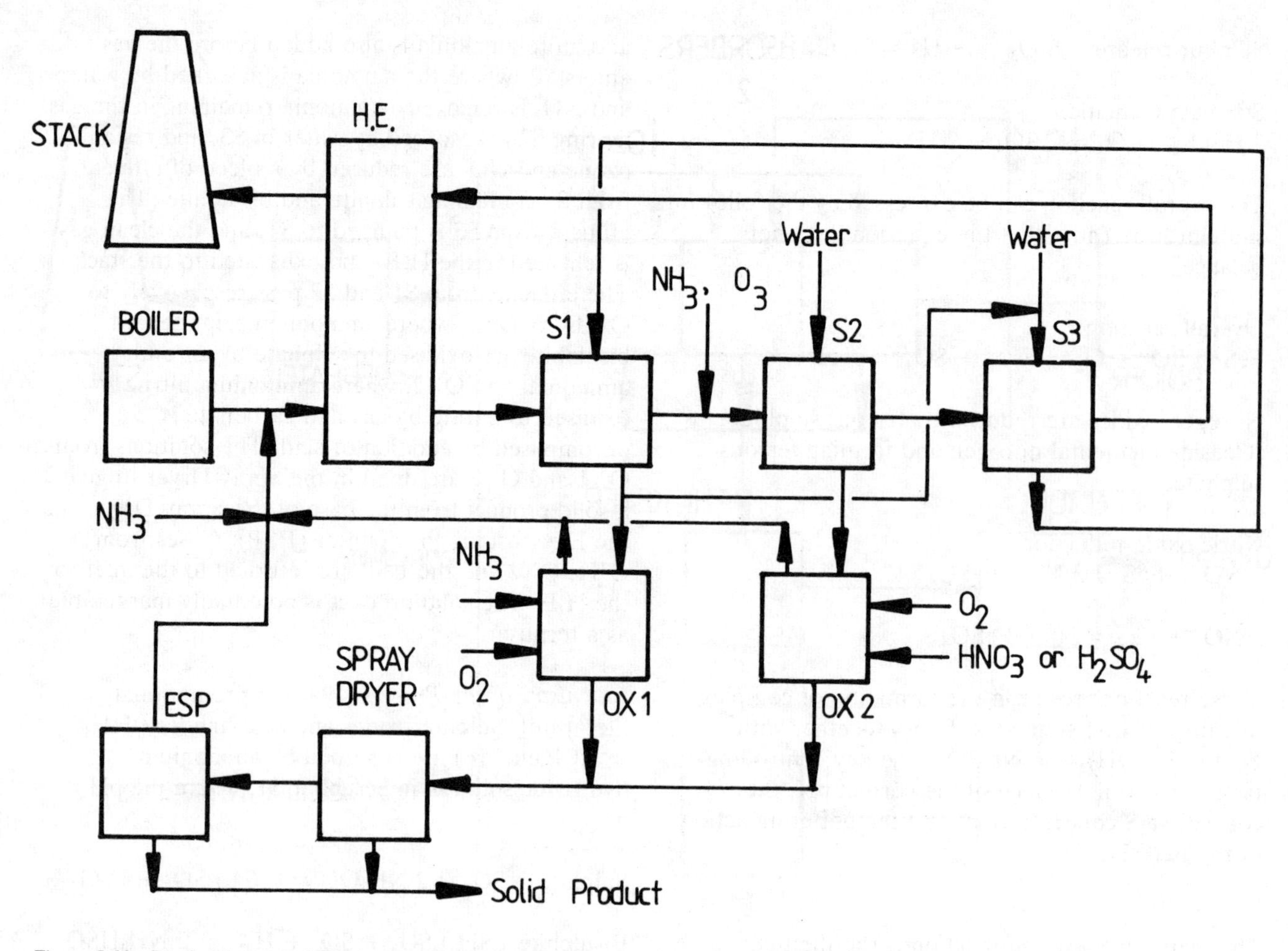

Figure 5.13 Block Diagram of Walther Combined Abatement Process

In Scrubber S3, residual ammonia in the gas is absorbed by water, and residual ozone and NO_2 are reduced by ammonium sulphite and bisulphite from the S1 effluent at pH 5; thus, with ammonium sulphite:

$$O_3 + (NH_4)_2SO_3 = (NH_4)_2SO_4 + O_2$$

$$2\ NO_2 + 4\ (NH_4)_2SO_3 = 4\ (NH_4)_2SO_4 + N_2$$

Process Code NS41.2–Kawasaki Process

Outline of Process [419]: A block diagram of the process is presented in Figure 5.14. Gas from the Boiler passes through an Absorber comprising three sections, in each of which the gas is scrubbed with a magnesium hydroxide slurry. Sulphur oxides are absorbed in Section 1, and NO and NO_2 are absorbed in Section 2 in equimolar proportions. Ozone is supplied to Section 3, oxidising NO to NO_2 which is absorbed by the slurry. The clean gas is exhausted to the Stack via a Reheater. Slurry leaving the Absorbers passes to a Thickener. The overflow, containing magnesium nitrite and nitrate in solution, is acidified with sulphuric acid in the reactor system, releasing NO which is oxidised with air to NO_2 and returned to Absorber 2. The magnesium nitrite, converted to nitrate and sulphate in the Reactor, passes together with the underflow from the Thickener, containing magnesium sulphite and sulphate, to the Oxidiser, where sulphite is oxidised to sulphate by treatment of the slurry with air. In the Regenerator system, the magnesium hydroxide is regenerated by reaction between calcium nitrate, to produce gypsum, followed by treatment with lime to produce calcium nitrate solution, some of which is bled off.

Chemistry of the Process [419]: The gas is cooled and humidified in Absorber 1, and sulphur dioxide is absorbed by the magnesium hydroxide slurry to form the sulphite hexahydrate:

$$SO_2 + Mg(OH)_2 + 5\ H_2O = MgSO_3.6H_2O$$

In Absorber 2, NO and NO_2 (including NO_2 from the Reactor) react together and are absorbed by the slurry to form magnesium nitrite:

$$NO + NO_2 + Mg(OH)_2 = Mg(NO_2)_2 + H_2O$$

This reaction ceases when the NO_x concentration falls below about 200 ppm, and the gas then has to be treated with ozone in Absorber 3, where NO is

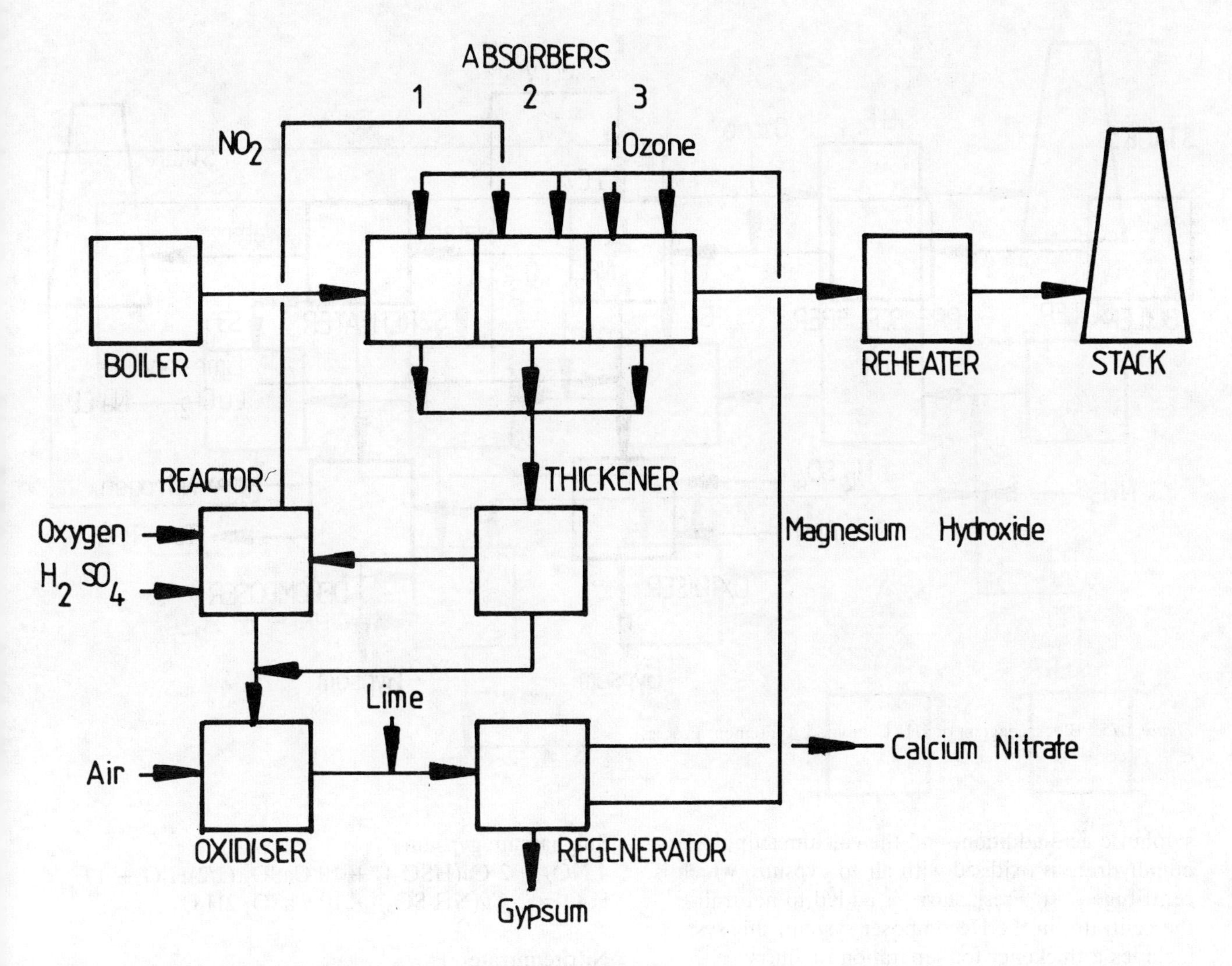

Figure 5.14 Block Diagram of Kawasaki Combined Abatement Process

oxidised to NO_2 which is absorbed by the slurry to form magnesium nitrate and nitrite:

$$NO + O_3 = NO_2 + O_2$$

$$4\ NO_2 + 2\ Mg(OH)_2 = Mg(NO_3)_2 + Mg(NO_2)_2 + 2\ H_2O$$

The effluents from all three Absorbers are treated in a Thickener; the nitrate and nitrite pass in the overflow to the Reactor system, where reaction with sulphuric acid converts nitrite to nitrate, sulphate and NO:

$$3\ Mg(NO_2)_2 + 2\ H_2SO_4 = 2\ MgSO_4 + Mg(NO_3)_2 + 4\ NO + 2\ H_2O$$

The NO released is oxidised with air, and passes to Absorber 2 where it reacts with NO and is absorbed as described above. The solution passes with the underflow from the Thickener, to the Oxidiser, where the slurry is treated with air to oxidise sulphite to sulphate. These reactions are:

$$2\ NO + O_2 = 2\ NO_2$$

$$2\ MgSO_3.6H_2O + O_2 = 2\ MgSO_4 + 6\ H_2O$$

The slurry then passes to the Regenerator, where reaction with calcium nitrate and make-up lime regenerates the magnesium hydroxide absorbent, and produces gypsum and excess calcium nitrate:

$$MgSO_4 + Ca(NO_3)_2 + 2\ H_2O = CaSO_4.2H_2O + Mg(NO_3)_2$$

$$Mg(NO_3)_2 + Ca(OH)_2 = Mg(OH)_2 + Ca(NO_3)_2$$

Process Code NS42.1–IHI Process

Outline of Process [419]: A simplified block diagram of the process is presented in Figure 5.15. Gas from the Boiler first enters a Prescrubber to remove acid halides and particulates, and to cool and humidify the gas. Ozone is mixed with the gas, to oxidise NO to NO_2, and the gas is then scrubbed in a Turbulent Contact Absorber (TCA) through which a slurry of lime or limestone is circulated. The slurry, which absorbs SO_2, contains copper and sodium chloride catalysts to promote absorption and reduction of NO_2. The clean gas is exhausted to stack via the Reheater. A side stream of the circulating slurry is treated in the Oxidiser system; the pH is reduced by

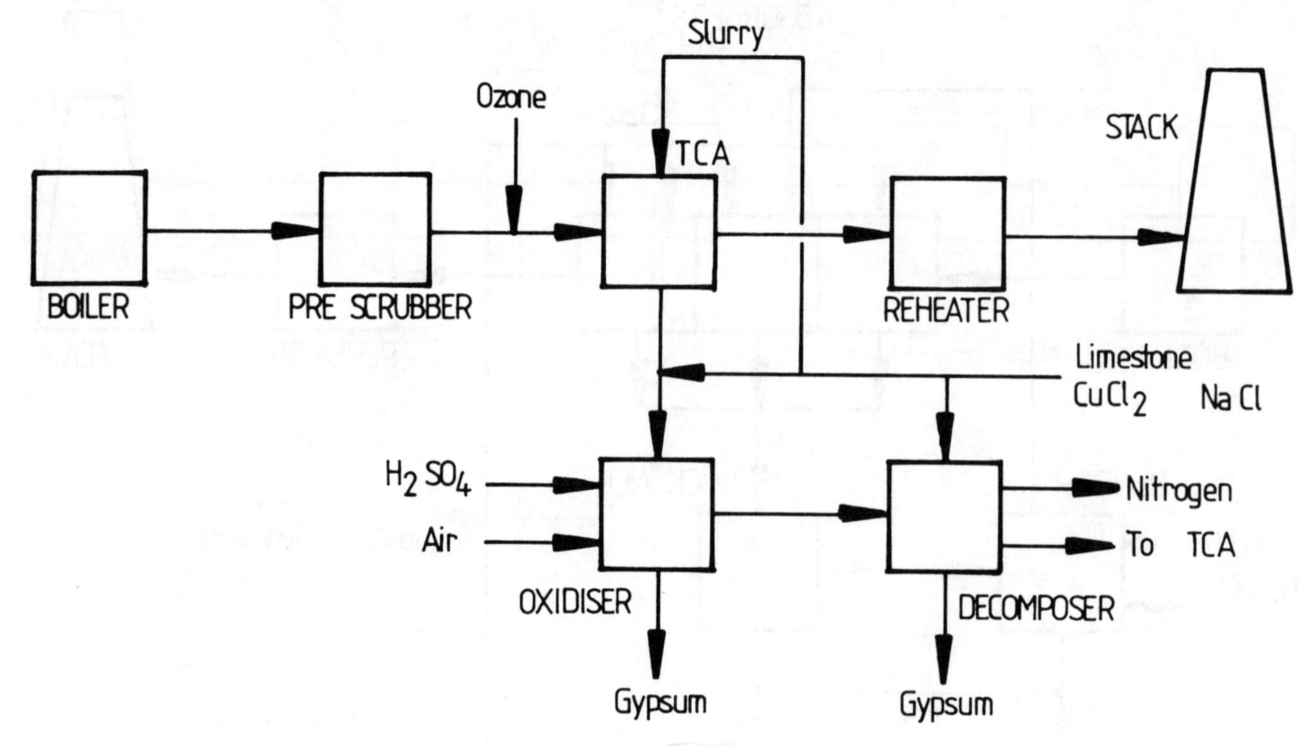

Figure 5.15 Block Diagram of IHI Combined Abatement Process

sulphuric acid addition, and the calcium sulphite hemihydrate is oxidised with air to gypsum, which is centrifuged off. Fresh slurry is added to neutralise the centrate, in the Decomposer system; this system includes a thickener for separation of slurry (returned to the TCA), an evaporator for the overflow, and a thermal decomposer for the evaporator residue. The decomposition stage is needed to eliminate nitrogen-sulphur compounds of calcium, and it produces further gypsum contaminated by copper and calcium chloride.

Chemistry of the Process [419]: SO_2 is absorbed by the slurry in the TCA to form calcium sulphite hemihydrate and bisulphite:

Sulphite
$2\ SO_2 + 2\ CaCO_3 + H_2O =$
$2\ CaSO_3.(1/2)H_2O + 2\ CO_2$

Bisulphite
$2\ SO_2 + 2\ CaSO_3.(1/2)H_2O + H_2O = 2\ Ca(HSO_3)_2$

Nitric oxide is oxidised to NO_2 by ozone, and the NO_2 is absorbed by the slurry and reduced to nitrogen by calcium sulphite and bisulphite to produce gypsum, calcium sulphamates and calcium nitrite and nitrate; the reactions include:

Oxidation $NO + O_3 = NO_2 + O_2$

Gypsum
$2\ NO_2 + 4\ CaSO_3.(1/2)H_2O + 6\ H_2O =$
$N_2 + 4\ CaSO_4.2H_2O$

Sulphamate/gypsum
$4\ NO_2 + 2\ Ca(HSO_3)_2 + 10\ CaSO_3.(1/2)H_2O + 17\ H_2O = 2\ Ca(NH_2SO_3)_2 + 10\ CaSO_4.2H_2O$

Nitrite/nitrate
$4\ NO_2 + 4\ CaSO_3.(1/2)H_2O =$
$Ca(NO_2)_2 + Ca(NO_3)_2 + 2\ Ca(HSO_3)_2$

In the thermal decomposition stage, the sulphamates, nitrite and nitrate are decomposed by reactions including:

Sulphamate: (The following equation, reproduced from Reference [33.12], does not balance):

$2\ (NH_2SO_3)_2Ca + 4\ CaCO_3 + 2\ Ca(HSO_4)_2 + 3\ O_2 + 8\ H_2O = 2\ N_2 + 8\ CaSO_4.2H_2O + 4\ CO_2$

Nitrite/nitrate:
$Ca(NO_2)_2 + Ca(NO_3)_2 + 2\ Ca(HSO_4)_2 =$
$2\ N_2 + 4\ CaSO_4.2H_2O + 2\ H_2O$

Process Code NS42.2–Moretana Calcium Process

Outline of Process [419]: A simplified block diagram of the process is presented in Figure 5.16. Gas from the Boiler is passed through a Prescrubber to remove acid halides and particulates, and to cool and humidify the gas. Chlorine dioxide is added to the gas, and it is then scrubbed with a limestone slurry, containing a catalyst, which is circulated through a plate tower Absorber. The clean gas is exhausted to stack via a Reheater. A side stream of the circulating

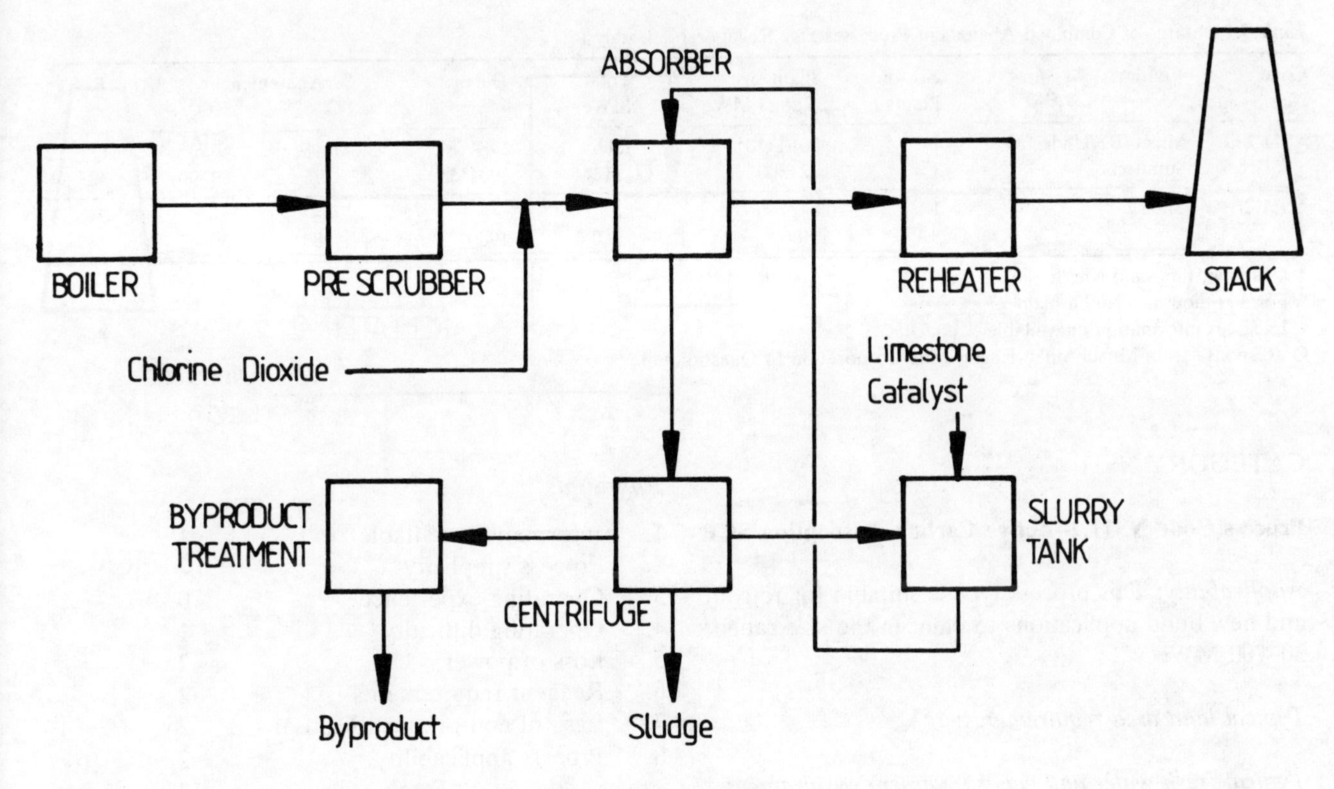

Figure 5.16 Block Diagram of Moretana Calcium Combined Abatement Process

slurry is centrifuged to remove gypsum, and the centrate is split into two streams; the larger passes to a Slurry Tank, where fresh limestone and catalyst is added for return to the Absorber, and the remainder is sent to a Byproduct Treatment system, where it is either evaporated to dryness, to produce calcium chloride and nitrate, or is treated with ammonium carbonate or sulphate to produce a liquid fertiliser.

Chemistry of the Process [419]: Chlorine dioxide oxidises NO to NO_2 and nitric acid:

Oxidation $2\ NO + C1O_2 + H_2O = NO_2 + HNO_3 + HCl$

In the Absorber, the SO_2 is absorbed by limestone to form calcium sulphite and bisulphite, and NO_2, HNO_3 and HCl are absorbed and undergo reactions with calcium sulphite and bisulphite (reduction of NO_2 in the presence of a catalyst), and with calcium carbonate:

Sulphite $2\ NO_2 + CaSO_3 + CaCO_3 + 2\ H_2O = Ca(NO_2)_2 + CaSO_4.2H_2O + CO_2$

Bisulphite: (The following equation, reproduced from Reference [33.12], does not balance):

$Ca(NO_2)_2 + Ca(HSO_3)_2 + H_2O = N_2 + 2\ CaSO_4.2H_2O$

Carbonate $CaCO_3 + 2\ HNO_3 = Ca(NO_3)_2 + CO_2 + H_2O$

$CaCO_3 + 2\ HCl = CaCl_2 + CO_2 + H_2O$

In the Byproduct Treatment system, the centrate, containing calcium nitrate and chloride, is either evaporated to dryness or treated with ammonium carbonate to produce calcium carbonate, together with a liquid fertiliser solution of ammonium chloride and nitrate:

Chloride $CaCl_2 + (NH_4)_2CO_3 = 2\ NH_4Cl + CaCO_3$

Nitrate
$Ca(NO_3)_2 + (NH_4)_2CO_3 = 2\ NH_4NO_3 + CaCO_3$

5.3 General Appraisal of Processes

This Section presents available information on the characteristics of the combined SO_2–NO_x abatement process types: status, applicability, space and typical land area requirements, fresh water and water treatment requirements, reagent consumption, end-product or waste materials disposal requirements, typical power consumptions and reductions in combustion plant efficiency, operating experience and process developments; see Section 1.5.

The status of Category NS11 processes is indicated in Table 5.2, which presents information on the size of plants built, year of completion, and percentage abatement of SO_2 and NO_x.

Table 5.2 Status of Combined Abatement Processes Dry Regenerable Catalyst

Code No.	Vendor	No. of Plants	Plant size Range MWe	Total MWe	Dates	% Abatement SO_2	NO_x	Ref.
NS11.1	Mitsui/BF/Uhde	4	30–1110*	1506*	1984–87	90–98	60	Q, 352
	Sumitomo	3	2–300*	312*	1984**	95	80–90	423,34
NS11.2	Shell	1	–	–	–	–	–	34
		+1	0.6	0.6	1980	88	50–65	137

* Gas flow, thousand Nm^3/h
**For 300 thousand Nm^3/h plant
– Indicates information unavailable
Q Response from Mitsui Miike Engineering Corporation to Questionnaire

CATEGORY NS11

Process Code NS11.1–Active Carbon Adsorption/SCR

Applicability: This process type is suitable for retrofit and new build applications to plant in the size range 30–700 MWe.

Typical land area requirements: ***

Typical fresh water and water treatment requirements: Estimates are presented in Table 5.3.

Consumption of reagents: Hydrogen consumption is 2.6–3.2 mol/mol sulphur captured; ammonia consumption is 0.5–1.4 mol per mol ingoing NO_x [656]. Further information in the form of published estimates is presented in Table 5.3.

End-product or waste disposal requirements: End product is elemental sulphur (99.9% purity), sulphuric acid (98% purity) or liquid sulphur dioxide; if elemental sulphur is produced the quantity is equivalent to 90–93% of the sulphur captured. Estimates are presented in Table 5.3.

Typical pressure losses: ***

Typical power requirements: ***

Typical reductions in combustion plant efficiency: ***

Operating experience: A demonstration plant (130,000 Nm^3/h from a coal-fired utility boiler) has operated for at least 16,000 hours, and a small commercial unit (30,000 Nm^3/h) was reported in 1986 to have operated trouble-free since October 1984 [352]. Estimated availability, based on experience of the 30,000 Nm^3/h plant, exceeds 95%, or equal to 100% if maintenance work on the abatement plant is carried out during scheduled boiler shut-down periods. It has been reported [656] that fouling by fly-ash in coal-fired systems is not a problem, but that there is some emission of ammonia.

Process developments: Development and use of lower-cost activated carbon is the most likely development to reduce the process operating cost.

Appraisal:

1.	Information available	1
2.	Process simplicity	2
3.	Operating experience	0
4.	Operating difficulty	1
5.	Loss of power	2*
6.	Reagent requirements	2
7.	Ease of end-product disposal	2
8.	Process applicability	2
	Total	12

*Assumed in absence of information

Table 5.3 Estimated Requirements

Reference		Q	352
MWe		60	460
Gas flow	'000 Nm^3/h	200	1200
Gas: Inlet SO_2 content	ppmv	373	1000
Inlet NO_x content	ppmv	200	240
Abatement: SO_2	%	90	98
NO_x	%	50	90
Water requirement per MWh:			
Process water		Nil	Nil
Cooling water	tonne	0.17	–
Consumption of reagents per MWh:			
Activated coke	kg	0.73	1.18
Ammonia	kg	0.42	0.67
Heavy oil	kg	–	0.65
Coke (for reduction)	kg	–	6.30
Nitrogen	Nm^3	0.22	–
Instrument air	Nm^3	0.50	–
End-product or waste disposal requirements per MWh:			
Activated coke fines	kg	–	0.49
Coal ash	kg	–	0.63
Sulphur	kg	1.3	3.41

– Indicates information unavailable
Q Response from Mitsui Miike Engineering Corporation to Questionnaire

Process Code NS11.2–Copper Oxide Absorption/SCR

Applicability: This process type is suitable for new build applications only.

Typical land area requirements: ***

Typical fresh water and water treatment requirements: ***

Consumption of reagents: Hydrogen consumption [282] 0.2 kg/kg sulphur captured. Ammonia consumption 0.52–0.57 kg/kg NO_x (as NO_2) abated (for 50.65% abatement) [137].

End-product or waste disposal requirements: End product is SO_2 which can be processed to produce sulphuric acid, elemental sulphur or liquid SO_2.

Typical pressure losses: Probably somewhat in excess of 20 mbar, the value (see Section 2.3) for a plant handling 125,000 Nm^3/h, 2500 ppmv SO_2, 90% sulphur capture.

Typical power requirements: ***

Typical reductions in combustion plant efficiency: ***

Operating experience: The process was applied to an industrial boiler in Japan for several years to remove 90% of the SO_2 and 40% of the NO_x, but the plant is no longer operated. The U.S. Environmental Protection Agency operated a pilot plant for 90 days on a side-stream (1600 m^3/h equivalent to 0.6 MWe) from a 400 MWe coal-fired boiler at the Big Bend station, North Ruskin, Florida; sulphur capture was 90% and NO_x abatement exceeded 70% [137, 352].

Process developments: The need for hydrogen makes this process more suitable for application at petroleum refineries than for utility or industrial boilers. Shell are no longer licensing the process.

Appraisal:

1.	Information available	1
2.	Process simplicity	1
3.	Operating experience	1
4.	Operating difficulty	0*
5.	Loss of power	1*
6.	Reagent requirements	2
7.	Ease of end-product disposal	1
8.	Process applicability	1
	Total	8

*Assumed in absence of information

The status of Category NS12 processes is indicated in Table 5.4, which presents information on the size of plants built, year of completion, and percentage abatement of SO_2 and NO_x.

CATEGORY NS12

Process Code NS12.1–Non-Selective Catalytic Reduction of SO_2 and NO_x (Ralph M. Parsons Co. Process)

Applicability: This process type is suitable for new build applications only.

Typical land area requirements: ***

Typical fresh water and water treatment requirements: ***

Consumption of reagents: ***

End-product or waste disposal requirements: ***

Typical pressure losses: ***

Typical power requirements: ***

Typical reductions in combustion plant efficiency: ***

Operating experience: A 200 m^3/h pilot plant has been operated in the FRG [352].

Process developments: ***

Appraisal:

1.	Information	0
2.	Process simplicity	2*
3.	Operating experience	0
4.	Operating difficulty	0*
5.	Loss of power	2*
6.	Reagent requirements	2*
7.	Ease of end-product disposal	2*
8.	Process applicability	1
	Total	9

*Assumed in the absence of information

Process Code NS12.2–Catalytic Reduction of NO_x and Catalytic Oxidation of SO_2

Applicability: This process is suitable only for new build applications.

Total land area requirements: ***

Table 5.4 Status of Combined Abatement Process Dry Non-Regenerable Catalysts

Code No.	Vendor	No. of Plants	Plant size Range MWe	Total MWe	Dates	% Abatement SO_2	% Abatement NO_x	Ref.
NS12.1	Parsons	1	0.2*	0.2*	–	–	–	419
NS12.2	–	–	–	–	–	–	–	352

* Thousand Nm^3/h
– Indicates information unavailable

Typical fresh water and water treatment requirements: Nil.

Consumption of reagents: *** Natural gas is needed to remove oxygen from the gas and to reduce the nitric oxide. Air is required to restore oxidising conditions for catalytic oxidation of SO_2.

End-product or waste disposal requirements: The end-product is concentrated (93%) sulphuric acid, equivalent in quantity to the SO_2 removed; this can be sold.

Typical pressure losses: ***

Typical power requirements: ***

Typical reductions in combustion plant efficiency: ***

Operating experience [352]: A 200 Nm^3/h pilot plant was reported in 1986 to be in operation to test the process.

Process development [352]: It was reported in 1986 that installation of a demonstration plant was planned.

Appraisal:

1.	Information available	0
2.	Process simplicity	1
3.	Operating experience	0
4.	Operating difficulty	0*
5.	Loss of power	2*
6.	Reagent requirements	1
7.	Ease of end-product disposal	2
8.	Process applicability	1
	Total	7

*Assumed in absence of information

CATEGORY NS21

Combustions techniques in Category NS21 processes are still at the development stage. The U.S. EPA have recently initiated a full-scale demonstration of the LIMB technique, which is expected to last for 4 1/2 years [351], whilst the pilot-scale development of a slagging burner was reported by Rockwell International in 1983 [427].

Process Code NS21.1–Limestone Injection into Multi-stage Burner (LIMB)

Applicability: This technique is suitable for retrofit and new build applications, although highest levels of NO_x and SO_x removal are to be expected from new boilers [431]. At present, the effects of the sorbent on the performance of an existing particulate collection system have not been determined, so upgrading of this equipment may be required in retrofit situations.

Space requirements: No information is available, but this technique requires sufficient space for limestone storage, preparation and injection, and for the disposal of the solid waste products and unreacted limestone.

Typical land area requirements: ***

Consumption of reagents: Limestone is normally used for SO_x absorption although alternative sorbents have been used in trials, e.g. calcite, dolomite or argonite [431]. Fine grinding of the limestone is necessary to improve the reactivity (surface area) of the sorbent and increase its utilisation. The Peabody Coal Co. found that limestone utilisation improved when the maximum particle size of the sorbent was reduced from 149 um to 74 um [431]. Calcium/sulphur molar ratios of 2.0 are typically required for SO_x reductions of 50%.

End-product or waste disposal requirements: ***

Typical pressure losses: ***

Typical power requirements: ***

Typical reductions in combustion plant efficiency: ***

Operating experience: The fore-runner of LIMB was the injection of dry limestone into the furnace above the flame zone. In early tests carried out for the EPA, low levels of SO_2 control were achieved. These levels were especially low when the limestone was added to the coal: e.g. less than 8% SO_2 reduction was reported by Babcock and Wilcox with this arrangement [431]. It was subsequently found that at high peak temperatures, e.g. in excess of 1250°C, deadburning of the limestone occurred leading to loss of reactivity.

With the development of low-NO_x coal burner designs, and their inherently lower operating temperatures, interest in limestone injection was revived. The EPA in the USA and several European and U.S. companies renewed their efforts to achieve acceptable SO_2 control levels by sorbent injection. For example, EPA-sponsored trials by Energy and Environmental Research Corporation in 1978 achieved 73% SO_2 removal with limestone injection (Ca:S ratio = 2) through an advanced, but experimental, 1.8 MWt low-NO_x burner. Two coals of 0.7% and 3.8% sulphur content were used in these trials and the limestone was mixed with the coal. With the lower sulphur coal only, SO_2 control rates of 50% and 88% were achieved for Ca:S ratios of 1.0 and 3.0, respectively [431]. These tests were aimed at minimising NO_x emission without optimising SO_2 removal.

Other companies which are known to be developing limestone injection techniques in conjunction with low-NO_x burner designs include the International Flame Research Foundation in the Netherlands [297], L. and C. Steinmuller in FRG [431], Foster Wheeler Energy Corporation (F.W.E.C.) [421, 368], Riley Stoker, TRW Inc. and Rockwell International [296].

Process development: The limited pilot-scale testing in the U.S.A. indicated that direct limestone injection into the furnace could not compete with other FGD processes because of its limited sulphur capture capabilities. There has, however, been an incentive to develop the LIMB technology as a combined SO_x-NO_x system because of its apparently high cost effectiveness where levels of control of the order of 70% for NO_x and 50% for SO_x are adequate [431].

The EPA have announced that their pilot unit testing of LIMB is essentially complete and a full-scale demonstration initiated in association with Ohio Edison, at the latter's 150 MWe Edgewater Station [351]. The programme is expected to last 4.5 years and has the prime objectives of demonstrating NO_x and SO_x reductions of 50–60% and determining the impact of the technique on the operation of the boiler.

Appraisal:

1.	Information	0
2.	Process simplicity	2
3.	Operating experience	0
4.	Operating difficulty	2
5.	Reagent requirements	1
6.	Ease of end-product disposal	1
7.	Process applicability	2
	Total	8

Process Code NS21.2–Slagging Combustion

Applicability: The burner under development is intended for gas or oil-to-coal retrofit applications only. The slagging burners would be fitted into the existing, and possibly enlarged, burner ports in the water wall in boilers where sufficient spacing exists between and behind the ports. An optional slag/fly-ash separator within the burner will allow conversion of a boiler designed for oil firing to pulverised coal firing, without significant de-rating.

In various other forms and modifications, the technique is expected to be suitable for new and retrofitted boilers designed for coal.

Space requirements: The burner is not small, especially when fitted with a fly-ash separator. Figure 5.7 shows a conceptual design and the dimensions of a 29MW burner. The separator and part of the burner would normally fit within the existing windbox, but part of the burner would protrude outside the windbox. Where the separator is not required, e.g. when installed in a boiler designed for coal, most or all of the burner would fit within the space occupied by the windbox.

Additional space is required for limestone storage, preparation and mixing with coal, prior to injection, and for the disposal of the solid waste products and unreacted limestone.

Typical land area requirements: ***

Consumption of reagents: Current data [427] indicate that adequate SO_2 control can be achieved with a Ca:S molar ratio of between one and two. The calcium is added in the form of limestone, which is pulverised with the coal prior to injection through the burner.

End-product or water disposal requirements: ***

Typical pressure losses: Very little information is available, although a combustion air pressure drop of not more than 75 mbar is expected through a 29MW burner, with an additional pressure loss of less than 7.5 mbar for the slag/fly-ash separator, if fitted [427].

Typical power requirements: ***

Typical reductions in combustion plant efficiency: ***

Operating experience: In 1978, Rockwell International initiated a programme to investigate other designs of and applications for a slagging, coal-fired burner previously developed for the U.S. DOE. Initial results suggested that the design was an effective means of SO_x control during combustion. Subsequent, and separate, tests verified that the burner was equally effective as a means of NO_x control. Testing was conducted on a 5 MW staged combustor which was later modified with a simple hot gas fly-ash separator.

In 1981, further development was geared to overcoming the problems associated with converting boilers designed for gas and oil fuels to pulverised coal firing, as well as achieving simultaneous control of SO_x and NO_x within the combustion process. A prototype burner, rated at 7MW, was constructed and operated in a test facility. This size of burner was considered to be pilot-scale for utility boiler applications, where burners would be at least four times its rated heat input, but more than full-scale for many industrial applications. Results from this test programme were used to prepare preliminary designs for a full-scale utility boiler burner rated at 29 MW. A full scale utility boiler demonstration was planned for the 1984–1986 period [427].

When the 7 MWt burner was operated for maximum SO_x control, and using lime, sulphur capture levels of 70–90% were achieved with high sulphur (2.8–3.9%) coals and up to 95% with low sulphur (0.5–1.0%) coals. When the same burner was operated for maximum NO_x control, flue gas NO_x levels were consistently below 88 ppm (measured as NO_2, dry, at 3% O_2) and frequently below 10 ppm. NO_x levels of the order of 500 ppm were consistently measured when the burner was operated like a conventional coal burner, suggesting that NO_x reductions of 84–98% can be achieved.

Process development: A major objective of the development work is to optimise the design and reduce the overall size and weight of the burner. It was the stated intention, during the pilot-scale testing, to operate the burner with continuous slag removal; this had not previously been attempted and no details are available on the results of this operation.

Appraisal:

1.	Information	0
2.	Process simplicity	2
3.	Operating experience	0
4.	Operating difficulty	0
5.	Reagent requirements	2*
6.	Ease of end-product disposal	1
7.	Process applicability	2
	Total	8

*Assumed in absence of information

CATEGORY NS22

The status of Category NS22 processes is indicated in Table 5.5, which presents information on the size of plants built, year of completion, and percentage abatement of SO_2 and NO_x.

Process Code NS22.1–Electron Beam Radiation Process

Applicability: This process type is suitable for retrofit and new build applications.

Typical land area requirements: ***

Typical fresh water and water treatment requirements: ***

Consumption of reagents: Ammonia: 0.53 kg/kg SO_2 removed; 0.74 kg/kg NO_x (as NO_2) abated.

End-product or waste disposal requirements: About 3% of equivalent electrical power generation of combustion system [34, 419].

Typical reductions in combustion plant efficiency: ***

Operating experience: Appears to be confined to a 6000 Nm^3/h pilot plant at the Paducah, Kentucky station of TVA [678]; and to a 10,000 m^3/h pilot plant in Japan, where over 95% sulphur capture and 80% NO_x abatement were obtained with gas containing 570 mg/m^3 SO_2 and 360 mg/m^3 NO_x [352].

Process developments: The Pittsburgh Energy Technology Center of the U.S. Department of Energy was reported [145, 685] to be co-sponsoring a development programme with Indianapolis Power & Light.

Appraisal:

1.	Information available	0
2.	Process simplicity	2
3.	Operating experience	0
4.	Operating difficulty	0*
5.	Loss of power	1
6.	Reagent requirements	2
7.	Ease of end-product disposal	2
8.	Process applicability	2
	Total	9

*Assumed in absence of information

CATEGORY NS23

The status of Category NS23 processes is indicated in Table 5.6, which presents information on the size of plants built, year of completion, and percentage abatement of SO_2 and NO_x.

Process Code NS23.1–Sodium Carbonate Adsorption Process (NOXSO Process)

Applicability: This process type is suitable for retrofit and new build applications.

Space requirements: The overall space requirements for a 125 MWe module are estimated [434] to be approximately: area 7.4 m^2; height 24.5 m.

Typical land area requirements: ***

Typical fresh water and water treatment requirements: ***

Consumption of reagents: A bed inventory loss of 0.2% per hour has been reported [673].

Table 5.5 Status of Combined Abatement Processes: Radiation Processes

Code No.	Vendor	No. of Plants	Plant size Range MWe	Total MWe	Dates	% Abatement SO_2	NO_x	Ref.
NS22.1	Ebara/JAERI	1	100	100	–	90	80	145, 34
		+2	6*–10*	16*	–			352, 678

* Thousand Nm^3/h
– Indicates information unavailable

Table 5.6 Status of Combined Abatement Processes: Adsorption Processes

Code No.	Vendor	No. of Plants	Plant size Range *	Total *	Dates	% Abatement SO_2	NO_x	Ref.
NS23.1	NOXSO	60		60	1982	90	90	434, 673

*Gas flow, Nm^3/h

End-product or waste disposal requirements: SO_2 and H_2S streams could be treated to produce elemental sulphur.

Typical pressure losses: ***

Typical power requirements: ***

Typical reductions in combustion plant efficiency: ***

Operating experience: Not yet tested beyond the bench scale, in which over 50 sorbent regenerations have been achieved [673].

Process developments: The Pittsburgh Energy Technology Center of the U.S. Department of Energy conducted development work on the process [685]. Testing on a large scale needed.

Appraisal:

1.	Information available	0
2.	Process simplicity	0
3.	Operating experience	0
4.	Operating difficulty	0*
5.	Loss of power	1
6.	Reagent requirements	1
7.	Ease of end-product disposal	2
8.	Process applicability	2
	Total	6

*Assumed in absence of information

CATEGORY NS31

The status of Category NS31 processes is indicated in Table 5.7, which presents information on the size of plants built, year of completion, and percentage abatement of SO_2 and NO_x.

Process Code NS31.1–Lime Spray Dryer Process

Applicability: This process type is suitable for retrofit and new build applications.

Typical land area requirements: ***

Typical fresh water and water treatment requirements: Water to prepare a slurry containing 20% suspended lime absorbent (see below). No waste water from the process.

Consumption of reagents [432]: Lime at a molar ratio of $Ca/(SO_2 + NO_x)$ of about 1.5. Sodium hydroxide at a Na/Ca molar ratio of about 0.56.

End-product or waste disposal requirements [352]: Solid product for disposal containing calcium and sodium sulphite, sulphate, nitrate, carbonate and hydroxide, with captured particulates.

Typical pressure losses: ***

Typical power requirements: ***

Table 5.7 Status of Combined Abatement Processes: Liquid Phase NO_x Oxidation

Code No.	Vendor	No. of Plants	Plant size Range MWe	Total MWe	Dates	% Abatement SO_2	NO_x	Ref.
NS31.1	Niro	2	20–96	116	1980–81	90–95	20–60	213, 352

Table 5.8 Status of Combined Abatement Processes: Liquid-Phase NO_x Reduction

Code No.	Vendor	No. of Plants	Plant size Range MWe	Total MWe	Dates	% Abatement SO_2	NO_x	Ref.
NS32.1	Asahi	1	0.05*	0.05*	–	99	80–85	419
NS32.2	SHL	1	5*	5*	–	–	–	352
NS32.3	PENSYS	3	1–1.5	3.5	1977–82	Over 95	Over 90	145

* Thousand Nm^3/h
– Indicates information unavailable

Typical reductions in combustion plant efficiency: ***

Operating experience: This process has been operated in short-term tests at the two plants indicated in Table 5.7, and in the laboratory. The possibility of emission of NO_2, formed by oxidation of NO, has been reported [685].

Process developments: The Pittsburg Energy Technology Center of the U.S. Department of Energy has conducted development work on the process [685]. Further full-scale testing needed.

Appraisal:

1.	Information available	1
2.	Process simplicity	2
3.	Operating experience	0
4.	Operating difficulty	0
5.	Loss of power	2*
6.	Reagent requirements	1
7.	Ease of end-product disposal	1
8.	Process applicability	2
	Total	9

*Assumed in absence of information

CATEGORY NS32

The status of Category NS32 processes is indicated in Table 5.8, which presents information on the size of plants built, year of completion, and percentage abatement of SO_2 and NO_x.

Process Code NS32.1–Sodium Sulphite Scrubbing Process with Chelating Compound (Asahi Chemical Process)

Applicability: This process type is suitable for retrofit and new build applications.

Typical land area requirements: ***

Typical fresh water and water treatment requirements: Published estimates are presented in Table 5.9.

Consumption of reagents: Published estimates are presented in Table 5.9.

End product or waste disposal requirements: Published estimates are presented in Table 5.9.

Typical power requirements: Published estimates are presented in Table 5.9.

Table 5.9 Published Estimated Requirements

Reference		419
MWe		500
Inlet gas: SO2 content		
NO_x content		
Cooling water per MWh	tonne	0.9
Steam (11 bar max.)	tonne	0.1
Reagents per MWh:		
Limestone	kg	26
$FeSO_4.7H_2O$	kg	0.1
EDTA	kg	0.1
Na_2CO_3	kg	3.4
or NaOH	kg	1.3
Residual oil (cracking)	litre	11
End-product and waste disposal requirement per MWh:		
Gypsum	kg	45
Na_2SO_4	kg	5.3
Waste water for treatment		Small quantity
Electric power:	kWh	54

Typical pressure losses: ***

Typical reductions in combustion plant efficiency: ***

Operating experience [419]: This process has been operated as an integrated system only in a small unit treating 40–60 Nm^3/h of gas from an oil-fired boiler, but the absorption, crystallisation and decomposition parts of the system have been confirmed, and scale-up data obtained, on a bench-scale unit treating 500–600 Nm^3/h of gas. The system has not been tested on flue gas from a coal-fired boiler.

Process development: Development to full scale needed.

Appraisal:

1.	Information available	1
2.	Process simplicity	0
3.	Operating experience	0
4.	Operating difficulty	0*
5.	Loss of power	1
6.	Reagent requirements	1
7.	Ease of end-product disposal	1
8.	Process applicability	2
	Total	6

*Assumed in absence of information

Process Code NS32.2–Limestone Slurry Scrubbing with Chelating Compound

Applicability: This process type is suitable for retrofit and new build applications.

Typical land area requirements: ***

Typical fresh water and water treatment requirements: ***

Consumption of reagents: ***

End-product or waste disposal requirements: ***

Typical pressure losses: ***

Typical power requirements: ***

Typical reductions in combustion plant efficiency: ***

Operating experience [352]: The process was reported in 1986 to be under test in a laboratory and on a 5000 m^3/h pilot plant.

Process development: The Pittsburgh Energy Technology Center of the U.S. Department of Energy has conducted development work on the process [685]. Development to full scale needed.

Appraisal:

1.	Information available	0
2.	Process simplicity	1
3.	Operating experience	0
4.	Operating difficulty	0*
5.	Loss of power	1
6.	Reagent requirements	1
7.	Ease of end-product disposal	1
8.	Process applicability	2
	Total	6

*Assumed in absence of information

Process Code NS32.3–Sulf-X Process

Applicability: This process type is suitable for retrofit and new build applications. See Section 2.3 for estimates of requirements for application of the process to sulphur capture alone. There appear to be no published estimates of requirements for combined abatement of SO_2 and NO_x.

Typical land area requirements: ***

Typical fresh water and water treatment requirements: ***

Consumption of reagents: ***

End-product or waste disposal requirements: ***

Typical pressure losses: ***

Typical power requirements: ***

Typical reductions in combustion plant efficiency: ***

Operating experience [145, 277]: Experience limited to 1.5 MWt scale, with some stages of the regeneration system omitted.

Process development: Integrated operation, and development to full scale, needed.

Appraisal:

1.	Information available	0
2.	Process simplicity	0
3.	Operating experience	0
4.	Operating difficulty	0*
5.	Loss of power	1*
6.	Reagent requirements	2
7.	Ease of end-product disposal	2
8.	Process applicability	2
	Total	7

*Assumed in absence of information

CATEGORY NS41

The status of Category NS41 processes is indicated in Table 5.10, which presents information on the size of plants built, year of completion, and percentage abatement of SO_2 and NO_x.

Process Code NS41.1–Oxidation plus Ammonia Scrubbing (Walther Process)

Applicability: This process type is suitable for retrofit and new build applications.

Typical land area requirements: ***

Table 5.10 Status of Combined Abatement Processes: Liquid-Phase NO_x Oxidation

Code No.	Vendor	No. of Plants	Plant size Range MWe	Total MWe	Dates	% Abatement SO_2	NO_x	Ref.
NS41.1	Walther	–	–	–	–	91	88	**
NS41.2	Kawasaki	2	0.1–5*	5.1*	1974	Over 95	91	419

* Thousand Nm^3/h
** Vendor's promotional literature
– Indicates information unavailable

Typical fresh water and water treatment requirements: Published estimates are presented in Table 5.11.

Consumption of reagents: Published estimates are presented in Table 5.11.

End-product or waste disposal requirements: Published estimates are presented in Table 5.11.

Typical power requirements: Published estimates are presented in Table 5.11.

Typical pressure losses: ***

Typical reductions in combustion plant efficiency: ***

Operating experience: Pilot plant operation only.

Process developments: Needs to be developed to full scale operation.

Appraisal:

1.	Information available	1
2.	Process simplicity	1
3.	Operating experience	0
4.	Operating difficulty	0*
5.	Loss of power	1
6.	Reagent requirements	1
7.	Ease of end-product disposal	2
8.	Process applicability	2
	Total	8

*Assumed in absence of information

Table 5.11 Published Estimated Requirements

Reference		Walther*
Furnace type		Slag tap
MWe		400
Gas flow	'000 Nm^3/h	1300
Inlet gas: SO_2 content	mg/Nm^3	2200
NO_x content	mg/Nm^3	1640
Abatement: SO_2	%	91
NO_x	%	88
Water and water treatment requirements per MWh:		
Cooling water	tonne	9.7
Water treatment		Nil
Consumption of reagents per MWh:		
Ammonia	kg	5.1
Ozone (see power consumption)	kg	4.0
End-product or waste disposal per MWh:		
$(NH_4)_2SO_4$	kg	13.4
NH_4NO_3	kg	8.2
Ammonium halides, particulates		Site-specific
Power requirements per MWh:		
Ozone production		42 kWh
Other		20 kWh

*Walther promotional literature

Process Code NS41.2–Kawasaki Process

Applicability: This process type is suitable for retrofit and new build applications.

Typical land area requirements: ***

Typical fresh water and water treatment requirements: ***

Consumption of reagents: ***

End-product or waste disposal requirements: ***

Typical pressure losses: ***

Typical power requirements: ***

Typical reductions in combustion plant efficiency: ***

Operating experience: Limited to operation on a pilot plant handling a flow of 5000 Nm^3/h from a coal-fired boiler.

Process developments: Development to full scale needed.

Appraisal:

1.	Information available	0
2.	Process simplicity	0
3.	Operating experience	0
4.	Operating difficulty	0*
5.	Loss of power	1*
6.	Reagent requirements	1
7.	Ease of end-product disposal	1
8.	Process applicability	2
	Total	5

*Assumed in absence of information

Table 5.12 Status of Combined Abatement Processes: Liquid-Phase NO_x Reduction

Code No.	Vendor	No. of Plants	Plant size Range MWe	Total MWe	Dates	% Abatement SO_2	NO_x	Ref.
NS42.1	IHI	2	5–27*	32*	1975	90	80	419
NS42.2	Fuji Kasui-Sumitomo	1	25*	25*	1976	95	90	419

*Thousand Nm^3/h

CATEGORY NS42

The status of Category NS42 processes is indicated in Table 5.12, which presents information on the size of plants built, year of completion, and percentage abatement of SO_2 and NO_x.

Process Code NS42.1–IHI Process

Applicability: This process type is suitable for retrofit and new build applications.

Typical land area requirements: ***

Fresh water and water treatment requirements: Published estimates are presented in Table 5.13.

Consumption of reagents: Published estimates are presented in Table 5.13.

End-product or waste disposal requirements: ***

Typical pressure losses: ***

Typical power requirements: Published estimates are presented in Table 5.13.

Typical reductions in combustion plant efficiency: ***

Table 5.13 Published Estimated Requirements

Reference		419
Gas flow	'000 Nm^3/h	100
Approximate equivalent MWe		40
Boiler fuel		Oil
Inlet gas: SO_2 content	ppmv	1500
NO_2 content	ppmv	180
Particulates content	mg/Nm^3	100
Abatement: SO_2 emission	%	90
NO_x emission	%	80
Water	tonne/MWh	1.4
Steam	tonne/MWh	0.11
Consumption of reagents, kg/MWh:		
Limestone		79
Sulphuric		4.5
Slaked lime		4.0
Additives		1.7
Fuel oil		53
Power requirements:	kWh	210

Operating experience: The process was tested on flue gas from an oil-fired boiler in a 5000 Nm^3/h pilot plant in a 3000 hour test run. A 27,000 Nm^3/h prototype unit was reported in 1979 [419] to be under test.

Process development: Development to full scale needed.

Appraisal:

1.	Information available	0
2.	Process simplicity	0
3.	Operating experience	0
4.	Operating difficulty	0*
5.	Loss of power	0
6.	Reagent requirements	1
7.	Ease of end-product disposal	1
8.	Process applicability	2
	Total	4

*Assumed in absence of information

Process Code NS42.2–Moretana Calcium Process

Applicability: This process type is suitable for retrofit and new build applications.

Space requirements: ***

Typical land area requirements: ***

Typical fresh water and water treatment requirements: ***

Consumption of reagents: ***

End-product or waste disposal requirements: ***

Typical pressure losses: ***

Typical power requirements: ***

Typical reductions in combustion plant efficiency: ***

Operating experience: Process has been tested on a pilot plant scale, treating 25,000 Nm^3/h gas from a sintering furnace, and on the bench scale with a synthetic coal-fired flue gas.

Process developments: Development to full scale needed.

Appraisal:

1.	Information available	0
2.	Process simplicity	0
3.	Operating experience	0
4.	Operating difficulty	0*
5.	Loss of power	0
6.	Reagent requirements	1*
7.	Ease of end-product disposal	1
8.	Process applicability	2
	Total	4

*Assumed in absence of information

5.4 Processes for Detailed Study

The selection of processes for detailed study in this Manual has been based upon their suitability for application in the U.K. for the three datum combustion systems (Section 1.3) considered:

- 2000 MWe (4 × 500 MWe) power station (Datum system 1)

- Large (450 tonne steam/h) industrial boiler (Datum system 2)

- Small (13 tonne steam/h) factory boiler (Datum system 3)

In principle, all of the processes listed in Section 5.1 can be applied to all combustion plant, but the attraction of many processes diminishes with factors such as decrease in plant operating scale, and increases in Combined Abatement process complexity, reagent costs, and end-product disposal difficulty.

Appraisal of processes

To evaluate some of these factors, a rough appraisal of each process type has been made in Section 5.3 by assigning 'merit points' for a number of features; merit points have been awarded according to the scale:

0 Below average merit

1 Average merit

2 Above average merit

The features to which these points have been assigned are described in Section 1.5: they are briefly:

1. Information available

2. Process simplicity

3. Operating experience–extent and difficulties encountered

4. Operating difficulty–availability, reliability

5. Loss of power sent out–by installation of the FGD process.

6. Reagent requirements–quantities

7. Ease of end-product disposal

8. Process applicability–e.g. for retrofit

All of the processes listed in Section 5.1 and outlined in Section 5.2 are appraised in Section 5.3. The merit points assigned to the process types for each of the above features are summarised in Table 5.14. It should be noted that the number of points in the merit point system adopted in other Sections of the Manual are not strictly comparable with those considered here.

Table 5.14 Summary of Combined Abatement Process Appraisals

Code No. NS	Name	Merit Points for Feature No: 1 2 3 4 5 6 7 8	Total Points
11.1	Active carbon/SCR	1 2 0 1 2 2 2 2	12
11.2	Copper oxide	1 1 1 0 1 2 1 1	8
12.1	Ralph M. Parsons	0 2 0 0 2 2 2 1	9
12.2	Cat. Redn/oxidn	0 1 0 0 2 1 2 1	7
21.1	LIMB	0 2 0 0 2 1 1 2	8
21.2	Slagging combustor	0 2 0 0 1 2 1 2	8
22.1	Electron beam	0 2 0 0 1 2 2 2	9
23.1	NOXSO	0 0 0 0 1 1 2 2	6
31.1	Spray dryer	1 2 0 0 2 1 1 2	9
32.1	Asahi	1 0 0 0 1 1 1 2	6
32.2	Limestone/Ferrous EDTA	0 1 0 0 1 1 1 2	6
32.3	Sulf-X	0 0 0 0 1 2 2 2	7
41.1	Walther	1 1 0 0 1 1 2 2	8
41.2	Kawasaki	0 0 0 0 1 1 1 2	5
42.1	IHI	0 0 0 0 0 1 1 2	4
42.2	Moretana	0 0 0 0 0 1 1 2	4

Features:
1. Information available
2. Process simplicity
3. Operating experience
4. Operating difficulty
5. Loss of power
6. Reagent requirements
7. Ease of end-product disposal
8. Process applicability

Processes suitable for the U.K.

The principal purpose of assigning merit points to each of the processes was to aid in the selection of processes that could be considered suitable for application in the U.K. It was arbitrarily assumed that suitable combined abatement processes would be those having more than 10 merit points.

Selection of processes for detailed study

All of the combined abatement process types are

Table 5.15 Applications of Combined Abatement Processes Considered in Detail

Code No.	Name	Application to Datum System			Section
		1	2	3	
NS11.1	Active carbon/SCR	Yes	Yes*	–	4.5
NS11.2	Copper oxide	–	–	–	–
NS12.1	Ralph M. Parsons	–	–	–	–
NS12.2	Cat.Redn./oxidn.	–	–	–	–
NS21.1	LIMB	–	–	–	–
NS21.2	Slagging combustion	–	–	–	–
NS22.1	Electron beam	–	–	–	–
NS23.1	NOXSO	–	–	–	–
NS31.1	Spray dryer	Yes**	Yes**	Yes**	4.5
NS32.1	Asahi	–	–	–	–
NS32.2	Limestone/Ferrous EDTA	–	–	–	–
NS32.3	Sulf-X	–	–	–	–
NS41.1	Walther	–	–	–	–
NS41.2	Kawasaki	–	–	–	–
NS42.1	IHI	–	–	–	–
NS42.2	Moretana	–	–	–	–

* In general, only if a centralised reagent reprocessing plant were available
**Tentative evaluation: holds promise for future application

shown in Table 5.15 with an indication of those considered to be suitable for application in the U.K.

None of the basic process types is considered to be suitable for immediate application to the smallest operating scale dealt with in this Manual (Datum System 3), as the only contender (Process Code NS11.1–Active Carbon/SCR), although rated as being a simple process, would be too complex for a small-scale boiler.

The otherwise attractive Spray Dryer Process (Process Code NS31.1–Merit Rating 9 points) fails to achieve the criterion of 10 Merit Points because of lack of information available and of operating experience. However, this process is considered to have promise for future application to all operating scales because of its simplicity, and is therefore tentatively evaluated in Section 5.5.

For utilities and large industrial boilers, the simplest process for consideration in the U.K. is: Active Carbon/SCR (Process Code NS11.1–merit rating 12 points), evaluated in Section 5.5. This process, though of above average simplicity, would be attractive only where reagent and by-product processing facilities were available.

Owing to the small number of plants built, and the very limited operating experience of combined abatement processes, and in particular of the two processes selected for evaluation, there are insufficient data available to enable reliable estimates of capital and operating costs to be made. The Active Carbon/SCR process (Code NS11.1) is an elaboration of the similar flue gas desulphurisation (FGD) process (S51.1; see Section 2) for which also there were inadequate data for cost estimates. The Lime Spray Dryer Absorber process (Code NS31.1) is an elaboration of the similar FGD process (Code S22.1) for which cost estimates have been made for application to Datum Systems 1 and 2 only. It can be assumed that costs for the combined abatement version of this process would be higher than those for the FGD version.

5.5 Evaluations of Selected Combined Abatement Processes

Evaluation of BF-Mitsui Active Carbon Combined Abatement Process (Process Code NS11.1)

See Section 5.2 for outline of the basic process; its chemistry; block diagram.

See Section 5.3 for general appraisal of the basic process; name of manufacturers offering this type of equipment.

See Section 5.4 for the reason for choosing the BF-Mitsui (Code NS11.1) basic process.

This basic process type is considered (Section 5.4) to be suitable for application in the UK to utility boilers, and to large industrial boilers having access to reagent reprocessing facilities, and hence it is evaluated here only for Datum Combustion Systems 1 and 2. The process, developed by Bergbau-Forschung GmbH (Essen, FRG) and Mitsui Mining Co. Ltd. (Tokyo, Japan), is offered by Uhde GmbH (Dortmund, FRG). See Appendix 2 for details of this manufacturer.

Process Description

Figure 5.17 shows a simplified flow diagram for application of the process to a coal fired boiler. It is assumed that the sulphur dioxide produced in the process is treated to recover elemental sulphur.

Dust and sulphur oxides removal: Gas at a temperature of 120–150°C from downstream of the electrostatic precipitator or baghouse, boosted by a fan, is cooled to 120°C if necessary by injection of water. The cooled gas enters the 1st Stage Adsorber containing activated carbon granules (prepared from coal in the form of extruded granules about 5 mm in length) from the 2nd Stage Adsorber. The carbon, which is contained in louvred channels, with the gas flowing across the bed, adsorbs sulphur dioxide, oxygen and water vapour to form adsorbed sulphuric acid. Nitrogen dioxide (forming 5–10% of the total nitrogen oxides content of the gas) and acid halides are also adsorbed, and the carbon bed filters out much of the particulates content of the gas. Ammonia

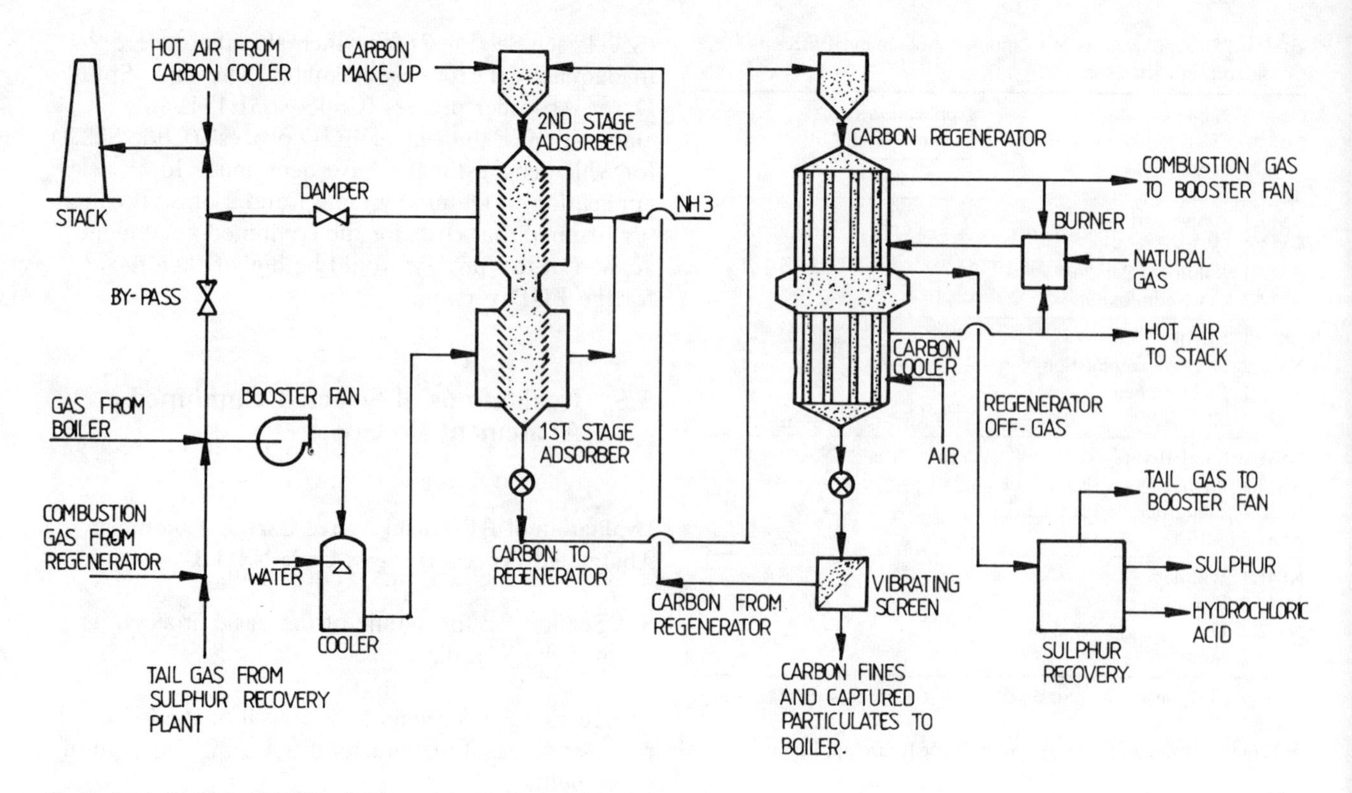

Figure 5.17 B.F-Mitsui Active Carbon Adsorption Combined Abatement Process

is then added to the gas before it enters the 2nd Stage Adsorber, which is fed with fresh and regenerated carbon; the carbon, which acts as a Selective Catalytic Reduction catalyst, is again contained within louvred channels with the gas flowing across, allowing the ammonia and nitric oxide to react forming water vapour and elemental nitrogen. Ammonia also reacts with residual sulphur dioxide to form ammonium sulphate and bisulphate which are adsorbed on the surface of the carbon. Carbon passes from the 2nd to the 1st Stage Adsorber. The gas temperature rises by 15–20°C across the Adsorber/SCR system; the cleaned gas passes via a Damper to Stack.

Adsorbent regeneration: The active carbon is removed continuously from the base of the 1st Stage Adsorber and conveyed to the Regeneration section, which can be either on- or off-site. The carbon flows slowly through vertical tubes, which are heated indirectly by hot combustion gas at 600°C derived from combustion of natural gas. The carbon temperature is raised to 400–450°C. The hot combustion gas leaves the heating section at 300°C; the Adsorber inlet. The adsorbed gases, sulphuric acid and nitrogen dioxide, react with the carbon, releasing sulphur dioxide, elemental nitrogen and carbon dioxide, and consuming some of the carbon. The ammonium sulphate and bisulphate decompose, giving elemental nitrogen, sulphur dioxide and water vapour. The heated carbon resides in a vessel to complete the regeneration process, and it then passes through vertical tubes in the cooling section, where it is indirectly cooled to 100°C by air; part of the air leaving the cooling section at 250°C supplies combustion air for the natural gas combustion, and the remainder is exhausted to stack. Carbon fines and particulates trapped from the gas are removed by a Vibrating Screen and burned in the Boiler. The oversize material is recycled to the 2nd Stage Adsorber; carbon losses are made up by adding fresh carbon.

Sulphur recovery: The off-gas from the Adsorber contains SO_2, CO_2, HCl and water vapour. HCl is removed from the gas, which then passes via a reduction reactor to a Claus unit to recover sulphur, which is condensed out; tail gas from the Claus unit is returned to the Adsorber.

Status and Operating Experience

The status of the BS-Mitsui process is indicated in Section 5.3, where it is seen that four installations have been (or are being) erected, in the size range 30,000–1,110,000 Nm³/h, totalling 1,505,000 Nm³/h. Three units of a similar process (Sumitomo) have been erected, totalling 312,000 Nm³/h.

It is seen in Section 5.3 that the process has been operated on one pilot plant for 16,000 hours, and on a small commercial unit for a similar period, with very high reliability, especially if maintenance work is carried out during scheduled boiler shut-down periods.

Variations and Development Potential

Design and operating variations can include:

- Adaptation of the process to either adsorption of SO_2 or removal of NO_x alone.

- Direct heating of carbon in the Regenerator with hot sand.

- Alternatives for sulphur recovery: as elemental sulphur by a variety of processes (Alliance process, Claus process, Foster-Wheeler 'Resox' process); as sulphuric acid (by conventional catalytic oxidation of SO_2); or as liquid SO_2.

Although complex, the process can achieve high sulphur capture (up to 99%) and high NO_x abatement (up to 90%) without excessive reagent make-up requirements or production of waste products. The process also removes acid halides and particulates.

The sulphur produced can be marketed as sulphuric acid, liquid SO_2 or as elemental sulphur; if it is produced as elemental sulphur, it can be disposed of safely. High operating flexibility is obtainable by installing large surge capacity for regenerated and unregenerated active carbon.

Development and use of lower-cost active carbon is the most likely development.

Process Requirements for Each Application Considered

These are shown in Table 5.16 for the two applications considered: Datum Combustion Systems 1 and 2.

It is assumed that the pollutant contents of the gas is to be reduced to the following levels:

- System 1:
  SO_2 250 mg/Nm³ (dry), equivalent to 94.1% capture of SO_2
  NO_x 200 mg/Nm³ (dry), equivalent to 80.6% abatement of NO_x
  HCl 10 mg/Nm³ (dry), equivalent to 97.1% capture of HCl

Table 5.16 Process Requirements

Datum Combustion System		1	2
Inlet Gas at full load			
Volume flow	'000 Nm³/h	1524*	551
Dry gas	'000 Nm³/h	1426*	513
Water Vapour	'000 Nm³/h	99*	38
Actual volume flow	'000 m³/h	2326*	854
Temperature	°C	150	150
Particulates content	mg/Nm³ (dry)	115	115
SO2 content	mg/Nm³ (dry)	4240	3520
NO_x content	mg/Nm³ (dry)	1025	1440
HCl content	mg/Nm³ (dry)	340	315
Exit Gas at full load			
Volume flow	'000 Nm³/h	1550*	561
Dry gas	'000 Nm³/h	1422*	512
Water Vapour	'000 Nm³/h	128*	49
Actual volume flow	'000 m³/h	2316*	838
Temperature	°C	135	135
Particulates content	mg/Nm³ (dry)	15 (25)	15 (25)
SO_2 content	mg/Nm³ (dry)	250 (649)	400 (712)
NO_x content	mg/Nm³ (dry)	200 (283)	750 (819)
HCl content	mg/Nm³ (dry)	10 (43)	10 (41)
Reaction temperature	°C	135	135
Particulates removal		Simult.	Simult.
Reagent		Active	Carbon
Requirements at full load			
Water	tonne/h	24.1*	8.7
Active carbon	tonne/h	0.64*	0.18
Ammonia	tonne/h	1.04*	0.30
Natural gas	GJt/h	30.5*	8.61
Air	tonne/h	59*	20
Electric Power	MWe	2.6*	1.0
Manpower	men/shift	***	***
Average load factor	%	***	***

*Per 500 MWe boiler
Figures in parentheses are annual average emissions for 90% combined abatement plant availability

Particulates 15 mg/Nm^3 (dry), equivalent to 87.0% capture of particulates

- System 2:
  SO_2 400 mg/Nm^3 (dry), equivalent to 88.7% capture of SO_2
  NO_x 750 mg/Nm^3 (dry), equivalent to 47.9% abatement of NO_x
  HCl 10 mg/Nm^3 (dry), equivalent to 96.8% capture of HCl
  Particulates 15 mg/Nm^3 (dry), equivalent to 86.9% capture of particulates.

It is further assumed that for both systems:

- The 1st Stage Adsorber captures 80%, and the 2nd Stage Adsorber 20%, of the total sulphur captured; the sulphur is captured in the 1st Stage as sulphuric acid, and in the 2nd Stage as ammonium sulphate.
- The sulphur is recovered as elemental sulphur in a Claus unit.
- The NO_x content is reduced by 8% by adsorption of NO_2 in the 1st Stage Adsorber; the remaining NO_x abatement occurs in the 2nd Stage Adsorber by SCR with ammonia.
- The mechanical breakdown of carbon leads to a carbon make-up rate 50% higher than the theoretical.
- The requirements for Regenerator off-gas treatment and sulphur recovery are not included.

Byproducts and Effluents

The byproducts from the system are elemental sulphur and hydrochloric acid, and there is a carbon waste in addition to the captured fly-ash. The rates of production of the material produced are summarised in Table 5.17 for the two Datum Combustion Systems considered.

Table 5.17 Estimated Rates of Output Solids Production

Datum Combustion System		1	2
Rate of production:			
Sulphur (99%)	tonne/h	11.5*	0.81
Hydrochloric acid (35%)	tonne/h	5.4*	0.45
Active carbon fines	tonne/h	0.86*	0.06
Fly-ash	tonne/h	0.57*	0.05

*Total from four 500 MWe boiler plants

Efficiency and Emission Factors

The efficiency and emission factors for the process are summarised in Table 5.18 for the two applications considered: Datum Combustion Systems 1 and 2.

In calculating the efficiency factors, it is assumed that at full load:

- For Datum System 1, the combined abatement unit for each 500 MWe set consumes:

  30.5 GJt/h of natural gas for the Regenerator, equivalent to an increase of 1.2 tonne/h of coal;

  and 2.6 MWe of electrical power, reducing the power sent out.

- For Datum System 2, the combined abatement unit consumes:

  8.61 GJt/h of natural gas for the Regenerator, equivalent to an increase of 0.3 tonne/h of coal;

  and 1.0 MWe of electrical power, equivalent to the combustion of a further 0.4 tonne/h of coal at a power station, assuming an overall power generation efficiency of 33%. This additional coal is arbitrarily assumed to be included with the coal burned in the boiler for calculating the efficiency factor.

For illustration purposes, the annual average emissions and emission factors for combined abatement plant availabilities of 100% and 90% are given in Tables 5.16 and 5.18. Further details are given in Section 1.6.

Effect of load variations: ***

Effect of design variations: ***

Limitations: ***

Costs

Capital Costs: ***

Annual running costs and cost factors: ***

Effects of design variations on costs: ***

Effects of annual load patterns on annual running costs: ***

Process Advantages and Drawbacks

The advantages are that the process can be retrofitted; process readily adaptable to either NO_x or SO_2 abatement alone; process also removes particulates and other air pollutants; no gas reheat needed; high sulphur capture efficiency (over 90% capture) and NO_x abatement efficiency (50–90%) can be attained; operating flexibility; no requirement for large quantities of reagent; no large quantities of waste products; by-products potentially marketable or (if elemental sulphur is produced) can be easily and safely disposed of.

Table 5.18 Efficiency and Emission Factors

Datum Combustion System		1	2
Coal heat (gross)	MWt	1292*	468
Coal fired	tonne/h	189*	60.2
Abatement plant power	MWe	2.6*	1.0
Equivalent coal input	tonne/h	–	0.4+
Natural gas for Regenerator	GJt/h	30.5*	8.61
Equivalent coal input	tonne/h	1.2*	0.3
Useful energy from system	GJe/h	1664*	–
	GJe/h	–	1468
Total equivalent coal input	tonne/h	190.2* (a)	60.9 (a)
Efficiency factor	GJe/tonne	8.75 (a)	–
	GJt/tonne	–	24.1 (a)
Emissions			
Sulphur in SO_2	kg/h	178* (463*)	103 (183)
Nitrogen in NO_x	kg/h	87* (123*)	117 (128)
Chlorine in HCl	kg/h	14* (60*)	5 (20)
Particulates	kg/h	21* (35*)	8 (13)
Emission factors (per tonne coal)		(a)	(a)
Sulphur	kg/tonne	0.94 (2.43)	1.69 (3.00)
Nitrogen	kg/tonne	0.46 (0.65)	1.92 (2.10)
Chlorine	kg/tonne	0.07 (0.32)	0.08 (0.33)
Particulates	kg/tonne	0.11 (0.18)	0.13 (0.21)

* For each 500 MWe boiler plant
+ Calculated assuming overall power generation efficiency = 0.33
(a) Based on coal fired to boiler plus coal equivalent to electric power and natural gas consumed
Figures in parentheses are annual average emissions for 90% combined abatement plant availability

The disadvantages are the complexity of the process; need for experienced labour.

Evaluation of Lime Spray Dryer Combined Abatement Process (Process Code NS31.1)

See Section 5.2 for outline of the basic process; its chemistry; block diagram.

See Section 5.3 for a list of manufacturers offering this type of equipment.

See Section 5.3 for general appraisal of the basic process.

See Section 5.4 for the reason for choosing the Code S31.1 basic process.

This basic process type is considered (Section 5.4) to be potentially suitable for future application in the UK to utility boilers, large industrial boilers and small factory boilers, and hence it is tentatively evaluated here for Datum Combustion Systems 1, 2 and 3. The process presented here is of the type using lime slurry, with an additive (sodium hydroxide) in a Spray Dryer Absorber, producing a dry end product. The process is offered in Europe by Flakt Industri AB, (Vaxjo, Sweden) in collaboration with A/S Niro Atomizer, (Copenhagen, Denmark), and in the USA by Joy-Niro. See Appendix 2 for details of these manufacturers.

Process Description

Figure 5.18 shows a simplified flow diagram for application of the process to a coal fired boiler. It is assumed that for all applications, the solid end-product will be disposed of to landfill, but because the end-product contains soluble components, the precise details of its disposal will be site-specific.

Gas from downstream of the electrostatic precipitator or baghouse enters a Spray Dryer Absorber (SDA) fed with a spray of calcium hydroxide slurry containing sodium hydroxide in solution. Operating conditions are chosen to give a gas temperature leaving the SDA of about 100°C, at which temperature the maximum extent of combined abatement occurs. The residence time of the slurry droplets in the SDA is about 10 seconds. The gas then passes to an Electrostatic Precipitator (ESP) or, more usually a Baghouse, to remove the gas-borne reaction products which have not been collected in the hopper base of the SDA. A significant proportion (10–20%) of the total abatement occurs in the ESP or Baghouse. The collection hoppers of the SDA and of the ESP or Baghouse are electrically heated to prevent solids build-up resulting from condensation. The cleaned gas then passes, via a Fan, to the Stack; reheating is not required. Part of the collected solids is returned via a Recycle Product Silo to the Feed Tank, where it is mixed with fresh lime/caustic soda slurry. The fresh lime is slaked with water in the Slaker, and the calcium hydroxide is slurried with

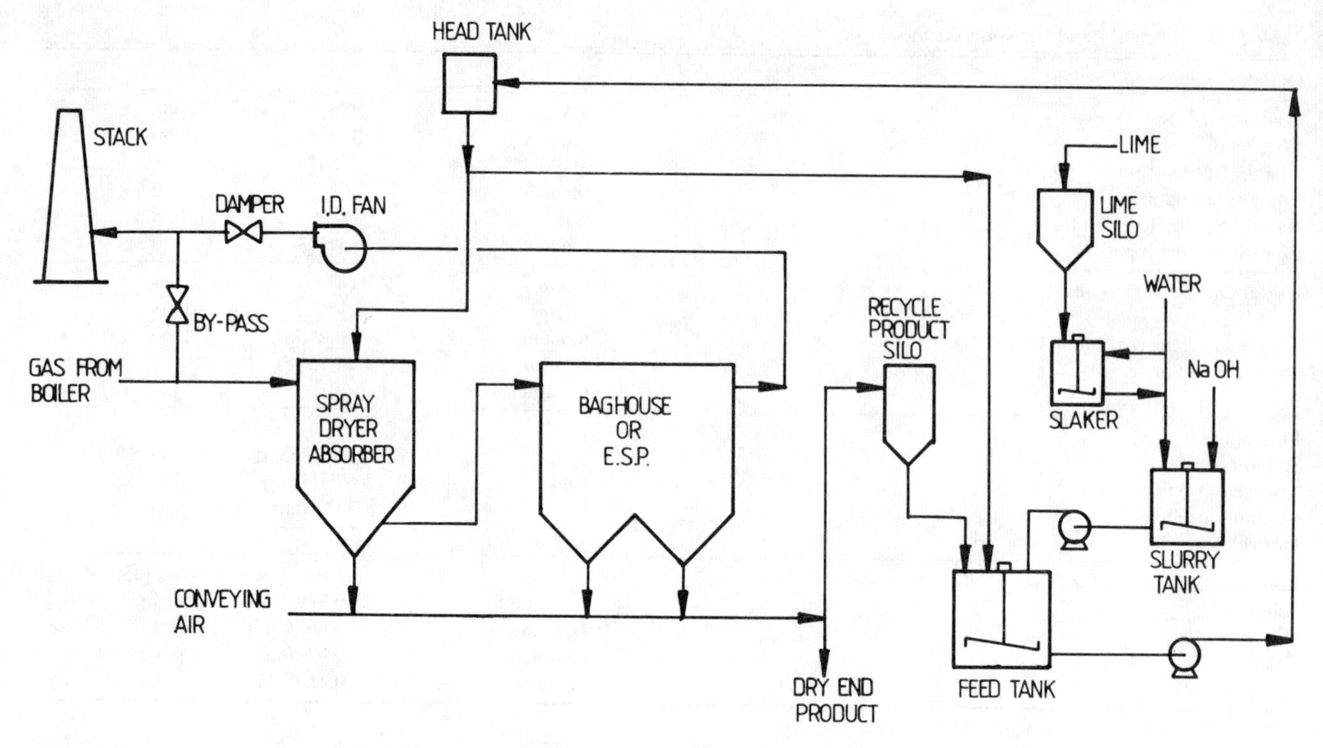

Figure 5.18 Lime Spray Dryer Combined Abatement Process

water and sodium hydroxide solution in the Slurry Tank before being pumped to the Feed Tank. The dry waste end-product contains fuel, ash, sodium and calcium sulphite, sulphate, nitrite, nitrate and chloride, and unreacted hydroxide.

Status and Operating Experience

The status of the process is indicated in Section 5.3, where it is seen that two units (70 MWt and 96 MWt) have been installed and operated as combined abatement plant.

Experience is limited to short-term; information is scanty. Experiences of operating lime spray dryers for FGD alone are mentioned in Section 2.5.

Variations and Development Potential

Design and operating variations can include:

- Number, type and location of slurry atomisers in the SDA.

- Alternatives of Electrostatic Precipitator and Baghouse for removal of gas-borne end-product.

- Removal of fly-ash from the boiler flue gas either upstream of the SDA, or in the ESP or Baghouse downstream of the SDA.

The simplicity of the process is its chief characteristic, making it suitable for application on even small-scale plant (e.g. System 3).

More operating experience is needed to establish operating data.

Process Requirements for Each Application Considered

Estimated requirements are shown in Table 5.19 for the three applications considered: Datum Combustion Systems 1, 2 and 3.

It is assumed that the pollutant contents of the gas are to be reduced to the following:

- System 1:
  SO_2 400 mg/Nm^3 (dry), equivalent to 90.6% capture of SO_2
  NO_x 400 mg/Nm^3 (dry), equivalent to 61.1% abatement of NO_x
  HCl 10 mg/Nm^3 (dry), equivalent to 95.6% capture of HCl
  Particulates 15 mg/Nm^3 (dry), equivalent to 87% capture of particulates

- System 2:
  SO_2 700 mg/Nm^3 (dry), equivalent to 80.1% capture of SO_2
  NO_x 700 mg/Nm^3 (dry), equivalent to 51.4% abatement of NO_x
  HCl 10 mg/Nm^3 (dry), equivalent to 95.2% capture of HCl
  Particulates 15 mg/Nm^3 (dry), equivalent to 87% capture of particulates

Table 5.19 Process Requirements

Datum Combustion System		1	2	3
Inlet Gas at full load				
Volume flow	'000 Nm^3/h	1524*	551	16.52
Dry gas	'000 Nm^3/h	1426*	513	15.46
Water vapour	'000 Nm^3/h	99*	38	1.06
Actual volume flow	'000 m^3/h	2362*	854	30.44
Temperature	°C	150	150	230
Particulates content	mg/Nm^3 (dry)	115	115	660
SO_2 content	mg/Nm^3 (dry)	4240	3520	3255
NO_x content	mg/Nm^3 (dry)	1025	1440	515
HCl content	mg/Nm^3 (dry)	340	315	290
Exit Gas at full load				
Volume flow	'000 Nm^3/h	1589*	573	18.11
Dry gas	'000 Nm^3/h	1423*	512	15.44
Water vapour	'000 Nm^3/h	166*	61	2.67
Actual volume flow	'000 m^3/h	2171*	783	24.74
Temperature	°C	100	100	100
Particulates content	mg/Nm^3 (dry)	15 (25)	15 (25)	15 (25)
SO_2 content	mg/Nm^3 (dry)	400 (784)	700 (982)	(80) (1226)
NO_x content	mg/Nm^3 (dry)	400 (463)	700 (774)	300 (322)
HCl content	mg/Nm^3 (dry)	10 (43)	10 (41)	10 (38)
Reaction temperature	°C	90	90	90
Particulates removal		Simult.	Simult.	Simult.
Lime supplied as:		CaO	CaO	$Ca(OH)_2$
Slurry composition				
Ca(OH)2	wt. %	15.4	13.7	4.8
NaOH	wt. %	4.2	3.7	1.3
Molar ratio				
Ca/(Captured SO_2 + NO_x)		1.5	1.5	1.5
Na/(Captured SO_2 + NO_x)		0.75	0.75	0.75
Requirements at full load				
Quicklime (94.8%)	tonne/h	9.31*	2.74	–
Slaked lime (96%)	tonne/h	–	–	0.071
Caustic soda (47%) liquor	tonne/h	6.70*	1.97	0.039
Water (slaking)	tonne/h	2.8*	0.83	–
Water (slurrying)	tonne/h	57.9*	19.8	1.33
Electric Power	MWe	6.6*	3.9	0.1
Manpower	men/shift	***	***	***
Average load factor	%	***	***	***

*Per 500 MWe boiler ***data unavailable
Figures in parentheses are annual average emissions for 90% combined abatement plant availability

- System 3:
  SO_2 1000 mg/Nm^3 (dry), equivalent to 69.3% capture of SO_2
  NO_x 300 mg/Nm^3 (dry), equivalent to 41.6% abatement of NO_x
  HCl 10 mg/Nm^3 (dry), equivalent to 94.9% capture of HCl
  Particulates 15 mg/Nm^3 (dry), equivalent to 97.7% capture of particulates

It is further assumed that for all three systems:

- The same molar ratio of (fresh lime)/(captured SO_2 + NO_x) is used (1.5 mol Ca per mol SO_2 + NO_x absorbed).

- The same molar ratio of (fresh Na)/(fresh Ca) is used (0.5 mol Na per mol Ca).

For System 3 it is assumed that the lime would be supplied as $Ca(OH)_2$.

By-products and Effluents

It is arbitrarily assumed that, in addition to captured particulates and to impurities present in the lime, the composition of the end-product is made up as follows:

- All of the input NaOH is converted to sodium sulphate decahydrate.

- All of the captured NO_x appears in the end-product as calcium nitrate.

- All of the captured HCl appears in the end-product as calcium chloride hexahydrate.

- The captured SO_2 not appearing in the end-product as sodium sulphate appears in equimolar proportions as calcium sulphate dihydrate (gypsum) and as calcium sulphite hemihydrate.

- Calcium hydroxide that has not combined with SO_2 or NO_x appears in the end-product as the unchanged hydroxide.

Table 5.20 Estimated Properties of Waste Solids Produced

Datum Combustion System		1	2	3
Rate of production:	tonne/h	123*	9.03	0.19
Composition of wet product:				
Moisture	wt. %	1.8	1.8	1.8
Coal ash	wt. %	0.5	0.6	5.1
Calcium chloride	wt. % (a)	4.6	5.2	6.6
Calcium hydroxide	wt. %	22.8	23.8	19.6
Calcium sulphate	wt. % (b)	12.9	10.5	13.9
Calcium sulphite	wt. % (c)	9.7	7.9	10.3
Calcium nitrate	wt. %	5.2	7.5	3.0
Sodium sulphate	wt. % (d)	41.0	41.2	38.3
Other impurities	wt. %	1.5	1.5	1.4

* Total from four 500 MWe boiler combined abatement plants
(a) As hexahydrate
(b) As gypsum
(c) As hemihydrate
(d) As decahydrate

The end-product has to be disposed of e.g. to a landfill, but with precautions taken to deal with the soluble components–sodium salts, and calcium nitrate and chloride–and the excess alkalinity.

The estimated rates of production and composition of the material produced are summarised in Table 5 20 for the three Datum Combustion Systems considered.

Efficiency and Emission Factors

The efficiency and emission factors for the process are summarised in Table 5.21 for the three applications considered: Datum Combustion Systems 1, 2 and 3.

In calculating the efficiency factors, it is assumed that at full load:

- The performance without the incorporation of the combined abatement plant would be as shown in Table 1.4.3.

- For Datum System 1, the combined abatement unit for each 500 MWe set consumes 6.6 MWe of electric power, reducing the power sent out.

Table 5.21 Efficiency and Emission Factors

Datum Combustion System		1	2	3
Coal heat input (gross)	MWt	1292*	469	13.1
Coal fired	tonne/h	189*	60.3	1.68
Abatement plant power	MWe	6.6*	3.9	0.1
Equivalent coal input	tonne/h	–	1.5+	0.04+
Useful energy from system	GJe/h	1650*	–	–
	GJt/h	–	1468	35.3
Total equivalent coal input	tonne/h	189*	61.8(a)	1.72(a)
Efficiency factor	GJe/tonne	8.73	–	–
	GJt/tonne	–	23.8(a)	20.5(a)
Emissions				
Sulphur in SO_2	kg/h	285*	180	7.73
		(559*)	(252)	(9.48)
Nitrogen in NO/	kg/h	173*	109	1.41
		(200*)	(121)	(1.51)
Chlorine in HCl	kg/h	14*	5.0	0.15
		(60*)	(20.1)	(0.57)
Particulates	kg/h	21*	7.7	0.23
		(35*)	(22.5)	(1.23)
Emission factors (per tonne coal)			(a)	(a)
Sulphur	kg/tonne	1.51	2.91	4.49
		(2.96)	(4.08)	(5.51)
Nitrogen	kg/tonne	0.92	1.76	0.82
		(1.06)	(1.96)	(0.88)
Chlorine	kg/tonne	0.07	0.08	0.09
		(0.32)	(0.33)	(0.33)
Particulates	kg/tonne	0.11	0.12	0.13
		(0.19)	(0.36)	(0.72)

* For each 500 MWe boiler combined abatement plant
\+ Calculated assuming overall power generation efficiency = 0.33
(a) Based on coal fired to boiler plus coal equivalent to electric power consumed
Figures in parentheses are annual average emissions for 90% combined abatement plant availability

- For Datum System 2, the combined abatement unit consumes 3.9 MWe of electric power, equivalent to the combustion of a further 1.5 tonne/h of coal at a power station, assuming an overall power generation efficiency of 33%. This additional coal is arbitarily assumed to be included with the coal burned in the boiler for calculating the efficiency factor.

- For Datum System 3, the combined abatement unit consumes 0.1 MWe of electric power, equivalent to the combustion of a further 0.04 tonne/h of coal at a power station, assuming an overall power generation efficiency of 33%. This additional coal is arbitrarily assumed to be included with the coal burned in the boiler for calculating the efficiency factor.

For illustration purposes, the annual average emissions and emission factors for combined abatement plant availabilities of 100% and 90% are given in Tables 5.18 and 5.20. Further details are given in Section 1.6.

Effect of load variations: ***

Effect of design variations: ***

Limitations: ***

Costs

Capital costs: ***

Annual running costs and cost factors: ***

Effects of design variations on costs: ***

Effects of annual load patterns on annual running costs: ***

Process Advantages and Drawbacks

The advantages are that the basic process is well-established in USA and Europe for FGD; potentially simple process allowing application to small scale without high labour demands; land area required is small; fine grinding not needed as caustic soda solution and lime are used as absorbent (calcium hydroxide is a fine powder); simple chemistry; fairly high sulphur capture efficiency (over 90% capture) and NO_x abatement (up to 60% abatement) has been demonstrated; waste product can be easily and safely disposed of if account is taken of its soluble salts content and excess alkalinity.

The disadvantages are uncertain, but (based on FGD experience) likely to include tendency for deposits to form on SDA walls.

6. Fluidised Bed Combustion

6. Fluidised Bed Combustion (FBC)

Fluidised bed combustion is currently an established technology for burning coal in plant generating up to about 200 tonne/h of steam, and several larger units (up to the output for System 2) will shortly come into operation. Scale-up to the size of modern utility power plant (e.g. 500–660 MWe units) can be envisaged towards the end of the century.

The majority of the 250 or thereabouts of the FBC's of all types now in operation in the western world (and of about a further 130 units on order or under construction) are, however, relatively small units, e.g. boilers producing less than 20 tonne/hour of steam, and despite the attributes of the technology as a means for limiting acidic emissions, the motivation for using FBC has more often been because of its other attributes, such as the ability to burn coal with high and variable ash content more efficiently than other systems and the capability for burning waste materials.

The inherent capability of also reducing acidic emissions will however become more important as stricter emission regulations come into force. A recent survey [578] of boiler orders placed in 1985 (mainly in the U.S.A.) notes that, as in the previous year, FBC is the preferred combustion system for boilers firing coal as the primary fuel where the steam output is more than about 90 tonne/h. The survey also notes that the environmental advantages of fluidised bed firing have virtually eliminated pulverised coal fired boilers from consideration for new industrial plant.

6.1 The Principles of FBC

BUBBLING BED FBCS

A fluidised bed is established when a gas flows upwards through a vessel containing granular material. The base of this vessel is formed by a horizontal or gently sloping plate containing a grid of injection nozzles which are designed to ensure uniform distribution of the combustion air. When the bed is static, or 'slumped', the base plate supports the weight of the bed. As the air is introduced, the bed of particles will remain stationary (or 'fixed') at low velocities, but as the gas velocity is increased there comes a point at which the bed material is raised, its weight then being supported by the gas flow. The volume occupied by the bed material expands and the bed as a whole exhibits many of the characteristics of a liquid. It is then said to be 'fluidised'.

At the minimum fluidisation velocity, the particles forming the bed remain in more or less the same position relative to each other, but as the velocity is raised still further, the excess of air over that required just to fluidise the bed rises through the bed as bubbles. This creates a stirring action and results in a turbulent, rapidly mixed bed in which temperatures are more or less uniform, and heat transfer coefficients to surfaces in contact with the bed are high compared with those encountered in fixed bed or pulverised fuel combustion appliances.

In FBC, the fuel is fed to a bed consisting mainly of incombustible particles–sand, sulphur sorbent or ash, or mixtures of these. Carbon concentrations in coal-fired combustors range from a fraction of 1% by weight of the bed up to a few per cent, depending mainly on coal particle size. The heat released as a particle burns is rapidly transferred to the surrounding incombustible particles and from them to cooling tubes immersed in the bed, so that the burning particles only reach temperatures 100–200 deg. C higher than the mean bed temperature. By keeping this in the range 800–900 deg. C, efficient combustion can be obtained without causing the ash to soften, and this greatly facilitates the removal of ash from the combustor.

The low concentration of fuel in the bed and the low combustion temperature confer a number of other benefits, such as lack of sensitivity to the caking characteristics and ash content of the coals. Because of the low gas temperatures, and the consequent absence of sticky ash particles, no further cooling is needed before the gases enter the convective sections of the boiler; this is in contrast to the requirements for pulverised fuel (PF) fired furnaces, and hence heights of bubbling bed FBC furnaces can be much lower.

The environmental benefits are also a consequence of the low combustion temperature. The most important of these–the ability to fix sulphur in the bed by feeding limestone or dolomite with the coal–results from the fact that sulphur dioxide is absorbed readily when limestone is fed to beds operating at the temperatures used in FBC systems. This is in contrast to the poor absorption generally obtained by limestone injection into PF fired furnaces. The limestone remains in the fluidised bed for periods of 10 minutes to an hour or more, and has time to absorb considerably more SO_2 than in PF firing.

The second environmental benefit of the low combustion temperature–that of low NO_x emissions–is that oxidation of atmospheric nitrogen is avoided. Most of the NO_x that is emitted from FB combustors comes from the conversion of nitrogen present in chemical combination in the coal, and ways of further reducing NO_x formation (e.g. by staged air admission) are now known.

The control of sulphur and NO_x emissions in FBC is described more fully in Sections 2.2 and 3.2. Possibilities of removing acid halides are discussed in Section 7.

Combustion efficiencies of over 98% can be achieved with most coals. For other than operation under pressure however, this usually entails recycling to the bed material that has been elutriated from it, or alternatively returning the material to a separate bed specially designed to encourage high burnout.

CIRCULATING FBCS

The development of FB Combustors was initially centred on the bubbling bed system, as described above. Subsequently, combustors of a somewhat different type have been developed, known as circulating fluidised beds (CFB's). In these, velocities are higher than in bubbling beds and a much higher rate of solids entrainment results. The solids are separated from the gas at the combustor outlet and are returned to the lower part of the combustor. One consequence of this is that a larger proportion of the combustion and sulphur capture occurs in the circulating zone than normally occurs in the space above the bed (the 'freeboard'), of a bubbling bed combustor. A further consequence is that the heat release takes place over a much larger volume than in bubbling beds, and the flow of cooled solids returning to the bottom of the combustion shaft obviates the need to have tubes immersed in the bed in the combustion zone in order to limit bed temperature. Having the heat transfer surfaces sited remotely from the zone in which combustion is occurring is potentially beneficial to the life of the tubing.

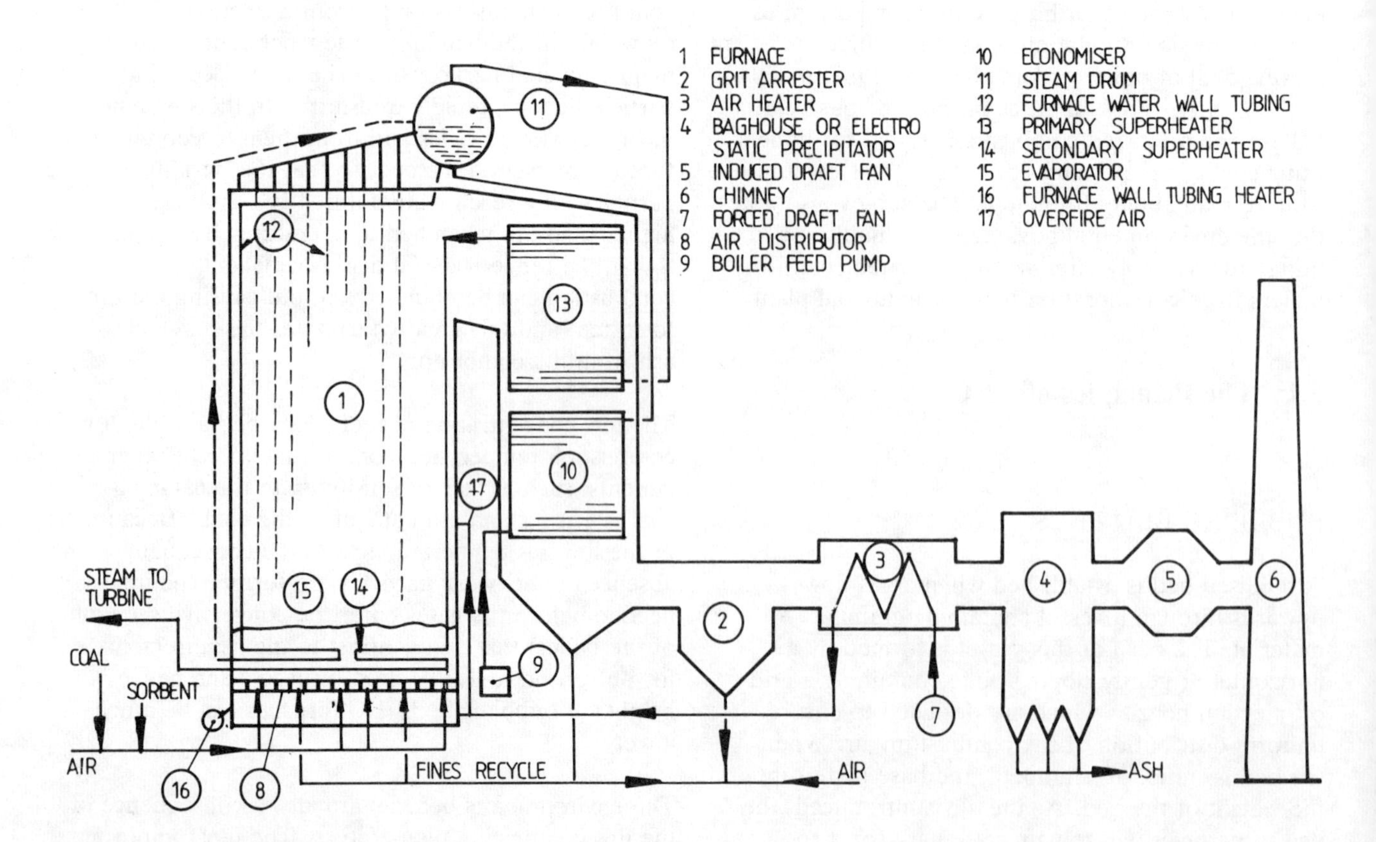

Figure 6.1 Deep Bubbling Bed AFBC Boiler (Water Circulating Pumps not shown)

6.2 Types of Fluidised Bed Combustor

BUBBLING BED FBCs

Bubbling bed FBCs may be of several types, as follows:

Deep bed: (Generally 0.6 to 1.5m fluidised depth) with pneumatic injection of crushed coal at a point (or points) just above the Air Distributor (Figure 6.1). Alternatively, coal with a top size of 20mm or even larger may be fed to the combustor at a point above the bed (overbed firing) into which it is allowed to fall.

Shallow bed: (Less than 0.3m fluidised depth) with overbed feeding of washed, lump coal (Figure 6.2). These are used mostly for small industrial boilers; the configuration is not suitable for efficient sulphur capture.

Dual bed: In these the gases leaving the first bed are caused to fluidise a second bed in which further combustion and sulphur absorption takes place (Figure 6.3).

All of the above are operated at around atmospheric pressure, and are examples of atmospheric pressure fluidised bed combustors (AFBCs).

Pressurised fluidised bed combustors (PFBC): At pressures up to 20 bar, with recovery of energy in a gas turbine operated by the cleaned, hot products of combustion. Deep beds (e.g. 3–4 m) are necessary to contain the large amount of heat transfer surface that is needed to match the high combustion intensity achieved when operating under pressure.

The ranges of steam/power output for the above types of plant for which manufacturers are currently prepared to supply fluidised boilers, and the numbers installed, are listed in Tables 6.1 and 6.2. Table 6.3 lists the organisations involved in the development/assessment of PFBC systems. The features of AFBCs are described in Section 6.3, and of PFBCs in Section 6.5. Selected types of bubbling bed FBCs are appraised in Section 6.9.

CIRCULATING FBCs

Circulating fluidised bed combustors (CFBCs) may also be of several different types. These differ principally in the disposition of the heat absorbing surface, but there are also differences in other features such as the way in which the circulating burden is separated from the combustion gases:

Integral Heat Exchanger Systems: (Figure 6.4) The heat transfer surface forms all or part of the walls of

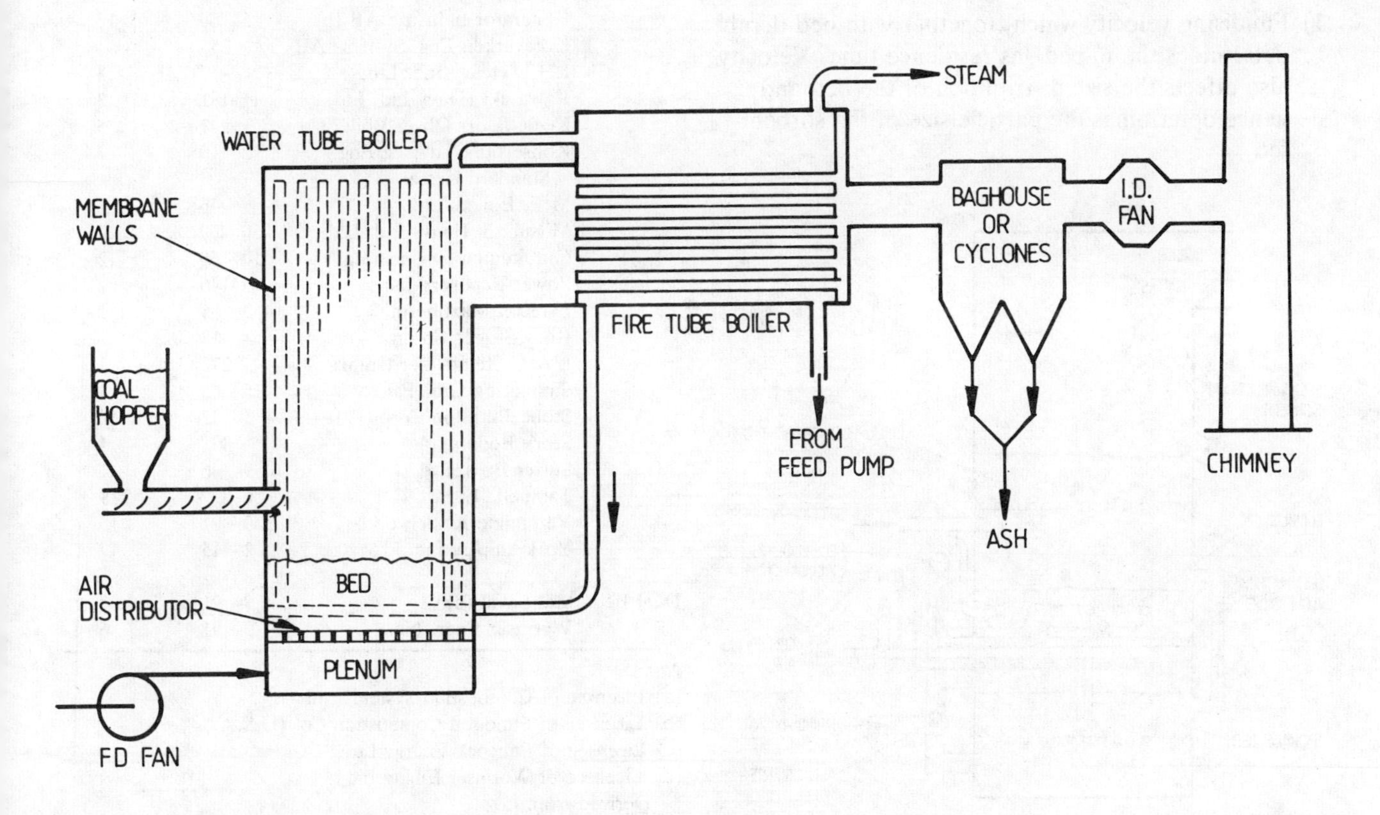

Figure 6.2 Shallow-Bed AFBC Boiler

the combustor shaft and may form part of the solids disengaging zone [579].

External Heat Exchanger Systems: (Figures 6.5 and 6.6) The major part of the heat transfer surface may be immersed in a bubbling bed external to the circulating bed through which the separated hot solids pass before returning to the combustor [580].

It would be possible to operate circulating bed combustors under pressure and there are potential benefits in doing so, but as of now there is no operating experience.

An appraisal of selected types of bubbling bed and circulating bed combustors is given in Section 6.9. Before this, however, attention will be focussed on the features of FBC systems that are most relevant to their use for reducing atmospheric emissions, and on other matters that have a bearing on the use of FBC for steam raising with SO_2 emission control.

Manufacturers of CFBCs are listed in Table 6.4 together with the numbers and capacity ranges of units installed or on order. The features of CFBCs are described in Section 6.4, and an appraisal of selected types is given in Section 6.9.

6.3 Features of Bubbling Bed Combustors

The main design factors affecting the performance of atmospheric bubbling bed FBC systems are as follows:

1) Fluidising velocity which, together with bed depth, determines the in-bed gas residence time. Velocity also affects the size distribution of the bed and hence determines the particle size of the sorbent fed.

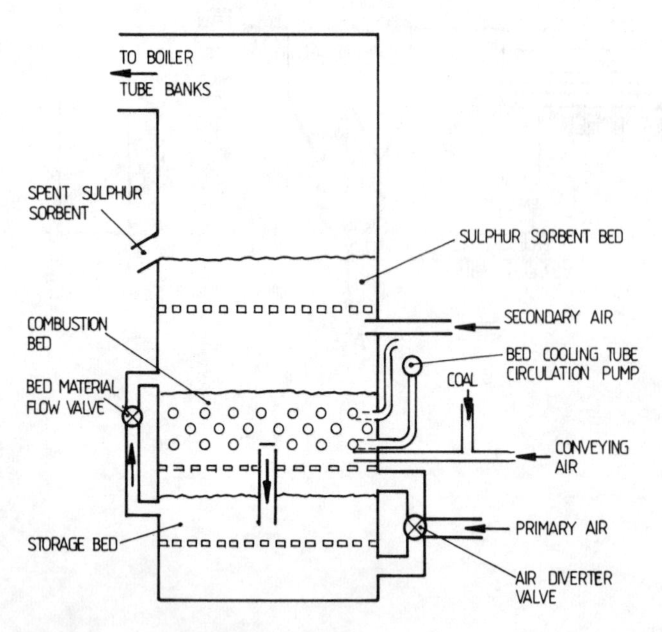

Figure 6.3 Dual Bed AFBC Combustor

2) Bed temperature, which affects the in-bed combustion rate and the reactivity of the calcined sorbent.

3) Sorbent air, and the points at which air is introduced, which are particularly relevant to combustion efficiency and to NO_x emission.

4) Height of the space above the bed before the gas exit (the freeboard) which, together with the gas velocity, determines the gas residence time and affects the solids entrainment rate at the furnace exit.

Table 6.1 Main Manufacturers of AFBC Plant

Type	Supplier	Capacity range tonne steam/h	Number of Boilers**
Deep Bed	ENEL/Ansaldo	180**	2
	Babcock & Wilcox Co.	23–680	2
	Babcock Energy Ltd. (a)	5–115	3
	Babcock Australia Ltd	2– 9	3
	Babcock Hitachi KK	5+	7
	Burmeister & Wain Energi	11	1
	C.E. Power Systems	23–460	2
	Cockerill Mech. Industries	3– 40	5
	Combustion Power Co.	7–115	8
	Crone, Holland	2	1
	Dedert Corpn.	2.5–46	6
	Deutsche Babcock Werke AG	9–318	15
	Energy Equip. Ltd.	1– 40	13
	Energy Products of Idaho	5–113	18
	Fluidyne Eng. Corpn.	3– 23	1–
	Foster Wheeler Corpn.	16–272	11
	Foster Wheeler Power Products Ltd. (b)	14–272	–
	Generator Industrie AB (b)	5– 50	16
	Gotaverken Eng. Systems AB	25	1
	I.H.I. Heavy Ind. Ltd.	–	3
	Kawasaki Heavy Ind. Ltd.	11–181	3
	Keeler/Dorr Oliver Boiler Co.	227+	5
	Konsortium Thyssen Eng. Standard Kessel	2– 16	17
	M.E. Boilers Ltd.	9– 46	2
	Mitsubishi Heavy Ind. Ltd.	30–400	4
	Outokumpu Engineering	10– 50	12
	Power Recovery Inc.	5– 46	–
	Pyrecon Maxitherm	3– 5	2
	Riley Stoker Corpn.	18	–
	L & C Steinmuller GmbH	272+	2
	Simmering Graz Pauker	15– 65	3
	Stone Johnston Corpn. (a)	1– 32	30
	Stork Boilers (c)	80	1
	Sulzer Bros. Ltd.	9– 46	1
	Tampella Ltd	6–227	19
	Vereinigte Kesselwerke	9– 91	15
	York Shipley Inc.	2– 15	17
Dual Bed	Stal-Laval (d)	9– 91	2
	Wormser Engr. Inc.	9– 91	6

Notes:
(a) Licensee of Combustion Systems Ltd
(b) Licensee of Fluidised Combustion Co. (U.S.A.)
(c) Licensee of Babcock Energy Ltd
(d) Licensee of Wormser Engng Inc
* Derived from [598]
** Installed or on order
\+ Up to

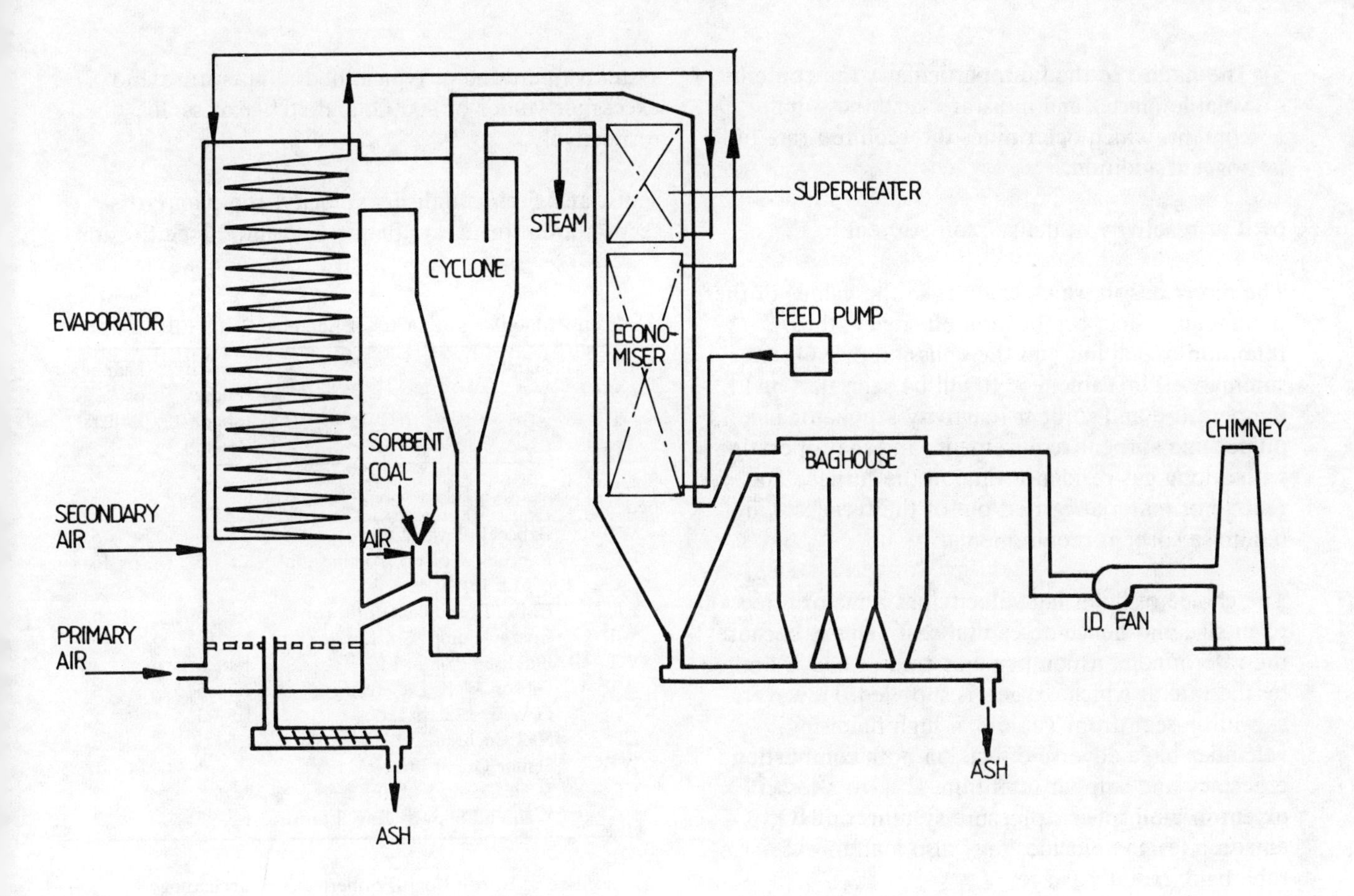

Figure 6.4 CFB Fired Boiler with Integral Heat Exchanger

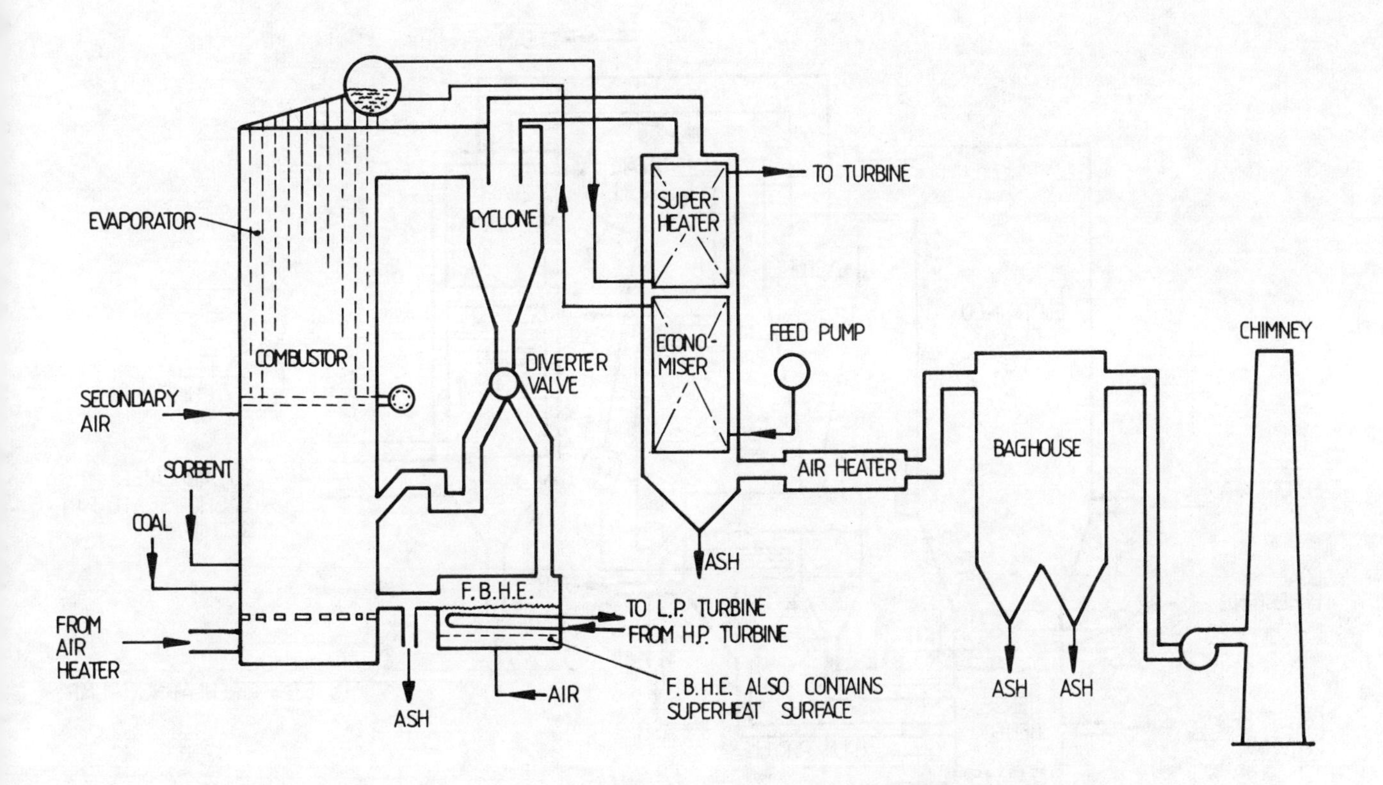

Figure 6.5 CFB Fired Boiler with External Heat-Exchanger–Lurgi System [F.B.H.E.]

5) The nature of the fuel, particularly the contents of volatile matter and moisture; also the sulphur content, which determines the required rate of sorbent addition.

6) The reactivity of the sulphur sorbent.

The directions in which changes in the values of these parameters affect combustion efficiency, the retention of sulphur and the emission of NO_x are summarised in Table 6.5. It will be seen that bed temperature and sorbent reactivity are major factors influencing sorbent requirements and that operation with a long gas residence time in the furnace and recycle of material carried out of the furnace will minimise sorbent requirements.

The choice of fluidising velocity has a major effect on plant size and hence on capital cost. This is because the rate of combustion per unit area of bed is limited by the rate at which oxygen is supplied. However, as will be seen from Table 6.5, high fluidising velocities have adverse effects on both combustion efficiency and sulphur retention, and, to a lesser extent in atmospheric pressure systems on NO_x emission. High velocities may also lead to excessive tube bank metal wastage.

The value chosen for fluidising velocity is therefore a compromise, and typical values are from 1 to 3 m/s, corresponding to maximum gross thermal inputs per square metre of bed of 0.9 to 2.7 MWt for AFBC systems operating at typical bed temperature and excess air values of 900°C, and 30% excess air, respectively.

Thus, at 2.5 m/s fluidising velocity, the plan cross-section area for datum plants 1, 2 and 3 (see Section

Table 6.2 Main Manufacturers of Bubbling Bed AFBC Plant*

Type	Supplier	Capacity range tonne steam/h	Number of Boilers**
Shallow bed			
WT	Babcock Power Ltd. (e)	5– 30	16
FT	Babcock Worsely Combustion Ltd. (e) (f)	1.8– 5	10
FT	EMS (Thermplant) Ltd. (e)	2– 5	6
WT	Energy Equip. Co. Ltd.	1– 50	10
WT	Fluidised Comb. Ltd.	1– 1.5	2
WT	Gibson Wells Ltd. (e) (g)	7.7–15	3
FT	G.W.B. Boilers Ltd.	1.5– 3	1
FT	NEI Cochran Ltd.	1.5– 5	6
WT	Senior Green Ltd. (e)	3	1
FT	Stone Danks Ltd. (e)	5	1
FT	Wallsend Slipway Eng. Ltd. (e)	1.5– 7	7

Notes:
(e) Licensee of British Coal (Formerly NCB) technology
(f) Includes Allied Boilers Ltd
(g) Subsidiary of Foster Wheeler (U.K.) Ltd
* Derived from [598]
** Installed or on order
WT Water Tube boiler
FT Fire Tube boiler

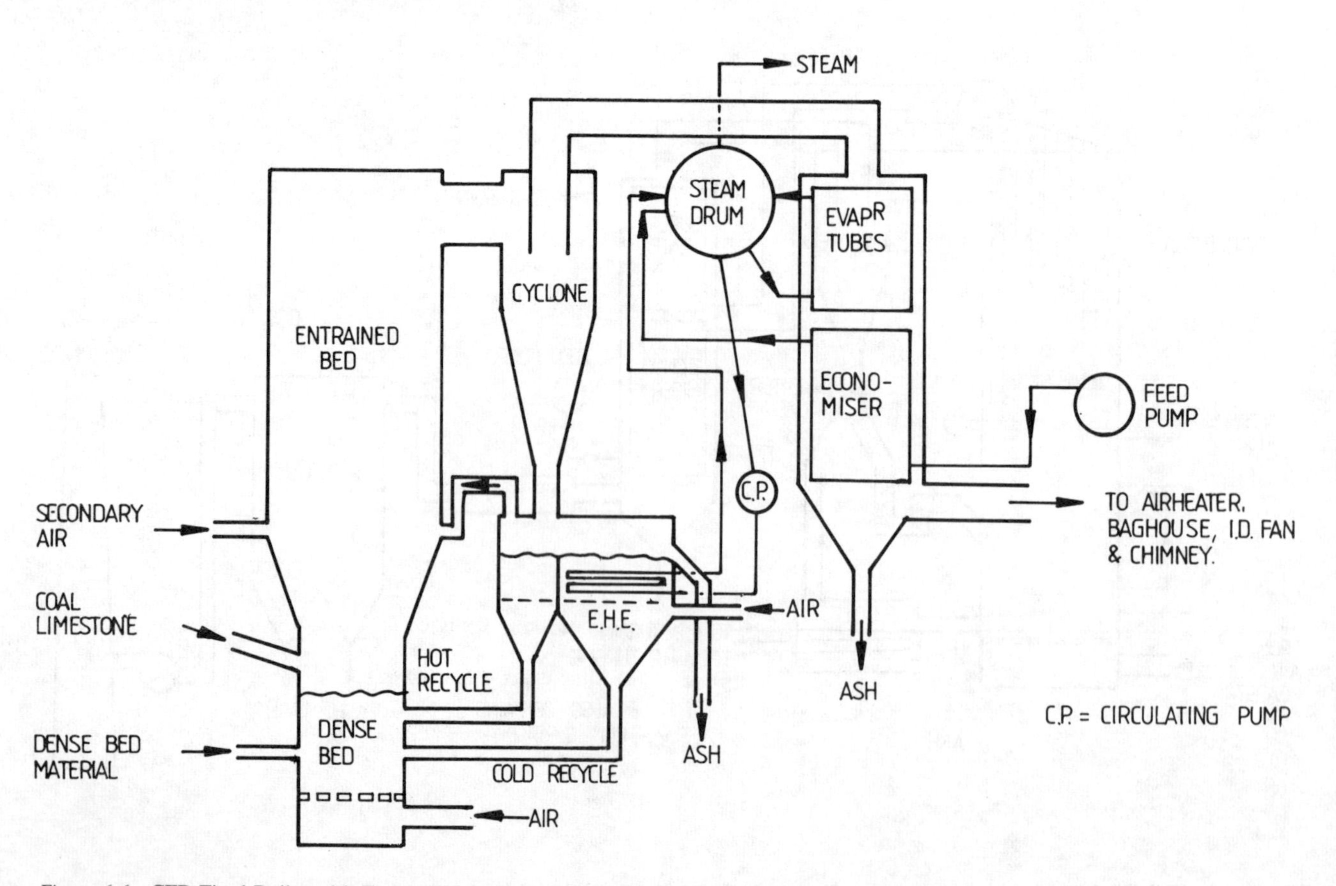

Figure 6.6 CFB Fired Boiler with External Heat Exchanger [E.H.E.] Multi-Solids FBC System

1.4) would be respectively 574 m² (per 500 MWe set) 208 m² and 5 m² (assuming no changes in efficiency factors). The plan areas for the large industrial and utility appliances are larger than for pulverised coal firing but, because the ash entrained with the outlet gas is below the melting temperature there is a considerable saving in plant height. In PFBC, for a given fluidising velocity, the bed area required is reduced in proportion to the increase in pressure.

In the large AFBC plant, the bed would comprise a number of smaller modules, varying numbers of which could be brought on stream as required to follow the variations in demand for steam. In small boilers, load following would probably be by intermittent slumping of the bed, sufficient heat being retained in the slumped bed to allow rapid return to full firing rate on restarting the air and fuel feeds.

For large industrial and utility plants, the bed will consist of the SO_2 sorbent together with adventitious ash associated with the coal (as distinct from inherent ash, which is fine and is quickly elutriated from the bed). Sand beds may be used in smaller plant operating with low ash, washed coals. An indication of the bed size likely to give satisfactory fluidisation can be obtained from reference [581]. Typical values for the top size of sorbent fed to the bed are 3 mm for a velocity of 1 m/s, 4.5 mm for 2 m/s and 6 mm for 3 m/s.

The SO_2 sorbent is normally fed through nozzles immersed in the bed. The coal may be fed through in-bed or through over-bed nozzles. Over-bed feeding has been shown to give poorer sulphur retention [582, 583] probably because some of the sulphur is released from the coal in the freeboard and is therefore less likely to be absorbed.

NO_x emission (expressed as NO_2) from AFBC systems with single stage combustion is typically in

Table 6.3 Manufacturers/Designers of Pressurised Fluidised Bed Plant

Type	Manufacturer	Capacity MWe	Status
Bubbling Bed (Deep)	ASEA PFBC	80	One on order
	ASEA PFBC	250	Design study
	Foster Wheeler USA	250	Design study
	Babcock Werke	300	Design study
	Babcock U.K.	660	Design study
	Babcock U.S.A. along with ASEA PFBC	80	One on order
	Steinmuller	100 MWt	One in operation (Aachen)

Table 6.4 Manufacturers of Circulating Fluidised Beds*

Type	Supplier	Capacity Range tonne/h	No. of Boilers Installed**
Integral Heat Exchange	A. Ahlstrom Oy.	14–454	15
	Babcock & Wilcox Co. (a)	23–680	4
	Babcock Power Ltd. (b)	5+	1
	Energie u. Verfahrenstechnik GmbH (c)	140+	4
	Foster Wheeler Corpn.	16– 22	1
	Gotaverken Energy Systems A.B.	23– 27	8
	Keeler/Dorr Oliver Boiler Co.	227+	9
	Pyropower Corpn. (d)	23–454	6
	Studsvik Energiteknik AB	25	–
External Heat Exchange	C.E. Power Systems (e)	23–460	2
	Foster Wheeler Power Products Ltd. (f)	14–272	1
	Lurgi GmbH	23–460	3
	Mitsui Eng. & Ship Building Co. Ltd. (f)	9–272	–
	Riley Stoker Corpn. (f)	18**	2
	Struthers Wells Corpn. (f)	34–227	2

Notes:
(a) Licensee of Studsvik Engergiteknik AB
(b) Licensee of Solids Circulation Systems Inc
(c) Licensee of Ahlstrom
(d) Subsidiary of Ahlstrom, offering Ahlstrom technology
(e) Affiliated with Lurgi Gmbh
(f) Licensee of Battelle Development Corpn
* Derived from [598]
** Installed or on order
\+ Up to

Table 6.5 Effect on Emission Performance of Main Operating Parameters: Bubbling Bed Systems

Operating Variable	Reduction of Loss/Emission Favoured by		
	Combustion Loss	Sulphur Emission	NO_x Emission
Excess Air	High (a)	Low (b)	Low
Bed Temperature	High	Optimum (c)	Moderate (d)
Fluidising Velocity	Low	Low	Low (e)
Bed Depth	High	High	High (f)
Sorbent Reactivity	No effect	High	No effect
Pressure	High (g)	High (h)	High (k)
Grit Recycle Ratio	High (l)	High (g)	No effect
Freeboard Residence time	High (m)	High (m)	Uncertain

Notes:
(a) Effect large below 20–30% excess air, otherwise small
(b) Effect small except near zero excess air
(c) Optimum at about 850 deg. C in AFBC; above 900 deg. C in PFBC
(d) Little effect below 900 deg. C
(e) Effect small, but significant in PFBC
(f) Evidence inconsistent, but in general deep beds are better than shallow
(g) Effect small
(h) Effect large at high temperatures but small at low temperatures
(k) Effect not large and diminishes above 6 bar
(l) Effect large up to about 2.8 times coal feed rate
(m) Effect small above about one second

the range 600–900 mg/Nm³ (300–450 vpm) measured at NTP with 6% oxygen in the dry flue gas. The level for a specific coal is primarily dependent on the excess air value at which the plant is operated. It is not usually feasible to use less than about 25% excess air, because this would result in an increase in combustible loss. However, staged combustion in AFBC systems has been found to result in NO_x reductions of up to 50% [584]. It is probable that optimum conditions are found when stoichiometric air is fed to the bed, and a further 25 to 30% is fed to the freeboard or to a second bed to complete combustion. The chemistry of NO_x formation and destruction in coal fired combustors is described in Section 3. Although the temperature/concentration fields in FBC differ from those in pulverised coal systems, the general effects of staged combustion can be explained in a similar way for both PF and FBC.

There appears to be little prospect for reduction in acid halide emissions, apart from the use of a separate bed containing an alkaline sorbent at a temperature several hundred degrees lower than that of the main bed (see Section 7). Some chloride may, however, be captured by dust deposited in the flues and dust removal equipment.

6.4 Features of Circulating Bed Combustors

The nature of a bubbling fluidised bed is such as to result in a degree of 'by-passing' of the fuel in the bed by the gas rising in the bubbles (although there is a flow of gas into and out of each bubble as it rises, which reduces the by-passing to an extent depending on the bed depth).

A significant feature of fast fluidisation, which is the name given to the process occurring in CFB combustors, is that gas-solid contacting is more thorough, and takes place over a longer period of time than in the bubbling bed. In addition, mass transfer of oxygen through the boundary layer of gas surrounding a burning particle has been shown to be more rapid than in bubbling bed combustion [585].

This confers the capability for higher combustion and sulphur capture efficiencies in CFB combustors.

Key plant items in CFB combustors are the hot cyclones for capturing the solids leaving the combustor, and the means employed for feeding the recirculated solids into the lower part of the combustor. This has to resist blow-back due to the higher pressure in the combustor, and must therefore not be of a type that can suffer from leakage due to the wear of moving parts. Pneumatically operated devices have been found most suitable for this application [586]. On the other hand, the feeding of fresh fuel and sorbent is simpler than in bubbling bed systems, fewer feed points being needed for a given combustor rating.

Combustor cooling surfaces either take the form of water-cooled walls, or of tubes suspended from above [587]. The former method is preferable, because suspended heat transfer surfaces are subject to erosion due to the high velocities employed (up to 10 m/s) and also they impede radial mixing of gas and solids, tending to reduce combustion and sulphur capture efficiencies.

Each of the broad categories referred to in Section 6.2 contain variants. These differ in such matters as design of the combustion chamber, particulate separation device, method of feeding the recycled solids and arrangement of the heat transfer surfaces. For example, CFB systems with integral heat transfer include Ahlstrom/Pyropower, Gotaverken (some of which employ inertial solids collection devices that are not conventional cyclones), Studsvik, Keeler/Dorr Oliver and the Babcock Power Solids Circulation Boiler, which incorporates both a bubbling bed chamber and a fast-fluidised chamber, the water-cooled wall of which comprises most of the primary heat transfer surface. Temperature control in this system is by proportioning the combustion air between the bubbling and the fast-fluidised beds.

There are two principal designs of system with external heat exchangers. These are the Lurgi and the Battelle Multisolids FBC systems.

Both utilise low-velocity bubbling beds for the external heat exchangers, but the Battelle system (licensed to a number of manufacturers) has a novel form of combustor in which the lower section, which has a smaller cross-sectional area, contains a non-entrained bed of larger-sized material.

Three types of CFB systems are considered in detail:

1) that with a wholly integral heat transfer surface, typified by the Ahlstrom Pyropower systems (Figure 6.4), developed initially for firing wood wastes and other fuels of low heating value, but now extended in range to use fuels of higher calorific value as well, (the Gotaverken CFB boiler is very similar),

2) that with an external fluidised bed heat exchanger for which the Lurgi system (Figure 6.5) is described,

and

3) the somewhat similar Multi-Solids FBC (MSFB) system (Figure 6.6) developed by Battelle, which used an external fluidised bed heat exchanger and also a combination of dense and circulating beds in the combustor.

Thomas [588] reports that about 20–25 CFB units which use coal as the major fuel were in operation at the end of 1985 and about 15 had commissioning dates in 1986.

6.5 Pressurised Fluidised Bed Combustion: 'Bubbling Beds'

The main motivations for Pressurised Fluidised Bed Combustion (PFBC) are the potentials for:

- Enabling coal instead of distillate oil or gas to be the fuel for combined cycle power plant, and hence offer the possibility of significantly higher power generating efficiency; the combination of gas and steam turbine driven power plant facilitates higher cycle efficiency than either system alone.

- More compact plant, and compared with both AFBC and conventional plant with FGD, lower capital cost.

- Inherently lower NO_x emissions than in AFBC.

- The capability of efficiently using dolomites (unlike AFBC) as the sulphur capture medium in the bed, in circumstances where limestones are less readily available.

- Coal can be fed as a coal-water mixture containing typically 70% coal, with less loss of efficiency than when this method of feeding is used with either AFBC or conventional pulverised fuel firing.

PFBC typically involves the combustion of coal at pressures in the range 10 to 16 bar. For the process to be economic, the power recovered by the expansion of the combustion gases must be at least adequate to supply the power requirements for pressurisation of the combustion air.

There are two main groups of combined cycles for which PFBC has been proposed:

- Steam cycles, in which most of the power is generated by steam turbines, and in which the gas turbines either generate additional power (high gas temperature cycles, Figure 6.7) or only pressurise the boiler (turbocharged boiler cycles, Figure 6.8).

- Air cycles, in which most of the power is generated by gas turbines, with exhaust gases passing to waste-heat boilers supplying steam to steam turbines.

Design studies [589] have shown that steam cycles have a greater potential than air cycles, and the further development of the latter has been largely discontinued. The main application for PFBC is ultimately for large units of 200 to 600 MWe. Because of this, and because of the inherent complexities of pressurised operation, development has been slower than for AFBC plant. Small-scale (up to 6 MWt) combustors have been operated for many years, leading to the pilot-scale operation of the combustor at Grimethorpe (60 MWt) and the CTF combustor at Malmö, Sweden (15 MWt). Most of the small-scale and the Grimethorpe information is available in open publications e.g. [593, 594].

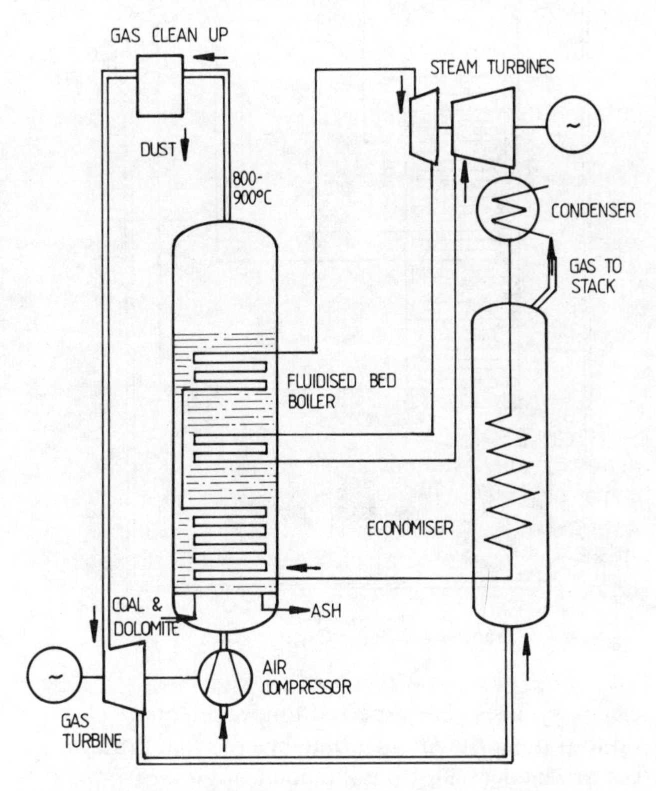

Figure 6.7 High Gas Temperature Steam Cycle (Supercharged Boiler Cycle)

However, no fully-integrated combined-cycle plant (i.e. operating with a gas turbine) has yet operated, although a small number of such plants are currently being built as demonstration units (of c.80 MWe output), and negotiations for commercial plant are in progress.

General descriptions of PFBC combined cycles are given in the following paragraphs. Examples of plant proposed for specific sites are discussed later.

High Gas Temperature Steam Cycle: This cycle (Figure 6.7) represents the most advanced of the first generation of PFBC combined cycles. It incorporates a basically conventional Gas Turbine operating at a full-load pressure in the range of 10–16 bar and a full-load turbine entry temperature of 800–950°C. Coal is burned in the bed at 850–950°C, the bed being fluidised by air from the Compressor; the bed is cooled to this temperature by immersed steam generation, superheating and reheating tubes. The combustion gas leaves the bed at the bed temperature, passes through hot Gas Cleaners, and is then expanded through the Turbine before passing through an Economiser to the stack. Currently, all the designs postulate conventional cyclones for hot-gas

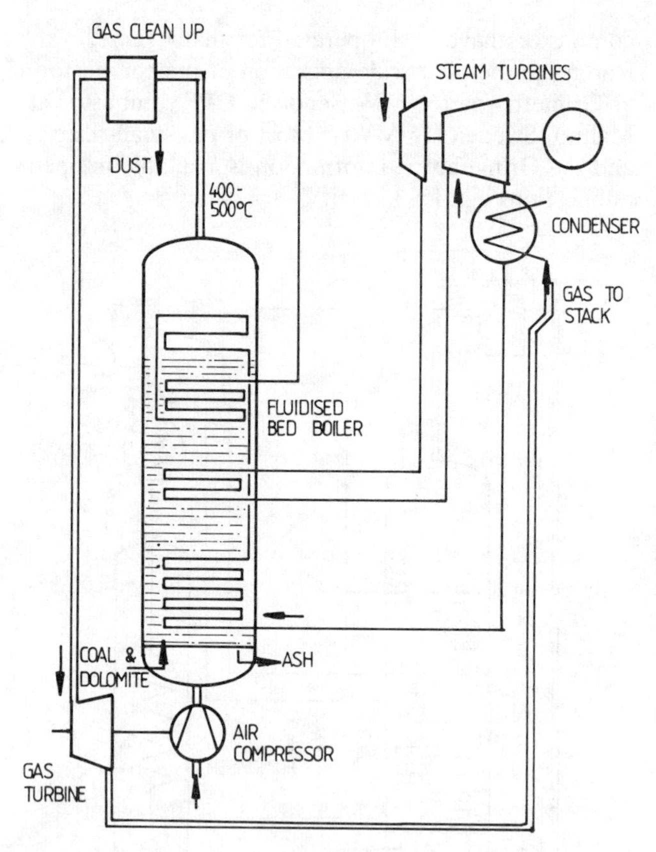

Figure 6.8 Turbocharged Boiler Cycle

clean-up. These are expected to give adequate clean-up to protect the Gas Turbine blades from erosion, but are inadequate from the point of view of particulate emissions to atmosphere. Consequently, a conventional bag-filter unit or electrostatic precipitator (ESP) is proposed between the Economiser and the stack. Filter units operating at the high temperature and pressure upstream of the Gas Turbine are expected to be smaller and cheaper than the combination of cyclones and bag-filter or ESP, and their development is actively being pursued. If, and when, such units become reliable, it will be possible to omit the final gas clean-up stage.

The proportion of total power output derived from the gas turbine depends on the level of excess air chosen for the cycle; it varies from about 25% of the total output at 25% excess air to about 30% of the total output at 100% excess air. Although a high design excess air leads to a higher cycle efficiency, it also leads to a higher capital cost, and designers have tended to finalise on a full-load excess air value of 20–30%. Design studies for utility power plant have of necessity been based on existing large industrial gas turbines of 60–100 MWe output. Consequently, to achieve a plant output of, say, 600 MWe involves two or three gas turbines in conjunction with a single steam turbine.

The actual cycle efficiency depends on a number of factors–such as coal characteristics, steam parameters, choice of gas turbine–but is generally computed to be in the range 39 to 41%. Where direct comparisons have been made [585] the efficiency has been 3 to 4 percentage points higher than a conventional coal-fired plant with FGD, and the capital costs about 10% lower. These have, however, been for installations in the U.S.A., and the results may be different for the conditions applying in the U.K. The results of a study jointly sponsored by the CEGB, British Coal and the U.K. Department of Energy are not available at the time of compiling this document.

Low Gas Temperature Steam Cycle (Turbocharged Cycles): This cycle (Figure 6.8) has convection passes, forming part of the steam generating/superheating/reheating system, installed in the gas path after the Fluidised Bed to cool the combustion gases down to a turbine entry temperature (TET) of 400–500°C. At this TET, the Gas Turbine generates no net power (except, possibly, a small amount for control purposes) but is acting as a turbocharger. At full load the pressure level is in the range 6–10 bar. The Turbine can be either a derated Gas Turbine or developed from a turbocharger.

This cycle is, in essence, a supercharged boiler in a conventional steam cycle. As such, it has a lower cycle efficiency than the high gas temperature steam cycle, but has the attraction of lower initial risk. It still has a higher cycle efficiency than a conventional coal-fired plant fitted with FGD.

6.6 Disposal of FBC Residues

Apart from a small amount of unburnt carbon, these residues contain coal ash, calcium sulphate, calcium oxide and calcium carbonate where the sorbent is limestone, with additional magnesium oxide when dolomite is used as, for example, in PFBC systems. Table 6.6 shows the estimated compositions and quantities of waste products from four notional FBC systems when burning the coals, having the analyses shown in Table 1.2 with the addition of sorbent at a Ca:S molecular ratio of 2.5:1 (sufficient to give from 85 to over 90% sulphur retention if the sorbent is of high reactivity). The compositions and flow rates of the residues are for the combined bed plus carry-over residues collected in the particulate removal equipment downstream of it.

An estimate [592] of land requirements for disposal of residues by tipping, suggests that for coal having the analysis of reference coal 1 shown in Table 1.2, the areas would range from about 0.006 to 0.010 ha.m/MWe.year as sulphur capture increases from 75% to 90%. This requirement is over and above that needed to accommodate the coal ash, which for a 16.7% ash coal and a load factor of 65% amounts to 0.044 ha.m/MWe.year. Assuming the depth of landfill is 10 m, dumping of the spent sorbent would result in an increase in land requirement for a 2000 MWe station burning 5 million tonnes of coal a year from 8.7 ha to 9.9 or 10.7 ha, depending upon the extent of sulphur removal.

The principal environmental concern when tipping wastes from combustors is the contamination of rivers and ground-waters by the leaching of soluble material from the wastes. Researchers at Westinghouse Research and Development Center and at Radian Corporation have studied the leaching properties of both AFBC and PFBC residues [593, 594] and the former have investigated methods for processing the residues in order to reduce the environmental impact of their disposal. The processing consisted of mild heat treatment to convert them into a cement-like material. Results of the leaching studies [593, 595] showed that the major potential contaminants in the leachate are the high pH, total dissolved solids, and sulphate levels. Heat released when FBC residues are initially contacted with water, due to calcium hydroxide hydration, may also cause a handling hazard. Trace metal concentrations in the leachates are very low, and do not generally exceed drinking water regulations. This is because of the alkaline nature of the residues. The actual levels of trace elements in the leachates, and in the residues themselves, tend to be very variable, and depend upon the composition of the mineral fraction of the coal and of the sorbent.

Table 6.6 Estimated Waste Products from Typical FBC Systems

1. System	1 AFBC	1 PFBC	2 AFBC	3 AFBC
2. Sorbent type	Limestone	Dolomite	Limestone	Dolomite
3. Sorbent Feed Rate tonne/h	23.9	44.2	7.06	0.18
4. Residues tonne/h	53.8	69.4	12.43	0.315
Estimated composition %wt:				
5. Ash	61.4	47.1	48.8	48.8
6. $CaSO_4$	19.6	15.9	24.9	24.9
7. $CaCO_3$	2.8	19.6	3.5	3.5
8. CaO	14.0	1.2	20.0	20.0
9. MgO	0	13.0	0	0
10. Other minerals excluding sorbent	2.2	3.2	2.8	2.8

Notes

Systems 1 = 500 MWe units for Power Utility.
2 = Large Industrial Boiler. 450 tonne/h of steam.
3 = Small Factory Boiler. 13 tonne/h of steam.

U.K. Limestone composition: 95% $CaCO_3$: 5% other minerals.

U.K. Dolomite composition: 51.6% $CaCO_3$: 43.4% $MgCO_3$; 5% other minerals.

Feed rate based on Ca/S ratio of 2.5 estimated to give 85% sulphur retention for AFBC and 90% for PFBC.

Residues: Material discharged from the bed and captured by the particulate removal equipment excluding moisture and unburned carbon.

When exposed to the weather, piles of FBC sulphated residues absorb water and carbon dioxide, causing swelling and recrystallisation. This leads to a loss of permeability, so that rain falling upon the pile tends to run off as surface water [596]. Thus, it appears likely that the main problem in the tipping of FBC wastes is likely to be the area of land occupied, rather than the environmental effects of leaching.

Possible uses for FBC solid residues [592, 593, 597, 598] include:

- as a raw material for cement manufacture, the unburnt carbon present in carry-over material contributing to a reduced energy usage on firing the cement;

- use in hydraulically-bonded building materials where a lower compressive strength than normal is tolerable;

- the production of artificial aggregate (either the sintered 'Lytag' type or the lime/heat-treated 'Aardelite' type);

- the production of ceramic and of sand-lime bricks; and

- as a possible future source of aluminium, should the import of bauxite become uneconomic.

In most of the applications producing materials of construction, the sulphate content may be limiting, and in some applications some further size reduction will probably prove necessary. Other potential applications include: the treatment of trade or municipal wastes; control of acid mine drainage; the replacement of hydrated lime in various applications including utilisation as a sorbent for SO_x scrubbers and as a reagent for stabilising flue gas desulphurisation sludges; as a raw material for asphalt fillers (in the mixing of asphalt concrete) where the carry-over material has roughly the right physical and chemical characteristics, although a problem may arise due to variability in composition of the material. As a general rule, if the product is to command a high market value it will be required to be of reliable, constant composition, usually within a fairly close range of properties. Because the residues from FBC are expected to vary quite widely, being affected by such things as coal properties, variation in coal-sorbent ratio, or in boiler load, it will be difficult to find an outlet for premium uses and therefore a low price may have to be accepted.

Liquid effluents are not, as a rule, a major problem in boiler operation and there is no reason why they should prove more so in FBR firing than in older established forms of coal firing.

Possible contributions to contaminated liquid effluents from FBC-fired boilers are:

- Run-off from coal stockpiles (suspended and dissolved solids; sometimes a low pH).

- Boiler blow-down (dissolved solids).

- Blow-down from hydraulic ash transport loop (suspended and dissolved solids: high pH).

- Accidental spillages caused by loss of control during conditioning of solid residues (suspended and dissolved solids; high pH).

- Water from plant cleaning operations (oil, suspended solids, possibly high C.O.D.)

- Sewage (suspended and dissolved solids, high C.O.D.)

Well-established treatment processes are available for ensuring that such effluents are rendered of acceptable quality for discharge. In the case of coal stockpile run-off for large boiler plants such as a power station, provision should be made for diverting run-off to a treatment area before final discharge.

In the case of some industrial plants, the boiler plant will only generate a small fraction of the total liquid effluent from the factory, and its treatment will be handled together with the bulk of the factory effluent.

6.7 Areas of Uncertainty

'Scaling-up': The largest boiler with a bubbling bed currently in operation generates about 70 tonnes/h steam, and the largest CFB boiler generates 122 tonne/h steam. A bubbling bed boiler for a 160 MWe demonstration plant is under construction, and design studies have been carried out for boilers for an 800 MWe plant.

Uncertainties exist on issues including:

- The number and size of FBC modules needed for applications of different types of bubbling and circulating FBCs.

- For bubbling beds: the best ways of distributing coal; and the arrangement of the beds (i.e. horizontal or vertical stacking)

- For circulating beds: the arrangement of heat transfer surfaces; air jet penetration in the combustor shaft; and the arrangement of hot cyclones.

None of these uncertainties appears to be of sufficient magnitude to impose technical limits on the application of fluidised bed combustion.

Metal Wastage: Metal wastage due to erosion/abrasion of in-bed tubes has been a serious problem in some AFBC and PFBC plant, while some combustors have operated for many thousands of hours without experiencing any metal wastage [599]. Where erosion does occur, it has often been associated with high velocity jets of air impinging on a tube or wall surface, resulting in a 'sand-blasting' effect. More general tube metal wastage has also occurred in a number of instances. The indications are, however, that it should be possible to achieve commercially acceptable tube life by measures such as:

- avoiding design features that lead to impingement of jets of particles on to tubes,

- minimising the distance between the air distributor and the tube bank,

- using parallel rather than staggered tube arrays,

- using horizontal or vertical tubes rather than inclined tubes,

- avoiding high fluidising velocities,

- the use of fins or studs to protect the tubes [600],

- chromising of tubes in sensitive areas.

In view of this problem and the lack of a full understanding of the mechanisms involved, special attention is being given by boiler designers to minimising downtime should tubes have to be replaced.

Corrosion of tubing has not so far been found to be a significant problem area. If it occurs, it is mostly at metal temperatures of 500°C and above, and is therefore only of concern where superheater/reheater tubes are immersed in the bed. Iron-based materials appear to be the most resistant [601]. Tubing in the free-board and down-stream heat exchange surfaces suffers little from corrosion, and it has been concluded that corrosion problems on the whole appear to be handled by choice of materials and by keeping within currently available design and operating guidelines [602]. High chlorine coals are recognised as potentially presenting more serious corrosion risks [603] but there is little experience of operation with these.

Use of residues: The excess of lime that has to be used to obtain the required sulphur removal presents both a potential environmental hazard and a potential benefit in the disposal of residues. The hazard results from the alkalinity of the material, should leaching occur, while the benefit is by utilising this alkalinity to neutralise unwanted acidity in solids, or in solid or liquid wastes. A second benefit resulting from the free lime content, is to make use of its contribution to particle bonding in the manufacture of blocks for construction purposes, or the consolidation of other solid wastes for disposal as landfill.

Because any positive use for the residues will have, for economic reasons, to be close to the point of origin, it is desirable to identify as wide a range of uses as possible, and further research to this end is currently in progress in many countries.

6.8 Future Developments

For small and medium-sized boilers, AFBC is now a proven commercial system, with over 380 units of all types in operation or on order. Prospects are good for application in the U.K. in view of the ability to reduce emissions of SO_2 and NO_x (except for the shallow bed type of AFBC).

It can be expected that there will be a successful outcome to the work that is currently in progress in the U.S.A. and elsewhere to prove the viability of AFBC for large scale power and steam generating plant; to simplify and improve the reliability of coal feeding and solids recycle systems; further to reduce metal wastage, NO_x emissions, capital costs, and sorbent requirements for SO_2 removal; and to eliminate doubts on disposing of the residues.

Advanced concepts that have been the subject of recent research projects include the following:

- Multi-stage bubbling bed systems with more than two beds and solids interchange between them using underflow downcomers [604, 605]. The main combustion bed would be surmounted by sulphur capture beds and above them would be heat transfer beds. Spent solids cooling beds would be below the main combustion bed. The potential benefits claimed as compared with current plant, include a saving of nearly 20% in capital cost, 14–17% saving in the cost of steam, and lower NO_x emissions.

- A CFB design [606, 607] where the main heat transfer surface is below the coal and sorbent feed points. Recycle solids are fed to the heat transfer zone, together with sufficient fluidising air to produce a gently bubbling bed, most of the air being introduced higher up the shaft, producing a fast fluidised bed. The objective would appear to be to obtain the advantages of an external heat exchanger without having the complex ducting.

- Another CFB design [608] in which a draft tube is used to produce internal recirculation in the bed. Fuel, sorbent and air are fed to the base of the narrow spouted bed section, which feeds at the top into a bubbling bed containing the heat transfer surface. Air is fed to the base of this bed and further air is introduced into the freeboard to burn up any elutriated carbon. The more compact layout compared with conventional CFB arrangements is claimed to reduce plant costs. Staging of the air, as in other arrangements, reduces NO_x emission.

- The development of custom-made regenerable sorbents, e.g. cement, or alumina based materials [609, 610, 611] with the objective of minimising environmental disturbance resulting from the extraction of limestone and disposal of sulphated material. Beneficial means found for reducing sorbent requirements include: crushing of spent bed material [612]; water treatment of spent bed material [613]; pre-treatment of sorbents by controlled calcination [614]; and the use of salts [615] or iron oxide [616]

Developments that lead to a higher power generating efficiency also contribute to a reduction in atmospheric emissions. Pressurised fluidised bed combustion in combined cycle power plant is a leading contender for a significant increase in power generating efficiency, whilst on the smaller scale, combined cycle plant in which part of the power is generated by a gas turbine using air heated in a fluidised bed combustion system, may also be worthy of consideration. PFBC therefore has attractions for large utility installations, but as it has yet to be proven at this scale, it is unlikely to be available before the year 2000.

Factors likely to delay the commercialisation of PFBC include: restricted public sector research and development budgets; reluctance of plant purchasers to order untried technologies; and tight capital constraints in the industrial and utility sectors. The important technical issue that needs to be demonstrated is the ability to expand combustion gases through gas turbines without causing uneconomic rates of fouling, corrosion and erosion of the blades.

Thus the prospects are as yet uncertain. The system is good for reducing SO_2 and NO_x emissions. If demonstrations, such as that at Grimethorpe, show that the system can be practically applied to commercial-scale combined cycle power generation, then the prospects for adoption of PFBC in about fifteen years time are good.

6.9 Evaluation of Selected Types of Fluidised Bed Combustors

COMBUSTORS OPERATING AT ATMOSPHERIC PRESSURE

This section contains detailed evaluations for representative plants of each basic process type described in Section 6.2, with the exception of shallow bed systems, which are not ideal for sulphur capture by the use of added sorbent. Some of the plants described (in particular, those for Datum System 1) are inevitably only conceptual, since plants of this size have yet to be built and tested. Smaller boilers for utility power plant are now under construction, such as the 160 MWe demonstration boiler for TVA at their Shawnee station in Kentucky; the 125 MWe boiler conversion at Northern States Power Company's Black Dog station near Minneapolis, and the 110 MWe circulating bed fired boiler at Colorado Ute Electrical Association's Nucla station, which will, within the next few years, provide much needed technical and economic data. In the meantime, designs for 500 MWe units have to be regarded as unproven, and technical and cost estimates must be treated with caution. Such costings as are available relate to sites in the USA, burning American coals and using local limestones for sulphur control. Site-specific factors for such design studies could result in significant differences in costs compared with a UK site.

The main basis for estimating operating costs has been the results from units such as the TVA 20-MWe AFBC unit at the Shawnee Power Station in Paduca, Kentucky, together with data from smaller experimental rigs.

There has been more experience in the construction and operation of small commercial AFBC plants, including CFB systems, but little has been published on the costs of these.

Evaluation of Large Bubbling Bed AFBC

The detailed description that follows is based on a conceptual design study for a 500 MWe bubbling bed boiler carried out by Bechtel for EPRI [617]. The technical data on which the design was based was obtained from the 20 MWe bubbling bed boiler operated by TVA at their Shawnee station, together with additional input from a 1.8m square experimental combustor at the Babcock & Wilcox Company's Alliance (Ohio) Research Centre.

Process Description

A flow diagram of the process is shown in Figure 6.1. The notional 500 MWe unit boiler comprises a single-level fluidised bed, 1.5m deep when fluidised, in which is submerged water and steam superheating tubes. The bed is divided into 24 compartments, each with several coal and limestone feed points, immersed in the bed and feeding solids by pneumatic transport. The coal and sorbent are crushed to a top size of 6 mm (coal) and 3 mm (limestone). Fuel oil fired burners feed hot combustion gases into the wind-box of each bed compartment for light-up.

The gases leaving the freeboard zone above the bed enter convection passes containing reheat tubes and additional evaporation and superheat tubing, and then pass through an Economiser to a multicyclone Grit Collector which separates relatively coarse particulate material from the gas stream. This grit is partly or wholly recycled to the combustion bed, the remainder being discharged. The gases subsequently pass through a combustion Air Heater and fabric filters (Baghouse) to an Induced Draught Fan and the Chimney.

Drains in the bed floor provide for bed removal in order to control bed level. Solid residues (quenched bed discharge, drop-out from the economiser, discharge from the multicyclone and bag filter dust) are water-conditioned and transported to a tipping area or to a commercial outlet.

The combustion efficiency is estimated to be 99%, giving a boiler efficiency of 87%. The sorbent is limestone of reasonably good reactivity, added at a calcium: sulphur mole ratio of 2.5:1, and effecting retention of 90% of the sulphur from the coal with a calcium utilisation of 36%. Assuming staged combustion is not used, NO_x emissions (as NO_2) will average 600 mg/Nm^3 (measured at 25% excess air). This is equivalent to 430 mg/Nm^3 at 6% oxygen in dry flue gas. Expressed as a volume fraction, this is 292 vppm, and as weight of NO_2 per kg of coal fired it is 4.4 g. If staged introduction of combustion air were to be used, NO_x emission might be reduced to about half of this.

Particulate emissions, where a baghouse is used for final gas clean-up, would be about 50 mg/Nm^3, or 0.37 g per kg of coal fired. An electrostatic precipitator may be used instead of a baghouse.

The plant envisaged is described in more detail in Reference [617].

Operating Experience

Experience of in-bed feeding of crushed coal and limestone in large AFBC plant is limited. Work has been carried out by Babcock & Wilcox Ltd (UK) at Renfrew beginning in 1975 in a 3.05 m square bed firing a converted water-tube boiler generating 21 tonne steam/h [617]. In Japan, a 20 tonne/h pilot plant has been in operation since 1981 [619]; and in the USA, a 20 MWe unit was commissioned by TVA in 1982 [620] and, together with a 2 MWt, 2 m square test rig operated by EPRI at the Babcock & Wilcox (USA) Alliance Research Centre [621] have given valuable design data for larger scale plant.

The work to date has provided information on the design of coal, sorbent and recycle feeding systems, on the spacing between coal feed points that should not be exceeded to avoid loss of combustion efficiency, on ways of minimising corrosion and erosion, light-up procedures, bed height control, and techniques for load following and general boiler automatic control. Overbed feeding has also been studied on many of these plants. Although much simpler and more reliable in operation, combustion and sulphur retention efficiencies are significantly lower than with underbed feeding, especially when the coal feed contains a high proportion of fines.

In over 5,600 hours of operation, the TVA 20 MWe boiler has operated well, with no major technological problems. It has been free from corrosion and has only experienced a small amount of tube erosion [622].

There is intrinsically no need for major items of stand-by equipment. There are, however, items which are expected to need fairly frequent maintenance, and the provision of stand-by plant could reduce the risk of boiler outage while repairs are effected. This applies particularly to components of the pneumatic solids feeding systems where abrasion can result in the need for more frequent replacements than in conventional firing equipment. The need for spares for equipment downstream of the boiler is unlikely to differ from that for conventional plant without acid emission control equipment.

Variations and Development Potential

Most manufacturers who offer AFBC plant are prepared to offer variations of the basic process type to suit customers' requirements and the composition and size distribution of fuel available. The design may also depend on the size of the boiler and on the form of the fluctuations in load.

Design variations include the method adopted for coal and sorbent feeding, the presence or absence of recycle facilities, techniques for light-up and load following, arrangement of the heat transfer surface in the bed and convection zones, the split of air between fluidising air and secondary air admitted to the freeboard, and the method adopted for final gas clean-up.

By introducing approximately 20–25% of the combustion air into the freeboard the NO_x emissions may be roughly halved, but care must be taken in design of the secondary air injection system to ensure thorough mixing with the gases leaving the bed, otherwise the combustion efficiency will suffer unacceptably. The use of staged combustion air also requires redesign of the heat transfer surfaces, because of the greater proportion of the heat release occurring above the bed when using staged combustion.

Development potential at present lies mainly in the field of engineering design, to develop reliable solids feeding equipment ensuring an equal flow of fuel to a large number of feed nozzles, to minimise tube and wall erosion, to develop effective automatic control systems and reliable sensors needed to provide input for them, and to select the most satisfactory equipment for handling and conditioning the solid wastes.

Because of the fundamental differences between the design of an AFBC power generating boiler and a PF-fired boiler, retrofitting is not, in general, an attractive proposition, although for economy, some of the development work has been done on existing boilers that have been modified to use fluidised bed combustion. Replacement of an existing PF-fired boiler at a power station is however worth considering as much of the ancillary equipment would be re-usable.

Process Requirements and Effluents

Table 6.7 shows estimated performance data for a 500 MWe plant including flue gas flowrate and composition at the stack, reagent quality and input rate, all at maximum load. Also shown are the annual requirements for fuel, sorbent, fuel oil, water and operating manpower, based on a load factor of

Table 6.7 Process Data for 500 MWe FBC Operating at MCR based on reference [617]

Coal feed rate, tonne/h	212.0
Gross heat input, MWt	1449
Steam cycle heat rate, kJ/kWh	8352
Boiler efficiency, %	88
Gross heat rate, kJ/kWh	9491
Gross efficiency factor, GJe/tonne	9.34
Turbine generator gross output, MWe	549.2
Auxiliary power requirements, MWe	36.4
Turbine generator net output, MWe	512.8
Net heat rate, kJ/kWh	10,177
Net efficiency factor, GJe/tonne	8.71
Dry flue gas flowrate Nm^3/s	444
Composition of dry flue gas:	
CO_2 % vol.	14.8
N_2 % vol.	80.8
O_2 % vol.	4.4
CO vppm	500
SO_2 vppm	150
NO_x vppm	290
Limestone feed rate, tonne/h	26.50
Limestone analysis:	
$CaCO_3$, % wt	95
$SiO_2 + Al_2O_3$, % wt	3
Moisture, % wt	2
Particle top size mm	3
Particle median size mm	1.8
(ii) Annual inputs, 500 MWe unit, 65% load factor	
Coal, thousand tonne	1,207
Limestone, thousand tonne	158.9
Water, thousand tonne	6,400
Fuel oil (for start-up), GJt	66,500
Fuel oil (for coal drying), GJt	316,500

65%. A breakdown of auxiliary power requirements is shown in Table 6.8.

The steam cycle is condensing, and the design is for an inland site, employing cooling towers. A large part of the water usage is therefore accounted for by evaporation in the cooling tower. Water consumption would be considerably less in a back-pressure turbine system, e.g. for use in combined heat and power (CHP) plant.

Coal drying has been found necessary in the TVA 20 MWe plant when using in-bed feeding, to avoid plugging of the coal feed lines. A fuel-oil fired dryer is used.

The rates of production and chemical composition of solid residues, together with the amount of liquid effluents per 500 MWe generating unit with a 65% annual load factor, are shown in Table 6.9. Upon contact with water, the CaO will be rapidly hydrated to $Ca(OH)_2$, with a 32% increase in mass, and the $CaSO_4$ to $CaSO_4.2H_2O$, with a 26% increase in mass.

Table 6.8 Comparison of Auxiliary Power Requirements, AFBC and PF with FGD

	Power per 500 MWe unit, kW	
	AFBC	*PF + FGD*
Coal and limestone handling (1)	2871	1482
Coal milling	0	1688
Recycle feed system (1)	58	0
Ash cooling and handling	1891	656
Forced draught fans (2)	14201	3333
Induced draught fans	5777	7339
Boiler circulation pumps	995	0
Cooling tower fans and pumps	5830	5830
Condensate pumps	890	890
Regenerative air heater drive	43	29
Baghouse reverse air fans	448	827
Flue gas desulphurisation	0	4019
FGD sludge handling	0	919
Miscellaneous	3400	3400
Total	36404	30412

Notes:

(1) Excluding conveying air

(2) Includes fluidising and secondary air; coal, limestone and recycle transport air fans for AFBC. Includes combustion and coal transport air fans for PF

For AFBC the coal is fed 'under bed'

For PF the FGD system is limestone slurry scrubbing

Table 6.9 Solid and Liquid Effluents from 500 MWe unit operating at 65% load factor

Production of solid wastes, tonne/year dry weight	345,800
Composition of dry solid wastes, % wt	
Ash	41.6
Carbon	3.1
Anhydrous $CaSO_4$	29.8
$CaCO_3$	3.9
CaO	19.7
Impurities from limestone	1.9

Table 6.10 Efficiency and Emission Factors for Large AFBC Operating at MCR

System No.	1	2
Gross heat input, MWt	1449	460
Coal fired, tonne/h	212	59.1
Efficiency factor, GJ/tonne	8.60	24.1
Ashes output, tonne/h	60.7	11.2
Carbon content of ashes, %	2.1	3.2
Calcium content of ashes, %	17.5	24.8
Sulphate content of ashes, %	15.1	20.6
Elements in acidic emissions:		
Sulphur, kg/h	339	88.7
Nitrogen (a) kg/h	292	81.6
Chlorine (b) kg/h	530	153.7
Emission factors kg/tonne coal		
Sulphur	1.60	1.51
Nitrogen (a)	1.32	1.38
Chlorine (b)	2.50	2.60

Notes

System 1 : One 500 MWe unit

System 2 : One 450 tonne steam/h boiler

(a) These values may be approximately halved if a staged combustion design is used.

(b) Maximum possible values, assuming no retention of chloride in cyclones and bag filters or electrostatic precipitators.

Table 6.11 Order of Magnitude Capital Costs 4 × 500 MWe Power Plant, $ million (EOY1982) (converted to £ sterling, DEC '86)

Based on information in [617]

Item	Cost £M '86 FBC	Cost £M '86 PF + FGD
Civil/structural/architectural	160	160
Steam generators, air heaters and feed water systems	325	245
Turbine generators, condensers and mechanical ancillaries	160	160
Coal and limestone reception, storage, handling & preparation	50	25
Ash handling	25	20
Baghouse, ID fans and stack	65	65
Piping	105	105
Controls/instrumentation	20	20
Electrical equipment, switchyard	80	75
Distributables	80	80
FGD plant	-	160
Total field cost	1070	1115
Engineering services, fees	90	90
Contingencies	210	180
Total plant cost	1370	1385
Advance for funds during construction plus escalation	105	105
Total plant investment	1475	1490
Total investment, £/kW	737	745

The liquid effluents arise from cooling tower and sluice-water blowdown, area rainfall run-off, waste water from plant cleaning operations, and sewage. The sewage disposal will be by methods directed by the relevant Water Authority; the other streams will be treated on-site for removal of oil and suspended solids, pH adjustment and dechlorination, if necessary, before disposal to a surface watercourse.

Table 6.12 Annual Operating and Maintenance Costs 4 × 500 MWe Power Plant; $ million (EOY 1982) (converted to £ sterling DEC '86)

Based on information in [617]

	Cost*	
	£M '86 AFBC	£M '86 PF + FGD
Operating labour (a)	7.74	7.49
Maintenance labour and materials:		
Steam Generator	13.85	4.84
Coal Feed	1.83	0.92
FGD	–	14.90
Rest of plant	9.07	8.75
Total maintenance	24.75	29.42
Administrative charge (b)	5.29	5.77
Total O & M Costs	37.78	42.68
Consumables costs:		
Coal (c)	193.14	190.05
Fuel oil (start-up) (d)	1.54	1.54
Fuel oil (coal dryer) (e)	7.28	–
Limestone (f)	7.04	2.66
General chemicals (allowance)	1.62	1.62
Water (g)	3.30	3.71
Solid waste disposal (h)	6.80	4.08
Crude gypsum disposal (i)	–	–
Total consumables costs	220.72	203.66
Total fixed + variable annual running costs	258.50	246.34
Costs, pence/kWh (j)	2.21	2.14

Notes

(a) 522,000 Man-hours for AFBC, 505,440 Man-hours for PF+FGD charged at £14.82 per man-hour
(b) 30% of labour cost (maintenance labour 40% of total maintenance cost)
(c) 4,828,500 tonnes for AFBC, 4,751,300 tonnes for PF; cost £40/tonne
(d) 266,000 GJ for both AFBC and PF; cost £5.75/GJ
(e) 1,266,000 GJ for AFBC, nil for PF; cost £5.75/GJ
(f) 620,800 tonnes for AFBC, 234,300 tonnes for PF + FGD; cost £4.86/tonne
(g) 25,600 tonnes for AFBC, 28,600 tonnes for PF + FGD; cost £129.58/tonne
(h) 1,400,000 tonnes for AFBC, 840,000 tonnes for PF; cost £4.86/tonne
(i) As $CaSO_4.2H_2O$ with 10% free moisture, 101,700 tonnes; cost £6.48/tonne
(j) At 65% annual load factor, exclusive of charges on capital

*Costs as originally given were in end-of-year 1982 U.S. Dollars With the exception of coal price, conversion to pounds sterling, December 1986, is on following basis:
$1.6175 = £ Dec. 1982
Inflation rate assumed:
1.31 for period December 1982 to December 1986

Emission and Efficiency Factors

Table 6.10 presents emission and efficiency factors for AFBC as applied to Systems 1 and 2.

Costs

Capital costs for the plant described above are shown in Table 6.11, along with costs for a PF fired plant of equal output firing the same coal and equipped with limestone slurry flue gas desulphurisation. The costs are order-of-magnitude estimated costs in end-of-year 1982 US dollars, and exclude cost of land, working stocks of fuel and sorbent and other pre-production costs.

The land area needed for the 4 × 500 MWe AFBC boiler and turbine buildings, complete with coal bays is approximately 27,000 m^2 and the total building volume is in the region of one million cubic metres. This compares with 17,000 m^2 area and 950,000 m^3 volume for the PF-fired plant, excluding the FGD system. This would be expected to occupy a further 10,000 m^2 for limestone slurry scrubbing. On top of this there would, for both plants, be baghouses of total plan area over 10,000 m^2.

The area needed for tipping of solid wastes from AFBC would be about 25% greater than that for pulverised coal firing with production of saleable gypsum, or 10% greater than for pulverised coal firing with tipping of crude gypsum from FGD.

The capital costs for AFBC are very similar to those for a PF-fired plant equipped with limestone slurry scrubbing for sulphur oxides removal.

Contingency allowances for AFBC are higher than those for PF because they include a process contingency charge (appropriate to a technology not yet operating on that scale) in addition to the normal project contingency. Running costs, both fixed and variable, but exclusive of annualised capital cost, are shown in Table 6.12. If overbed feeding were to be used, coal drying would be unnecessary, and coal and limestone preparation, handling and feeding costs would be lower. As pointed out earlier this would be at the cost of reduced sulphur retention efficiency and reduced combustion efficiency.

In the PF plant with FGD it is assumed, for purposes of costing, that no market exists for the gypsum produced. If there were a market, the cost of crude gypsum disposal would be replaced by a credit for sales, but this benefit would be reduced by the additional costs of treatment to give a product that meets the sales specification (see Section 2.3).

Even without gypsum sales, electricity generation by PF is slightly cheaper than by AFBC. This is mainly due to the slightly lower boiler efficiency assumed for the AFBC plant (88.0% compared with 88.3% for the PF plant) and the higher auxiliary power (36.3 MWe per 500 MWe unit compared with 30.5 MWe for the PF plant) in circumstances where the cost of fuel is a high proportion of the cost of power generation. There are a number of uncertainties in the data, hence the outcome of the comparison should not be taken as being final.

Process Advantages and Drawbacks

The main advantages of AFBC, *vis a vis* PF with limestone slurry FGD are as follows:

(a) Combustion and emissions control are carried out in a single unit, as compared with two or more interacting units in PF firing. This simplifies both plant and operation.

(b) Coal and sorbent preparation costs are lower, because a larger sized particle feed is required for FBC.

(c) The use of wet scrubbing is avoided, resulting in energy savings (no need for stack gas reheat) and freedom from cool-side corrosion.

(d) A wide range of coal types and quality can be burned without serious complications.

Disadvantages are the following:

(a) For the highest combustion and sulphur retention efficiencies, underbed feeding is necessary. This has the disadvantage, for large plants, of requiring a very large number of coal feed points (one for every two square metres of bed).

(b) Satisfactory operation of coal, limestone and recycle feeding depends on careful attention to details of design. Reliable solutions may add significantly to the cost of the boilers.

(c) Considerable skill in plant design is needed in order to avoid metal wastage due to erosion.

(d) The solid residues are strongly alkaline, and care must be exercised in their handling and dumping.

Evaluation of Small Bubbling Bed AFBC

Boilers in the 13 tonne/h size range (Datum System 3) may be either water tube or fire tube but, particularly in the UK, fire tube designs predominate, and these are normally available as packaged units.

Currently oil and gas firing predominate in this market, and for coal firing to be generally attractive, the equipment used must be reasonably competitive as regards space requirements, controllability, and amenity considerations without incurring undue penalty as regards capital costs.

Fluidised bed boilers with shallow beds burning cleaned coal with low fines content can be expected to meet most of these requirements. In the event of there being strict limitations on sulphur emissions, deep beds would be needed.

Space and cost considerations will however usually limit the application of features (under-bed feeding, tall hot freeboard, and high fines recycle) in small FBCs that enable the optimum combustion and sulphur retention performance to be obtained in large units. In Tables 6.13 and 6.14 comparisons are made between the performances of four relatively small deep bed FBCs, and the TVA test boiler used as a basis for the design of large units. Small plants will in general have thermal efficiencies 2 to 3 percentage points lower than large plants, and require higher Ca:S ratios to attain the same sulphur retention.

The plant selected for evaluation is a firetube boiler, supplied by the Stone Johnston Corporation [623]. About 30 boilers of this design are in operation, 23 of which burn coal only, five can burn coal, gas or oil, and two burn other fuels. The total installed steam generating capacity of the Johnston boilers is around 400 tonne/h.

Process Description

The boiler is a horizontal firetube boiler with the fluidised combustion bed contained in a downward extension of the steam/water pressure shell. Crushed coal and limestone are fed by a screw-feeder discharging beneath the surface of the bed, which is tapered towards the bottom.

The bed is cooled by natural circulation of water through inclined tubes passing through the bed and fixed into the steam/water pressure shell at each end. Ash and spent sorbent are discharged through drain pipes at the base of the bed, and light-up is by the use of gas or oil burners passing upward through the plenum chamber and firing into the lower part of the bed.

The boiler shape necessitates a low freeboard, and carryover of bed material into the smoke tubes is reduced by positioning a baffle beneath the entrance to the smoke tubes.

The gases leaving the boiler pass first through a separated mechanical grit collector, the grits separated from the gas stream being optionally recycled to the bed in a stream of conveying air, and then through a baghouse or electrostatic precipitator, an induced draught fan and to the chimney. A damper at the I.D. fan outlet is used to control the firebox pressure at approximately −150 Pa gauge (−15 mm of water).

Operating Experience

The total operating hours and comments on problems experienced are shown in Table 6.14 for the Johnston boiler and for the other boilers for which performance data were given in Table 6.13.

Most of the Johnston boilers are able to meet sulphur dioxide emission limits by burning compliance fuels, but the ability to use limestone sorbent effectively on this type of boiler has however been demonstrated [623]. It should be noted that sulphur retention performance is strongly dependant upon the reactivity of the sorbent, and unless the same sorbent

Table 6.13 AFBC Bubbling Bed Operating Results

		US NAVY Grt. Lakes	Central Soya, Marion, Ohio	Canadian Dept. of Defence	Georgetown University	TVA Test Boiler
Type of Boiler		Package W/T	Package F/T	Package W/T	Package W/T	Package W/T
Manufacturer		Comb. Engng	Stone/Johnstone	Foster Wheeler	Foster Wheeler	B & Wilcox
Rating, tonne/h steam		23	18	2×18	45	77 (equiv.)
Method of coal feeding		Pneumatic	Screw-feed	Spreader	Spreader	Pneu. or Spreader
Position of coal feed		In-bed	In-bed	Overbed	Overbed	In/Overbed
Bed depth (fluidised), m		0.75–0.9	0.9	1.2	1.35	1.0–1.2
Boiler efficiency, %		(80 design)	82	(81.6 design)	77 (a)	–
Combustion efficiency, % At Recycle to coal feed ratio	0	90	96	92–93.5	–	92 (b)
	0.3	94	–	–	–	
	1.0	96	–	–	–	96 (b)
	2.5	–	–	–	–	97.5 (c)
Sulphur retention, % At Ca:S mole ratio	2.0	76–85 (d)	70–75	–	–	79
	2.5	–	–	71	–	92 (e)
	3.0	85–91 (d)	–	81	–	96 (e)
	3.6	–	–	–	86	–
	4.0	91–85 (d)	–	–	–	–
NO_x emission, mg (as NO_2) per Nm^3 dry flue gas at 6% O_2 At Excess Air	20%	500	–	–	–	260
	25%	–	760	–	–	330
	30%	–	855	–	–	410
	75%	–	–	–	360 (f)	–
	110%	–	–	715 (g)	–	–

Notes
(a) Efficiency 80% when recycling grits
(b) With underbed feeding; about four percentage points lower for overbed feeding
(c) With underbed feeding; about three percentage points lower for overbed feeding
(d) Range of values as recycle ration is increased from zero to 1.2
(e) At a recycle ratio of 2.0
(f) Excess air reduced to between 25% and 50% in later operation
(g) High excess air levels were in use during test period to assist in bed temperature control

WT = Water Tube
FT = Fire Tube

Table 6.14 AFBC Bubbling Bed Operating Experience

	US Navy Gt.Lakes	Central Soya	Can. Dept of Defence	Georgetown University	TVA Test Boiler
Date Commissioned	June 1981	April 1979	December 1982	July 1979	May 1982
Problems encountered	Bed tube + wall erosion (a)	Clinkering, Tube erosion (controlled by use of fins & metal spray)	High excess air levels necessary for comb. efficiency	Wall tube erosion. Start-up difficult. Blinding of filter bags.	Some clinkering with overbed feeding. Erosion at welds on wall tubing.
Hours of Operation	5,300 (b)	25,000	3,000	19,000	10,000 (b)

Notes: (a) Worst at vertical surfaces, otherwise low
(b) Up to June 1985

is being used in each boiler under review the results are not strictly comparable.

Combustion efficiencies for the better boilers are comparable with those obtained in stoker-fired boilers and, as mentioned earlier, higher efficiencies can be obtained in the FBC systems by grit refiring.

Variations and Development Potential

Most manufacturers who offer AFBC plant are prepared to offer variations of the basic process type to suit customers' requirements and the composition and size distribution of fuel available. The design may also depend on the size of the boiler and on the form of the fluctuations in load. There are a wide range of designs that can be applied on the 13 tonne/h scale of operation.

Design variations include the method adopted for coal and sorbent feeding, techniques for light-up and load following, arrangement of the heat transfer surface in the bed and convection zones, the split of air between fluidising air and secondary air admitted to the freeboard, and the means embodied to minimise tube erosion.

Development potential at present lies mainly in the field of engineering design, to develop reliable solids feeding equipment capable of dealing with a wide range of fuel sizes and moisture contents, to minimise tube and wall erosion, to develop effective automatic control systems and reliable sensors needed to provide input for them, and to select the most satisfactory equipment for handling and conditioning the solid wastes.

Process Requirements and Effluents

Table 6.15 shows performance data, flue gas flowrate and composition at the boiler outlet, reagent quality and input rate (all at MCR), together with the annual requirements for fuel, sorbent and operating manpower based on a load factor of 65%. The coal fired is the second coal in Table 1.2.

A sulphur removal efficiency of 80% is obtained by adding limestone sorbent at a Ca:S mole ratio of 2.3:1 using a moderate degree of solids recycle. Combustion efficiency is 96% with 25% excess air, giving a boiler efficiency of 82%.

The NO_2 concentration of 430 vppm is equivalent to 400 vppm or 820 mg/Nm^3 (as NO_2) at 6% oxygen in dry flue gas.

The annual production and the chemical composition of solid wastes from a 13 tonnes steam an hour boiler at a load factor of 65% are shown in Table 6.16. As

Table 6.15 Process Data for Small Bubbling Bed AFBC 13 tonne/h Steam Boiler at MCR

Coal Feed rate, tonne/h	1.68
Gross heat input MWt	13.1
Combustion efficiency, %	96.0
Boiler efficiency, %	82.0
Efficiency factor, GJ/tonne	21.0
Electric power useage	Not known
Dry flue gas flow rate, Nm^3/s	4.02
Composition of dry flue gas	
CO_2, % vol.	14.1
N_2, % vol.	81.0
O_2, % vol.	4.9
CO, vppm	500
SO_2, vppm	244
NO_x, vppm	430
Limestone feed rate, kg/h	190
Limestone analysis:	
$CaCO_3$, % wt.	95
SiO_2 + Al_2O_3, % wt.	3
Moisture, % wt.	2
Particle top size, mm	4
Particle median size, mm	2
Annual Inputs at 65% Load Factor	
Coal, tonne	9,566
Limestone, tonne	1,085
Auxiliary fuel, GJt	Not known
Manpower for operation (a), Man-hours	10,300
Water for ash conditioning, tonne	234

Note: (a) Based on one operator per shift, 4 shifts, plus one man on days; 6 day week

Table 6.16 Solid Residues from 13 tonne/h Steam Boiler at 65% Load Factor

Production of solid wastes, tonne/annum dry weight:	1948
Production of Conditioned wastes: tonne/annum	2620
Composition of dry wastes, % wt.:	
Ash:	39.3
Carbon	13.3
Anhydrous $CaSO_4$:	25.0
CaO:	17.3
$CaCO_3$:	3.4
Impurities from limestone:	1.7

Table 6.17 Efficiency and Emission Factors (at MCR)

Gross heat input, MWt	13.1
Coal fired, tonne/h	1.68
Efficiency factor, GJ/tonne	21.0
Ashes output, tonne/h	0.34
Carbon content of ashes, %	13.3
Calcium content of ashes, %	21.1
Sulphate content of ashes, %	17.7
Elements in acidic emissions:	
Sulphur, kg/h	5.0
Nitrogen kg/h	3.9
Chlorine kg/h	4.4
Emission factors kg/tonne coal	
Sulphur	3.0
Nitrogen	2.3
Chlorine	2.6

mentioned previously, the calcium sulphate and oxide will be hydrated on contact with water, and the mass will absorb further water to give a moist, easily-handled consistency.

Efficiency and Emission Factors

Data for the system described are shown in Table 6.17. The figures for chlorine emissions are calculated on the assumption that there is no retention of chloride anywhere in the system. Confirmatory data is lacking.

Costs

Firm costs for the smaller AFBC plant are difficult to obtain, but an IEA Coal Research Study [588] reviews the information available.

For water-tube boilers in the output range 100–140 MWt (150–210 tonne/h steam), capital costs for AFBC firing appear to be a little lower than for PF firing without flue gas desulphurisation or measures for NO_x control, indicating that if used for emission control, fluidised-bed firing would have a decided advantage in capital investment in this size range and also down to about 50 MWt (the lowest size generally accepted as suitable for PF firing).

For packaged firetube boilers using deep enough fluidised beds to permit efficient sulphur capture, capital costs in the size range 5–15 MWt (7.5 to 22.5 tonne/h steam) are reported to be almost double those for underfeed or chain grate stoker firing. Shallow-bed fired boilers are considerably cheaper, being comparable in costs to chain grate fired boilers.

The IEA report also compares estimates of non-fuel operating costs. A survey of manufacturers and users of FBC boilers showed that the labour requirements for FBC units are little different from those for other types of solid fuel firing.

Electrical power requirements are about three times as great as for stoker firing. The difference is principally accounted for by the high fan power required for FBC.

The cost of sorbent would be greater than if wet FGD were to be used, since the calcium utilisation is only about half as much. The cost of crushing will, however, be less because the particle size required for bubbling bed FBC is much larger than for wet FGD processes. If the comparison were with the spray dry FGD process, the limestone required for FBC would be considerably less. It may be open to doubt as to whether the operators of 13 tonne steam/h boilers would find it attractive to install new stoker fired boilers and FGD plant.

Ash disposal costs are increased by the addition of sulphur sorbent, and potential uses for the ash may be affected. Care may have to be taken in the tipping of residues to avoid environmental harm and this also may increase disposal costs.

Allowing for capital allocation on a two year simple pay-back period, total non-fuel operating costs for a deep bubbling bed fired low-pressure steam boiler of c.22 tonne steam output/h are roughly 50–55% higher than for a chain grate stoker fired boiler of equal rating and without FGD. If the stoker-fired boiler is equipped with limestone slurry scrubbing for SO_2 control, the non-fuel operating costs of the FBC fired boiler are 10–15% lower.

For larger (100 MWt) water tube boilers, the deep bubbling bed mode of firing gives a 0–5% higher non-fuel operating cost than a PF fired boiler of the same rating without FGD, and a 15–20% lower non-fuel operating cost than a similar boiler equipped with limestone slurry FGD.

Boiler thermal efficiencies for the two plants are equal, therefore fuel costs when burning the same quality of fuel will be the same. The FBC system has however greater flexibility as regards fuel quality.

Process Advantages and Disadvantages

Compared with stoker firing and FGD the main advantages of FBC are as follows:

- Combustion, SO_2 control and NO_x control are carried out in a single unit, as compared with two or more interacting units. This simplifies both plant and operation.

- Freedom from cool-side corrosion compared with using wet FGD, and better plume buoyancy.

- A wide range of coal types and quality can be burned without serious complications.

- Simpler and probably cheaper to dispose of residues than from stoker firing using FGD.

- Less ground space needed than for a stoker fired boiler with scrubbers.

The disadvantages are as follows:

- The solid residues are strongly alkaline, and care must be exercised in their handling and dumping.

- Considerable skill in plant design is needed in order to avoid metal wastage due to erosion.

Evaluation of Dual Bed AFBC

As presently marketed, this process is applicable to datum combustion system 3, and to plants up to about seven times the output capacity of this. Vendors offering dual bed AFBC systems are shown in Table 6.1.

In basic design, the process is similar to the deep bubbling bed AFBC, the principal differences being in the design of the actual fluidised bed unit. The two bubbling beds are positioned one over the other; the lower, combustion bed containing sand as bed material and operating at about 900 deg. C; the upper, sulphur-capture and carbon burn-out bed consisting of limestone and operating at about 845 deg. C. A third bed, beneath the combustion bed, is used for sand storage, the bed material being transferred pneumatically between this and the combustion bed, as required.

Process Description

The process here described is that offered by Wormser Engineering, Inc. [624]. A flow diagram of the process is shown in Figure 6.3.

The 13 tonne steam per hour boiler uses rectangular, non-compartmented beds, each 6.5 m^2 in area. Each bed is less than 0.3 m deep when unfluidised and the fluidising velocity in the upper bed is about 2.2 m/s.

Coal is dried and crushed, then fed pneumatically via a stream splitter to eight in-bed feed points in the combustion bed. Crushed limestone is fed to the upper bed and spent sorbent is removed by an overflow drain positioned at the bed's surface. Excess air is typically 22%, the fluidising air being 1.07 of stoichiometric, and a further 0.15 of stoichiometric is injected into the space above the combustion bed.

Turn-down to 31% of full load is achieved by varying the fluidising velocity–the bed cooling surface is so arranged that at low velocities much of the tubing is above the bed surface. Turn-down to lower levels is obtained by intermittant slumping of the bed.

The combustion gases leaving the furnace chamber pass through a boiler/economiser unit, then to a baghouse and I.D fan to the stack. The overall height of the boiler is 4.3 m.

Operating Experience

Four pilot plants and eight full-sized dual bed AFBCs were in operation in July 1985 (under licence from Wormser Engineering, Inc.). Two of these are in Sweden, one in Japan and the rest in the U.S.A. Their capacities range from 13.6 to 68 tonnes steam per hour. The most fully tested, a 32 tonnes/h 650 psig saturated steam boiler at Amarillo in Iowa, has been in use since February 1983.

Problems were encountered initially in feeding wet coal, due to water evaporated in the dryer condensing downstream and leading to blockages. These difficulties have been overcome by modifications to the dense-phase coal feed system, including preheating of the conveying pipe.

The plant otherwise operated satisfactorily until routine shut-down in autumn of 1984, when examination of the tube bank in the combustor showed that there had been no erosion. Some tubes that had been damaged due to overheating during the start-up phase when the water flowrate was too low were replaced. There was also some evidence of water-side corrosion, and this necessitated improvement of the water treatment system.

The FBC availability in the first 21 months of operation was 95%, most of the down-time being attributed to failure of thermocouples, the coal rotary valve and wind-box seal, together with overheating of the control panel [625].

Variations and Development Potential

The ASEA group have also developed a dual bubbling bed combustor, available in a range of sizes from 10 MWth to 100 MWth. Advantages claimed for this type of system are compactness and suitability for modular fabrication, saving erection time and costs on site, easy operation with a high degree of automatic control, and low SO_2 and NO_x emissions.

Descriptions of the ASEA system show that there are only two beds, material being transferred between the combustion and sulphur control beds as required to regulate heat transfer by altering the tube surface area immersed in the combustion bed. Both beds therefore contain limestone. Like the Wormser combustor, a sophisticated automatic control system carries out the operations required for combustion control and bed material transfer. Ash recycle is optional.

ASEA are understood now to have a marketing arrangement with Wormser in regard to the installation of plants in different areas and, no doubt, their technologies will tend to converge.

Process Requirements

Table 6.18 shows performance data, flue gas flowrate and composition of dry flue gas at the stack, reagent quantity and quality, all at maximum load; also annual requirements for fuel, sorbent, water, auxiliary fuel, and operating manpower, based on a load factor of 65% and one shut-down per week.

Table 6.18 Dual Bed AFBC Process Data 13 tonne/h steam boiler at MCR

(a)	*Inputs per hour*	
	Coal feed rate, tonne/h	1.43
	Gross heat input, MWt	11.1
	Combustion efficiency, %	97
	Boiler efficiency, %	84
	Efficiency factor	23.4
	GJ/tonne	
	Electric power usage, kWe	160
	Dry flue gas flow rate, Nm^3/s	3.34
	Composition of dry flue gas;	
	CO_2, % vol.	14.5
	N2, % vol.	81.1
	O_2, % vol.	4.4
	CO, vppm	500
	SO_2, vppm	175
	NO_x, vppm	300
	Limestone feed rate. kg/h	176
	Limestone analysis:	
	$CaCO_3$, % wt.	95
	$SiO_2 + Al_2O_3$, % wt.	3
	Moisture, % wt.	2
	Particle top size, mm	3–4
	Particle median size, mm	2
(b)	*Inputs per year*	
	Coal, tonne	8,142
	Limestone, tonne	1,004
	Water, tonne	Variable
	Auxiliary fuel for light-up, GJt	660
	Auxiliary fuel for coal drying, GJt	2,275
	Manpower for operation, Man-hours	10,000

The rate of limestone addition represents a Ca/S mole ratio of 2.5 and effects an 86% sulphur retention efficiency. The NO_x concentration of 300 vppm is equivalent to 615 mg/Nm^3 (as NO_2) at 22% excess air, or 550 mg/Nm^3 at 6% oxygen in dry flue gas.

Expressed as mass of NO_2 per unit of heat input it is 194 ng/J. Some improvement on this may be obtained by optimising the proportion of secondary air. Emissions as low as 100 ng/J have been reported, using staged combustion in dual bed AFBC [626], but details of how this was achieved are not available.

The annual production and the chemical composition of solid wastes and the amount of liquid effluent from a 13 tonnes steam per hour dual bed AFBC boiler, operating at a load factor of 65%, are shown in Table 6.19.

On contact with water, the CaO will be rapidly hydrated to $Ca(OH)_2$ with a 32% increase in mass, and the $CaSO_3$ will slowly hydrate to gypsum, $CaSO_4.2H_2O$, with a 26% increase in the mass. Additional water will be absorbed to give a moist, crumbly consistency required for easy handling without risk of producing airborne particulate matter.

There may be no liquid effluent as the principal source of effluent (boiler blow-down), can be used for ash conditioning. Assuming the water required for conditioning is 20% of the mass of residues, over and above that required for hydration, the annual water requirement for solid wastes disposal is about 600 tonnes. This is about 0.8% of the total steam production.

The quantity of blowdown produced will depend on the feed rate of fresh boiler water (which will be less than the evaporation rate if some of the steam condensate is recycled) and on the quantity of dissolved solids in the feedwater.

Other liquid effluents will arise from site drainage, plant cleaning, and sewage. These may require treatment for oil and suspended solids removal, after which they can be discharged in a manner acceptable to the Water Authority. In many cases, the boiler plant will only generate a small fraction of the total liquid effluent from the factory, and its treatment will be handled together with the bulk of the factory effluent.

Table 6.19 Solid and Liquid Effluents

Operation at Annual Load Factor of 65%.	
Production of solid wastes, tonne/annum dry weight	1,693
Composition of dry solid wastes, % wt.	
Ash	38.5
Carbon	11.0
Anhyd.$CaSO_4$	26.4
CaO	18.6
$CaCO_3$	3.7
Impurities from limestone	1.8
Production of liquid effluent, tonne/annum (Excluding site run-off, wash-water, sewage, etc.)	0 to 3000

Efficiency and Emission Factors

Table 6.20 shows efficiency and emission factors for dual bed AFBC as applied to datum system 3.

The calculations are for a 13t/h steam plant operating at the conditions stated above, and are based on measured operating data [624] for a 32t/h boiler at Amarillo, Texas. The sulphur retention at any plant will differ, for given operating conditions, according to the reactivity of the limestone supplied. It may be possible, by adjusting the primary:secondary air ratio, to reduce NO_x emissions or to increase the combustion efficiency. It is unlikely, however, that both will be optimised at the same primary:secondary air ratio.

Table 6.20 Efficiency and Emission Factors

Operation at MCR	
Gross heat input, MWt	11.1
Coal fired, tonne/h	1.43
Efficiency factor, GJ/tonne	23.3
Ashes output, tonne/h	0.30
C content of ashes, %	11.0
Ca content of ashes, %	22.5
SO_4 content of ashes, %	18.6
Elements in acidic emissions, kg/h	
Sulphur	3.00
Nitrogen	2.26
Chlorine (a)	3.72
Emission factors (kg per tonne coal)	
Sulphur	2.10
Nitrogen	1.58
Chlorine	2.60

Note: (a) Maximum possible value, assuming no retention of chloride in baghouse or ESP.

The figure for chlorine emissions is calculated on the assumption that all the chlorine in the coal is emitted from the stack as HCl. There could well be some absorption of HCl in the dust collected on the fabric filter, because this operates in the temperature range in which reaction of HCl with limestone or lime to produce calcium chloride is possible. No reports of measurements of such chlorine absorption in the baghouse appear to have been published.

Costs

The capital cost of the 32 tonne/h plant at Amarillo has been given as $6.3 million (1983 US dollars). This included, in addition to the boiler, baghouse, stack and storage silos, the cost of a railway spur to the plant, and reception facilities. It is anticipated that a second, similar unit at the same site would cost $1 million less.

No operating costs have been reported.

Process Advantages and Drawbacks

The concept of dual fluidised beds permits the use of relatively shallow beds (less than 0.3 m slumped depth) to obtain good combustion and sulphur capture efficiency. Most of the sulphur is released into the gas stream in the lower bed and this is able to react throughout the whole depth of the sorbent bed.

The injection of part of the combustion air at a point just below the upper bed ensures that oxygen is available to complete the burn-out of carbon elutriated from the lower bed which is trapped in the upper bed. Both functions of sulphur capture and carbon burn-out in the upper bed are accomplished sufficiently effectively to avoid the necessity for recycling the grit contained in the flue gases.

It would appear that the manufacturers have been able to evolve a design of distributor for the second bed that is resistant to damage by the passage of hot, dust-containing gases at high temperatures.

The problems in accommodating thermal stresses may limit the size of boiler using this technology.

Evaluation of CFB with Integral Heat Transfer

Suppliers of CFB systems in which the heat transfer surface is principally inside the combustion chamber include Ahlstrom/Pyropower [627–631], Gotaverken [632], Studsvik [633] and Keeler/Dorr Oliver [634]. The system described in this section is the Ahlstrom, which has been widely installed in several countries and which has been used as the basis for design studies for 150 MWe and 74 MWe (equivalent) boilers commissioned by Nova Scotia Power Corporation and by Colorado-Ute Electric Association respectively.

The plant described is a dual boiler system, each boiler generating 225 tonne/h of steam at 100 bar pressure and 480 deg. C temperature to supply the requirements of Datum Combustion System 2. Plants are also available for the smaller Datum System 3.

Process Description

The general arrangement of the plant is shown diagrammatically in Figure 6.4. Coal is stockpiled on site and conveyed to an intermediate storage hopper from where it is fed into the combustor via a weigh-belt. A second storage hopper is for prepared limestone, and the limestone feed rate is controlled to maintain a preset limestone:coal ratio, with adjustment about this ratio by means of a signal from an SO_2 analyser to compensate for variations in coal sulphur content. The coal and limestone are injected pneumatically into the combustion chamber (C.C.) and primary air is introduced through the base of the C.C. Secondary air is introduced at two higher levels, both in the lower part of the C.C. Three hot cyclones in parallel separate entrained particles from the exit gases, and the separated particles drop into a non-mechanical loop seal before being fed back to the C.C. Flue gas leaving the cyclones passes to the convection zone of the boiler, thence to an air heater and a baghouse (or ESP), finally, via an induced draught fan, to the stack. The primary air distribution grid also carries a bottom ash removal pipe; other solids removal points being at the economiser base, air heater and bag filter. Light-up is by means of auxiliary fuel burners firing into the C.C.

A turn-down ratio of 3 to 1 is possible by changing the air and fuel flow rates, and a rapid load-following ability has been demonstrated. Portions of the bed may de-fluidise at low load. The C.C. temperature is controlled closely at about 850°C in order to optimise sulphur retention. Temperature control is effected by altering the proportions of primary and secondary air flows so as to alter the heat release pattern within the chamber. The water-wall design of the chamber is such that the total heat flux is a function of heat release pattern, thus enabling the rate of heat generation to be balanced by the total heat flux. The overall excess air remains constant with variations in load.

Operating Experience

The first Ahlstrom systems to be installed, about 8 in number commissioned between 1979 and 1982, were in Finland and Sweden, burning peat, wood-waste, coal or oil, and ranging from 6 to 65 MWt. There were, in addition, two sludge incinerator plants, one of them burning sewage sludge (this one in Czechoslovakia). Following the setting up of the Pyropower subsidiary in the U.S.A., several more plants have been commissioned from 1983 onwards, in the U.S.A. and Europe, burning a wide range of fuels.

Such operational problems as have been reported have mainly been associated with plant auxiliary equipment. A 73 tonne/h steam boiler at Oulu in Finland, firing mainly peat but also coal, oil, wood-waste, gasifier dust and waste gas, had logged over 20,000 hours of operation up to June 1985 [629]. There was apparently no evidence of serious damage to the combustion chamber or hot cyclones during this period of operation, but it is not clear whether limestone was fed during the whole of the period.

Combustion efficiency on coal is reported to be between 98 and 99.5% when operating in the range 2–3% oxygen in dry flue gas (10–15% excess air) [629]. Sulphur dioxide removal, when burning low-sulphur coal (0.6% S) and with a Ca:S mole ratio of 2.2 is stated to be 96.4% and the concentration of NO_x in dry flue gas is 148 vpm (240 mg NO_2 per Nm^3 of flue gas at 6% oxygen content).

The sulphur removal efficiency is exceptionally high, even for CFB combustors. It is probable that this is due to the low sulphur content of the fuel fired, because it is sometimes found that higher sulphur removal efficiencies are obtained when burning low sulphur coals. This is because the contribution of ash to the sulphur capture is proportionately higher, and at the same time, these coals often contain highly reactive, alkaline mineral contents that react readily with sulphur dioxide produced during combustion. When burning an average British coal, such as the reference coals shown in Table 1.2, a sulphur dioxide removal of about 90% at a Ca:S molar ratio of 2.25 is thought to be more likely.

Variations and Development Potential

CFB has been widely used for the combustion of low grade coals such as high ash coals, peat and combustible industrial waste materials. Basu [635] has stated two main requirements for the design of a CFB combustor. They are:

(a) The furnace height should allow adequate retention time for the finer fuel particles (i.e. those escaping collection in the hot cyclones) to complete their combustion.

(b) The furnace walls should absorb enough heat to maintain the bed temperature within the desired range (800–900°C) over the given ranges of excess air and solids circulation rate.

In cases where the heat transfer surface is located mainly within the combustor (i.e. CFB with integral heat transfer), little control over combustion temperature can be exerted by the solids circulation rate and, for the combustion of high calorific value fuels, additional heat transfer capacity has to be supplied within the C.C. This necessitates increased furnace height, and furnaces with heat release rates above about 40–100 MWt will probably have to be equipped with some suspended tube cooling in the absence of an external heat transfer bed [636].

Another problem that arises in the scale-up of CFB combustors to larger sizes is that of ensuring good penetration of secondary air jets. The suspensions of solid particles have densities of over 16 kg/m^3 and there is a lack of knowledge about the penetration of air jets into such media [603].

The cyclones used to separate entrained solids from the gases leaving the C.C. are very large in order to handle the high volumes of gas. They also have to operate under severe conditions of high temperature and the impact of abrasive particles. There have been attempts to develop internal recycling devices consisting of conical baffles positioned towards the top of the C.C. [637, 638], but these have yet to be proved capable of withstanding the severe erosion that might be anticipated as a result of the high velocities and heavy loadings of suspended particles.

CFB has gained rapidly in popularity in recent years because of its flexibility in operation and its favourable environmental features. It is to be expected that further progress will be made in engineering design to increase the durability of components, and to extend the range of unit sizes offered to include large power station boilers.

Process Requirements

Table 6.21 shows performance data, flue gas flowrate

and composition at the stack, reagent quality and input rate at MCR, and annual requirements for fuel, sorbent, water and operating manpower, based on a load factor of 65%.

The boiler thermal efficiency, based on a combustion efficiency of 99%, is 86.5%. An excess air level of 20% is assumed, which is somewhat higher than quoted [629]. Limestone utilisation is 40%.

Table 6.21 Ahlstrom Process Data

2 × 225 tonne/h 100 bar, 480 deg. C steam boiler	
(a) *Hourly Input at MCR*	
Coal feed rate, tonne/h	60.2
Gross heat input, MWt	468
Combustion efficiency, %	99
Efficiency factor, GJ/tonne	24.2
Electric power requirement, MWe	–
Dry flue gas flowrate, Nm^3/s	138
Composition of dry flue gas:	
CO_2, % vol.	15.2
N_2, % vol.	81.1
O_2, % vol.	3.7
CO, vppm	50
SO_2, vppm	127
NO_x vppm	150
Limestone feed rate, tonne/h	6.68
Limestone analysis:	
$CaCO_3$, % wt.	95
$SiO_2 + Al_2O_3$, % wt.	3
Moisture, % wt.	2
Particle top size, micron	1000
Particle median size, micron	200
(b) *Annual Input at 65% Load Factor*	
Coal, tonne	342,800
Limestone, tonne	38,055
Water, tonne	23,400*
Auxiliary fuel, GJt	–
Manpower requirement, man-hours	17,500

*Excluding boiler feed water and water used for plant cleaning

No firm data are to hand for auxiliary power requirements, but it has been reported [Battelle Memorial Institute confidential report] that the 14.6 MWt boiler operated by Gulf Oil Co. at Bakersfield is equipped with electric motors totalling 600 h.p. for operation of the three fans and a Roots blower. Assuming 50% over capacity at MCR, the electric power consumption (adjusting for plant size) would be about 8 MWe for the 2 × 225 tonne/h steam boilers. This excludes other demands for power such as coal, limestone and solid residues handling machinery. The rates of production and the chemical composition of solid wastes from the conceptual 2 × 225 tonne/h plant operating at an annual load factor of 65%, are shown in Table 6.22.

No analyses of the discharged solids have been published, and it has been assumed here that any reduced forms of sulphur are fully oxidised to sulphate before the material has cooled.

Table 6.22 Solid and Liquid Effluents (65% annual load factor)

Production of solid wastes, tonne/annum dry weight	62,420
Composition of dry solid wastes, % wt.	
Ash	43.9
Carbon	3.3
Anhydrous $CaSO_4$	31.5
CaO	19.5
$CaCO_3$	0
Impurities from limestone	1.8
Production of liquid effluents, tonne/annum	Variable

On contact with water, the CaO will be rapidly hydrated to $Ca(OH)_2$ with a 32% increase in mass, and the $CaSO_4$ will slowly be hydrated to gypsum, $CaSO_4.2H_2O$ with a 26% increase in mass. Additional water will be absorbed to give the material a moist, crumbly consistency for easy, dust-free handling. At 20% free moisture, the water required for treatment of the solid residues will be 23,300 tonne/annum and the weight of conditioned residues will be 85,700 tonne/annum.

The conditioned waste will gradually harden over a period of weeks, especially if compacted during tipping. This will reduce the risk of leaching from the tip.

Efficiency and Emission Factors

Table 6.23 presents efficiency and emission factors for the Ahlstrom/Pyropower combustor described in previous sections. The sulphur dioxide emissions are those that would be expected when using a reasonably reactive limestone and may be less or more (for the same Ca:S ratio) for the use of sorbent from different sources.

Table 6.23 Efficiency and Emission Factors

Operation at MCR	
Gross heat input, MWt	468
Coal feed rate, tonne/h	60.2
Efficiency factor, GJ/tonne	24.2
Ashes output, tonne/h	10.96
Carbon content of ashes, % wt.	3.3
Calcium content of ashes, % wt.	23.2
Sulphate content of ashes, % wt.	22.2
Elements in acidic emissions, kg/h	
Sulphur	90.3
Nitrogen	46.7
Chlorine (a)	156.5
Emission factors (kg per tonne coal)	
Sulphur	1.50
Nitrogen	0.78
Chlorine	2.60

Note: (a) Maximum possible value, assuming no HCl absorption in baghouse or ESP

The chloride emissions are calculated on the assumption that there is no absorption of HCl in the system. Conditions in the baghouse are favourable for absorption of HCl by the layer of dust on the bags. Since this dust contains CaO and $CaCO_3$ it is quite possible that the layer will absorb a significant amount of HCl with conversion of the oxide or carbonate to calcium chloride or oxychloride, so leading to a reduction in acid chloride emissions and to the presence of chlorides in the solid residues. No confirmatory measurements are, however, available.

Costs

The following discussions of costs relates to CFB firing generally and includes costs of other types of CFB systems.

In reviewing the costs of FBC firing, Thomas [588] states that most sources agree that a CFB fired boiler is more capital intensive than either a pulverised coal or a bubbling bed fired boiler. The CFB system costs between 10 and 36% more than a deep bubbling bed system, reported costs being from 170 to 220 $/kWt for a 30 MWt boiler, falling to 120 to 150 $/kWt for a 120 MWt boiler.

Electric power consumption for CFB systems is higher than for deep bubbling beds. For slow CFB (velocities of about 5–6 m/s), consumption is about 30% higher than for deep bubbling beds and 50% higher than for PF firing (without FGD). For fast CFB (velocities of about 9–10 m/s), consumption is considerably greater, at about 85% greater than for the deep bubbling bed and 115% greater than for PF.

Labour requirements for boiler operations are similar to those for deep bubbling bed and for conventional firing systems. Thomas notes evidence of high availability for CFB systems, indicating that maintenance costs could be relatively low.

The high capital costs and auxiliary power consumption of CFB systems can be offset partly by higher combustion efficiency (as compared with bubbling bed systems without recycle) and partly by their ability to burn cheaper fuel. Thomas calculates that, for a 100 MWt boiler, the required coal savings that would make CFB cost the same as a bubbling bed are $11 per equivalent tonne of 26 MJ/kg coal for a 2 year payback and $3 per equivalent tonne for a 4 year payback.

Process Advantages and Drawbacks

Mass transfer rates between solids and gas are high in fast-fluidised (CFB) systems because of the high relative velocity between gas and solids. Another factor contributing to high combustion and calcium utilisation efficiency is the increased time during which the solids and gas are in contact, compared with bubbling beds. There is a high degree of solids back-mixing in the CFB combustor and this results in a once-through mean solids residence time considerably longer than that of the gas, and the gas is in contact with a high concentration of solids throughout the whole height of the C.C.–much greater than the depth of a bubbling bed.

It seems probable that the high velocities and large mass of solids supported by the gas flow requires high fan-power. If the estimate of 8 MWe for fan power at MCR (see above) is correct, it means that the CFB fired boiler is a fairly heavy consumer of auxiliary power in comparison with a pulverised coal fired boiler of comparable output, for which the total electricity consumption by primary and secondary air fans together with an induced draught fan would be about 3 MWe at MCR.

The principal advantages of the system are its flexibility in regard to fuel type, its high combustion and sulphur capture efficiencies and its low NO_x emissions.

Evaluation of Multisolids FBC System

For the combustion of premium fuels, the CFB systems incorporating an external bubbling bed heat exchanger are probably the most satisfactory, because they call for a smaller amount of heat transfer surface within the combustor and, therefore, have less potential for corrosion and erosion. They also make it easier to compensate for changes in combustor heat release such as can occur due to changes in boiler load or fuel properties. One of the main difficulties in bubbling bed combustors is to keep heat transfer rates in the combustor in step with heat release rates, so as to avoid wide fluctuations in bed temperature. The external heat transfer bed offers better heat balance control than the CFB system with an integral heat transfer surface. The external bed also operates under fully oxidising conditions, at somewhat lower temperatures than the combustor, and at relatively low fluidising velocities so that erosion and corrosion are virtually eliminated.

At present, CFB systems are available in a range of sizes starting at rather larger than Datum Combustion System 3, up to a size rather smaller than Datum Combustion System 2. The demonstration 420 tonne/h, 105 bar/540 deg. C steam boiler being built by Colorado-Ute Electric Association at its Nucla Station, could open the way for large CFB-fired power plants. The Nucla plant will fire low-grade coals.

The process selected for description in the following

sections is the Multisolids FBC system (MSFB) developed by Battelle Development Corporation and offered under licence by firms in the U.K. and U.S.A. (Table 6.4).

The other major supplier, Lurgi, uses a more conventional CFB combustor, without the dense bed (see below). The chemistry of sulphur capture is described in Section 2.2, and a general appraisal of the process, in comparison with other FBC systems, and the reason for its selection for detailed consideration, are given previously.

Process Description

The MSFB employs a dense fluidised bed of large, non-entrained particles occupying the lowest part of the combustor, which is narrower than the upper part and is operated under reducing conditions. Passing through this, and occupying the rest of the combustor, is a circulating bed which serves as the primary heat transfer medium (see Figure 6.6). Plants employing the MSFB process are described in references [639, 640].

The dense bed of large particles aids mixing of the fuel and limestone and delays the passage of the circulating solids through the combustor, increasing the residence time. Coal is fed towards the top of the dense bed, and limestone at a point near the bottom. The dense bed also receives recycle solids from the heat-exchange bed. Secondary air and off-gas from the heat-exchange bed are introduced at the bottom of the expanded zone of the combustor, so that the velocities in the two zones are approximately equal. Primary air represents only about 50% of the stoichiometric air requirements, ensuring very low NO_x emissions.

The limestone feed contains a high proportion of fine particles, because the rate of attrition is not always rapid enough to ensure a sufficiently high concentration of fresh sorbent in the circulating bed if lump stone is used. Limestone can also form part of the dense bed, if sufficiently resistant to breakage; or the bed can be wholly of a durable gravel, large enough to remain in place at fluidising velocities in the range 5 to 10 m/s.

Combustion temperature is maintained between 800°C and 815°C, giving a sulphur retention efficiency of 90% at a limestone addition rate corresponding to a Ca:S mole ratio of 2.25. Spent sorbent and ash are removed through an overspill output at the surface of the heat-exchange bed, ensuring the complete absence of calcium sulphide in the discharged solids. The heat-exchange bed operates at a fluidising velocity of about 0.5 m/s and a temperature of 550–650°C.

The combustion gases leaving the primary cyclones enter a convective heat transfer section which includes an economiser, before passing to an airheater (optional) or baghouse or other clean-up appliance, I.D. fan and stack. Well over half of the total heat transfer takes place in the external fluidised bed heat exchanger. Start-up is by firing combustion gases from an auxiliary fuel burner into the dense bed section of the combustor, while recycling only hot material.

The plant evaluated consists of twin 225 tonne/h steam boilers based on the MSFB system.

Review of Multisolids Fluid Bed Process and Operating Experience

Foster Wheeler Power Products Limited is a Licensee of the multisolids fluid bed (MSFB) process. This process was developed by Battelle Laboratories to meet the need for an advanced combustion process, capable of burning a wide variety of fuels efficiently and also to meet increasingly stringent environmental standards [568].

The MSFB process incorporates a circulating fluid combustion system and steam generator and consists of four main components:

1) Combustor vessel.
2) High efficiency hot gas cyclones.
3) An external heat exchanger.
4) A compact boiler and economiser.

Combustion of fuel takes place within the combustor vessel with the temperature of combustion controlled by recycled entrained bed material. This material enters at a controlled rate and maintains an even temperature throughout the combustor. Flue gas and entrained bed material is separated with high efficiency cyclones. The flue gas passes to the convective boiler where up to 50% of the heat input is recovered. The entrained bed material enters an external heat exchanger, which comprises a fluid bed and heat transfer surface to generate saturated or superheated steam.

A plant designed on these principles has been operating in the Irish Republic since June 1984, and to-date over 14,000 operating hours have been logged. The unit generates 53,000 kg/h of saturated steam at 24 bars and fires a variety of fuels including coal, peat and wood wastes. The availability of this plant has been in excess of 94% and has met or exceeded the original design requirements. These include combustion efficiencies generally greater than 99%, high turndown ratios in the order of 7.1 on automatic control, low levels of noxious flue gas emissions and fuel flexibility [568].

Variations and Development Potential

CFB has been extensively used for the combustion of low grade fuels, such as high-ash coals or wood-waste from sawmills. For these applications, combustor temperatures can be kept within required limits by using a membrane-wall construction for the combustor (integral heat exchange). The higher temperature, resulting from combustion of premium fuels, necessitates some additional cooling, provided in systems with external heat exchangers by recycling cooled solids to the combustor.

The cyclones used to separate entrained solids from the gas leaving the combustor have to operate under exceptionally severe conditions of high temperature, handling abrasive particles. Designs of cyclones installed in plant currently operating are based on conventional high efficiency cyclone designs and have proved capable of capturing 99% of particles larger than 20 microns.

In the system here described, temperature control is affected by varying the proportions of hot and cold recycle, the former drawn from an uncooled compartment of the external fluidised bed and the latter from the cooled compartment.

In one variation of the process, cool recycle is introduced to the oxidising zone as well as to the reducing zone, so increasing still further the degree of temperature control throughout the combustor.

CFB has gained rapidly in popularity in recent years because of its flexibility in operation and its favourable environmental features. It is to be expected that further progress will be made in engineering design to increase the durability of components and extend the range of unit sizes on offer to include large, power station boilers.

Process Requirements

Table 6.24 shows process data, flue gas flow rate and composition at the stack, reagent quality and input rate at MCR, and annual requirements for fuel, sorbent, water, auxiliary fuel and operating manpower, based on a load factor of 65%.

Combustion efficiency, at 97.5%, is lower than is reported for the Ahlstrom/Pyropower systems. This is probably accounted for by the shorter combustion chamber in the MSFB (15m compared with 34m in an Ahlstrom combustor designed for bituminous coal firing).

The excess air level is approximately 20%, which is greater than the optimum for control of NO_x emissions, but is a compromise value allowing a reasonably good combustion efficiency together with

Table 6.24 MSFB Process Data

2 × 225 tonne/h 100 bar, 480 deg. C steam boiler	
(a) *Input per hour at MCR*	
Coal feed rate, tonne/h	61.2
Gross heat input, MWt	476
Combustion efficiency, %	97.5
Efficiency factor GJ/tonne	23.8
Electric power requirement:	
Dry flue gas flowrate, Nm^3/s	141
Composition of dry flue gas:	
CO_2, % vol	14.8
N_2, % vol	81.0
O_2 % vol	4.2
CO, vppm	250
SO_2, vppm	127
NO_x, vppm	150
Limestone feed rate, tonne/h	6.79
Limestone analysis:	
$CaCO_3$, % wt.	95
$SiO_2 + Al_2O_3$, % wt.	3
Moisture, % wt.	2
Particle top size, mm	1
Particle medium size, mm	0.2
(b) *Annual Inputs at 65% Load Factor*	
Coal, tonne	348,500
Limestone, tonne	38,700
Water, tonne	***
Auxiliary fuel, GJt	***
Manpower requirement	***

an acceptable NO_x emission level. 115 vppm SO_2 is equivalent to 330 mg/Nm^3 (296 mg/Nm^3 at 6% oxygen) and 160 vppm NO_x is equivalent to 308 mg (as NO_2) per Nm^3 (277 mg/Nm^3 at 6% oxygen). The limestone utilisation is 40%.

The rates of production and the chemical composition of solid wastes, together with the amount of liquid effluent from the conceptual 2 × 225 tonne/h steam plant operating at an annual load factor of 65%, are shown in Table 6.25. On contact with water, the CaO will be rapidly hydrated to $Ca(OH)_2$ with a 32% increase in mass, and the $CaSO_4$ will slowly hydrate to gypsum, $CaSO_4.2H_2O$ with a 26% increase in mass. Additional water will be absorbed to give a moist, crumbly consistency required for easy handling without production of airborne dust. At 20% free moisture, the water required for treatment of the solid residues will be 24,750 tonne/annum, and the weight of conditioned residues will be 97,900 tonne/annum.

The conditioned waste will gradually harden over a period of weeks, especially if compacted during tipping. This will reduce the risk of leaching from the tip.

Efficiency and Emission Factors

Table 6.25 presents efficiency and emission factors for the MSFB circulating bed combustor described in previous sections.

The sulphur dioxide emissions are those that would be expected when using a reasonably reactive limestone, and may be less or more (for the same Ca:S ratio) for different sources of sorbent.

The NO_x emissions could be reduced further by using a lower excess air value, but this would be at the cost of reduced combustion efficiency.

The chloride emissions are calculated on the assumption that there is no absorption of HCl in the system. Conditions in the baghouse are favourable for absorption by the layer of dust, containing calcium oxide and carbonate, on the bags, and it is therefore quite possible that the actual emissions of acid chloride will be appreciably less than the figure given. No confirmatory measurements are, however, available.

Costs

Capital costs for a 38 MWt boiler are detailed in Table 6.26. Operating costs cannot at present be specified.

Process Advantages and Drawbacks

Mass transfer rates between solids and gas are high in fast-fluidised (CFB) systems because of the high relative velocity between gas and solids and the longer time during which the gas is in contact with solids, compared with bubbling beds. This results in good combustion efficiency at relatively low operating temperatures, and high sulphur retention efficiencies. The staged introduction of air makes possible low NO_x emissions.

The use of an external heat exchange bed gives more flexibility in temperature control with fluctuating load and reduces the potential for deterioration of the heat transfer surface.

Table 6.25 Solid and Liquid Effluents (65% Annual Load Factor)

Production of solid wastes, tonne/annum dry weight	68,300
Composition of dry solid wastes, % wt.	
Ash	40.8
Carbon	10.2
Anhydrous	29.3
CaO	18.0
$CaCO_3$	0
Impurities from limestone	1.7
Production of liquid effluent, tonne/annum	***

Against this must be set the need to employ highly efficient primary cyclones, operating at high temperature and handling abrasive solids larger in size than those handled by the hot cyclones in PFBC. Any wear occurring may seriously reduce the efficiency of the cyclones, leading to a fall-off in combustion efficiency and overloading of the downstream gas passages and equipment with solids. Regular inspection and maintenance of these cyclones is therefore essential. A further disadvantage is the relatively high fan power needed.

Evaluation of CFB with External Heat Transfer: Lurgi System

This is one of the two main forms of the CFB process using an external fluidised bed to perform part of the combustor circuit heat transfer duty, the other being the Battelle Multi-Solids FBC system.

In the Lurgi system [580, 641, 642, 331, 332] the combustor design is basically similar to that of the Ahlstrom/Pyropower combustor, but the requirement for cooling surface in the combustion chamber is reduced by the use of the external heat transfer fluidised bed which receives solids from the hot cyclones at about 850°C and cools them to about 400°C before they are recharged to the combustor. In addition to the FBHE (Fluidised Bed Heat Exchanger) and the water-cooled walls in the combustion chamber, heat is recovered from the gases leaving the hot cyclones in a conventional convection pass, with economiser and combustion air heater, before they enter a baghouse or ESP for particulate removal, induced draught fan and stack.

The proportion of the total primary heat transfer taking place in the external bed decreases from about 40% at MCR to less than 20% at low load (about

Table 6.26 Efficiency and Emission Factors Operation at MCR

Gross heat input, MWt	476
Coal feed rate, tonne/h	61.2
Efficiency factor, GJ/tonne	23.8
Ashes output, tonne/h	12.0
C content of ashes, %	10.2
Ca content of ashes, %	21.5
SO_4 content of ashes, %	20.7
Elements in acidic emissions, kg/h	
Sulphur	91.8
Nitrogen	47.5
Chlorine (a)	159.1
Emission factors (kg/tonne coal)	
Sulphur kg/tonne	1.50
Nitrogen kg/tonne	0.78
Chlorine kg/tonne	2.60

Note:
(a) Maximum possible value, assuming no HCl absorption in baghouse or ESP

one third of MCR). The term primary heat transfer is here used to denote the combined heat removal in the combustion chamber and in the FBHE.

The plant described here is a conceptual 500 MWe power generating boiler described in [641]. It is therefore, to some extent, untried technology although it must be noted that there is considerable operating experience on smaller Lurgi systems, including units sized up to about 100 MWe.

Process Description

The boiler has twin combustion chambers each rectangular in cross section, 8.1 m × 16.2 m and approximately 30 m high. The chamber walls consist of a lower, refractory-coated section, 6.1 m in height, the rest being of membrane-tube heat transfer surface. Each combustion chamber feeds gases into two cyclones, each over 9 m in diameter (four cyclones in all), and the gases leaving the cyclones enter the convective sections of the boiler (see Figure 6.5).

The hot solids from each pair of cyclones enter an FBHE and the cooled solids are returned to the lower part of the combustion chamber. Each of the two FBHEs has an area of 112 m^2 and a height of 4.3 m.

Table 6.26 MSFB Costs–December 1986
38 MWt (120 × 10^3 lb/hr)

	£ '000s
Combustor, steam generator, economiser, air heater, cyclones, EHE burners	460
Coal and limestone handling and bed makeup vessel	315
Ducting and expansion joints	70
Ash handling	66
Dust collector, fans and stack	145
De-aerator, feedwater system, chemical dosing and pumps	105
Support steel–ladders and galleries	105
Piping, valves, safety valves, blowdown vessel and firefighting equipment	80
Control & Instrumentation, sample coolers and air receiver	120
Electrics	85
Refractory and insulation	245
Miscellaneous and transport	42
Civils	190
Construction	425
Design, draughting and contract management	700
Total	3,153

The feed coal is crushed to a top size of 6–10 mm; the limestone top size is not more than 1 mm and may be considerably less. Fuel and limestone are fed through four feed points into the lower part of each of the combustors. Preheated primary air, representing about 50% of the stoichiometric air requirement is supplied to the base of the combustor, with secondary air inlets higher up. Air leaving the FBHEs is also fed to the combustors as secondary air.

The total air supplied to the combustor (primary, secondary, FBHE off-gas, solids feed and recycle transport air and air used for fluidisation of siphon seals) amounts to 1.2 times stoichiometric (i.e. 20% excess air).

Ash is extracted from the system by diverting part of the flow of cooled solid from the FBHE through another fluidised cooler, which reduces the temperature from 400°C to about 80°C. Ash is also discharged from the base of the economiser and from the bag filter hopper.

The FBHE is used for all of the reheat and part of the superheat, the rest of the superheating surface is wholly in the combustion chamber wall. The FBHE is compartmented, with superheater and reheater tube bundles in different compartments, as required for boiler control.

Three air input fans are required; for air to the base of combustor; for secondary air; and for FBHE fluidising air. There are two induced draught fans, in parallel. It is stated in [641] that electrical usage for 'equipment within the boiler island' amounts to approximately 17 MWe. It is not clear what this includes, but it can be assumed that the I.D. fans are excluded, also coal crushing, and coal and limestone handling remote from the boiler. If it is assumed that 90% of this 17 MWe is accounted for by fan power, the rest being for such things as coal and limestone feeders, boiler recirculation pump, etc. and if the I.D. fan power is taken to be 5.8 MWe (as in the equivalent bubbling bed boiler), the total fan power is 21.2 MWe.

The design is for a conventional steam cycle with single reheat. Further design studies are needed to assess the feasibility of using CFB units at advanced steam conditions (240 bar pressure, temperatures above 540°C).

Operating Experience

Lurgi have been supplying fluidised bed heaters for industrial heating applications for over 20 years, and the Lurgi CFB boiler technology represents a development of this expertise. In straightforward boiler applications, heat transfer surfaces in the combustion chamber, FBHE and convection pass are all used for steam generators and superheating. Lurgi have also installed plants in which FBHE tube circuits are used for special heating duties, e.g. for heating process-heating fluids.

Six units of 70 MWt input or more, incorporating external FBHEs, have been or are being installed in West Germany, the USA and Canada. Those in the Western hemisphere are installed by Combustion-Engineering Power Systems Inc., who are licensees of the Lurgi process. The largest plant, at Duisburg in West Germany, is rated at 208 MWth input and has twin FBHEs in parallel: one containing evaporative surface and the other, reheat surface.

The plant for which most information is available is a boiler at Lunen in West Germany which has a thermal input of 84MWt. In this the FBHE is used to heat molten salt as a source of process heat. This plant was started up in July 1982, firing a coal washery waste containing up to 55% ash and 2% sulphur, with a gross C.V. of 14,000 to 18,600 kJ/kg. Lime kilm baghouse rejects are used for SO_2 emission control at a rate of 1.6 tonne/hour, representing a Ca:S mole ratio of 1.5 and effecting 90% removal of the sulphur input to the combustor. Combustion efficiency is 99%, most of the loss being due to incomplete burn-out of carbon (the carbon-in-ash is 0.5 to 1.5%).

Problems encountered have principally been associated with the feeding of moist, high ash coals containing a high proportion of fine particles. These have tended to compact in the conveyors and silos. The problem was overcome by maintaining the moisture content of the feed coal below 14%. The roller mill first installed for coal crushing did not give the required particle size distribution for combustion of the high ash fuel, being too fine with a relatively flat distribution curve. Modifications to the mill and classifier corrected this. The grinding roller had, by late 1985, reached a service life of 8000 hours without renewal. Initial problems of wear in the coal conveying system and in the pneumatic feeding equipment were overcome by using Ni-hard in the worst affected areas. There have been no combustion problems and the level of burn-out is considered very satisfactory.

The refractory lining in the combustor showed no significant wear after more than two years of operation, and the tubes in the FBHE, which are made of high-alloy steel, have proved highly reliable. It is claimed that rapid changes of load are possible, response times of up to 20% per minute being achieved. The availability of the plant during two years of operation was over 90%.

Variations and Development Potential

Variations are mainly concerned with the arrangement of heat transfer surface in the combustor, FBHE and convection pass. In two plants being installed in the U.S.A. to burn respectively low grade lignite and anthracite culm, FBHEs will not be used because both boilers are to be used as base load units with dedicated fuel supplies, so that the control capability offered by an FBHE will not be necessary. In both boilers, an adequate cooling surface for temperature control can be built into the combustion chamber without having to resort to an external heat exchanger. In most cases where an FBHE is used as part of the boiler heat transfer surface, the tube banks consist of evaporative surfaces and either superheater or reheat tubing. An exception to this could be the case of very large units, such as the one described above, where all the evaporative surface is in the combustion chamber. Two of the German plants produce low pressure steam for district heating in addition to generating electric power.

The focus of attention in future development is on scaling up to the sizes used in modern power stations, the design described above being one such conceptual scheme. Equipment and process information requiring additional development for scale-up have been identified in [641]. They include: the cyclones, which will be about 30% larger diameter and height than the largest cyclones currently being used in Lurgi technology: ensuring adequate fuel distribution and good combustion in the larger combustion chambers that will be needed; and the adaption of designs to utilise supercritical steam cycles.

Process Requirements

Table 6.27 shows the estimated performance data, flue gas flowrate and composition at the stack, reagent quality and input rate for a single 500 MWe at MCR, and annual requirements of a 4 × 500 MWe station for fuel, sorbent and water, based on an annual load factor of 65%. No figures are available for auxiliary fuel or operating manpower. The fuel fired is the first of the two coals in Table 1.2.

The combustion efficiency, 99%, is typical of those reported for the majority of the Lurgi plants now in operation [580]. Values as low as 98% can occur [643] but it is probable that at least 99% will be achieved when using 20% excess air, as in the plant described here.

The high calcium utilisation of 60% is obtained by using very fine limestone (90% less than 90 micron). If such a material is not readily available (Vereinigte Aluminium Werke use lime kiln baghouse rejects at their Lunen plant), additional power will be required for grinding the limestone. Use of a coarser material is possible, but would probably result in a lower SO_2 removal for a given rate of sorbent addition.

The total auxiliary power requirement without limestone grinding, 37.5 MWe, is slightly higher than that required for bubbling bed AFBC (see Table 6.27), the difference arising from the higher total power of the forced draught fans for the Lurgi plants.

The rates of production and the chemical composition of solid wastes from the conceptual 4 × 500 MWe unit station operating at an annual load factor of 65% are shown in Table 6.28. The risk of discharge of calcium sulphide formed in the reducing zone in the lower part of the combustion chamber is avoided by discharging ash from the FBHE. This ensures that discharged material has had an opportunity of being fully oxidised in the upper part of the combustor, the cyclones, and the FBHE, before reaching the discharge point.

Table 6.27 Lurgi CFB Process Data

(a) *Single 500 MWe Unit at MCR*	
Coal feed rate, tonne/h	210
Gross heat input, MWt	1435
Steam cycle heat rate, kJ/kWh	8352
Boiler efficiency factor, GJe tonne	9.35
Turbine generator gross output, MWe	545.5
Auxiliary power, MWe	
Forced draught fans	15.3
Induced draught fans	5.8
Coal, limestone and ash handling	4.8
Boiler pumps and cooling tower fans	7.7
Other equipment	3.9
Total power	37.5
Turbine generator net output, MWe	508.0
Net heat rate, kJ/kWh	10,170
Net efficiency factor, GJe/tonne	8.71
Wet flue gas flow rate, Nm^3/s	453
Dry flue gas flow rate, Nm^3/s	422
Composition of dry flue gas;	
CO_2, % vol.	15.4
N_2, % vol.	80.9
O_2, % vol.	3.7
SO_2, vppm	152
NO_x, vppm	150
CO, vppm	250
Limestone feed rate, tonne/h	16.58
Limestone analysis:	
$CaCO_3$, % wt.	95
$SiO_2 + Al_2O_3$, % wt.	3
Moisture, % wt.	2
Particle size	90% less than 90 micron
(b) *Annual Inputs, 4 × 500 MWe Units at 65% Load Factor*	
Coal, thousand tonne	4,783
Limestone, thousand tonne	378
Water, thousand tonne	363 (a)

Note: (a) For conditioning of solid residues only.

On contact with water, the CaO will be rapidly hydrated to $Ca(OH)_2$ with a 32% increase in mass, the $CaSO_4$ will slowly hydrate to gypsum, $CaSO_4.2H_2O$, with a 26% increase in mass.

Additional water will be absorbed to give a moist, crumbly consistency required for easy handling without production of airborne dust. The material will gradually harden after tipping and, provided it has been properly compacted, will produce a firm, impermeable mass suitable for eventual development of the site.

At 20% free moisture in the hydrated product (based on dry weight of hydrated material), the water required for treatment of the solid residues is 365,200 tonne/annum, and the weight of conditioned residues 1,571,000 tonne/annum.

Efficiency and Emissions Factors

Table 6.29 shows efficiency and emission factors for the conceptual 4 × 500 MWe Lurgi CFB fired power plant, operating at full output. The chloride emissions have been calculated on the assumption that there is no absorption of HCl in the system. Conditions in the

Table 6.28 Solid Effluents from Lurgi CFB, 4 × 500 MWe Units at 65% Load factor

Production of solid wastes, thousand tonne/annum dry weight	1206
Composition of solid wastes, % wt.	
Ash:	66.2
Carbon:	1.9
Anhydrous $CaSO_4$:	24.3
CaO:	6.7
$CaCO_3$:	0
Impurities from limestone	0.9

Table 6.29 Efficiency and Emission Factors 4 × 500 MWe Lurgi CFB Fired Power Plant at MCR

Gross heat input, MWt	5,740
Coal feed rate, tonne/h	840
Efficiency factor, GJe/tonne	8.71
Ashes output (dry), tonne/h	212
Ashes output (conditioned), tonne/h	277
Carbon content of dry ashes, %	1.9
Calcium content of dry ashes, %	11.9
Sulphate content of dry ashes, %	17.1
Elements factors (kg per tonne coal)	
Sulphur	1,344
Nitrogen	570
Chlorine (a)	2,100
Emission factors (kg/tonne coal)	
Sulphur	1.60
Nitrogen	0.68
Chlorine	2.50

Note: (a) Maximum possible value, assuming no absorption of HCl in the baghouse (or ESP)

baghouse are favourable for the absorption of HCl by the layer of dust collected on the bags, which contains CaO and $CaCO_3$, and it is therefore possible that the actual emissions of acid chloride will be appreciably less than the value shown, while the baghouse rejects will contain some calcium chloride. No confirmatory measurements are, however, available.

Costs

Costings for the conceptual 500 MWe boiler described in [641] have not been published yet.

Process Advantages and Drawbacks

The Lurgi systems possesses the advantages of high sulphur removal, good combustion efficiency at relatively low excess air levels, low NO_x emissions, and the ability to burn a wide range of fuels. It has, in addition, the advantage over CFB systems with integral heat transfer that it is possible to scale up to large boilers without having to resort to internal cooling tubes in the combustion chamber, with their associated risks of corrosion and erosion damage.

The use of an external heat transfer bed also introduces greater flexibility in boiler control because not only is it possible to vary the amount of heat transferred in the external bed by-passing solids, but the proportions of heat transferred to the evaporate, superheat and reheat surfaces can be varied. This is done by placing the various categories of tubing in different compartments of the bed, each with a controllable hot solids input from a solids flow divider.

Experience has shown that there is insignificant erosion or corrosion of tubes in the FBHE.

The principal drawback is that of all fluidised combustion systems, namely the high power requirements for fans on the input side of the combustor. The total fan power (including the I.D. fan) of 21.1 MWe compares with 10.7 MWe for a pulverised coal fired boiler of equal rating.

A further drawback is the size of the cyclones used for separating the entrained solids at the combustion chamber outlet. On scaling-up to power station boiler size, these become very large, and may dictate the height of the combustion chamber.

EVALUATION OF 'BUBBLING BED' PRESSURISED FLUIDISED BED COMBUSTION

Combustion of coal under pressure in a fluidised bed was first accomplished in 1969, and as described later, several successful combustors have been in operation in the intervening years. For the process to be commercially viable it is necessary at least to recover the energy needed to drive the compressor for the combustion air by expanding the combustion gases through a gas turbine. The feasibility of so doing using an industrial design of gas turbine at the operating temperatures (850–900°C) relevant to combined cycle power plant has yet to be demonstrated. As mentioned later, the first combined cycle demonstration plant is scheduled to come into operation in 1990. Experience will, however, be gained on operation of the supercharged boiler cycle, where the gas turbine operates at temperatures in the range 400–500°C, in 1987–88.

Process Description

Pressurised fluidised bed combustors have deep beds–typically 3 to 5 metres–in order to accommodate the heat transfer surface that is needed to match the much higher heat release per unit bed cross section than occurs at atmospheric pressure. Fluidising velocities are low–0.8 to 1.5 m/s–in the interests of better combustion and sulphur capture performance; lower bed particle size and better heat transfer; lower rates of particle breakdown, elutriation and tube erosion. At pressure these considerations greatly outweigh the penalty of a greater plan area. As with all fluidised beds, the operating conditions and particle size distribution of the feedstocks must be kept within limits to avoid excessive elutriation on the one hand, and segregation of large particles on the other. The maximum coal and sorbent particle size is usually in the range 1.6 to 3.2 mm. The bed, as with AFBC, is a mixture of coal-ash and partially-sulphated sorbent; unless certain high-ash coals are burned, the sorbent is the main constituent of the bed. Coal combustion rates are such that the carbon content of the bed is small–usually less than 1%.

The pressure vessel containing the combustor can be either a sphere, a horizontal cylinder or a vertical cylinder. The latter two can be refractory-lined to form the containment of the fluidised bed. Alternatively, the combustor can be suspended from the inside of the pressure vessel; in this case the walls of the combustor are likely to be membrane walls forming part of the steam circuitry.

This is the only practical arrangement if the pressure vessel is a sphere. Air from the compressor enters the space between the combustor and the pressure vessel so that the membrane walls have to withstand only a small pressure differential. With this arrangement, the gas cleaning equipment (cyclones) can be located between the combustor and the walls of the pressure vessel. An example of this arrangement is shown in Fig. 6.9. Alternatively (and always if the pressure

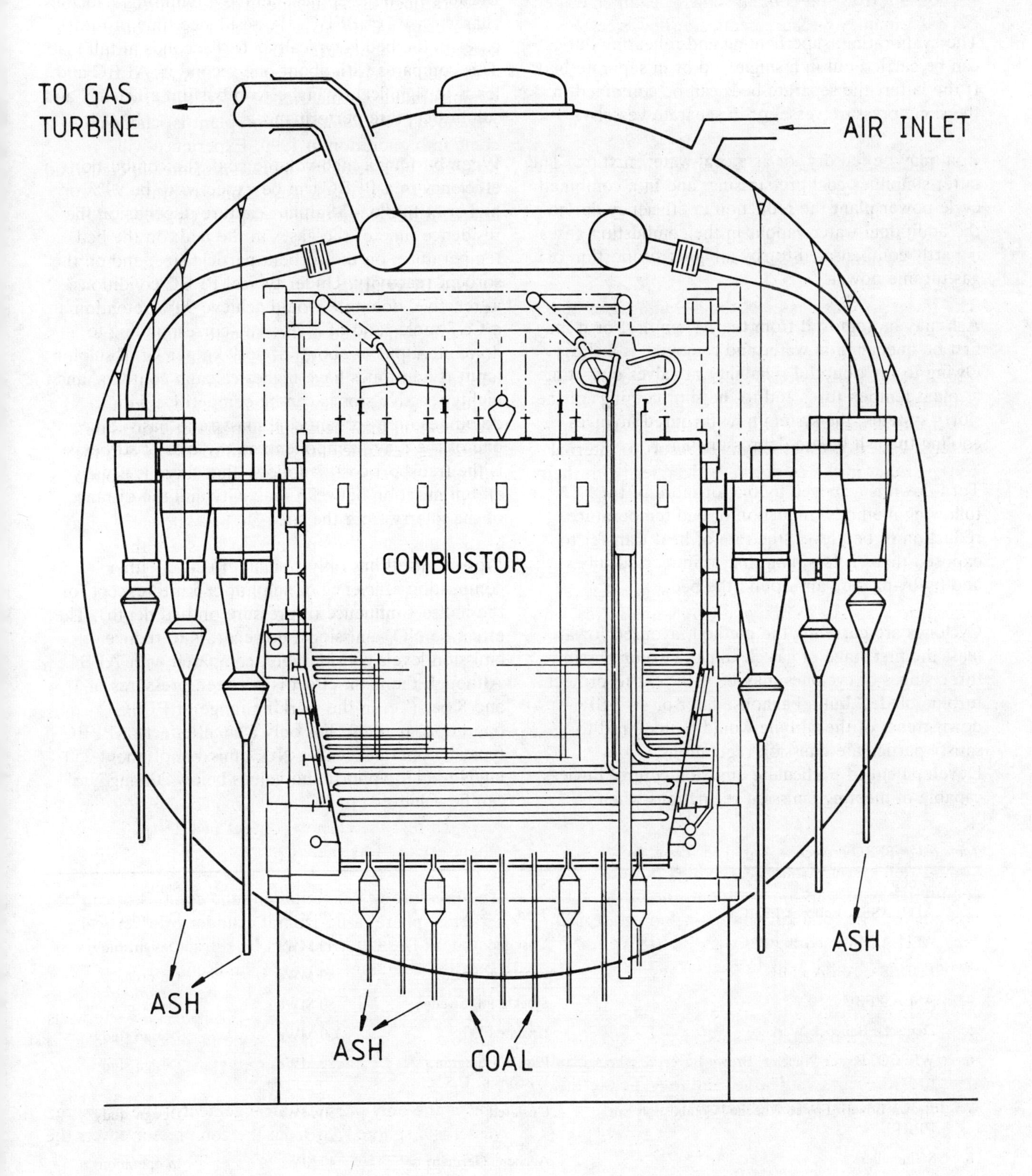

Figure 6.9 A Pressurised Fluidised Bed Boiler Arrangement

vessel forms the combustor walls) gas cleaning equipment can be installed in separate pressure vessels. Pressure vessels can be built to a maximum size equivalent to about 300 MWe. Thus, a plant of 600 MWe is usually designed to have at least two combustors (i.e. boilers).

The evaporating, superheating and reheating duties can be carried out in a single bed or in separate beds. If the latter, the separate beds can be contained in the same pressure vessel or in separate vessels.

Coal may be fed dry, or as a coal-water mixture. The latter simplifies coal pressurising, and in a combined cycle power plant the reduction in efficiency due to the additional water vapour in the combustion gases is partly compensated for by a higher proportion of gas turbine power.

Ash may be removed from the system dry, or it may first be quenched in water and removed as a slurry. Owing to the doubtful reliability of valves operating at plant temperatures and to the shortcomings of the slurry systems, the system now favoured involves cooling the ash before depressurising.

Turn-down is achieved by one or more of the following methods: reduction of bed temperature, reduction of bed level (the rate of heat transfer to exposed tubes is less than that to immersed tubes) and by by-passing air around the bed.

Cyclones are currently the method favoured for at least the first stage of gas cleaning. Two or possibly three stages of cyclones may be sufficient to protect turbine blades, but a baghouse filter or an ESP downstream of the turbine would be required to satisfy particulate emission regulations.
Development of particulate emission control devices, capable of meeting emission regulations when operating at gas turbine inlet temperature, is in progress.

Operating Experience

Because of the deep beds and low fluidising velocities characteristic of PFBC, the residence time of the gases in the bed is typically 3 to 4 seconds at full load. This compares with about one second in AFBC and leads to significantly higher combustion efficiency and sulphur capture performance.

When burning a high-volatile coal, the combustion efficiency in a PFBC can be expected to be 99% or higher at full-load. Sulphur capture depends on the residence time of the gases in the bed; on the bed temperature; on the sorbent particle size; and on the sorbent reactivity. Under typical PFBC conditions, a 'reactive' dolomite would achieve 90% retention at a Ca/S mole ratio of 1.2 to 1.5–corresponding to a dolomite input of about 7.2 to 9 kg per kg of sulphur input. Limestones have higher calcium contents, and highly reactive stones can be competitive with dolomites on a sorbent/coal mass ratio basis. Since one of the main components in the cost of sorbents is the transport cost, it follows that there is usually an optimisation between reactivity and the distance of the quarry from the site.

Pressure itself has only a small effect on either combustion efficiency or sulphur capture (except for the indirect influence of pressure on bed depth). The effect on NO_x emissions, however, is to reduce emission levels considerably, compared with AFBC. Although the main effect is between pressures of 1 and 8 bar (i.e. in the working range of PFBC operation, pressure has only a small effect) a PFBC typically operates with a NO_x emission of about 250 mg/Nm^3 (125 vppm), and values below 200 mg/Nm^3 can be achieved.

Table 6.30 PFBC Projects

	Group	Site	Capacity	Status
1.	AEP, ASEA Babcock PFBC	Tidd, U.S.A.	80 MWe	On order
2.	AEP, ASEA Babcock PFBC	Unstated	640 MWe	Design study
3.	ENDESA, ASEA PFBC	Escatron, Spain	80 MWe	On order
4.	ASEA PFBC	Stockholm, Sweden	80 MWe	Order pending
5.	Deutsche Babcock Werke	Unstated	330 MWe	Design study
6.	WEPCO, Foster Wheeler, Brown Boveri & others	Port Washington, Wisconsin, U.S.A.	80 MWe	Design study
7.	Brown Boveri, Foster Wheeler, Combustion Eng. EPRI	Unstated	250 MWe	Design study
8.	Steinmuller	Aachen, Germany	40 MWt	In operation
9.	Steinmuller	Unstated	100 MWt	Design study
10.	Babcock U.K.	Unstated	3 × 6600MWe	Design study

Items 1, 2, 3, 4, 5, 9 and 10 are combined cycle plants. The others are turbocharged boilers. References to Items 1–9 are contained in [646]. The three 80 MWe plants of items 1, 3 and 4 are demonstration units being retrofitted to existing (i.e. old) steam turbines and utilise the 15 MWe ASEA STAL GT-35P gas turbine.

A total of well over 16,000 operating hours have been amassed on PFBC equipment operating at: the Leatherhead Coal Utilisation Research Laboratory, British Coal's Coal Research Establishment, the Grimethorpe PFBC facility (initially sponsored jointly by the Governments of the U.S.A., the U.K. and the F.R.G. and latterly by British Coal, and the U.K. Central Electricity Generating Board), the ASEA-PFBC test facility at Malmö in Sweden, the Curtiss-Wright pilot plant, the General Electric Malta rig, and the New York University test facility.

The foregoing data obtained have formed the basis for the design activities listed in Table 6.30.

Variations and Development Potential

The main scope for improving performance lies, (1) in adopting advanced steam conditions (supercritical pressure and double reheat), and (2) in developing a hybrid cycle where the combustor is proceeded by a gasification stage. After cleaning, the fuel gas produced in the first stage is burned in the products of combustion from the second stage, thus raising their temperature. In this way it may be possible to operate at higher turbine temperatures (e.g. 1100°C) than is possible where the combustion gases come directly from the bed. Investigation of (2) is in progress under a contract with the USDOE [647]. The target is a power generating efficiency of 45%.

Process Requirements and Costs

The design study for PFBC combined cycle plant comprising three 660 MWe units (Item 10 Table 6.30), is jointly sponsored by the CEGB, British Coal and the U.K. Department of Energy. The results of this study are not available at present. Table 6.31 summarises the results of a study [645] carried out by American Electric Power for a 500 MWe commercial PFBC combined cycle plant and gives comparative data for a conventional plant with FGD.

Uncertainties and Future Development

Two uncertainties which have to be resolved before commercial exploitation can proceed are:

- The ability of a gas turbine or turboexpander to operate reliably in a dirty, potentially corrosive atmosphere (the uncertainty is less in the case of turboexpanders). Pilot-scale investigations have indicated that this should be possible, but only experience with demonstration plant will be convincing.

- Erosion of tubes immersed in the bed. This has

Table 6.31 Comparative Data for PFBC Combined Cycle Plant

	Conventional plant with an FGD system	Combined cycle PFBC plant
Net capacity (MWe)	640	640
Net plant efficiency (%)	36.6	40.1
Sorbent	Lime	Dolomite
Ca/S molar ratio	1.05	1.6
Water consumption (gpm)	13000	11250
Sorbent consumption (ton/h)	20	81
Waste production (ton/h)	149 (sludge)	128 (dry)
Waste disposal costs (dollars/ton)	6.25	4.00
Total capital cost (dollars/kW)	1640	1505
Levelised busbar power cost (mill/kWh)	100.0	92.0

Steam cycle: Supercritical single reheat 250 bar/537°C
Coal: High volatile, 16% ash, 4% sulphur

Table 6.32 Comparative Data for Turbo-Charged CFB Plant

	Conventional Plant with FGD	Bubbling Bed Turbo-Charged	Circulating Bed Turbo-Charged
No. of units	4	4	4
Unit size (MWe)	250	250	250
Plant efficiency (%)	33.5	34.2	36.2
Sorbent	Limestone	Dolomite	Limestone
Ca/S molar ratio	0.99	1.5	1.2
Coal feed rate, tonne/h*	114.6	112.3	105.5
Solid Waste, tonne/h*	54.4	51.0	36.3
Efficiency factory, (GJe nominal/tonne)	7.85	8.01	8.53
Elements in acidic emissions, kg/h*			
Sulphur	457	465	433
Nitrogen	211	72	59
Emission factors (kg per tonne coal)			
Sulphur	3.99	4.14	4.10
Nitrogen	1.84	0.64	0.56

Note:
*denotes per unit

been a problem with some pilot-scale plant and solutions are being actively pursued.

EVALUATION OF PRESSURISED FLUIDISED CIRCULATING BED COMBUSTORS

In the combined cycles described in the previous section, the 'bubbling' beds could, in principle, be replaced by circulating fluidised beds. Such beds would be pressurised versions of those described in Section 6.2.

Although, as yet, there has been no experimental work on pressurised circulating beds, a design study [647] for a 250 MWe turbocharged boiler cycle has been carried out. The design of the combustor was based on Lurgi experience with their atmospheric units. As with its 'bubbling' bed counterpart (Item 7 of Table 6.30) the plant was designed to have the maximum number of shop-fabricated units which could be shipped to site by barge. As part of the design study, the 'bubbling' bed and conventional coal plant were independently compared [648] with the following main results:

The higher plant efficiency for the circulating bed is mainly because the coal is fed dry (crushed to minus 6.4 mm) whereas in the 'bubbling' bed the coal is fed as a coal/water mixture. This advantage would be less marked in a high-gas-temperature combined cycle.

From the above it would appear that the circulating bed could be attractive for this particular application (highly modularised). On the debit side, however, the capital costs of the circulating bed plant were estimated to be 19% higher than those of the 'bubbling' bed plant [649].

The benign environment for tubes in the external heat exchangers of this combustion system may reduce tube metal wastage to insignificant levels, and more readily facilitate application of the advanced steam conditions system referred to in the previous section.

The true status of pressurised circulating beds will only become apparent when operational experience with pilot-scale units becomes available.

The potential advantages of the circulating bed system would appear to make this worthwhile.

7. Processes for Removal of Halides

7. Processes for Removal of Halides

Compounds of the halogen elements: fluorine, chlorine, bromine and iodine, are, by and large, volatile and highly soluble in water. The member of the series occurring most abundantly in fossil fuels is chlorine, with smaller amounts of fluorine and trace amounts of the two other halogens.

Concentrations of chlorine in U.K. coals vary from a trace up to about 1.2%, averaging 0.32% (probably rising to about 0.34% by the year 2000) [36].

Fluorine concentrations lie in the range 98 to 130 ppm (U.K. coalfield averages), with a national average of 114 ppm [35].

Because of their volatility, the halogens are almost completely discharged from the furnace in the form of gaseous compounds, principally the hydrogen halides which dissolve readily in water to form strongly acid solutions. Solid halides are not stable at furnace temperatures, but can exist at the lower temperatures occurring in the downstream sections of the boiler, e.g. the air heater, particulate removal equipment and flues. Some absorption of halides may therefore be possible in these regions if an alkaline sorbent is present.

7.1 Classification of Processes

Processes can be divided into three groups: wet/dry (spray drying processes) and dry. In the first two groups, the halide removal occurs as a secondary process to sulphur oxide removal in standard FGD processes. In the third group, halides may not be removed by the FGD process and their removal, if required, necessitates an additional stage. Dry halide removal is also of interest in certain operations that generate gases containing an appreciable concentration of acid halides but only small amounts of sulphur dioxide, such as municipal incinerator operation.

7.2 Process Descriptions and Evaluations

WET SCRUBBING PROCESSES

Because of the high solubility of the acid halides, they are efficiently removed by wet scrubbing processes, even when the scrubbing liquor is so acid that absorption of sulphur dioxide is negligible [447]. When a feed coal contains more than about 0.1% of chlorine, most wet FGD systems only operate satisfactorily if a prescrubber is used to ensure prior removal of the HCl.

Operation of the prescrubber liquid circuit and the isolation and subsequent disposal of the chloride are discussed in Section 2.5. Lime requirements and the quantities of waste products generated for combustion of the two standard coals (Table 1.2) are shown in Table 7.1. There is virtually no market for calcium chloride and, being highly soluble, inland tipping of solid calcium chloride is environmentally unacceptable. Similarly, the discharge of calcium chloride solution into rivers will not be permitted in the quantities produced by a 2000 MWe power station burning the majority of U.K. coals.

Table 7.1 Process Data for HCl Removal in FGD Prescrubbing: 65% Load Factor

	System 1 Coal 1	System 2 Coal 2	System 3 Coal 2
Lime usage, tonne/yr	9,400	3,120	87
$CaCl_2$ prodn. tonne/yr			
As 30% solution	62,000	20,570	573
As $CaCl_2.6H_2O$	36,700	12,170	339

Note: Lime usage is expressed as weight of CaO (pure) and is the quantity required for HCl neutralisation only.

Possible methods of disposal are tipping of the salt at sea, or discharge of a concentrated solution by pipeline into the sea. As explained in Section 2.5, the prescrubber effluent is expected to contain a proportion of the minor and trace elements present in the coal, including heavy metals. In any disposal to the sea, international agreements may require the prior removal of certain elements from the wastes, using techniques described in Section 2.5.

SPRAY DRYING PROCESSES

In all spray drying FGD processes, whatever the sorbent, hydrogen halides will react as well as the sulphur oxides, to give a mixture of halide, sulphite, sulphate and unreacted excess sorbent. The reaction of sorbent with HCl is believed to result in almost total absorption of the pollutant, if not in the spray chamber, then downstream in the flues and baghouse (or ESP).

Problems posed by the presence of chlorides in the product from lime slurry spray drying FGD systems are discussed in Section 2.5. The quantities of calcium chloride produced for the three datum systems are the same as shown for $CaCl_2.6H_2O$ in Table 7.1, although hydration of the salt may not always be complete unless the residue is water-conditioned after discharge.

If the coal ash is removed from the flue gases before they enter the spray dryer, heavy metals from the coal will not be present in the residues but if, as is more likely, the ash is collected together with the residues, then heavy metals and other trace elements such as selenium and arsenic will be present, associated mainly with the coal ash.

DRY PROCESSES

Removal of acid halides may be required in systems employing dry FGD processes which do not also remove halides; also in fluidised bed combustion, where the bed temperature is generally too high to allow the retention of halides. The case of FBC will be dealt with first.

Fluidised Bed Combustion

During fluidised bed combustion, even when limestone is present in the bed, most of the chlorine and from 10 to 60% of the fluorine (increasing as bed temperature is raised from 880 to 950°C) are released into the flue gas as the hydrogen halides [517].

The fate of chlorine during the combustion of a high-chlorine coal (0.7–0.8%, dry basis) in a 4.3 MWt industrial FB boiler has been investigated [532]. The bed temperature was in the region of 900°C and results were presented for three mass balance test periods. There was no sulphur sorbent present. It was found that 90% of the chlorine was emitted as hydrogen chloride in the flue gas, 3% occurred as chloride in the particulate material collected by the grit arrestor and 2% as chloride in the dust suspended in the flue gas leaving the grit arrestor. Five per cent was unaccounted for. This distribution of the element agrees closely with that determined by Munzner and Schilling [517], who found 95% emission in the flue gas at 900°C, falling to 87% at 800°C and to 73% at 750°C bed temperature. The bound chlorine was present mostly in the cyclone ash and filter dust and there was practically none in the bed material.

A similar picture was shown for fluorine but, in this case, bed retention was a little higher, at 7% for a bed temperature of 750°C, falling to 3% at 800°C and to 1–2% at 950°C. Munzner and Schilling also found that the retention of fluorine in the ashes could be increased by adding increasing amounts of fine limestone (1 to 2 μm median particle size). The biggest change in retention occurred between Ca:S mole ratios of 1 and 2, in other words, when the amount of limestone present was in excess of that needed to capture all the sulphur (assuming complete conversion to sulphate). Retention of fluorine increased from 50% to over 80% in this range of Ca:S ratios. The bed temperature at which this was observed is not stated in the paper.

The capture of chlorine by limestone did not occur under the conditions in which fluorine was retained, but the authors suggest that a second bed operating at a lower temperature (300°C optimum) could effect a worthwhile reduction in hydrogen chloride emissions.

Work on chlorine retention in FBC of chlorine-containing waste material has been done in Japan [449]. It was reported that in a packed bed of 1.0 to 2.4 mm particles of uncalcined limestone or dolomite at 800°C, 80 to 95% of the inlet HCl (inlet concentration 500 vppm) was removed from the gas. At the same time, it was noted that the chloride promoted the SO_2 absorption efficiency. The Cl absorption efficiency, determined from inlet and outlet HCl concentrations, remained high, at 70%, even after the absorption of enough HCl to react with 40% of the limestone. This is in contrast with the behaviour on absorption of SO_2, where absorption drops off rapidly due to the formation of an impermeable film of calcium sulphate on the particle surface. X-ray diffraction analysis of the sorbent at the end of the experiment showed the presence of calcium hypochlorite, $Ca(ClO)_2.4H_2O$, and the authors were of the opinion that this compound, rather than $CaCl_2$, was the principal chlorine-containing product.

When starting with calcined limestone or dolomite, the HCl absorption rate was initially high, but it fell more rapidly than when using uncalcined stones. Other sorbents tested were nickel ore and olivine sand, but neither was nearly so effective as limestone or dolomite.

Because of the encouraging nature of these results, tests were then carried out in a 250 mm square fluidised bed combustor in which was burned granular PVC using a bed of sand and limestone [533]. The stone, which had a top size of 2 mm, was premixed with the PVC and sand in the appropriate ratios to give, in a series of tests, Ca:Cl atomic ratios of 1, 2 and 3.

Hydrogen chloride absorption efficiencies increased from zero, at a Ca:Cl ratio of 0, to 40% at Ca:Cl of 1, and to 65% at Ca:Cl of 2 (bed temperature 850°C). At first sight, these results appear to contradict those of the German workers discussed above. The difference can, however, be explained. Fortunately, the Japanese team examined the particles in the bed at the end of each test and also those collected in a cyclone connected to the combustor outlet, using an X-ray diffraction technique. Chlorine-containing compounds were detected in the cyclone material but not in the bed material. It is probable that the cyclone was unheated, and it may be assumed that it operated at a considerably lower temperature than that of the bed–low enough for chlorine absorption to occur as suggested by Munzner and Schilling.

This is clearly a field for further study.

Other Dry Processes

Very little has been published regarding processes for the dry removal of acid halides, but the results quoted above indicate that, given suitable temperature conditions, contacting of the flue gas with certain sorbents downstream of the combustor could lead to substantial reductions in the emissions of acid halides.

Mitsubishi [450] have patented a process in which a slurry of slaked lime or $CaCO_3$ powder is injected into flue gases (especially those from municipal incinerators, glass furnaces and aluminium refining plants, which tend to have high HCl contents), followed by the injection of a 'reaction aid' powder to accelerate the neutralisation which takes place mainly in a bag filter, which is used to collect the injected material. The reaction aid material is a mixture of 90% diatomaceous earth and 10% perlite. It is claimed that the HCl concentration can be reduced from 700 vppm to 10 vppm, using 100 kg of $Ca(OH)_2$ per 30,000 m^3 of flue gas plus 10 kg of reaction aid powder.

A similar process has been patented by Nippon Kokan [448].

Battelle report [651] that Steag AG, at their Kellerman Power Station in West Germany, remove fluorine and chlorine from boiler flue gases in a fixed bed of slaked lime at a temperature of under 200°C before the gases pass to a Resox SO_2 reduction process.

In all absorption processes of this type, the spent sorbent will contain a substantial proportion of calcium chloride and care must be exercised in its disposal (see Section 2.5).

COSTS OF ACID HALIDE REMOVAL PROCESSES

In wet scrubbing or spray dry processes, halide removal is an integral feature of the flue gas desulphurisation process, and separate costing is not necessary or indeed possible. When the fuel contains a large amount of chlorine, the costs of FGD are increased, because prescrubbing is necessary whereas it might be dispensed with if the chlorine content is very low. Even if prescrubbing is to be used in any case, costs will be increased because of the provision that has to be made for isolating large amounts of calcium chloride from the prescrubber liquid circuit, and for its disposal. Costing of the additional expense for this has not been attempted.

The cost of dry processes is likely to be similar to that of the flue gas lime-injection processes for SO_2 removal (Process Category S61).

Appendices

Appendix 1

Bibliography

GENERAL PAPERS ON ACID EMISSIONS ABATEMENT

General Approach to Acidic Emissions Abatement

1 Farrell R.J. and Ziegler E.N. 'Processes for reducing emissions.'

2 Mori A. (Hitachi Ltd.), Yamada H. (Hitachi P.E. and C. Co. Ltd) and Kuroda H. (Babcock-Hitachi K.K.) 'Flue gas treatment system for coal-fired power plants'. *Hitachi Review*, 1980, **29** (6) 303–306.

3 Kyte W.S., Bettelheim J. and Cooper J.R.P. (CEGB). 'Possible fossil fuel developments within the electric power generation industry and their impact on other industries'. Symposium Series 78, 1983.

4 Van der Brugghen F.W. (Kema). 'Technological measures to reduce air pollution during coal firing'. *Resource and Conservation*, 1981, **7**, 133–143.

5 Meagher J.F., Stockburger L., Bananno R.J., Bailey E.M. and Luria M. (Tennessee Valley Authority). 'Atmospheric oxidation of flue gases from coal fired power plants–a comparison between conventional and scrubbed plumes'. *Atmospheric Environment*, 1981, **15** (5), 749–762.

6 Jahnig C.E. and Shaw H. (Exxon). 'A comparative assessment of flue gas treatment processes, Part 1–Status and design basis.' *JAPCA*, 1981 (Apr), **31** (4), 421–428.

7 Jahnig C.E. and Shaw H. (Exxon). 'A comparative assessment of flue gas treatment processes, part 2–Environmental and cost comparison'. *JAPCA*, 1981 (May), **31** (5), 596–604.

8 Kyte W.S. (CEGB) 'Some implications of possible emission control technologies in the electric power generation industry'. Paper to I. Chem. E/SCI Symposium, Control of Acid Emissions in the UK, November 1985.

9 Catalano L., Elliott T.C. and Makansi J. 'Control technologies mature as policy debate lulls'. *Power*, 1985 (May), 13–20.

10 Environmental Resources Ltd. 'Acid rain–a review of the phenomenon in the EEC and Europe', Report for the Commission of the European Communities. Graham and Trotman Ltd., 1983.

11 Szabo M., Shah Y. and Abraham J. (PEDCO) 'Acid rain: control strategies for coal fired utility boilers. Volume 1'. Report No. DOE/METC-82-42 (Vol. 1) to US Department of Energy, May 1982.

12 Institution of Chemical Engineers/Society of Chemical Industry. 'Control of acid emissions in the UK' London, November 1985.

13 Dudley N., Barret M. and Baldock D. (Earth Resources Research). 'The acid rain controversy', 1985.

14 Anonymous. 'Acid rain. Its causes, effects and abatement'. *Sulphur* 1983, (Nov–Dec), (169), 32–37.

15 Cope D.R. (IEA Coal Research). 'Acid rain: The available control technology'. Paper to Acid Rain Inquiry, Edinburgh, 29 Sept, 1984.

16 Cope D.R. (IEA Coal Research). 'Control of acidic emissions from static plant.' Conference, 1985.

17 Wallin S.C. (Warren Spring Laboratory). 'Abatement systems for SO_x, NO_x particles–Technical options'. Paper to Institution of Environmental Sciences Seminar, An Update on Acid Rain, London, November 1984.

18 Elliott T.C. and Schwieger R.G. (editors) 'The acid rain source book'. New York: McGraw-Hill Inc, 1984.

19 Schwieger R.G. and Elliott T.C. (editors). 'Acid rain engineering solutions, regulatory aspects'. New York: McGraw-Hill Publications Co., 1985.

20 Goklany I.M. and Hoffnagle, G.F. (TRC Environmental Consultants). 'Trends in emissions of PM, SO_x and NO_x and VOC: NO_x ratios and their implications for trends in pH near industrialised areas'. *JAPCA.*, 1984, 844–846.

21 Beer J.M. 'Clean combustion of coal, research and applications. An overview of recent developments in the USA.' *J. Inst. Energy*, 1986 (March), **59** (438), 3–19.

22 Truchot A. et al. 'Contribution of petroleum refineries to emissions of nitrogen oxides.' Report No. 9/84 to CONCAWE, May 1984.

23 De Meulemeester A. et al. 'Sulphur dioxide emissions from oil refineries and combustion of oil products in Western Europe in 1979 and 1982.' Report No. 10/84 to CONCAWE, May 1985.

24 Nakabayashi Y. (EPDC). 'A future forecast of the research and evaluation on the integrated flue gas treatment system'. Paper to EPDC/IIP/VDI, NO_x Symposium, Karlsruhe, West Germany, February 1985. Paper I.

25 Hovey H.H., Davis E., Sistla G., Glavin P., Twaddell R. and Rao S.T. (New York State Department of Environmental Conservation). 'Evaluation, selection and economic assessment of control strategies for acidic deposition.' Paper to APCA, 78th Annual Meeting, Detroit, USA, June 1985. Paper 85–1B.3.

26 Miller M.J. (EPRI). 'Retrofit SO_2 and NO_x control technologies for coal-fired power plants'. Paper to APCA, 78th Annual Meeting, Detroit, USA, June 1985. Paper 85-1A.2.

27 Rubin E.S. (Carnegie-Mellon University). 'Air pollution constraints on increased coal use by industry–an international perspective.' *JAPCA*, 1981 (Apr), **31** (4), 349–360.

28 Sedman C.B. (US EPA) and Ellison W. (Ellison Consultants). 'German FGD/deNO_x experience'. Paper to Third Annual Pittsburg Coal Conference, Pittsburgh, USA, September 1986.

29 Koch H. (VEBA Kraftwerke Ruhr). 'Effects of environmental protection measures on planning, operation and costs of conventional thermal power stations.' *VGB Kraftwerkstechnik* 84, 1984 (Dec), (12), 935–941.

30 Anonymous. 'CEGB claims credit for acid rain move as international pressure intensifies.' *ENDS Report* 140, 1986 (Sept), 3–5.

31 Anonymous. 'Acid emissions: an opportunity for British industry?' *ENDS Report* 140, 1986 (Sept), 13–15.

32 Kimura T., Nakabayashi T. and Mouri K. (Electric Power Development Co.). 'Overall flue gas treatment technology from coal-fired power plants in Japan.' Paper to 5th International Conference on Coal Research, Düsseldorf, FRG, September 1980, Paper D-5, 313–333.

33 Barsin J.A. (Babcock & Wilcox). 'Options for reducing NOx and SO emissions during combustion'. Paper to *Power Magazine*, First International Conference on Acid Rain, Washington, USA, March 1984.

34 Ando J. (Chuo University). 'Recent developments in SO_2 and NO_x abatement technology in Japan.' Paper to EPA/EPRI, Symposium on Flue Gas Desulphurisation, Cincinnati, USA, June 1985.

35 National Coal Board. 'Study into the emission of air pollutants coming from the use of coal within the United Kingdom.' Report No. EUR 6853 EN to Commission of the European Communities, Part 1, p. 14, 1980.

36 National Coal Board (British Coal), Coal Research Establishment. CRE Technology Review No. 2, 1982 (December).

37 Chem Systems International Ltd. 'Reducing pollution from selected energy transformation sources'. A study for the Commission of the European Communities, Environment and Consumer Protection Service. London: Graham & Trotman Ltd., 1976.

38 Buckley-Golder D.H. (ETSU). 'Acidity in the environment'. ETSU Report No. R.23, June 1984.

39 Derwent R.G. (ETSU). 'The nitrogen budget for the UK and NW Europe'. ETSU Report No. R.37, April 1986.

40 Department of the Environment. 'Digest of environmental protection and water statistics: 1986, No.9. HMSO.

41 Walker D.S., Galbraith R. and Galbraith

J.M. (Warren Spring Laboratory). 'Survey of nitrogen oxides, carbon monoxide and hydrocarbon emissions from industrial and commercial boilers in Scotland'. WSL Report No. LR524 (AP)M.

Emission Limitation Regulations

42 Evans P. (Department of the Environment, UK). 'The EC Directive on smoke and SO_2: The future for smoke control'. Paper to National Society For Clean Air, 47th Annual Conference, Bournemouth, September 1980.

43 Parkinson G. 'SO_2–Removal techniques ready for tighter curbs.' *Chemical Engineering*, 1983 (25 July), 17–20.

44 Catalano L. and Makansi J. 'Acid rain: New SO_2 controls inevitable.' *Power*, 1983 (Sept), 25–33.

45 Remirez R., Hoppe R., McQueen S. and Smith J. 'Acid rain: Europe, Canada, Act; the US Dithers'. *Chemical Engineering*, 1983 (11 July), 29–31.

46 Anonymous. 'An administration backed acid rain bill may be forthcoming'. *Chemical Engineering*, 1983 (22 August), 20.

47 Anonymous. 'An acid rain study is slated in Europe; action is expected in US as well'. *Chemical Engineering*, 1983 (19 Sept), 10.

48 Siegfriedt W.E. and Ludwig M. (Fluor). 'Desulphurisation processes in West Germany–an overview'. Power conference 1984.

49 Select Committee on the European Communities. 'Air pollution–22nd Report, Session 1983–84'. Report No. HL 265 to House of Lords, 26 June 1984.

50 Department of the Environment. 'Acid rain: The Government's reply to the fourth report from the Environment Committee, Session 1983–84'. Report No. HC446-1 to Parliament, December 1984.

51 Environment Committee. 'Acid rain. Vol. 1. Report together with the proceedings of the committee relating to the report.' Fourth report to the House of Commons, Session 1983–84 (HMSO 446-1).

52 Westaway M.T. and McKay J. (British Petroleum). 'The impact of legislative requirements and legislative change on industry.' National Society for Clean Air, Oxford, March 1984.

53 Brady G.L. (President's Council on Environmental Quality, Washington) and Conway G. (Imperial College). 'Market approaches for sulphur dioxide management: A comparative analysis of Great Britain and the United States.' Proposal to Economic and Social Research Council, September 1984.

54 Clarke A.J. 'European legislative position'. Watt Committee Acid Rain Working Group.

55 Means C.S. (Associated Electric Cooperative Inc.) and Landwehr J.B. (Burns and McDonnell). 'Estimated economic impact of proposed acid rain control regulations on Associated Electrics Cooperative Inc., Springfield, Missouri'. Paper to Coal Technology '83, 6th International Coal and Lignite Utilisation Conference, Houston, USA, November 1983, **3**, 269–280.

56 Trisko E.M. (Stern Brothers). 'Potential impacts of acid rain control legislation.' Paper to Coal Technology '83, 6th International Coal and Lignite Utilisation Conference, Houston, USA, November 1983, **3**, 233–257.

57 Commission of the European Communities. 'Proposal for a Council Directive on limitation of emissions of pollutants into the air from large combustion plants'. Report COM (83) 704 Final, Brussels, 15 December 1983. Amended Proposal, COM (85) 47 Final, 18 February 1985.

58 Dacey P.W. (IEA Coal Research). 'An overview of international NO_x control regulations.'

59 Organisation for Economic Co-operation and Development. 'Emission standards for major air pollutants from energy facilities in OECD member countries'. Paris, 1984.

60 Aniansson B. 'A firm commitment'. *Acid Magazine 3*, Autumn 1985.

61 Haigh N. 'EEC environmental policy and Britain'. London: Environmental Data Services Ltd, 1984.

62 Short H., Herd J., McQueen S., and Smith J. 'EC producers grapple with a plan that limits air emissions'. *Chemical Engineering*, 1984 (14 May), 20E–20H.

63 Catalano L., 'Acid rain controls defeated in surprising sub-committee vote.' *Power*, 1984 (June), 9–10.

64 Catalano L., 'Environmentalists sue EPA to

issue industrial NSPS'. *Power*, 1984 (May), 9–10.

65 Catalano L., 'SO_2 standard not included in EPA's proposed NSPS for industrial boilers'. *Power*, 1982 (June), 9–10.

66 Clean Air (Emission of grit and dust from furnaces) Regulations 1971, (Statutory Instrument 1971 No. 162).

General Papers on Acidic Emissions Abatement Costs

67 Leggett A., Rubin E. and Torrens M. (OECD). 'Comparing the costs of flue gas treatment systems internationally'. Paper to ECE, Fourth Seminar on the Control of Sulphur and Nitrogen Oxides from Stationary Sources, Graz, Austria, May 1986.

68 Eriksson S., Forrester R., Johnston R. and Teper M. (IEA Coal Research). 'Economic and technical criteria for coal utilisation plant. Part 1: Economic and financial conventions'. Report No. A1/77, December 1977.

69 Torrens I.M. (OECD). 'Coal pollution abatement costs'. Paper to Coal Technology '83, 6th International Coal and Lignite Utilisation Conference, Houston, USA, November 1983, **3**, 257–267.

70 Maxwell J.D., Humphries L.R. (Tennessee Valley Authority), and Mobley J.D. (US Environmental Protection Agency). 'Economics of NO_x, SO_2 and ash control systems for coal-fired utility power plants.' Joint Symposium on Stationary Combustion NO_x Control, Boston, USA, May 1985. Paper 8a–7.

71 Remirez R. 'Looking into cheaper ways to deal with acid rain'. *Chemical Engineering*, 1986 (7 July), 17–19.

72 Ireland P.A. and Keeth R.J. (Stearns-Roger). 'Economic comparison of wet vs dry FGD'. Paper to National Lime Association Conference, Effective Use of Lime for Flue Gas Desulphurisation, Denver, USA, September 1983.

73 Organisation for Economic Co-operation and Development, Environment Directorate. 'Understanding pollution abatement cost estimates'. Report No. W.0067, March 1986.

74 Rubin E.S. and Torrens I.M. (Editors). 'Cost of coal pollution abatement: results of an international symposium'. Paris, OECD, 1983.

75 Ponder T.C., Yerino L.V., Katari V., Shah Y. and Devitt T.W. (PEDCO). 'Simplified procedures for estimating FGD system costs'. Report No. EPA-600/2-76-150 to US EPA, June 1976.

76 Kaplan N., Lachapelle D.G. and Chappell J. (US Environmental Protection Agency). 'Control cost modelling for sensitivity economic comparison'. Paper ID 5A.

77 Reisdorf J.B., Keeth R.J., Miranda J.E., Scheck R.W. (Stearns-Roger) and Morasky T.M. (Electric Power Research Inst.). 'Economic evaluation of FGD systems'. Paper to EPA/EPRI Flue Gas Desulphurisation Symposium, New Orleans, USA, November 1983.

78 Bobman M.H., Weber G.F. (University of North Dakota) and Dorchak T.P. (US Department of Energy). 'Comparative costs of flue gas desulphurisation: advantages of pressure hydrated lime injection'. Paper to EPA/EPRI Joint Symposium on Dry SO_2 and Simultaneous SO_2/NO_x Control Technologies, Raleigh, USA, June 1986. Report DOE/FE/60181-177.

79 Mora R.R., Ireland P.A. (Stearns-Roger) and Morasky T. (Electrical Power Research Institute). 'Estimating procedure for retrofit FGD costs'. Paper to EPA/EPRI Flue Gas Desulphurisation Symposium, New Orleans, USA, November, 1983. Paper 2B.

80 Damon J.E., Scheck R.W. (Stearns-Roger Engineering Co.) and Cichanowicz J.E. (Electric Power Research Institute). 'Economics of SCR post combustion NO_x control processes'. Paper to Joint Symposium, Stationary Combustion NO_x Control, 1982.

81 Ireland P.A., Brown G.D., Sebesta J.J. (Stearns Catalytic Corp.) and McElroy M.W. (Electric Power Research Institute). 'Economics of furnace sorbent injection for SO_2 Emission control'. Paper to EPA/EPRI Symposium on Dry SO_2 and Simultaneous SO_2/NO_x Control Technologies, Raleigh, USA, June 1986. Paper 5B.

82 McMahon T.C. and Rigsby L.S. (Ashland Coal). 'The high road to sulphur dioxide reduction'. Paper to Coal Technology Europe, 3rd European Coal Utilisation Conference, Amsterdam, The Netherlands, October 1983, **4**, 137–147.

83 Lachapelle D.G., Kaplan N. and Chappell J. (US Environmental Protection Agency).

'EPA's LIMB cost model: development and corporative care studies'. Paper 6H.

84 Storm J. (Niro Atomizer). 'Economics of dry FGD and byproducts handling'. Paper to OECD International Symposium, The Economic Aspects of Coal Pollution Abatement Technologies, Petten, The Netherlands, May 1982.

85 Naulty D.J. (Stearns-Roger), Muzio L.J. (KVB Corp.) and Hooper R. (Electric Power Research Institute). 'Economics of dry FGD by sorbent injection'. Paper to Coal Technology '83, 6th International Coal and Lignite Utilisation Conference, Houston, USA, November 1983, **3**, 209–229.

86 Krieg J.P., Mortimer G.W. and Weiss L.H. (Cogit Consulting Group). 'Economics of sulphur removal using "dry"-carbonated trona ore FGD sorbent'. Paper No. 5D.

87 Musgrove J.G. and Donnelly J.R. (Bechtel Power Corp.) 'Dual-alkali flue gas desulphurisation system cost versus operating availability'. Paper to Coal Technology '83, 6th International Coal and Lignite Utilisation Conference, Houston, USA, November 1983, **3**, 139–153.

88 Hollinden G.A., Stephenson G.A. (Tennessee Valley Authority) and Stensland J.G. (FMC Corp). 'An economic evaluation of limestone double-alkali flue gas desulphurisation systems'. Paper to Coal Technology '83, 6th International Coal and Lignite Utilisation Conference, Houston, USA, November 1983, **3**, 155–187.

89 Scharer B. and Haug N. 'The cost of flue gas desulphurisation and denitrification in the Federal Republic of Germany'. Paper to Fourth Seminar, Control of Sulphur and Nitrogen Oxides from Stationary Sources, Austria, May 1986.

90 Torrens I.M. (OECD). 'Coal pollution abatement costs'. Paper to Coal Technology Europe, Third European Coal Utilisation Conference, Amsterdam, October 1984, **4**, 213–223.

91 Sutherland H. et al. 'Cost of control of sulphur dioxide, nitrogen oxides and particulates emissions from large combustion plants in oil refineries'. Report No. 7/84 to CONCAWE, September 1984.

92 Samish N.C. (Shell Development Company). 'The cost of FGD'. Paper to 14th Annual Meeting of the Battelle Stack Gas Assessment and Technologies Survey Programme, London, September 1986.

93 Bakke E. (Peabody Process Systems). 'Economical retrofit of wet scrubbers to coal fired boilers'. 1986.

94 Morasky T.M., Dalton S.M. and Preston G.T. (EPRI). 'Economic assessment and operating and maintenance costs of FGD systems'. Paper to Seminar on Flue Gas Desulphurisation, Ottawa, Canada, September 1983.

95 Keeth R.J., Miranda J.E., Reisdorf J.B. and Scheck R.W. (Stearns-Roger). 'Economic evaluation of FGD systems. Volumes 1–3.' Report No. CS-3342 to EPRI, December 1983.

96 Shattuck D.M. et al. (Stearns Catalytic Corp.). 'Retrofit FGD cost-estimating guidelines'. Report No. CS-3696 to EPRI, October 1984.

97 Rubin E.S. 'Contribution of pollution costs to total annual revenue requirements, US, Germany and Japan (1982 US $)'. *Env. Sci. Technol.*, 1983, **17** (8), 366A–377A.

98 Highton N.H. and Webb M.G. 'Pollution abatement costs in the electricity supply industry in England and Wales'. *Journal of Industrial Economics*, 1981 (September), **3–** (1), 49–65.

99 Anonymous. 'Will pollution control costs hinder the return of coal?' (Summary of OECD Symposium). *ENDS Report*, 1983 (October), (105), 10–13.

100 Scharer B. and Haug N. 'On the economics of flue gas desulphurisation–measures, costs and effectiveness'. Paper to OECD, International Symposium on the Economic Aspects of Coal Pollution Abatement Technologies, Petten, The Netherlands, May 1982. OECD Proceedings, 1983. (Editors: Rubin E.S. and Torrens I.M.).

101 Wijdeveid H.W.J. (ESTS). 'Recent developments in reduction of cost of FGD systems in the Netherlands'. Paper to OECD/ENEA Enclair '86 Symposium, Taormina, Italy, October 1986.

102 Leggett J.A. (OECD). 'Costs and cost-effectiveness of techniques to reduce emission of NO_x and VOC: development of the OECD compendium'. Paper to OECD/ENEA Enclair '86, Taormina, Italy, October 1986.

103 Smith T.F. (CEGB). 'Factors affecting costs of power station emission controls with particular reference to the UK' Paper to OECD/ENEA Enclair '86, Taormina, Italy, October 1986.

Effects of Acidic Gases on Equipment and Materials of Construction

104 Verhoff F.H. and Choi M.K. (West Virginia University). 'Effects of sulphuric acid condensation on stack gas equipment'. *Journal of Institute of Energy*, 1980 (Jun), 92–99.

105 Berger D.M., Trewella R.J., and Wummer C.J. (Gilbert/Commonwealth). 'Evaluating linings for power plant SO_2 scrubbers'. *Power Engineering*, 1980 (Nov), 71–75.

106 Ellis P.F., Anliker D.M., Jones G.D. (Radian Corporation), and Steward D.A. (EPRI). 'FGD system failure analyses of metallic components'. *Materials Performance*, 1986 (Mar), 15–23.

107 Beavers J.A. and Koch G.H. 'Review of corrosion related failures in FGD systems'. *Materials Performance*, 1982 (Oct), 13–25.

108 Ellison W. (consultant) and Lefton S.A. (Aptech). 'FGD's reliability: What's being done to achieve it?' *Power*, 1982 (May), 71–76.

109 Kyte W.S. (CEGB). 'Corrosion in FGD plant'. Chapter 6, 'Dewpoint corrosion'. (Holmes, Editor). Chichester: Ellis Horwood Ltd., June 1985.

110 Javetski J. 'Solving corrosion problems in air-pollution control equipment, Part I'. *Power*, 1978 (May), 72–77.

111 Javetski J. 'Solving corrosion problems in air-pollution control equipment, Part II'. *Power*, 1978 (June), 80–97.

112 Dene C.E., Syrett B.C., Koch G.M. and Beavers J.A. 'Alloys and coatings for SO_2 scrubbers'. Paper to American Power Conference, Chicago, USA, April 1982.

113 Forsythe R.C., Hirt F.K. and Richards W.E. 'Chimney liner experience at the Bruce Mansfield plant'. Paper to American Power Conference, Chicago, USA, April, 1982.

114 Lee T.S. (LaQue Centre for Corrosion Technology) and Lewis R.O. (Montana State University). 'Evaluation of corrosion behaviour of materials in a model SO_2 scrubber system'. *Materials Performance*, 1985 (May), 25–32.

115 Pitt W.G. and Andersen T.N. (Kennecott Minerals Co.). 'Corrosion of alloys in simulated smelter FGD scrubber solutions'. *Materials Performance*, 1982 (May), 26–29.

116 Dille E.R., Froelich D.A. and Weilert C.V. (Burns & McDonnell) 'Tame the latest FGD-system corrosion pest: fluorides'. *Power*, 1983 (Aug) 41–42.

117 Rosenberg H.S., Koch G.H. (Battelle Columbus), Meadows M.L. (Black & Veatch) and Steward D.A. (EPRI). 'Materials for outlet ducts in wet FGD systems'. *Materials Performance*, 1986 (Feb), 41–55.

118 Deleted

119 Johnson C.A. (Peabody Process Systems). 'Evaluation of materials of construction for Alabama Electric Co-operative's limestone FGD system'. Paper to National Association of Corrosion Engineers, Denver, USA, August 1981.

120 Anonymous. 'Why put a fan on the wet side of a scrubber'. *Power*, 1986 (Sept.), 151–152.

PAPERS ON FLUE GAS DESULPHURISATION

General Approach to Flue Gas Desulphurisation and Recovery Processes

121 Makansi J. 'Optimizing today's processes for utility and industrial power plants. A special report'. *Power*, 1982 (Oct), S.1–S.24.

122 Kyte W.S. (CEGB). 'Some chemical and chemical engineering aspects of FGD'. *Trans. I. Chem. E.*, 1981, **59**, 219–228.

123 United Nations, Economic Commission for Europe. 'Air pollution studies 1. Air-borne sulphur pollution: effects and control'. Report No. GE.84–40823, 1984.

124 Moser R.E. (Brown and Root). 'FGD options offer environmental trade-offs'. *Hydrocarbon Processing*, 1981 (Oct), 88–92.

125 Melia M.T., McKibben R.S. and Pelsor B.W. (PEDCo). 'Project summary. Utility FGD survey: October 1983–September 1984'. Report No. EPA–340/1-85-014 to US Environmental Protection Agency, October 1984.

126 Rittenhouse R.C. 'Equipment retrofits add conformance to emissions control'. *Power Engineering*, 1986 (Sept), **90** (9), 18–24.

127 Kyte W.S., Bettelheim J. and Cooper J.R.P. (CEGB). 'Sulphur oxides control options in the UK electric power generation industry'. *I.*

Chem. E. Symposium Series, (77), Loughborough, April 1983.

128 Marx J.A. and Nagaraja M.L. (M.W. Kellogg). 'Structural engineering vital to FGD design'. *Electrical World*, 1983 (July), 93–95.

129 Martel G. (Northeast Utilities Service) and Veratti T. (Nalco). 'Reduce impact of acid emissions from your oil-fired boiler'. *Power*, 1983 (Oct), 105–106.

130 Wall J.D. 'Control FCC SO_x emissions'. *Hydrocarbon Processing*, 1984 (Oct), 45–46.

131 Sheppard S.V. (Ceilcote). 'Tailor air pollution control equipment to applications and requirements'. *Power Engineering*, 1986 (Feb), 32–35.

132 Ellison W. (Ellison), Leimkuhler J. (GEA) and Makansi J. 'West Germany meets strict emission codes by advancing FGD'. *Power*, 1986 (Feb), 29–33.

133 Mobley J.D. (US EPA) and Dickerman J.C. (Radian). 'Commercial utility FGD systems'. *Mechanical Engineering*, 1984 (July), 62–71.

134 Beals J. (Pennwalt), Cannell L. and Hengel J. (Black and Veatch). 'How FGD reagent quality affects system performance'. *Power*, 1984 (Mar), 27–30.

135 Anonymous. 'Product guide: flue gas desulphurisation'. *Modern Power Systems*, 1986 (June), 99–101.

136 Schwieger R., and Hayes A. 'Reliability concerns, regulations lead to virtual standardisation of air pollution control systems'. *Power*, 1985 (April), **129** (4), 81–93.

137 Morrison G.F. (IEA Coal Research). 'Control of sulphur oxides from coal combustion'. Report No. ICTIS/TR21, IEA Coal Research, London, November 1982.

138 CONCAWE. 'SO_2 emission trends and control options in Western Europe'. Report No. 1/82, 1982 (Jan).

139 The Watt Committee on Energy. 'Acid rain'. Report No. 14, August 1984.

140 Meyer C.E. 'Flue gas heat exchanger increases efficiency, reduces pollution'. *Power Engineering*, 1986 (Mar), 30–31.

141 Gillette J.L. and Chiu S.Y. (Argonne National Laboratory). 'FGD: Review of selected commercial and advanced technologies'. Report No. ANL/FE-81-51 for US Department of Energy, February 1981.

142 NATO Committee on the Challenges of Modern Society. 'FGD. First follow-up report: Control of air pollution from coal combustion.' Report No. 138, Ottawa, Canada, June 1982.

143 NATO Committee on the Challenges of Modern Society. 'FGD. Second follow-up report: Control of air pollution from coal combustion–Focus on NO_x and Limestone injection multistage burner NO_x/SO_x control technology'.

144 United Kingdom contribution to NATO-CCMS FGD Study Group. 'Control of air pollution from coal combustion'. Vienna, Austria, May 1986.

145 Lunt R.R. and Mackenzie J.S. (United Engineers & Constructors). 'Longer-term options for reducing SO_2 emissions'. Paper to Power Magazine, First International Conference on Acid Rain, Washington, USA, March 1984.

146 Pruce L.M. 'Why so few regenerative scrubbers'. *Power*, 1981 (June), 73–76.

147 VEW Special Steels (UK) Ltd. 'Materials for FGD Plant'. February 1986.

148 Steiner P. (Foster Wheeler), Dalton S.M. (EPRI) and Knoblauch K. (Bergbau Forschung). 'Capture and conversion of SO_2 RESOX prototype demonstration in Germany'. *Combustion*, 1980 (Jan), 28–31.

149 Gutterman C. and Steiner P. (Foster Wheeler). 'Continuous testing of the RESOX process–Final Report'. Report No. FWC/FWDC/TR-84/28 to Electric Power Research Institute, August 1984.

150 Gutterman C., Steiner P., Aiello M. and Violante D. (Foster Wheeler). 'RESOX process for urban FGD system–Final Report'. Report No. FWC/FWDC/TR-85/9 to Empire State Electric Energy Research Corporation, June 1985.

151 Ellis R.J. et al. 'Sulphur emissions from combustion of residual fuel oil based on EEC energy demand and supply, 1980–2000'. Report No. 5/86 to CONCAWE, July 1986.

152 Rittenhouse R.C. 'Additives: a lower cost alternative to hardware retrofits'. *Power Engineering*, 1986 (June), **90** (6), 18–24.

153 Anonymous. 'Growing FGD usage brings its own problems'. *Process Engineering*, 1985 (May), 41–42.

154 Neukam H. 'Flue gas desulphurisation with Ljungstrom heat exchanger'. *Chem. Tech.*, 1983, **12** (1), 18–20.

155 Rosenberg H.S. and Choi P.S.K. (Battelle). 'Energy aspects of FGD and stack gas reheat'. *AIChE Symp. Ser.*, 1980, **76** (196), 28–37.

156 Bettelheim J., Kyte W.S. and Littler A. (CEGB). 'Fifty years' experience of FGD at power stations in the UK'. *The Chemical Engineer*, 1981 (June), 275–278.

157 Kohl H. and Riesenfeld F. 'Gas Purification'. Gulf Publishing Co., Third Edition, 1979.

158 McIlvaine Scrubber Manual, Vol. IV, 1979.

Non-Regenerable Solution-Based Wet Processes (Category S10)

159 Reason J., Baur P. and Makansi J. 'Finch, Pruyn cleans air and water while increasing steam production'. *Power*, 1981 (Nov), 73–83.

160 Brady J.D. (Andersen 2000). 'Particulate and SO_2 removal with wet scrubbers'. *CEP*, 1982 (June), 73–77.

161 Anonymous. 'NaOH scrubbing process most attractive for small-scale SO_2 pollution abatement'. *TI Chem.E.* 1971 (Jan), **16** (1), 7–9.

162 Ponder W.H. (US EPA), Fischer W.H. (Gilbert Associates) and Zaharchuk R. (Firestone). 'Environmental assessment of the dual alkali FGD system applied to an industrial boiler firing coal and oil'. *A.I.Ch.E. Symposium Series*, 1980, **76** (201), 80–95.

163 Henry J.R., Wrobel B.A. (Northern Indiana Public Service Company), Ellefson D.W., Katzberger S.M., Predick P.R. (Sargent and Lundy). 'Lime handling and preparation for two double-alkali FGD systems'. Paper to Coal Technology '83, 6th International Coal and Lignite Utilisation Conference, Houston, USA, November 1983, **2**, 37–48.

164 Kirchgessner D.A., Gullett B.K. (US Environmental Protection Agency) and Lorrain J.M. (Acurex). 'Physical parameters governing the reactivity of $Ca(OH)_2$ with SO_2'. Paper to EPA/EPRI Joint Symposium on Dry SO_2 and Simultaneous SO_2/NO_x Control Technologies, Raleigh, USA, June 1986. Paper 2D.

165 Deleted

166 Anonymous. 'FMC announces pre-engineered double alkali scrubber system'. ??? 1981 (8 Jul), 4.

167 Brady J.D. (Andersen 2000). 'Sulphur dioxide removal from exhaust gases'. *CEP*, 1984 (Sept), 59–62.

168 Anonymous. 'Dry scrubber overcomes scale-up problems at Coyote'. *Power*, 1983 (Apr), 114–115.

169 Lewis M.F. (Montana-Dakota Utilities), and Gehri D.C. (Rockwell). 'Atomisation–The key to dry scrubbing at the Coyote Station'. Paper to EPA/EPRI Symposium, Flue Gas Desulphurisation, May 1982.

170 Stern J.L. (Joy). 'Dry scrubbing for FGD'. *CEP*, 1981 (Apr), 37–42.

171 Francis D.V. (Arco Chemical Co.), Biolchini R.J. and Coons J.D. (FMC). 'Operating experience with high sulphur coal in an industrial double alkali FGD system'. Paper to Coal Technology '81, 4th International Coal Utilisation Conference, Houston, USA, November 1981, **3**, 215–229.

172 Tamaki A. (Chiyoda Chemical Engineering & Construction Co. Ltd). 'Commercial application of dilute sulphuric and/gypsum (the Chiyoda Thoroughbred 101) FGD process for large power plant boilers'. Paper to 66th Annual AIChE Meeting, Washington, USA, December 1974.

Non-Regenerable Slurry-Based Wet Processes (Category S20)

173 Sugita Y., Oguri H. and Sakamoto Y. (IHI). 'State of the art in desulphurisation system for preventing environmental pollution caused by flue gas from coal fired power stations.' *IHI Engineering Review*, 1984 (Oct), **14** (4), 29–35.

174 Esche M. (Saarberg-Holter). 'Stack gas desulphurisation without reheating'. Paper to Coal Technology Europe, 3rd European Coal Utilisation Conference, Amsterdam, Netherlands, October 1983, **4**, 93–102.

175 Johnson C.A. (Peabody Process Systems Inc.). 'Flyash alkali technology–Low cost FGD'. Paper to Coal Technology '80, Houston, USA, November 1980, 569–588.

176 Kunzweiler V.L., Landwehr J.B., Collier C.W. and Froelich D.A. (Burns and McDonnell). 'Start-up experience of five FHD systems'. Paper to American Power Conference, Chicago, USA, April 1980.

177 Van Ness R.P. (Louisville Gas and Electric Co), Kingston W.H. and Borsare D.C. (Combustion Engineering). 'Operation of C-E FGD system for high sulphur coal at Louisville Gas and Electric Co., Cane Run 5'. *Combustion*, 1980 (Feb), 10–16; and Vol. 41 of Proceedings of the American Power Conf. 1979, pp. 656–664.

178 Makansi J. 'Wet venturi doubles as an SO_2 scrubber'. *Power*, 1981 (Dec), 71–72.

179 Yeager K. (EPRI). 'Advanced SO_2 control'. *EPRI Journal*, 1981 (Mar), 38–39.

180 Johnson C.A. (Peabody Process Systems). 'Minnesota Power's operating experience with integrated particulate and SO_2 scrubbing'. *Journal of the Air Pollution Control Association*, 1981 (Jun), **31** (6), 701–705.

181 Hoffman D.C. (Dravo Lime). 'Thiosorbic lime for FGD processes'. *Mining Engineering*, 1981 (Nov), 1628–1631.

182 Catalano L. (compiler). 'FGD improves with adipic acid'. *Power*, 1982 (Jul), 84–85.

183 Nesbit W. 'Scrubbers: The technology nobody wanted'. *EPRI Journal*, 1982 (Oct), 8–15.

184 Ellison W. (consultant) and Kutemeyer P.M. (Bischoff). 'New developments advance forced-oxidation FGD'. *Power*, 1983 (Feb), 43–45.

185 Anonymous. 'Research boosts thiosulphate for FGD systems'. *Chemical Engineering*, 1983 (8 Aug), 11–12.

186 Chang J.C.S. (Acurex) and Mobley J.D. (US EPA). 'Testing and commercialisation of by-product dibasic acids as buffer additives for limestone FGD systems'. *JAPCA*, 1983 (Oct), (10), 955–956.

187 Esche M. and Igelbuscher, H. (Saarberg-Holter). 'Technical solutions for the new SO_2 legislation in West Germany–FGD without reheating'. Paper to Air Pollution Control Association, 77th Conference, San Francisco, USA, June 1984. Paper No. 84–971.

188 Makansi J. 'A limestone FGD system'. *Power*, 1985 (Sept), 107–109.

189 Kojima T., Shikishima S., Kanamori A. and Torii M. (IHI). 'Operating results of FGD system for unit No.3 (700MW) at Takehara thermal power station of the Electric Power Development Co. Ltd.' *IHI Engineering Review*, **17** (2), 1–5.

190 Murphy K.R., Shilling N.Z. (General Electric Environmental Services Inc) and Pennline B.H. (US Department of Energy). 'Low cost in-duct scrubbing system will be tested at Muskingum River'. *Modern Power Systems*, 1986 (Jun), 79–83.

191 Kirchner R.W. (Cabot Corp.). 'Materials of construction for flue gas desulphurization systems'. *Chemical Engineering*, 1986 (19 Sept), 81–86.

192 Rock K.L. (D.M. International), Glamser J.H. (Davy McKee) and Esche M. (Saarberg-Hoelter). 'Commercial operating history and latest developments to the Davy S-H process'. Paper to Coal Technology '81, 4th International Coal Utilisation Conference, Houston, USA, November 1981, **3**, 261–280.

193 Rader P.C., Hansen R.W. and Borsare D.C. (Combustion Engineering). 'Design of lime/-limestone flue gas desulphurisation systems for high chlorides'. Paper to Coal Technology 1981, 4th International Coal Utilisation Conference, Houston, USA, November 1981, **3**, 231–261.

194 Anonymous. 'High sulphur coal tests demonstrate the successful operation of dry scrubbing'. *Chemical Engineering*, 1982 (22 Mar), 18.

195 Karlsson H.T., Klingspor J., Linne M. and Bjerle I. (Chemical Centre). 'Activated wet-dry scrubbing of SO_2'. *JAPCA*, 1983 (Jan), **33** (1), 23–28.

196 Blythe G.M., Burke J.M., Kelly M.E., Rohlack L.A. (Radian) and R.G. Rhudy (EPRI). 'EPRI spray drying pilot plant status and results'. Paper to EPA/EPRI Symposium.

197 Mudgett J.S. (Strathmore paper), Sadowski R.S. (Riley Stoker), West W.W. (Mikropul) and Mutsakis M. (Koch). 'Dry SO_2 scrubbing achieved with spray dryer and fabric filter'. Reprint from the 'Eighth Annual Industrial Plant–Energy Systems Guidebook', McGraw-Hill, 1982.

198 Rainauer, T.V. (Mikropul), Monat J.P. (Abcor) and Mutsakis M. (Koch). 'Dry FGD on an industrial boiler'. *CEP*, 1983 (Mar), 74–81.

199 Emerson R.D. (Sunflower Electric). 'Dry FGC system: Start-up, performance and acceptance tests'. *Power Engineering*, 1984 (Oct), 50–52.

200 Kaplan S.M. and Felsvang K. (Niro Atomizer). 'Spray dryer absorption of SO_2 from industrial boiler flue gas'. *A.I.Chem. E. Symposium Series*, 1980, **76** (201), 23–30.

201 Meyler J.A. (Joy). 'Case history of a dry scrubber application at Northern States Power Company'. Paper to Coal Technology '80, 1980, 589–596.

202 Felsvang K. (Niro Atomizer). 'Results from operation of Riverside dry scrubber'. Paper to Riverside Dry FGD Symposium, Minneapolis, USA, June 1981.

203 Gude K.E. 'The spray dryer absorber concept for FGD'. Paper to Symposium on Danish Know-How and Technology on Energy and Pollution Control, Peking, November 1983.

204 Hansen S.K., Felsvang K.S. (Niro Atomizer), Morford R.M. and Spencer H.W. (Joy). 'Status of the Joy/Niro Atomizer dry FGD system and its future application for the removal of high sulphur, high chloride and NO_x from flue gases'. Paper to ASME Joint Power Generation Conference, September 1983.

205 Thousig J.T., Jorgensen C. and Fallenkamp B. (Niro Atomizer). 'Dry scrubbing of toxic incinerator flue gas by spray absorption'. Paper to ENVITEC 83, Düsseldorf, February 1983.

206 Schwartzback C. (Niro Atomizer). 'The science and art of spray dryer design for FGD'. Paper to Coal Technology '82, 5th International Coal Utilisation Conference, Houston, USA, December 1982.

207 Felsvang K. (Niro Atomizer). 'Desulphurization of low rank, high sulphur coal by dry flue gas desulphurisation'. Paper to 8th International Congress of Chemical Engineering, Chemical Equipment Design and Automation, Praha, Czechoslovakia, September 1984.

208 Jacobson P. (Flakt Industri AB) and Madhuk R. (Niro Atomizer). 'Flakt dry FGD Systems'. Paper to Flue Gas Desulphurisation Seminar, Bombay and New Delhi, Jan/Feb 1986.

209 Horn R.J. (Ecolair Environmental Co.). 'Installation and operation of a retrofit dry flue gas desulphurisation system'. Paper to Coal Technology '83, 6th International Coal and Lignite Utilisation Conference, Houston, USA, November 1983, **3**, 115–139.

210 Downs W., Sanders W.J. and Miller C.E. (Babcock and Wilcox). 'Control of SO_2 emissions by dry scrubbing', 262–271.

211 Burnett T.A., Threet G.E., Humphries L.R., Robards R.F. and Runyan R.A. (Tennessee Valley Authority). 'Spray dryer/baghouse flue gas desulphurisation–evaluation for high-sulphur utility applications'. Paper to American Institute of Chemical Engineers, Winter Annual Meeting, Chicago, USA, November 1985.

212 Robards R.F., DeGuzman J.S., Runyan R.A. and Flora H.B. (Tennessee Valley Authority). 'Spray Dryer/ESP testing for utility retrofit applications on high-sulphur coal'. Paper to American Power Conference, Chicago, USA, April 1986.

213 Livengood C.D. and Farber P.S. (Argonne National Laboratory). 'Performance and economics of a spray-dryer FGD system used with high-sulphur coal'.

214 Hammer P.R.R. (Niro Atomizer) 'Desulphurisation of flue gases from coal burning by spray absorption'. Paper to Coal Technology Europe, 2nd International Coal Utilisation Conference, Copenhagen, Denmark, September 1982, **3**, 297–311.

215 Colley J.D. (Radian Corporation), Donaldson T. (Central Illinois Light Co) and Stewart D.(Electric Power Research Institute). 'Process troubleshooting at a utility limestone F.G.D. system'. Paper to Coal Technology '83, 6th International Coal and Lignite Utilisation Conference, Houston, USA, November 1983, **3**, 187–207.

216 Hargrove O.W., Colley J.D. (Radian Corp.) and Mobley J.D. (US Environmental Protection Agency). 'Adipic acid-enhanced limestone flue gas desulphurisation system commercial demonstration'. Paper to Coal Technology '81, 4th International Coal Utilisation Conference, Houston, USA, November 1981, **3**, 201–213.

217 Felsvang K. (Niro Atomizer). 'Acid rain control through dry scrubbing'. Paper to Power Magazine, First International Conference on Acid Rain, Washington, USA, March 1984.

218 Makansi J. 'New processes enhance the in-duct emissions-control option'. *Power*, 1986 (July), **139** (7) 27–29.

219 Buschmann J.C., D'Ambrosi F.D. and Mezner M. (Flakt). 'Start-up and operating experience of the University of Minnesota dry FGD system'. Paper to APCA, 78th Annual Meeting, Detroit, USA, June 1985, Paper 85-58.3.

220 Widico M.J. and Dhargalkar P.H. (Research-Cottrell). 'Dry FGD process for various coals'. Paper to APCA, 78th Annual Meeting, Detroit, USA, June 1985. Paper 85–58.4.

221 Davidson L.N., Goffredi R.A. and Wedig C.P. (Stone and Webster). 'The importance of maintenance for lime FGD systems'. Paper to APCA, 78th Annual Meeting, Detroit, USA, June 1985. Paper 85–58.6.

222 Cannall A.L. and Meadows M.L. (Black and Veatch). 'Effects of recent operating experience on the design of spray dryer FGD systems'. Paper to APCA, 78th Annual Meeting, Detroit, USA, June 1985. Paper 85–58.8, and *JAPCA*, 1985, **35** (7), 782–788.

223 Burnett G.F. and Basel B.E. (Burns and McDonnell). 'The status of dry scrubbing in the United States'. Paper to APCA, 78th Annual Meeting, Detroit, USA, June 1985. Paper 85–58.1.

224 Mobley J.D. (US EPA), Cassidy M. and Dickerman J. (Radian). 'Organic acids can enhance wet limestone flue gas scrubbing'. *Power Engineering*, 1986 (May) 32–35.

225 Ashley M. (Lodge-Cottrell). 'Spray dry desulphurisation plant requires lower capital investment'. *Modern Power Systems*, 1985 (May).

226 Murphy K.R., Shilling N.Z. (GEESI) and Pennline H. (US Department of Energy). 'In-duct scrubbing pilot study'. *JAPCA.*, 1986 (Aug), **36** (8), 953–958.

227 Ashley M.J. (Lodge-Cottrell). 'The Lodge-Cottrell spray dry desulphurisation system'. Paper to 14th Annual Meeting of the Battelle Stack Gas Assessment & Technology Survey Programme, London, September 1986.

228 Dhargalkar P.H. (Research-Cottrell), and Ford P.G. (Davy McKee). 'Performance of two established FGD processes'. Paper to Coal Tech '85, 5th International Conference on Coal Utilisation and Trade, London, December 1985, **2**, 307–334.

229 Richman M. (Research-Cottrell Inc.). 'Advanced FGD technology for V.Y. Dallman Station'. Paper to Coal Technology '80, 1980, 597–607.

230 Yeargan R.D. (TVA). 'Paradise Fossil Plant: Units 1 & 2 scrubber operating experience'. Paper to EPRI, FGD Users Conference, Farmington, USA, June 1986.

231 Wallenwein E.H. (Bischoff). 'Desulphurisation plant developed by West Germany utility'. *Modern Power Systems*, 1985 (May), 33–37.

232 Hargrove O.W., Colley J.D. (Radian Corp.), Wadlington M. (Texas Utilities Generating Co.) and Stewart D.A. (EPRI). 'FGD system and water balance improvements at Texas Utilities Generating Company's Martin Lake station'. Paper to Symposium on Advances in Fossil Power Plant Water Management, Orlando, Florida, February 1986.

233 Mori T., Matsuda S., Nakajima F., Nishimura T. and Arikawa Y. (Hitachi). 'Effect of Al^{3+} and F^- on desulphurisation re-action in the limestone slurry scrubbing process'. *Ind. Eng. Chem. Process Des. Dev.*, 1981, **20**, 144–147.

234 Anonymous. 'Forced-O_2 FGD system achieves 99.8% availability in first year'. *Power*, 1986 (April), 19–20.

235 Martin J.R., Ferguson W.B. and Frabotta D. 'C-E dry scrubber systems: application to Western coals'. *Combustion*, 1981 (February), 12–20.

236 Crowe R.B. (Celanese Fibres), Lane J.F. (Rockwell International) and Petti V.J. (Wheelabrator-Frye). 'Early operation of the Celanese Fibres Company coal-fired boiler using the dry flue gas cleaning system'. *Combustion*, 1981 (February), 34–37.

237 Borgwardt R.H. (US EPA). 'Combined flue gas desulphurisation and water treatment in coal-fired power plants'. *Env. Sci. Technol.*, 1980 (March), **14** (2), 294–298.

238 Shattuck D.M., Stenby E.W., Lacey J.N. and Layton K.F. 'Utah Power and Light's experience with wet scrubbing of SO_2 at the Huntington and Hunter plants'. Paper to 42nd American Power Conference, Chicago, USA, April 1980.

239 Massey C.L., Moore N.D., Munson G.T., Runyan R.A. and Wells W.L. 'Forced oxidation of limestone scrubber sludge at TVA's Widows Creek Unit 8 steam plant'. Paper to US EPA, 6th Symposium on Flue Gas Desulphurisation, Houston, USA, October 1980.

240 Chan P.K. and Rochelle G.T. 'Limestone dissolution: effects of pH, CO_2 and buffers modelled by mass transfer'. Paper to ACS National Meeting, Atlanta, USA, March 1981.

241 Burke J.M., Metcalfe R.P., Cmiel R. and Mobley J.D. 'Technical and economic evaluation of organic acid addition to the San Miguel FGD system'. Paper to EPA's Industry Briefing on the Organic Acid Enhanced Limestone FGD Process, San Antonio, USA, July 1984.

242 Benson L.B. 'The role of magnesium in increasing SO_2 removal and improving the reliability in magnesium-enhanced FGD systems'. Paper to 2nd Annual Coal Conference, Pittsburgh, USA, September 1985.

243 Mobley J.D. (US EPA) and Chang J.C.S. (Accurex Corp.). 'The adipic acid enhanced limestone flue gas desulphurisation process: an assessment. *JAPCA*, (December), **31** (12), 1249–1253.

244 Chang J.C.S., Kaplan N. and Brna T.G. (Accurex Corp.). 'Effects of Mg^{++} and Cl^- ions on limestone dual alkali system performance'. *ACS Div. Fuel Chem.*, 1985, **30** (2), 1145–161.

245 Smith T. (Consultant), Colley D. (Radian Corp.) and Steward D. (Electric Power Research Institute). 'Apply process-chemistry know-how to your FGD system'. *Power*, 1985 (Sept), 35–37.

246 Dharmarajan (Central & South West Services Inc.). 'Stirred mill proves its worth for FGD lime-slaking duties'. *Power*, 1985 (Oct), 61–63.

247 Friedlander G.D. 'Huge scrubber retrofitted at four corners'. *Electrical World*, 1984 (Mar), 71–72.

248 Beals J. (Pennwalt), Cannell L. and Hengel J. (Black & Veatch). 'How FGD reagent quality affects system performance'. *Power*, 1984 (Mar), 27–30.

249 Ellison W. (Consultant) and Egan R. (Munters Corp.) 'Incorporate the latest FGD trends into mist-eliminator design'. *Power*, 1984 (Mar) 35–37.

250 Makansi J. 'Particulate and SO_2 scrubbers that require no wetted-surface internals'. *Power*, 1983 (Mar) 119.

251 Anonymous. 'SO_2 scrubber makes saleable gypsum'. *Oil & Gas Journal*, 1979 (5 Mar) 180.

252 Stowe D.H., Henzel D.S. and Hoffman D.C. (Dravo Lime Co.) 'The FGD reagent dilemma: lime, limestone or thiosorbic lime'. Report No. EPA-600/7-79-16TB to US EPA, July 1979.

253 Anonymous. 'Dry scrubber overcomes scale-up problems at Coyote'. *Power*, 1983 (Apr), 114–115.

254 Anonymous. 'Bechtel offers partial desulphurisation to reduce costs'. *Process Engineering*, 1986 (Oct), 15.

255 Laslo D. and Bakke E. (Peabody Process Systems). 'State-of-the-art design applications on a closed-loop FGD system'. Paper to EPA/EPRI, FGD Symposium, Cincinnati, USA, June 1985.

256 Laslo D. (Peabody Process Systems), Chang J.C.S. (Acurex) and Mobley J.D. (US EPA). 'Pilot plant tests on the effects of dissolved salts on lime/limestone FGD chemistry'. Paper to EPA/EPRI, Symposium on FGD, New Orleans, USA, November 1983.

257 Laslo D. and Bakke E. (Peabody Process Systems). 'The effect of dissolved solids on limestone FGD scrubbing chemistry'. Paper to ASME, 1983 Joint Power Generation Conference, Indianapolis, USA, September 1983.

258 Bakke E. (Peabody Process Systems). 'Cost effective wet FGD systems on medium to high sulphur coals'. Paper to 1985 Joint Power Generation Conference, Milwaukee, USA, October 1985.

259 Anonymous. 'Catenany-grid scrubber'. *Chemical Engineering*, 1986 (13 Oct), 39.

260 Fahlenkamp H. (Deutsche Babcock Anlagen). 'Recent developments in West Germany's limestone-based FGD technology'. Paper to Joint ASME/IEEE, Power Generation Conference, Portland, USA, October 1986.

261 Chang J.C.S. (Acurex Corp.) and Laslo D. (Peabody Process Systems). 'Chloride ion effects on limestone FGD system performance'. Paper to EPA/EPRI, FGD Symposium, Hollywood, Florida, USA, May 1982.

262 Tearney J.F., Froelich D.A. and Graves G.M. (Burns & McDonnell). 'SO_2 control of non-regenerable wet FGD systems'. Paper to Power Magazine, First International Conference on Acid Rain, Washington, USA, March 1984.

263 Wataya K., Hon A. (Toyama Hyoda), Hashimoto N., Koshizuka H. (Chiyoda) and Clasen D.D. (Chiyoda International). 'Operating results of Toyama Kyoda Electric Powers' Chiyoda Thoroughbred 121 FGD system'. Paper to EPA/EPRI, 9th FGD Symposium, Cincinnati, USA, June 1985.

264 Wiitala W.W. (Marguette Board of Light & Power), Arello J. (Lutz, Daily & Brain), Martinelli R. and Lapp D. (GEESI). 'Spray dry scrubbers at Marquette's Shiras power plant'. Paper to American Public Power Association, 28th Annual Engineering & Operations Workshop, Toronto, Canada, March 1984.

Regenerable Solution-Based Wet Reagent Processes (Category S30)

265 Bettelheim J., Cooper J.R.P., Kyte W.S. and Rowlands D.T.H. (CEGB). 'The integration of a regenerable FGD plant on to a 2000 MW coal fired power boiler station site in the UK'. *I. Chem. E. Symposium Series*, 1981, (72).

266 Dhargalkar P.H. (Research-Cottrell) and Ford P.G. (Davy McKee). 'Performance of two established FGD processes'. Paper to Coal Tech '85, 5th International Conference on Coal Utilisation and Trade, London, December 1985, **2**, 307–334.

267 Madenburg R.S. and Seesee T.A. (Morrison-Knudsen Co.). 'H_2S reduces SO_2 to desulphurise flue gas'. *Chemical Engineering*, 1980 (14 Jul), 88–89.

268 Farrington J. and Bengtsson S. (Flakt Inc.) 'Citrate solution absorbs SO_2'. *Chemical Engineering*, 1980 (16 Jun), 88–89.

269 Makansi J. (compiler). 'Regenerative FGD: progress is slow but steady'. *Power*, 1983 (Aug), 36–37.

270 Walker R.J., Wildman D.J. and Gasior S.J. (US Department of Energy). 'Evaluation of some regenerable SO_2 absorbents for FGD'. *JAPCA*, 1983 (Nov), **33** (11), 1061–1067.

271 Munson R.A., Fitch W.N., and Nissen W.I. '50 MW power plant demonstration of the removal of sulphur oxides from stack gases using the Bureau of Mines citrate process'. 95–98.

272 Langenkamp H. and Van Velzen D. (CEC). 'FGD by the Mark 13A process'.

273 Van Velzen D., Langenkamp H. and Ferrari, A. 'The Mark 13A Process for FGD'. Excerpts from Programme Progress Reports, Hydrogen Production Energy Storage and Transport, Jan 1983–Jun 1984.

274 Anonymous. 'The European Commission to co-finance a project against acid deposition'. *Biomass News International*, 1986 (Jun), (18), 7.

275 US Bureau of Mines, Report of Investigations 8638 (1982). 'FGD: evaluation of the modified citrate process draws important conclusions'. *Sulphur*, 1983 (Jan/Feb), (164), 43–45.

276 Neumann U., Vangala R. and Giovanetti A. (Davy McKee). 'Wellman-Lord SO_2 recovery operating experience serving coal fired boilers'. Paper to Coal Technology Europe '82, 2nd International Coal Utilisation Conference, Copenhagen, Denmark, September 1982, **3**, 281–296.

Regenerable Slurry-Based Wet Reagent Process (Category S40)

277 Makansi J. 'New regenerative FGD system demonstrated at state hospital'. *Power*, 1983 (Oct), 131–132.

278 Anonymous. 'Regenerable scrubber meets EPA limits, utility requirements'. *Power*, 1984 (Apr), 33–34.

279 Marawczyk C., MacKenzie J.S. (United Engineers and Constructors), and Bitsko R. (Philadelphia Electric). 'The outlook for regenerative magnesium oxide FGD' *CEP*, 1984 (Sep), 62–68.

280 Makansi J. 'MgO scrubber links utility to chemical firms'. *Power*, 1981 (Dec).

281 PETC Quarterly Progress Report, 30 September 1985.

Regenerable Dry Reagent Processes (Category S50)

282 Ploeg J.E.G. (Shell Internationale), Akagi E. (Showa) and Kishi K. (Japan Shell). 'How Shell's FGD unit has worked in Japan'. *Petroleum International*, 1974 (Jul) **14** (7), 50–58.

283 Steiner P. (Foster Wheeler Energy Corp.). 'Pollution control system and method for the removal of sulfur oxides'. UK Pat. Appl. GB 2,009, 117A, 13 June 1979 (filed 7 September 1978).

284 Townley D. and Winnick J. 'Flue gas desulphurisation using an electrochemical sulfur oxide concentrator'. *Ind. Eng. Chem. Process Des. Dev.*, 1981, **20**, 435–440.

285 Bee, R., Reale R. and Walls A. (Mitre Corp.). 'Demonstration/evaluation of the Cat-Ox flue gas desulphurisation system–final report'. Report No. EPA-600/2-78-063 to US Environmental Protection Agency, March 1978.

286 Steiner P. (Foster Wheeler Development Corp.), Dalton S.M. (EPRI) and Knoblauch K. (Bergbau Forschung). 'Capture and conversion of sulphur dioxide at the ReSOx prototype demonstration in Germany'. Proceedings of the American Power Conference, 1979, **41**, 719–723.

Non-Regenerable Dry Reagent Applied to Flue Gas (Category S60)

287 Yeager K. (EPRI). 'SO_2 control by dry sorbent injection'. *EPRI Journal*, 1983 (Mar), 36–37.

288 Samuel E.A., Furlong D.A. (Envirotech), Brna T.G. (US EPA) and Ostop R.L. (City of Colorado Springs). 'SO_2 removal using dry sodium compounds'. *AIChE Symposium Series*, 1981, **77** (211), 54–60.

289 Anonymous. 'Dry capture of SO_2'. *EPRI Journal*, 1984 (Mar), 14–21.

290 Hamala S. (Tampella Ltd.). 'LIFAC cuts SO_x in Finland'. *Modern Power Systems*, 1986 (June), 87–91.

291 Forsythe R.C. (Dravo Lime Co.) and Kaiser R.A. (Ohio Edison Co.). 'Hydrate addition at low temperature: SO_2 removal in conjunction with a baghouse'. Paper to 2nd Annual Pittsburgh Coal Conference, Pittsburgh, USA, September 1985.

292 Yoon H., Ring P.A. and Burke F.P. (Conoco). 'Coolside SO_2 abatement technology–1 MW field tests'. Paper to Coal Technology '85, Pittsburgh, USA, November 1985.

293 Graf R., 'Lurgi dry FGD processes based on the circulating fluid bed principle and the spray absorber system'. Paper to Technical Academy, Wuppertal, 1983. (In German.)

Non-Regenerable Dry Reagent Applied in Furnace (Category S70)

294 Maulbetsch J. (EPRI). 'Status of furnace sorbent injection technology'. Paper to 47th American Power Conference, Chicago, USA, April 1985.

295 Chughtai M.Y. and Michelfelder S. 'Direct desulphurisation through additive injection in the vicinity of the frame'. Paper to EPA/EPRI Flue Gas Desulphurisation Symposium, New Orleans, USA, November 1983. Paper 4C.

296 Parkinson G. and McQueen S. 'A shot of limestone may cure SO_2-removal woes'. *Chemical Engineering*, 1984 (20 Feb), 30–35.

297 Bortz S, and Flament P. (International Flame Research Foundation). 'Recent IFRF fundamental and pilot scale studies on the direct sorbent injection process'. Papers of First Joint Symposium, Dry SO_2 and Simultaneous SO_2/NO_x Control Technologies, San Diego, USA, November 1984, and Symposium, Schone Verbranding van Steenkool, Noordwijkerhout, Netherlands, January 1985.

298 Case P.L., Ho L., Clark W.D., Kau E., Pershing D.W., Payne R. and Heap M.P. (Energy and Environmental Research Corporation). 'Testing of wall-fired furnaces to reduce emissions of NO_x and SOx. Volume 1. Final report'. Report No. EPA/600/7-85/026a for US EPA, June 1985.

299 Doyle J.B. and Jankura B.J. (Babcock and Wilcox). 'Furnace limestone injection with dry scrubbing of exhaust gases'. Paper to 1982 Spring Technical Meeting of the Central States Section of the Combustion Institute, Columbus, USA, March 1982.

300 Gallaspy D.T. (Southern Company Services, Inc.). 'Dry sorbent emission control prototype conceptual design and cost study'. Paper 6G.

301 Kokkinos A., Lewis R.D., Borio D.C., Plumley A.L. (Combustion Engineering) and McElroy M.W. (Electric Power Research Institute). 'Feasibility of furnace injection of limestone for SO_2 Control'. Paper to Joint Symposium on Stationary Combustion NO_x Control 1982. EPRI, Proceedings, CS 3182, July 1983.

302 Burdett N.A., Cooper J.R.P. (CEGB), Dearnley S. (UK Department of Energy), Kyte W.S. (CEGB) and Tunnicliffe M.F. (Health & Safety Executive). 'The application of direct limestone injection to UK power stations'. *J. Inst. Energy*, 1985 (June), **58** (435), 64–69.

303 Ness H., Dorchak T.P. (US Dept. of Energy) and Reese J.R. (Energy & Environmental Research Corp.). 'Experience with furnace injection of pressure hydrated lime at the 50 MW Hoot Lake Station'. *Inside R&D*, 1985 (20 Mar).

304 Anonymous. 'SO_2 scrubbing: more work for sodium'. *Chemical Week*, 1984 (18 July), 34–35.

Waste Product Disposal

305 Haynes L.H. (Central Illinois Light Co.), Ansari A.H., and Owen J.E. (Gilbert/Commonwealth). 'Ash/FGD waste disposal options: A comparative study for CILCO Duck Creek site'. *Combustion*, 1980 (Jan), 21–27.

306 Kyte W.S. and Cooper J.R.P. (CEGB). 'The disposal of products from FGD processes'. Paper to Second International Conference, Ash Technology and Marketing, London, September 1984.

307 Kyte W.S. and Cooper J.R.P. (CEGB). 'The disposal of products and wastes from FGD processes'. *I. Chem. E. Symposium Series*, (96), 1986, 233–247.

308 Donnelly J.R., Jons E. (Niro Atomizer) and Webster W.C. (Webster and Associates). 'Synthetic gravel from dry FGD end-products'. Paper to 6th International Ash Utilisation Symposium, Reno, USA, March 1982.

309 Donnelly J.R. (Niro Atomizer), Webster W.C. (Webster and Associates), Duedall I.W., Hsu J., Parker J.H. and Woodhead P.M.J. (NY State University). 'Ocean disposal of consolidated spray dryer FGD wastes'. Paper to International Conference, Coal-fired Power Plants and The Aquatic Environment, Copenhagen, Denmark, August 1982.

310 Donnelly J.R. (Niro Atomizer). 'Disposal and utilisation of spray dryer FGD end-products'. Paper to Canadian Electrical Association Seminar, SO_2 Removal by Dry Process, Ottawa, Canada, October 1982.

311 Weis J.G., Hendry D.W. (Burns and McDonnell) and Baumgardner D. (Plains Electric Generation and Transmission Co-operative). 'Centrifuging FGD sludge can eliminate thickening step'. *Power*, 1985 (Oct), 67–69.

312 Johnson C.A. (Peabody Process Systems). 'Alternative methods of handling waste from flue gas desulphurisation systems'. Paper to Coal Technology Europe, 3rd European Coal Utilisation Conference, Amsterdam, The Netherlands, October 1983, **4**, 81–92.

313 Smith C.L. and Rau E. (IU Conversion Systems). 'Stabilised FGD sludge goes to work'. Paper to Coal Technology '81, 4th International Coal Utilisation Conference, Houston, USA, November 1981, **2**, 247–258.

314 Adams D.F. and Farwell S.O. (University of Idaho). 'Sulphur gas emissions from stored flue gas desulphurisation sludges'. *JAPCA*, 1981 (May), **31** (5), 557–564.

315 Goodwin R.W. (General Electric Environmental Services). 'Effect of auto-oxidation on treatment and disposal properties of lignite derived flue gas desulphurisation sludge'. Paper to Coal Technology '83, 6th International Coal Utilisation Conference, Houston, USA, November 1983, **6**, 263–284.

316 Johnson C.A. (Peabody Process Systems). 'FGD sludge stabilisation and fixation: an alternative disposal technique to produce a commercial gypsum'. Paper to Coal Tech '85, 5th International Conference on Coal Utilisation and Trade, London, December 1985, **2**, 349–369.

317 Ellison W. (Ellison Consultants). 'F.G.D. Gypsum: Utilisation vs disposal'. Background paper for 8th FGD Symposium, New Orleans, USA, November 1983.

318 Cope D.R. and Dacey P.W. (IEA Coal Research). 'Solid residues from coal use–disposal and utilisation'. Report No. ICEAS/B3, IEA Coal Research, London, July 1984.

319 Ellison W. (Consultant) and Luckevich L.M. (Ontario Research Foundation). 'FGD waste: Long-term liability or short-term asset?' *Power*, 1984 (June), 79–82.

320 Bengtsson S., Ahman S., Lillestolen T. (Flakt), and Koudijs G. (Dorr-Oliver). 'Thermal oxidation of spray dryer FGD waste product'. Paper for EPA/EPRI Symposium, Flue Gas Desulphurisation, Cincinnati, USA, June 1985.

321 Goodwin R.W. (Chemico). 'Waste treatment and disposal aspects: combustion and air pollution control processes'. *JAPCA*, 1981 (July), **31** (7), 744–747.

322 Anonymous. 'Environmental impacts of a flue gas desulphurisation programme'. ENDS Report No. 117, October 1984, 9–11.

323 Goodwin R.W. (GEESI). 'Resource recovery from flue gas de-sulphurisation systems'. *JAPCA*, 1982 (September), **32** (9), 986–989.

324 Mzyk D. (Texas Utilities Generating Co.) and Zmuda J. (Research Cottrell). 'By-product gypsum production at a 2300 MW power plant'. Paper to US EPA, 9th Symposium on Flue Gas Desulphurisation, Cincinnati, USA, June 1985.

325 Rosenstiel T.L. and Debus A.A.G. (US Gypsum Co.). 'Process for preparing wastes for non-pollutant disposal'. UK Patent GB 2,097,990B, 2 January 1986. (Appl. 8204421 filed 15 February 1982).

326 Boldt K.R., Tusa W. and Streets D. (Fred C. Hart Associates and Argonne Nat. Lab.). 'Analysis of industrial boiler solid waste impacts'. *JAPCA*, 1981 (July), **31** (7), 753–760.

327 Weeter D.W. 'Utilisation of dry calcium based flue gas desulphurisation waste as a hazardous waste fixation agent'. *JAPCA*, 1981 (July), **31** (7), 751–753.

328 Wirsching F., Poch W., Huller R. and Hamm

H. 'Environmentally safe disposal of coal-fired power station waste'. Eur. Pat. Appl. EP 139,953 (Cl. A62D3/00), 8th May 1985. DE Appl. 3,329,972, 19 August 1983.

329 Krueger B. and Kraus M. (Bischoff). 'Converting the residue from FGD installations to alpha-calcium sulphate hemihydrate crystals'. Ger. Pat. DE 3,331,838, 21 March 1985. Appl. 3 September 1983.

330 Hoelter H., Ingelbuescher H., Gresch H. and Dewert H. 'Making residues from coal-fired power plants environmentally favourable'. Ger. Pat. DE 3,322,539, 17 January 1985. Appl. 23 June 1983.

331 Mitsubishi Heavy Industries Ltd. 'Treatment of FGD wastewater'. Japan Pat. 60 60,886, 5 January 1985. Appl. 83/106,768, 16 June 1983.

332 Jons E. (Niro Atomizer) 'Properties of stabilised desulphurisation products from spray-dry process'. Niro Atomizer A/S, Soeberg, Denmark. Report No. NP-5750388 (Order No. T185750388), 1984. (In Danish.)

333 Aggour M.S. and Stanbro W.D. (Univ. of Maryland). 'Field ageing of fixed sulphur dioxide scrubber waste'. *J. Energy Eng.*, 1985, **111** (1), 62–73.

334 Sayre W.G. 'Selenium: a water pollutant from FGD'. *JAPCA*, 1980 (October), **30** (10), 1134.

335 Thompson C.M. (Radian Corp.). 'Chemical and physical characterisation of Western low-rank coal waste materials. Part 1: By-products from sodium-based dry scrubbing systems, final report'. Report No. DOE/FC/10200-T2 (DE83001167) to US Department of Energy, August 1982.

336 Rittenhouse R.C. 'Additives: the answer to freezing, dust and sludge instability'. *Power Engineering*, 1986 (July), **90** (7), 38–41.

337 Jons E. (Niro Atomizer), 'The use of spray drying absorption FGD products in building materials'. Paper to FGD Symposium, Leningrad, USSR, July 1986.

338 Jons E. (Niro Atomizer). 'SDA-ash as the only residue from flue gas cleaning'. Paper to ACI/RILEM Joint Seminar, Monterrey, Mexico, March 1985.

PAPERS ON NITROGEN OXIDES ABATEMENT

General Approach to Nitrogen Oxides Emissions Abatement

339 Ogunsola O.I. and Reuther J.J. (Pennsylvania State University). 'Relationship between fuel-nitrogen-to-NO_x conversion efficiency and boiling range for coal-derived liquid fuel combustion'. Report No. PSU-FCL-C-80-78.

340 Artem'ev Y.P., Verbovetskii E.K. and Kozhanov D.S. (VTI). 'The effect of air pre-heat temperature on the formation of nitrogen oxides'. *Thermal Engineering*, 1980, **27** (9), 527–528.

341 Parkinson G. 'NO_x controls: Many new systems undergo trials'. *Chemical Engineering*, 1981 (9th March), 39–43.

342 McCartney M.S. and Cohen M.B. (Combustion Engineering). 'Techniques for reducing NO_x emissions from coal fired steam generators'. Paper to Power Magazine, First International Conference on Acid Rain-Regulatory Aspects and Engineering Solutions, Washington, USA, March 1984.

343 Morrison G.F. (IEA Coal Research). 'Nitrogen oxides from coal combustion–abatement and control'. Report No. ICTIS/TR 11, IEA Coal Research, London, November 1980.

344 Dacey P. (IEA Coal Research). 'Developments in NO_x control for coal-fired boilers'. Working paper 67, IEA Coal Research, London, November 1984.

345 Siddiqi A.A. and Tenini J.W. (ARCO). 'NO_x controls in review'. *Hydrocarbon Processing*, 1981 (Oct), 115–124.

346 Parker L.B. and Trumbule R.E. 'Opportunities for increased control of nitrogen oxides emissions from stationary sources: Implications for mitigating acid rain'. Report No. 82–217 ENR to Congressional Research Service, December 1982.

347 Yanai M. (Kawasaki Heavy Industries Ltd). 'Kawasaki's technology on NO_x abatement'. Paper to EPDC/IIP/VDI, NO_x Symposium, Karlsruhe, West Germany, February 1985. Paper R.

348 Ishimoto R. and Miyamae S. (IHI). 'NO_x abatement technologies in IHI'. Paper to EPDC/-IIP/VDI, NO_x Symposium, Karlsruhe, West Germany, February 1985. Paper N.

349 Kuroda H. and Masai T. (Babcock Hitachi). 'Babcock Hitachi NO_x abatement technology'. Paper to EPDC/IIP/VDI/NO_x Symposium, Karlsruhe, West Germany, February 1985. Paper L.

350 Ando J. (Chuo University). 'Review of Japanese NO_x abatement technology for stationary sources'. Paper to EPDC/IIP/VDI, NO_x Symposium, Karlsruhe, West Germany, February 1985. Paper A.

351 Jones G.D. (Radian) and Mobley J.D. (US EPA). 'Review of US NO_x abatement technology'. Paper to APCA, 78th Annual Meeting, Detroit, USA, June 1985. Paper 85-55.2.

352 ECE NO_x Task Force. 'Technologies for controlling NO_x emissions from stationary sources'. Report No. IIP4/1986, April 1986.

353 Davids P., Oels H.J. and Rosenbusch K. (Umweltbundesamt). 'Technical consequences of NO_x emission limits in West Germany'. *Gaswarme Int.*, 1986 (May–June), **35** (4), 178–86.

354 Kircher U. 'NO_x emissions and reduction measures in the glass industry'. *Gaswarme Int.*, 1986 (May–June), **35** (4), 207–212.

355 Bergsma F. (TNO) 'Abatement of NO_x from coal combustion. Chemical background and present state of technical development'. *Ind. Eng. Chem. Process Des. Dev.*, 1985, 24 (1), 1–7.

356 Moore T. (EPRI). 'The retrofit challenge in NO_x control'. *EPRI Journal*, 1984 (Nov), 26–33.

357 Mason H.B. *et al.* 'Environmental assessment of stationary source NO_x control technologies'. Third Stationary Source Combustion Symposium, Vol. IV. US Environmental Protection Agency, EPA-600/7-79-050d, February 1979.

358 Ferrari L.M. *et al.* 'Nitrogen oxides emissions and emission factors for stationary sources in New South Wales'. Proceedings, International Clean Air Conference, Brisbane, Australia, May 1978. (Ann Arbor Science, 1978.)

359 US Environmental Protection Agency. 'Compilation of air pollutant emission factors, 2nd edition'. AP-42, US EPA, Research Triangle Park, N.C., 1975.

360 MacCurley W.R., Moscowitz C.M., Ochsner J.C. and Reznik R.B. 'Source assessment; dry bottom industrial boilers firing pulverised bituminous coal'. Report No. EPA-600/2-79019e to US Environmental Agency, June 1979.

Abatement by Combustion Modifications

361 Sekinguchi Y., Okigami N., Taninaka I. and Sakai S. (Hitachi Zosen). 'Development of new NO_x combustion control method'. Report Number UDC 661.5: 662.9, 67–77.

362 Kawamura T. (Mitsubishi) and Frey D.J. (Combustion Engineering). 'Current developments in low NO_x firing systems'. Paper to EPA/EPRI Joint Symposium, Stationary Combustion NO_x Control, Denver, USA, October 1980.

363 Wheeler W.H. (Urquhart). 'Chemical and engineering aspects of low NO_x concentration'. *The Chemical Engineer*, 1980 (Nov), 693–699.

364 Coe W.W. (CEA). 'How burners influence combustion'. *Hydrocarbon Processing*, 1981 (May), 179–184.

365 Parkinson G. 'Catalytic burning tries for NO_x control jobs'. *Chemical Engineering*, 1981 (15 June), 51–55.

366 Bell C.T. and Warren S. (Airoil-Flaregas). 'Experience with burner NO_x reduction'. *Hydrocarbon Processing*, 1983 (Sept), 145–147.

367 Ando J. (Chuo University) and Mobley J.D. (US EPA). 'Low NO_x burners for pulverised-coal-fired boilers in Japan'. Paper to FGD Pilot Study Group of NATO Committee on the Challenges of Modern Society, York, May/June 1984.

368 Vatsky J. (Foster Wheeler). 'Industrial and utility boiler NO_x control'. EPA/EPRI Symposium.

369 Vatsky J. (Foster Wheeler). 'High capacity low NO_x coal burner for retrofit and new units'. *Power Engineering*, 1982 (Jan).

370 Vatsky J. (Foster Wheeler). 'Modern combustion systems for coal-fired steam generators'. Paper to Pacific Coast Electric Association Conference, San Francisco, USA, March 1980.

371 Pruce L. 'Reducing NO_x emissions at the burner, in the furnace, and after combustion'. *Power* 1981 (Jan), **125** (1), 33–40.

372 Phelan W.J. (International Flame Research Foundation). 'The effect of pulverised coal type

and burner parameters when staging air combustion for NO_x reduction'. Paper to Coal Technology Europe, 3rd European Coal Utilisation Conference, Amsterdam, Netherlands, October 1983, **1**, 85–110.

373 Bancel P.L. and Massoudi M.S. (Kaiser Engineers Inc). 'Gas turbine NO_x controlled with steam and water injection'. *Power Engineering*, 1986 (June), **90** (6), 34–37.

374 Moore T. 'The retrofit challenge in NO_x control'. *EPRI Journal*, 1984 (Nov), **9** (9), 26–33.

375 Campbenedetto E.J. (Babcock & Wilcox) and Schuster H. (Deutsche Babcock). 'Development of low-NO_x pulverised coal firing system'. Paper to Coal Technology Europe '81, Cologne, West Germany, June 1981.

376 Takahashi Y., Tokuda K., Sengoku T., Nakashima F. and Kaneko S. (Mitsubishi). 'Evaluation of Tangential fired low NO_x burners'. Paper to EPA/EPRI Joint Symposium on Stationary Combustion NO_x Control, Dallas, USA, November 1982.

377 Phelan W.J. (IVO), 'The influence of P.F. burner design parameters on the NO_x-emission and char burnout when staging the combustion air'. Paper to Symposium, Noordwijkerhout, January 1985.

378 Masai T., Morita S., Akiyama I. (Babcock-Hitachi) and Ohtsuka K. (Hitachi). 'Low NO_x combustion technology for pulverised coal fuel'. *Hitachi Review*, 1985, **34** (5), 207–212.

379 Mason H.B. (Acurex). 'Survey of control techniques for nitrogen oxide emissions from stationary sources'. *AIChE Symposium Series*, 1979, **75** (188), 1–13.

380 Hunter, S.C. and Carter W.A. (KVB Inc). 'Application of combustion technology for NO_x emissions reduction on petroleum process heaters'. *AIChE Symposium Series*, 1979, **75** (188), 14–26.

381 Sakai M., Fujima Y., Namiki T. and Okada M. (Mitsubishi). 'Development on low NO_x combustion technology'. Paper to EPDC/-IIP/VDI, NO_x Symposium, Karlsruhe, West Germany, February 1985. Paper P.

382 Mahjoob A.L., Singh S.N. and Yokosh S.M. (Aqua-Chem Inc). 'An experimental investigation of the effects of flue gas recirculation on NO_x formation'. Paper to APCA, 78th Annual Meeting, Detroit, USA, June 1985. Paper 85–55.3.

383 Lisauskas R.A., Snodgrass R.J. (Riley Stoker), Johnson S.A. (Physical Sciences Inc.) and Eskinazi D. (EPRI). 'Experimental investigation of retrofit low-NO_x combustion systems'. Paper to EPA/EPRI, 1985 Symposium on Stationary Combustion NO_x Control, Boston, USA, May 1985. EPRI, Proceedings, CS-4360, January 1986.

384 Folsom B., Abele A. and Reese J. (Energy & Environmental Research Corp.). 'Field evaluation of the distributed mixing burner'. Paper to 1985 Symposium on Stationary Combustion NO_x Control, Boston, USA, May 1985. EPRI, Proceedings, CS4360.

385 Lisauskas R.A., Itse D.C. (Riley Stoker) and Masser C.C. (EPA). 'Extrapolation of burner performance from single burner tests to field operation'. Paper to EPA/EPRI, 1985 Symposium on Stationary Combustion NO_x Control, Boston, USA, May 1985. EPRI, Proceedings. CS-4360, January 1986.

386 Mulholland J.A. and Hall R.E. (US EPA). 'The effect of fuel nitrogen in reburning application to a firetube package boiler'. Paper to EPA/EPRI, 1985 Symposium on Stationary Combustion NO_x Control, Boston, USA, May 1985. EPRI, Proceedings, CS-4360, January 1986.

387 Yang R.J., Garacia F.J. and Hunter S.C. (KVB Inc). 'Screening and optimisation of in-furnace NO_x-reduction processes for refinery process heater applications'. Paper to EPA/EPRI, 1985 Symposium on Stationary Combustion NO_x Control, Boston, USA, May 1985. EPRI, Proceedings, CS-4360, January 1986.

388 England G., Kwan Y. and Payne R. (Energy & Environmental Research Corporation). 'Development and field-demonstration of a low-NO_x burner for TEOR steamers'. Paper to EPA/EPRI, 1985 Symposium on Stationary Combustion NO_x Control, Boston, USA, May 1985. EPRI, Proceedings, CS-4360, January 1986.

389 Hunter S.C. and Benson R.C. (KVB Inc). 'Reduction of nitric oxide emissions on a full-scale cement kiln using primary air vitiation'. Paper to EPA/EPRI, 1985 Symposium on Stationary Combustion NO_x Control, Boston, USA, May 1985. EPRI, Proceedings, CS-4360, January 1986.

390 Fleming D.K. (Institute of Gas Technology) and Kurzynske F.R. (Gas Research Institute). 'NO_x control for glass-melting tanks'. Paper to EPA/EPRI, 1985 Symposium on Stationary Combustion NO_x Control, Boston, USA, May

1985. EPRI, Proceedings, CS-4360, January 1986.

391 Suzuki T., Morimoto K., Ohtani K., Odawara R., Kohno T., Matsuda Y. and Suyari M. (Kobe Steel). 'Development of Low-NO_x combustion for industrial applications'. Paper to EPA/EPRI, 1985 Symposium on Stationary Combustion NO_x Control, Boston, USA, May 1985. EPRI, Proceedings, CS-4360, January 1986.

392 Kesselring J.P. and Krill W.V. (Alzeta Corp.). 'A low-NO_x burner for gas-fired firetube boilers'. Paper to EPA/EPRI, 1985 Symposium on Stationary Combustion NO_x Control, Boston, USA, May 1985. EPRI, Proceedings, CS-4360, January 1986.

393 Wendt J.O.L. (University of Arizona). 'Fundamental coal combustion mechanisms and pollutant formation in furnaces'. *Prog. Energy Combust. Sci.*, 1980, **6**, 201–222.

394 Waibel R. and Nickeson D. (John Zink Co.) 'Staged fuel burners for NO_x control'. Paper to International Flame Research Foundation, 8th Members Conference, Noordwijkerhout, The Netherlands, May 1986.

395 Takahashi Y., Sakai M., Junimoto T., Haneda H., Hawamura T. and Kaneko S. (Mitsubishi). 'Development of MACT: In-furnace NO_x removal process for utility steam generators'. Paper to American Power Conference, Chicago, USA, April 1982. Proceedings 1982, **44**, 402–412.

396 Penterson C.A. (Riley Stoker Corp.). 'Development of an economical low-NO_x firing system for coal fired steam generators'. ASME Paper 82-JPGC-Pwr-43, 1982.

397 Hunter S.C. (KVB Inc.). 'Refinery process heater NO_x control by staged combustion air lances'. Paper to 38th Petroleum Mechanical Engineering Workshop Conference, Philadelphia, USA, 1982.

398 Schaedel S.V. (GRI). 'Pyrocore–radiant burner with a bright future'. *Gas Research Institute Digest*, 1984 (July/Aug), **7** (4), 4–9.

399 Whitehead D.M. and Butcher R.W. (British Petroleum). 'Forced draft burners compared'. *Hydrocarbon Processing*, 1984 (July), 51–55.

400 Anonymous. 'Oil power–how they're making it more acceptable'. *Achievement*, 1986 (June), 19–20.

401 Makansi J. (compiler). 'Low-NO_x burners can play key role in retrofits, upgrades'. *Power*, 1986 (Sept), 61–62.

402 Lim K.J. *et al.* 'Technology assessment report for industrial boiler applications: NO_x combustion modifications'. Report No. EPA-600/7-79-178f to US Environmental Protection Agency, December 1979.

403 Gabrielson J.E., Langsjoen P.L. and Kosvic T.C. 'Field tests of industrial stoker coal-fired boilers for emission control and efficiency improvement'. Report No. EPA-600/7-79-130a, May 1979.

404 British Gas–Private communication.

405 Fenumore

406 Martenay

407 Clark A.G.

Abatement by Flue Gas Treatment

408 Makansi J. 'Controlling NO_x emissions from utility power plants'. *Power*, 1985 (Sept), 107–109.

409 Makansi J. (Compiler). 'Meeting future NO_x caps goes beyond furnace modifications'. *Power*, 1985 (Sept), 45–46.

410 Anonymous. 'Flue gas treatment aims for process simplicity, NO_x control'. *Power*, 1985 (May), **129** (5), 31–32.

411 Karlsson H.T. and Rosenberg H.S. (Battelle). 'Flue gas denitrification. Selective catalytic oxidation of NO to NO_2'. *Ind. Eng. Chem. Process Des. Dev.*, 1984, **23** (4), 808–814.

412 Sengoku T., Miyake J., Suzuki T., Seto T., Nishimoto Y., Lida K., Sera T. and Mitsuoka S. (Mitsubishi). 'A consideration on NO_x reduction catalysts for coal-fired boilers'. *Mitsubishi Heavy Industries Technical Review*, 1983 (Feb), **20** (1), 1–7.

413 Iwata K., Nishimoto Y. and Muraishi K. (Mitsubishi). 'Selective catalytic reduction'. *Modern Power Systems*, 1985 (Dec), 33–51.

414 Hurst B.E. (Exxon Research and Engineering Co.). 'Thermal denox technology update'. 1985 Joint Symposium on Stationary Combustion NO_x Control, Boston, USA, May 1985.

415 Nagai K. and Tanaka S. (Hitachi Zosen). 'NO_x abatement systems developed by Hitachi

Zosen'. Paper to EPDC/IIP/VDI, NO_x Symposium, Karlsruhe, West Germany, February 1985, Paper M.

416 Hurst B.E. (Exxon Research & Engineering Co.). 'Thermal deNO_x: the practical approach to deep NO_x reduction'. Paper to 32nd Canadian Chemical Engineering Conference, Vancouver, Canada, October 1982.

417 Hurst B.E. (Exxon Research & Engineering Co.) 'Exxon thermal deNO_x process for stationary combustion sources'. Paper to US–Dutch International Symposium on Air Pollution by Nitrogen Oxides, Maastricht, The Netherlands, May 1982.

418 Kerry H.A. and Weir A. 'Catalytical DeNO_x demonstration system at Huntington Beach Generating Station Unit 2'. Paper to Joint Symposium on Stationary Combustion NO_x Control, Dallas, USA, November 1982.

419 Faucett H.L., Maxwell J.D. and Burnett T.A. (Tennessee Valley Authority). 'Technical assessment of NO_x removal processes for utility application'. Report No. AF-568 to EPRI, March 1978.

PAPERS ON COMBINED FGD AND/OR NO_x ABATEMENT AND/OR HALIDES ABATEMENT

Combined Flue Gas Desulphurisation and Nitrogen Oxides Abatement

420 Knoblauch K., Richter E. and Juntgen H. (Bergbau-Forschung). 'Application of active coke in processes of SO_2- and NO_x-removal from flue gases'. *Fuel*, 1981 (Sept), **60**, 832–838.

421 Vatsky J. and Schindler E.S. (Foster Wheeler). 'Limestone injection with an internally-staged low-NO_x burner'. Paper to EPA/EPRI, 1st Joint Symposium on Dry SO_2 and Simultaneous SO_2/NO_x Control Technologies, San Diego, USA, November 1984.

422 Felsvang K., Morsing P. and Veltman P. (Niro Atomizer). 'Acid rain prevention through new SO_x/NO_x dry scrubbing process'. Paper to Eighth Symposium, Flue Gas Desulphurisation, New Orleans, USA, November 1983.

423 Takenouchi S., Takahashi K., Atsumi T. and Tanaka H. (Sumitomo). 'Simultaneous NO_x/SO_x removal from sinter waste gas by dry process'. *Transactions ISIJ*, 1983, **23**, 1076–1084.

424 Hoffmann V. (Uhde). 'Activated coke will reduce emissions in Arzberg'. Flue Gas Desulphurisation, 1986 (June), 71–77.

425 Rosenberg H.S. (Battelle). 'Combined NO_x/SO_2 removal for flue gases'. *CME*, 1985 (Jan), 48.

426 Drehmel D.C., Martin G.B. and Abbott J.H. (US EPA). 'Results from EPA's development of limestone injection into a low NO_x furnace'.

427 Dykema O.W. (Rockwell). 'SO_x and NO_x control in combustion'. Paper to Coal Technology '83, 6th International Coal and Lignite Utilisation Conference, Houston, USA, November 1983, **3**, 321–343.

428 Richter E. and Knoblauch K. (Bergbau-Forschung). 'BF-Process for SO_2- and NO_x-removal from flue gases'. Paper to Coal Tech. '85, 5th International Conference on Coal Utilisation and Trade, London, December 1985, **2**, 335–348.

429 Marshall A.R., Goldsack J.S. and Gray J.S. (Babcock Power). 'Reduction of sulphur and nitrogen oxide emissions from utility and industrial boilers'. Paper to Coal Technology Europe '84, 4th European Coal Utilisation Conference, Messe Essen, FRG, September 1984, **2**, 87–125.

430 Furusawa T., Koyama M. and Tsujimura M. (University of Tokyo). 'Nitric oxide reduction by carbon monoxide over calcined limestone enhanced by simultaneous sulphur retention'. *Fuel*, 1985 (March), **64**, 413–415.

431 Drehmel D.C., Martin G.B., Milliken J.O. and Abbott J.H. (US EPA). 'Low NO_x combustion systems with SO_2 control using limestone'. Paper to APCA Annual Meeting, Atlanta, USA, June 1983. Paper No. 83-38.7.

432 Drummond C.J., Markussen J.M., Plantz A.R. and Yeh J.T. (US Department of Energy). 'Advanced environmental control technologies for the simultaneous removal of sulphur dioxide and nitrogen oxides from the flue gas'.

433 Ito Y., Fujimoto T. and Nagaoka, O. (Mitsui). 'Mitsui-BF simultaneous SO_x and NO_x removal system'. Paper No. 8C.

434 Haslbeck J.L., Neal L.G. and Wang C.J. (NOXSO). 'The NOXSO process: a dry simultaneous SO_2/NO_x control technology'. Paper to EPA/EPRI, First Joint Symposium, Dry SO_2 and Simultaneous SO_2/NO_x Control Technologies, San Diego, USA, November 1984.

435 Richter E. and Knoblauch K. (Bergbau-

Forschung). 'BF Process for SO_2- and NO_x-removal from flue gases. Paper to Coal Tech. '85, 5th International Conference on Coal Utilisation and Trade, London, December 1985, **2**, 335–348.

436 Barnes H.L. and Shapiro E. (Pittsburgh Environmental & Energy Systems). 'Process for removing sulphur and/or nitrogen oxide or oxides from other gases containing such oxide or oxides'. UK Pat. Appl. No. GB 2,003,126A, 7 March 1979 (filed 21 August 1978).

437 Flament G. 'The simultaneous reduction of NO_x and SO_2 in coal flames by direct injection of sorbents in a staged mixing burner'. International Flame Research Foundation, Document No. G19/a/10, September 1981.

438 Chang S.G., Littlejohn D. and Lyon S. 'Effects of metal chelates on wet flue gas scrubbing chemistry'. *Env. Sci. Technol.*, 1983, 17 (11), 649–653.

439 Ploeg J.E.G. (Shell International Research). 'A process for the simultaneous removal of nitrogen oxides and sulphur oxides from a gas stream'. Eur. Pat. Appl. No. 80200733.6, 18 February 1981 (filed 31 July 1980).

440 Staudinger G. and Schrofelbauer H. 'Laboratory tests, field trials and application of furnace limestone injection in Austria'. Paper to EPRI/EPA, 1st Joint Symposium on dry SO_2 and simultaneous SO_2/NO_x Control Technologies, San Diego, USA, November 1984.

441 Dalton S.M. 'Current status of dry NO_x–SOx emission control process'. Paper to Joint Symposium on Stationary Combustion NO_x control, Dallas, USA, November 1982.

442 Anonymous. 'Process scrubs both sulphur and nitrogen oxides'. *Chemical Engineering*, 1983 (31 Oct), 21–22.

443 Anonymous. 'Wet-type simultaneous SOx, NO_x removing process developed'. *IHI Bulletin*, 1976 (Nov), 1.

444 Gleason R.J. and Helfritch D.J. (Cottrell Environmental Sciences). 'Alternative electron beam SO_x and NO_x control systems.' Paper to AIChE Spring National Meeting, Houston, USA, March 1985.

Combined Flue Gas Desulphurisation and Acid Halides Abatement

445 Anonymous. 'Gas cleaning: Combined removal of sulphur, dust and fluorine'. *Sulphur*, 1983 (March/April), (165), 42–43.

446 Uchida S. and Tsuchiya K. (Shizuoka University). 'Simulation of spray drying absorber for removal of HCl in flue gas from incinerators'. *Ind. Eng. Chem. Process Des. Dev.*, 1984, **23** (2), 300–307.

447 Kyte W.S., Bettelheim J. (CEGB), Nicholson N.E. and Scarlett J. (Davy McKee). 'Selective absorption of hydrogen chloride from flue gases in the presence of sulphur dioxide'. *Environmental Progress*, 1984 (Aug), **3** (3), 183–187.

448 Nippon Kokan K.K. 'HCl removal from flue gas'. Jap. Pat. JP 60 38,024 (85 38,024), 27 February 1985. Appl. 83/146,078, 10 August 1983.

449 Deguchi A., Kochiyama Y., Hosoda H., Miura M., Hirama T., Nishizaki H. and Horio M. 'The search for an absorbent for HCl and SO_2 removal at high temperature'. *Nenryo Gakkaishi*, 1982, **61** (668), 1105–1108.

450 Mitsubishi Heavy Industries. 'HCl removal'. JP 60 90,028 (85 90,028), 21 May 1985 (Appl. 83/196,128, 21 October 1983).

PAPERS ON FLUIDISED BED COMBUSTION

General Papers on Acid Emissions Abatement by Fluidised Bed Combustion

451 Doyle J.B., May M.P. and Sanders W.J. (Babcock and Wilcox). 'Fluid bed combustion with dry scrubbing of exhaust gases'. Paper to Coal Technology '81, 4th International Coal Utilisation Conference, Houston, USA, November 1981, **5**, 69–86.

452 Gunzelman C.P. (Baltimore Gas and Electric) and Causila H. (Bechtel). 'Methodology and results of a site-specific alternative generation technologies study'. Paper to Coal Technology '83, 6th International Coal and Lignite Utilisation Conference, Houston, USA, November 1983, **3**, 281–297.

453 Gamble R.L. (Foster Wheeler), Krippene B.C. (Stone & Webster) and Zylkowski J.R. (Northern States Power Co.). 'Fluidised bed is Black Dog's new trick'. *Modern Power Systems*, 1985 (Dec), 49–51.

454 Friedrich, F.D. (Energy Mines & Resources Canada). 'A review of major Canadian activities in fluidised bed combustion'. Paper to US

Department of Energy, Proceedings of the 7th International Conference on Fluidised Bed Combustion, Philadelphia, USA, October 1982. Report No. DOE/METC/83–48, 1983 (January), **2**, 644–658.

455 Chui S. (Argonne National Laboratory). 'An evaluation of the environmental and technical aspects of pressurised fluidised bed combustion'. Report No. ANL/EES-TM-257, October 1983.

456 Stringfellow T.E., Nolte F.S. and Sage W.L. (Stearns-Roger). 'Fluidised-bed retrofit a practical alternative to FGD'. *Power*, 1984 (Feb), 78–81.

457 Buck V.E. and Claypoole G. (Pope, Evans and Robbins). 'Controlling emissions with fluidised bed combustion'. Pages 407–413, 'Coal: Phoenix of the '80s' (Traweel A.M. Al Editor). Ottawa: Canadian Society for Chemical Engineering, 1982.

458 Dennis J.S. and Hayhurst A.N. (Univ. of Cambridge). 'Simplified analytical model for the rate of reaction of SO_2 with limestone particles'. *Chem. Eng. Sci.*, 1986, **41** (1), 25–36.

459 Fieldes R.B., Burdett N.A. and Davidson J.F. (Univ. of Cambridge). 'Reaction of sulphur dioxide with limestone particles: the influence of sulphur trioxide'. *Trans. I. Chem. E.*, 1979, **57**, 276–280.

460 Bhatia S.K. and Perlmutter D.D. (Univ. of Pennsylvania). 'The effect of pore structure on fluid–solid reactions: application to the SO_2–lime reaction'. *AIChE Journal*, 1981 (March), **27** (2), 226–234.

461 Munzner H. (Bergbau-Forschung). 'Sulphur bonding to lime in fluidised-bed combustion'. *VDI-Ber.*, 1979, **346**, 319–322.

462 Baker D.C. and Attar A. (Univ. of Houston, Texas). 'Sulphur pollution from coal combustion. Effect of the mineral components of coal on the thermal stabilities of sulphated ash and calcium sulphate'. *Env. Sci. Technol.*, 1981, **15** (3), 288–293.

463 Burdett N.A., Gliddon B.J., Hotchkiss R.C. and Squires R.T. (CEGB). 'SO_3 in coal-fired fluidised bed combustors'. *J. Inst. Energy*, 1983 (September), **56** (428), 119–124.

464 Walsh P.M., Chaung T.Z., Dutta A., Beer J.M. and Sarofim A.F. (MIT). 'Particle entrainment and nitric oxide reduction in the freeboard of a fluidised coal combustor'. *ACS Div. Fuel Chem.*, 1982, **27** (1), 243–261.

465 Martens F.J.A., van Koppen C.W.J. and Boersma D. (Delft Univ. of Technology, The Netherlands). 'The effect of coal type on the CO conversion and NO_x reduction in the freeboard'. Paper DISC/19 to Institute of Energy, 3rd International Fluidised Conference, London, October 1984.

466 Shaw J.T. (British Coal). 'Emissions of nitrogen oxides'. Chapter 6, pages 227–260, 'Fluidised beds: combustion and applications'. (Howard, J.R. Editor). London: Applied Science Publishers, 1983.

467 National Coal Board (British Coal). 'Fluidised bed combustion of coal'. Second edition, 1985. London: British Coal.

468 Stantan J.E. (NCB CURL). 'Sulphur retention in fluidised bed combustion'. Chapter 5, pages 199–225. 'Fluidised beds–combustion and applications'. (Howard, J.R. Editor). London: Applied Science Publishers, 1983.

Atmospheric Pressure Fluidised Bed Combustion

469 Highley J. (NCB Coal Research Establishment). 'Atmospheric FCB and CFB: The European approach'. Paper to Symposium, Advanced Coal Power Technology, Malmo, Sweden, November 1985.

470 Hutchinson B.R. and Virr M.J. (Johnston Boiler Co.). 'Operating experience with industrial package FBC boilers'. Paper to US Department of Energy, 7th International Conference, Fluidised Bed Combustion, Philadelphia, USA, October 1982. US Department of Energy, Proceedings, DOE/METC/83–48, 1983 (January), 1, 26–37.

471 Modrak T.M., Tang J.T., Dugum J.N. (Babcock and Wilcox), Aulisio C.J. (Electric Power Research Institute), Gorrell R.L. (Babcock & Wilcox Co) and Divilio R.J. (Pope, Evans and Robbins). 'The use of velocity turndown for AFBC boiler control'. Paper to US Department of Energy, 7th International Conference, Fluidised Bed Combustion, Philadelphia, USA, October 1982. US Department of Energy, Proceedings, DOE/METC/83–48, 1983 (January), **1**, 111–119.

472 Richards H.W. (Dorr-Oliver Inc.), Laukaitis J.F. (Shamokin Area Industrial Corp.), Lockman H. (Curtiss-Wright Corp.), Fisher B.L. (E. Keeler Co.) and Gmeindl, F.D. (US Department of Energy). 'Operating and maintenance experiences at the Shamokin culm burning boiler plant'. Paper to US Department

of Energy, 7th International Conference, Fluidised Bed Combustion, Philadelphia, USA, October 1982. US Department of Energy, Proceedings, DOE/METC/83–48, 1983 (January), **1**, 133–138.

473 Comparato J.R. and Vroom H.H. (Combustion Engineering). 'Fluidised Bed combustion operating experience at the Great Lakes demonstration plant'. Paper to US Department of Energy, 7th International Conference, Fluidised Bed Combustion, Philadelphia, USA, October 1982. US Department of Energy, Proceedings, DOE/METC/83–48, 1983 (January), **1**, 139–149.

474 Fisher M.J., Highley J., Willis D.M., Vickers M.A. (NCB) Coal Research Establishment) and Jenkins F. EMS Thermplant Ltd.). 'Operating experience with prototype shell boilers'. Paper to US Department of Energy, 7th International Conference, Fluidised Bed Combustion, Philadelphia, USA, October 1982. US Department of Energy, Proceedings, DOE/METC/83–48, 1983 (January), **1**, 150–163.

475 Moore J.P. (Wallsend Slipway Engineers Ltd.), Benson R.A.C. and Fisher M.J. (NCB Coal Research Establishment). 'Operating experience with horizontal and vertical shell boilers'. Paper to US Department of Energy, 7th International Conference, Fluidised Bed Combustion, Philadelphia, USA, October 1982. US Department of Energy, Proceedings, DOE/METC/83–48, 1983 (January), **1**, 168–179.

476 Struthers J.S. (Babcock Power Ltd.), Flett M., Highley J. and Tringham D. (NCB Coal Research Establishment). 'FBC package boilers–operation of 7 tonne/h prototype'. Paper to US Department of Energy, 7th International Conference, Fluidised Bed Combustion, Philadelphia, USA, October 1982. US Department of Energy, Proceedings, DOE/METC/83–48, 1983 (January), **1**, 180–189.

477 Gibson T., Ellis F. (Gibson Wells Engineering Ltd.), Highley J. and Tringham D. (NCB Coal Research Establishment). 'Design and operation of a 15 MW modular boiler).' Paper to US Department of Energy, 7th International Conference, Fluidised Bed Combustion, Philadelphia, USA, October 1982. US Department of Energy, Proceedings, DOE/METC/83–48, 1983 (January), **1**, 190–197.

478 Tang J.T., Duqum J.N., Modrak T.M. (Babcock & Wilcox) and Aulisio C.J. (Electric Power Research Institute). 'An overall review of the EPRI/B&W 6′ × 6′ fluidised bed combustion test facility'. Paper to US Department of Energy, 7th International Conference, Fluidised Bed Combustion, Philadelphia, USA, October 1982. US Department of Energy, Proceedings, DOE/METC/83–48, 1983 (January), **1**, 373–380.

479 Pelser J. (Netherlands Energy Research Foundation), Borgne K.G. (National Swedish Board for Energy Source Development), Friedrich F.D. (Canadian Centre for Mineral and Energy Technology), Hoult B.D. (Applied Combustion Systems Ltd.), Inoue T. (New Energy Development Organisation), Kausche R. (Sulzer Brothers Ltd.), Kollerup V. (Burmeister and Wain Energi), Osterbo E. (Technical University of Norway) and Saccenti G. (ENL-CRTN). 'The IEA atmospheric fluidised bed combustion programme; an international co-operation'. Paper to US Department of Energy, 7th International Conference, Fluidised Bed Combustion, Philadelphia, USA, October 1982. US Department of Energy, Proceedings, DOE/METC/83–48, 1983 (January), **1**, 390–405.

480 Cosar P. (Fives-Cail Babcock). 'Operating experience with the new fluid bed boiler of the Paris district heating system'. Paper to US Department of Energy, 7th International Conference, Fluidised Bed Combustion, Philadelphia, USA, October 1982. US Department of Energy, Proceedings, DOE/METC/83–48, 1983 (January), **1**, 420–428.

481 Fennelly P.F., Young C., Tucker G., Peduto E. (GCA Corp.) and Milliken J.O. (US Environmental Protection Agency). 'Long-term emission monitoring at the Georgetown University fluidised-bed boiler'. Paper to US Department of Energy, 7th International Conference, Fluidised Bed Combustion, Philadelphia, USA, October 1982. US Department of Energy, Proceedings, DOE/METC/83–48, 1983 (January), **1**, 506–513.

482 Leckner B., Jansson B., Lindqvist O. and Nielsen B.M. (Chalmers University of Technology). 'Emission from a 16 MWth FBC-boiler'. Paper to US Department of Energy, 7th International Conference, Fluidised Bed Combustion, Philadelphia, USA, October 1982. US Department of Energy, Proceedings, DOE/METC/83–48, 1983 (January), **1**, 514–524.

483 Chiplunker D.G., Kwon H.S., Richards H.W. (Dorr-Oliver Inc.) and Garver D.L. (E. Keeler Co.). 'Performance of a fluidised bed steam generator burning anthracite culm'. Paper to

US Department of Energy, 7th International Conference, Fluidised Bed Combustion, Philadelphia, USA, October 1982. US Department of Energy, Proceedings, DOE/METC/83–48, 1983 (January), **1**, 567–572.

484 Svensson G. (Generator Industri). 'Multifuel fluidised bed boilers in district heating system'. Paper to US Department of Energy, 7th International Conference, Fluidised Bed Combustion, Philadelphia, USA, October 1982. US Department of Energy, Proceedings, DOE/METC/83–48, 1983 (January), **1**, 573–586.

485 Delessard S., Puff R. and Kita J.C. (CERCHAR). 'The utilisation of carbonaceous shales by the CERCHAR process of fluidised bed combustion: the Drocourt plant'. Paper to US Department of Energy, 7th International Conference, Fluidised Bed Combustion, Philadelphia, USA, October 1982. US Department of Energy, Proceedings, DOE/METC/83–48, 1983 (January), **1**, 587–599.

486 Svensson G. (Generator Industri) and Leckner B. (Chalmers University of Technology). 'The fluidised bed boiler at Chalmers University of Technology'. Paper to US Department of Energy, 7th International Conference, Fluidised Bed Combustion, Philadelphia, USA, October 1982. US Department of Energy, Proceedings, DO/METC/83–48, 1983 (January), **2**, 625–636.

487 Spitsbergen U., Vincent C.J. (Twente Univ. of Technology, The Netherlands) and Longe T.A. (Ahmadu Bello Univ., Nigeria). 'Comparison of selected European limestones for desulphurisation of gases from atmospheric fluidised bed combustion'. *J. Inst. Energy*, 1981 (June), **54** (419), 94–99.

488 Verhoeffer F. (Stork Boilers) and van Gasselt M. (TNO). 'TNO/Stork fluidised bed combustion development'. Paper to US Department of Energy, 7th International Conference, Fluidised Bed Combustion, Philadelphia, USA, October 1982. US Department of Energy, Proceedings, DOE/METC/83–48, 1983 (January), **2**, 659–669.

489 Kaden M. and Huschauer H. (Vereinigte Kesselwerke). 'Experience with commercial scale AFBC and its influence on the design of the Hameln and Cebu plants'. Paper to US Department of Energy, 7th International Conference, Fluidised Bed Combustion, Philadelphia, USA, October 1982. US Department of Energy, Proceedings, DOE/METC/83–48, 1983 (January), **2**, 681–690.

490 Boyd T.J. (Electric Power Research Institute) and Gottschalk C. (Tennessee Valley Authority). 'Operating experience at TVA's 20 MW(e) AFBC pilot plant'. Paper to US Department of Energy, 7th International Conference, Fluidised Bed Combustion, Philadelphia, USA, October 1982. US Department of Energy, Proceedings, DOE/METC/83–48, 1983 (January), **2**, 755–760.

491 Lessig W.S., Callahan S.F., Rickman W.S. (General Atomic Co.), Manaker A.M. and Herness J.L. (Tennessee Valley Authority). 'Fines recycle in a fluidised bed coal combustor'. Paper to US Department of Energy, 7th International Conference, Fluidised Bed Combustion, Philadelphia, USA, October 1982. US Department of Energy, Proceedings, DOE/METC/83–48, 1983 (January), **2**, 761–780.

492 Terada H. (Babcock Hitachi K.K.) and Kawashima I. (Sumitomo). 'Utilisation of sedimented coal sludge in fluidised bed boiler'. Paper to US Department of Energy, 7th International Conference, Fluidised Bed Combustion, Philadelphia, USA, October 1982. US Department of Energy, Proceedings, DOE/METC/83–48, 1983 (January), **2**, 840–846.

493 Terada H. (Babcock Hitachi K.K.), Takagi M. (New Energy Development Organization), Shimizu T. (Electric Power Development Co. Ltd.), Tamanuki S. (Coal Mining Research Centre) and Tatebayashi J. (Kawasaki Heavy Industries Ltd.). 'Current topics on testing of the 20th fluidised bed boiler'. Paper to US Department of Energy, 7th International Conference, Fluidised Bed Combustion, Philadelphia, USA, October 1982. US Department of Energy, Proceedings, DOE/METC/83–48, 1983 (January), **2**, 876–885.

494 Ganesh A., Thirunavukkarasu R., Bhaskaran C. and Shanmugam S. (Bharat Heavy Electricals Ltd.). 'Operating experience of first commercial co-generation AFBC boiler in India'. Paper to US Department of Energy, 7th International Conference, Fluidised Bed Combustion, Philadelphia, USA, October 1982. US Department of Energy, Proceedings, DOE/METC/83–48, 1983 (January), **2**, 902–914.

495 Inoue T. (New Energy Development Organisation), Tamanuki S. (Coal Mining Research Centre), Shimizu T. (Electric Power Development Co.), Kawada, S. (Babcock-Hitachi K.K.) and Takada T. (Kawasaki Heavy Industries Ltd.). 'Operation of the 20t/h AFBC boiler pilot plant'. Paper to US Department of

Energy, 7th International Conference, Fluidised Bed Combustion, Philadelphia, USA, October 1982. US Department of Energy, Proceedings, DOE/METC/83–48, 1983 (January), **2**, 915–928.

496 Anthony E.J. (Energy Mines & Resources Canada), Becker H.A., Code R.K. (Queen's University, Canada), Liang D.T. (Energy Mines & Resources Canada) and Stephenson J.R. (Queen's University, Canada). 'Combustion of high-sulphur Eastern Canadian coals by AFBC'. Paper to US Department of Energy, 8th International Conference on Fluidised Bed Combustion, Houston, USA, March 1985. US Department of Energy, Proceedings, DOE/METC-85-6021, 1985 (July), **1**, 32–42.

497 Meulink J.P. (Stork Boilers), van Haasteren A.W.M.B. and Temmink H.M.G. (TNO). 'Operating experience with a 4MWth AFBB research facility'. Paper to US Department of Energy, 8th International Conference on Fluidised Bed Combustion, Houston, USA, March 1985. US Department of Energy, Proceedings, DOE/METC-85-6021, 1985 (July), **1**, 179–195.

498 Castleman J.M. (Tennessee Valley Authority). 'Process performance of the TVA 20-MW atmospheric fluidised-bed combustion (AFBC) pilot plant'. Paper to US Department of Energy, 8th International Conference on Fluidised Bed Combustion, Houston, USA, March 1985. US Department of Energy, Proceedings, DOE/METC-85-6021, 1985 (July), **1**, 196–207.

499 Shimizu T. (Electric Power Development Co.), Tatebayashi J. (Kawasaki Heavy Industries), Terada H. (Babcock-Hitachi K.K.), Furusawa T. (University of Tokyo) and Horio M. (Tokyo University of Agriculture & Technology). 'The combustion characteristics of different types of coal in the 20t/h fluidised-bed boiler'. Paper to US Department of Energy, 8th International Conference on Fluidised Bed Combustion, Houston, USA, March 1985. US Department of Energy, Proceedings, DOE/METC-85-6021, 1985 (July), **1**, 231–240.

500 Anthony E.J., Desai D.L., Friedrich F.D. and Razbin V.V. (CANMET, Energy, Mines and Resources Canada). 'The fluidised-bed combustion of a high-sulphur Maritime coal in a pilot-scale rig and industrial FBC boiler'. Paper to US Department of Energy, 8th International Conference on Fluidised Bed Combustion, Houston, USA, March 1985. US Department of Energy, Proceedings, DOE/METC-85-6021, 1985 (July), **1**, 241–254.

501 Duqum J.N., Tang J.T., Morris T.A., Esakov J.L. (Babcock & Wilcox Co.) and Howe W.C. (Electric Power Research Institute). 'AFBC performance comparison for underbed and overbed feed systems'. Paper to US Department of Energy, 8th International Conference on Fluidised Bed Combustion, Houston, USA, March 1985. US Department of Energy, Proceedings, DOE/METC-85-6021, 1985 (July), **1**, 255–278.

502 Takagi H., Koga M. (Coal Mining Research Centre), Nakabayashi Y., Abe H., Shimizu T. and Fujita M. (Electric Power Development Co. Ltd.). 'Development of atmospheric fluidised-bed combustion boiler for electric utility in Japan'. Paper to US Department of Energy, 8th International Conference on Fluidised Bed Combustion, Houston, USA, March 1985. US Department of Energy, Proceedings, DOE/METC-85-6021, 1985 (July), **1**, 279–288.

503 Oki K., Maeda M., Koga S., Yamamoto Y. and Takami N. (Babcock-Hitachi K.K.). 'Development of a large scale AFBC utility boiler'. Paper to US Department of Energy, 8th International Conference on Fluidised Bed Combustion, Houston, USA, March 1985. US Department of Energy, Proceedings, DOE/METC-85-6021, 1985 (July), **1**, 338–349.

504 D'Acierno J.P., Garver D.L. and Fisher, B. (Keeler/Dorr-Oliver). 'Design concepts for industrial coal-fired fluidised-bed steam generators'. Paper to US Department of Energy, 8th International Conference, Fluidised-Bed Combustion, Houston, USA, March 1985. US Department of Energy, Proceedings, DOE/METC-85-6021, 1985 (July), **1**, 406–414.

505 Dooley M.J. and Vroom H.H. (CE Power Systems). 'Operating evaluation of the Great Lakes fluidised-bed demonstration plant'. Paper to US Department of Energy, 8th International Conference on Fluidised-Bed Combustion, Houston, USA, March 1985. US Department of Energy, Proceedings, DOE/METC-85-6021, 1985 (July), **2**, 564–573.

506 Kikuzawa K., Tanabe T., Yano K., Takada T. (Kawasaki Heavy Industries Ltd.), Shimano S. and Motono S. (Mitsui Aluminium Co. Ltd.). 'Industrial Fluidised-Bed boiler cogeneration system at Mitsui Aluminium Co. Ltd., Wakamatsu Works, Japan'. Paper to US Department of Energy, 8th International Conference on Fluidised Bed Combustion, Houston, USA, March 1985. US Department of Energy, Proceedings, DOE/METC-85-6021, 1985 (July), **2**, 584–593.

507 Boden B.G. (Foster Wheeler Energy Corp.) and

Damon W.H. (JelCom/Commonwealth Associates). 'Commercial application of an atmospheric fluidised-bed steam plant'. Paper to US Department of Energy, 8th International Conference on Fluidised Bed Combustion, Houston, USA, March 1985. US Department of Energy, Proceedings, DOE/METC-85-6021, 1985 (July), **2**, 605–613.

508 Toth D.A. (Foster Wheeler Energy Corp.). 'Fluidised bed at Midwest Solvents Company–the economic edge'. Paper to US Department of Energy, 8th International Conference, Fluidised-Bed Combustion, Houston, USA, March 1985. US Department of Energy, Proceedings, DOE/METC-85-6021, 1985 (July), **2**, 614–618.

509 Sadowski R.S. (Wormser Engineering). 'The first 10,000 hours of operation of the Wormser grate at Amarillo'. Paper to US Department of Energy, 8th International Conference on Fluidised-Bed Combustion, Houston, USA, March 1985. US Department of Energy, Proceedings, DOE/METC-85-6021, 1985 (July), **2**, 634–643.

510 Stracener W.R., Wolfe M. (Griffin Industries), Wallish J.W. and Virr M.J. (Stone Johnston). 'Griffin Industries operating experience with an FBC "gob"-burning boiler'. Paper to US Department of Energy, 8th International Conference, Fluidised-Bed Combustion, Houston, USA, March 1985. US Department of Energy, Proceedings, DOE/METC-85-6021, 1985 (July), **2**, 924–936.

511 Trivett G.S., Field R.S. and MacKay G.D. (Technical University of Nova Scotia). 'Coal-limestone-water slurry testing in atmospheric fluidised-bed combustion'. Paper to US Department of Energy, 8th International Conference, Fluidised-Bed Combustion, Houston, USA, March 1985. US Department of Energy, Proceedings, DOE/METC-85-6021, 1985 (July), **2**, 948–961.

512 Rowley D.R. (Babcock & Wilcox), Lau I.T. and Friedrich F.D. (Department of Energy, Mines and Resources Canada). 'Combustion of coal-water slurries and coal tailings in a fluidised bed'. Paper to US Department of Energy, 8th International Conference, Fluidised-Bed Combustion, Houston, USA, March 1985. US Department of Energy, Proceedings, DOE/METC-85-6021, 1985 (July), **2**, 962–980.

513 Fisher B.L. (Keeler/Dorr-Oliver), LeBlanc F. (OTV) and Kwon H.S. (Keeler/Dorr-Oliver). 'Fluidised-bed steam generators burning low-grade fuels: operating experience'. Paper to US Department of Energy, 8th International Conference, Fluidised-Bed Combustion, Houston, USA, March 1985. US Department of Energy, Proceedings, DOE/METC-85-6021, 1985 (July), **2**, 1003–1018.

514 Haug R.T. (Consultant), Lewis F.M. (Consultant) and Sizemore H.M. (City of Los Angeles). 'Gasification and staged combustion for low air emissions from sludge-derived fuels'. Paper to US Department of Energy, 8th International Conference, Fluidised-Bed Combustion, Houston, USA, March 1985. US Department of Energy, Proceedings, DOE/METC-85-6021, 1985 (July), **2**, 1030–1049.

515 Rajan S. and Taylor E.W. (Southern Illinois University). 'Combustion and emissions characteristics of gob pile wastes in fluidised-bed combustors'. Paper to US Department of Energy, 8th International Conference, Fluidised-Bed Combustion, Houston, USA, March 1985. US Department of Energy, Proceedings, DOE/METC-85-6021, 1985 (July), **2**, 1075–1084.

516 Molayem B. and Bardakci T. (Benmol Corp.). 'Simultaneous removal of SO_2, NO_x, CO and CH_4 using Sorcat systems'. Paper to US Department of Energy, 8th International Conference, Fluidised-Bed Combustion, Houston, USA, March 1985. US Department of Energy, Proceedings, DOE/METC-85-6021, 1985 (July), **3**, 1197–1207.

517 Munzner H. and Schilling H.D. (Bergbau-Forschung). 'Fluorine and chlorine emissions from FBC enrichments in fly ash and filter dust'. Paper to US Department of Energy, 8th International Conference, Fluidised-Bed Combustion, Houston, USA, March 1985. US Department of Energy, Proceedings, DOE/METC-85-6021, 1985 (July), **3**, 1219–1226.

518 Ohki K., Takami N. and Maeda M. 'Development of AFBC utility boiler without FGD unit'. *Hitachi Review*, 1985, **34** (2), 87–90.

519 D'Acierno J.P. and Garver D.L. (Dorr-Oliver). 'Operating experience and design features for Keeler/Dorr-Oliver fluidised bed boilers'. Paper DISC/7 to Institute of Energy, 3rd International Fluidised Conference, London, October 1984.

520 Pomeroy M., Bannard J. (National Inst. for Higher Education, Limerick, Ireland) and Coakley S. (Electricity Supply Board, Dublin). 'The burning of low-grade Irish coals by fluidised-bed combustion'. Paper DISC/23 to Institute of Energy, 3rd International Fluidised Conference, London, October 1984.

521 Ekinci E., Pogson B. and Fells I. (Univ. of Newcastle upon Tyne). 'Sulphur dioxide capture by the inorganic matrix of a low grade fuel in a fluidised bed combustor'. *J. Inst. Energy*, 1984 (September), **57** (4332), 368–372.

522 Minchener A.J. (British Coal). 'Review of AFBC materials studies undertaken by the NCB Coal Research Establishment'. Paper to EPRI Workshop on Materials Issues in FBC, Nova Scotia, Canada, July/August 1985.

523 Krischke H.G. and Langhoff J. (Ruhrkohle). '6 MW AFBC test and demonstration facility "Konig Ludwig": 15,000 hours operational experience'. Paper KN/1B/3 to Institute of Energy, 3rd International Fluidised Conference, London, October 1984.

524 Pomeroy M.J., Bannard J.E. and Mulligan T.F. (National Inst. for Higher Education, Limerick, Ireland). 'Some operating experiences of the fluidised bed combustion of a high sulphur anthracite'. Paper DISC/2 to Institute of Energy, 3rd International Fluidised Conference, London, October 1984.

525 Bass J.W. and High M.D. (TVA). 'Operating and performance summary for TVA's 20-MW AFBC pilot plant'. Paper KN/1B/2 to Institute of Energy, 3rd International Fluidised Conference, London, October 1984.

526 Wormser A. (Wormser Engineering) and Beckwith W. (Iowa Beef Processors). 'The Wormser Grate installation at Iowa Beef's plant in Amarillo, Texas'. Paper to US Department of Energy, 7th International Conference, Fluidised Bed Combustion, Philadelphia, USA, October 1982. US Department of Energy, Proceedings, DOE/METC/83-48, 1983 (January).

527 Zielinski E.A. (Combustion Engineering). 'Atmospheric fluidised bed combustion–initial operating results from the facility at Great Lakes'. Paper to American Power Conference, Chicago, USA, April 1982.

528 Schwieger R. 'Coal-fired AFB boiler advances state-of-the-art'. *Power*, 1984 (June), 106–109.

529 Zylkowski J.R. (Northern States Power Co.), Gamble R.L. (Foster Wheeler Energy Corp.) and Krippene B.C. (Stone & Webster). Paper to 1984 Joint Power Generation Conference, Toronto, Canada, September, October 1984.

530 St. John B. (NUS Corp.) 'Economics of atmospheric fluidised-bed boilers'. *Chemical Engineering*, 1986 (8 Dec), 157–159.

531 Thomas J.F., Gregory R.W. and Takayasu M. (IEA Coal Research). 'Atmospheric fluidised bed boilers for industry'. Report No. ICTIS/TR35 to IEA, November 1986.

532 Bernhardt R.S., Hodges N.J., Newman J.O.H. and O'Brien D.G. 'Chlorine in coal: fate during fluidised bed combustion'. Private communication, British Coal, Coal Research Establishment, Committee Document, 1982 (December).

533 Deguchi, A., Kochiyama, Y., Hosoda, H., Miura, M., Hirama, T., Nishizaki, H. and Horio, M. 'Study on the HCl removal process in a fluidised bed combustor'. *Nenryo Gakkaishi*, 1982, **61** (668), 1109–1112.

Pressurised Fluidised Bed Combustion

534 Rubow L.N., Borden M. and Buchanan T.L. (Gilbert/Commonwealth). 'Cost and performance evaluation of air and steam cooled PFBC power plants'. Paper to US Department of Energy, 7th International Conference, Fluidised Bed Combustion, Philadelphia, USA, October 1982. US Department of Energy, Proceedings, DOE/METC/83-48, 1983 (January), **1**, 198–210.

535 Smith D., Anderson J.S., Atkin J.A.R., Bekofske K.L., Brown R.A., Cavanna J., Christianson S., Failing K.H., Friedman M.A., Glenn J.C., Herbden D.J., Mainhardt P.J., Schuetz M., Wheeldon J.M. and Carls E.L. (NCB). 'IEA Grimethorpe 2m × 2m pressurised fluidised bed combustion project–experimental performance results and future plans'. Paper to US Department of Energy, 7th International Conference, Fluidised Bed Combustion, Philadelphia, USA, October 1982. US Department of Energy, Proceedings, DOE/METC/83-48, 1983 (January), **1**, 439–452.

536 Hoy H.R., Roberts A.G., Phillips R.N. (NCB Coal Utilisation Research Laboratory) and Carpenter L.K. (US Department of Energy). 'Performance of a small combustor at pressures up to 20 atm'. Paper to US Department of Energy, 7th International Conference, Fluidised Bed Combustion, Philadelphia, USA, October 1982. US Department of Energy, Proceedings, DOE/METC/83-48, 1983 (January), **1**, 473–481.

537 Roberts A.G., Pillai K.K., Barker S.N. (NCB Coal Utilisation Research Laboratory) and Carpenter, L.K. (US Department of Energy). 'Combustion of "run-of-mine" coal and coal-water mixtures in a small PFBC'. Paper to US Department of Energy, 7th International

Conference, Fluidised Bed Combustion, Philadelphia, USA, October 1982. US Department of Energy, Proceedings, DOE/METC/83-48, 1983 (January), **1**, 482–489.

538 Zaharchuk R., Buchanan T.L., Rubow L.N. and Borden M. (Gilbert/Commonwealth). 'Cost evaluation of four PFBC hot gas cleanup systems'. Paper to US Department of Energy, 7th International Conference, Fluidised Bed Combustion, Philadelphia, USA, October 1982. US Department of Energy, Proceedings, DOE/METC/83-48, 1983 (January), **1**, 491–505.

539 Hoy H.R., Roberts A.G. (Hoy Associates) and Scott R.L. (US Department of Energy). 'Operation of a small combustor on dry coal and on coal-water mixtures at pressures up to 20 atm'. Paper to US Department of Energy, 8th International Conference, Fluidised-Bed Combustion, Houston, USA, March 1985, US Department of Energy, Proceedings, DOE/METC-85-6021, 1985 (July), **1**, 291–306.

540 Pillai K.K., Kreij S.E. and Paulsson R. (ASEA PFBC). 'Further experience on the Malmo PFBC pilot plant'. Paper to US Department of Energy, 8th International Conference, Fluidised-Bed Combustion, Houston, USA, March 1985. US Department of Energy, Proceedings, DOE/METC-85-6021, 1985 (July), **1**, 307–316.

541 Wheeldon J.M., Swift W.M., Anderson J.S., Bekofske K.L., Cavanna J., Christianson S.R., Friedman M.A., Glenn J.C., Hebden D.J., Mainhardt P.J., Mountford R.A., Schmidt D., Snow G.C. and Carls E.L. (NCB). 'Experimental results from the Grimethorpe PFBC facility'. Paper to US Department of Energy, 8th International Conference, Fluidised-Bed Combustion, Houston, USA, March 1985. US Department of Energy, Proceedings, DOE/METC-85-6021, 1985 (July), **1**, 317–335.

542 Olesen C., Pillai K.K., Wickstrom B. and Jansson S.A. (ASEA PFBC). 'The commercial status of ASEA PFBC technology'. Paper to US Department of Energy, 8th International Conference, Fluidised-Bed Combustion, Houston, USA, March 1985. US Department of Energy, Proceedings, DOE/METC-85-6021, 1985 (July), **1**, 512–521.

543 Wheeldon J.M. (NCB), Burdett N.A. (CEGB), Friedman M.A. (NCB) and Stantan J.E. (Hoy Associates Ltd.). 'SO_3 measurements at the Grimethorpe PFBC facility'. Paper to US Department of Energy, 8th International Conference, Fluidised-Bed Combustion, Houston, USA, March 1985. US Department of Energy, Proceedings, DOE/METC-85-6021, 1985 (July), **2**, 715–729.

544 Stantan J.E. (Hoy Associates Ltd.), Olen K.R. (Florida Power & Light Co.) and Scott R.L. (US Department of Energy). 'The performance of candidate PFB sorbents for use in South Florida'. Paper to US Department of Energy, 8th International Conference, Fluidised-Bed Combustion, Houston, USA, March 1985. US Department of Energy, Proceedings, DOE/METC-85-6021, 1985 (July), **3**, 1125–1155.

545 Smock R. (editor). 'PFBC aims for utility acceptance'. *Power Engineering*, 1986 (Sept), **90** (9), 31–34.

546 Minchener A.J., Stringer J., Brooks S., Lloyd D.M., Swift W.M., Anderson J.S. and Mainhardt P.J. 'Materials studies at the IEA Grimethorpe PFBC experimental facility'. Paper to US Department of Energy, 8th International Conference, Fluidised Bed Combustion, Houston, USA, March 1985. US Department of Energy, Proceedings, DOE/METC-85-6021, 1985 (July).

547 Stringer J. (EPRI) and Minchener A.J. (British Coal). 'High temperature corrosion in fluidised bed combustion systems'. Paper to AIME/-Amer. Soc. of Metals meeting, 'High Temperature Corrosion in Energy Systems', Detroit, USA, September 1984.

548 Swift W.M., Wheeldon J.M., Anderson J.S., Bekofske K.L., Cavanna J., Christianson S.R., Friedman M.A., Glenn J.C., Hebden D.J., Mainhardt P.J., Mountford R.A., Schmidt D., Snow G.C. and Carls E.L. 'IEA Grimethorpe 2m × 2m PFBC facility: test series 2 experimental results'. Paper KN/IV/2 to Institute of Energy, 3rd International Fluidised Conference, London, October 1984.

549 Burdett N.A. (CEGB). 'The prediction of SO_3 emissions from pressurised fluidised bed combustors'. Paper DISC/4 to Institute of Energy, 3rd International Fluidised Conference, London, October 1984.

550 Hoy H.R., Roberts A.G. and Scott R.L. 'Pressurised fluidised bed combustion: Past, present and future'. Paper DISC/13 to Institute of Energy, 3rd International Fluidised Conference, London, October 1984.

551 Markowsky J.J. (American Electric Power Service Corp.). 'Long term option for reducing SO_2 emissions'. Paper to Power Magazine, First International Conference on Acid Rain, Washington, USA, March 1984.

Circulating Fluidised Bed Combustion

552 Brereton C. (University of British Columbia) and Stromberg (Studsvik Energiteknik). 'Some aspects of the fluid dynamic behaviour of fast fluidised beds'.

553 Petersen V., Daradimos G., Serbent H. and Schmidt H.W. (Lurgi). 'Combustion in the circulating fluid bed: an alternative approach in energy supply and environmental protection'.

554 Engstrom F. (A. Ahlstrom Co.) and Yip, H.H. (Pyropower Corp.). 'Operating experience of commercial scale Pyroflow circulating fluidised bed combustion boilers'. Paper to US Department of Energy, 7th International Conference, Fluidised Bed Combustion, Philadelphia, USA, October 1982. US Department of Energy, Proceedings, DOE/METC/83–48, 1983 (January), **2**, 1136–1143.

555 McIntyre J. (Pyropower Corp.). 'Nucla upgrades to 110 MW'. *Modern Power Systems*, 1985 (December), 37–44.

556 Lund T. (Lurgi Corporation). 'Lurgi circulating fluid bed boiler: its design and operation'. Paper to US Department of Energy, 7th International Conference, Fluidised Bed Combustion, Philadelphia, USA, October 1982. US Department of Energy, Proceedings, DOE/METC/83–48, 1983 (January), **1**, 38–46.

557 Jones O. (Conoco) and Seber, E.C. (Struthers Thermo-Flood Corp.). 'Initial operating experience at Conoco's South Texas multi-solids FBC steam generator'. Paper to US Department of Energy, 7th International Conference, Fluidised Bed Combustion, Philadelphia, USA, October 1982. US Department of Energy, Proceedings, DOE/METC/83–48, 1983 (January), **1**, 381–389.

558 Kim B.C., Litt, R.D. and Nack, H. (Battelle). 'Multiple fuels emissions control'. Paper to US Department of Energy, 7th International Conference, Fluidised Bed Combustion, Philadelphia, USA, October 1982. US Department of Energy, Proceedings, DOE/METC/83–48, 1983 (January), **2**, 859–869.

559 Stromberg L. (Studsvik AB). 'Fast fluidised bed combustion of coal'. Paper to US Department of Energy, 7th International Conference, Fluidised Bed Combustion, Philadelphia, USA, October 1982, US Department of Energy, Proceedings, DOE/METC/83–48, 1983 (January), **2**, 1152–1162.

560 Jones O. (Conoco), Litt R.D. (Battelle) and Davis J.S. (Struthers Wells Corp.). 'Performance of Conoco's prototype MS-FBC oil field steam generator'. Paper to US Department of Energy, 8th International Conference, Fluidised-Bed Combustion, Houston, USA, March 1985. US Department of Energy, Proceedings, DOE/METC-85-6021, 1985 (July), **2**, 555–563.

561 Stromberg L., Kobro H., Brereton C. (Studsvik Energiteknik AB), Morris, T.A. and Walker D.J. (Babcock & Wilcox). 'The fast fluidised bed – a true multifuel boiler'. Paper to US Department of Energy, 8th International Conference, Fluidised-Bed Combustion, Houston, USA, March 1985. US Department of Energy, Proceedings, DOE/METC-85-6021, 1985 (July), **1**, 415–422.

562 Jones O. (Conoco), Litt R.D. (Battelle) and Davis J.S. (Struthers Wells Corp.). 'Performance of Conoco's prototype MS-FBC oil field steam generator'. Paper to US Department of Energy, 8th International Conference, Fluidised-Bed Combustion, Houston, USA, March 1985. US Department of Energy, Proceedings, DOE/METC-85-6021, 1985 (July), **2**, 555–563.

563 Kullendorff A., Jansson B. and Olofsson J. (Gotaverken Energy Systems AB). 'Operating experience of circulating fluidised-bed boilers'. US Department of Energy, 8th International Conference, Fluidised-Bed Combustion, Houston, USA, March 1985. US Department of Energy, Proceedings, DOE/METC-85-6021, 1985 (July), **2**, 574–583.

564 Beisswenger H. (Lurgi Corp.), Darling S. (Combustion Engineering), Plass L. (Lurgi GmbH) and Wechsler A. (Lurgi Corp.). 'Burning multiple fuels and following load in the Lurgi/Combustion Engineering circulating fluid-bed boiler'. Paper to US Department of Energy, 8th International Conference, Fluidised-Bed Combustion, Houston, USA, March 1985. US Department of Energy, Proceedings, DOE/METC-85-6021, 1985 (July), **2**, 619–633.

565 Fraley L.D., Keh-Hsien H., Solbakken A., Sadhukhan P. and Yung-Yi Lin (M.W. Kellogg Co.). 'Circulating multistage fluidised bed combustor design concept'. Paper to US Department of Energy, 8th International Conference, Fluidised-Bed Combustion, Houston, USA, March 1985. US Department of Energy, Proceedings, DOE/METC-85-6021, 1985 (July), **3**, 1260–1269.

566 Hintz D.B. (Keeler Dorr-Oliver). 'CFBC proves economic in gas to coal conversions'. *Modern Power Systems*, 1985 (December), 45–47.

567 Oakes E.J. (Pyropower Corp.). Javen J. and Engstrom F. (Ahlstrom). 'Start-up and operating experience at the Kautta 22-MW(e) co-generation circulating fluidised bed combustion plant'. Paper to American Power Conference, Chicago, USA, April 1982.

568 Stockdale W. and Stonebridge R. (FWPPL). 'The design and commercialisation of multi-solids fluidised bed combustors'. Paper to Coaltech '85, London, December 1985.

Waste Products Disposal

569 Shirley W.W. (Gilbert/Commonwealth). 'Fly ash revegetation – a case history in reclaiming Eastern and Western fly ash'. Coal Technology '83. 6th International Coal and Lignite Utilisation Conference, Houston, USA, November 1983, **3**, 7–17.

570 Grimshaw T.W., Little W.M. (Radian Corp), Minear R.A. (University of Tennessee) and Milliken J.O. (US EPA). 'Generation and attenuation of leachate from PFBC solid residues in simulated landfill conditions'. Paper to US Department of Energy, 7th International Conference, Fluidised Bed Combustion, Philadelphia, USA, October 1982. US Department of Energy, Proceedings, DOE/METC/83–48, 1983 (January), 534–543.

571 Minear R.A. (University of Tennessee), Grimshaw T.W. and Eklund A.G. (Radian Corp). 'Stepwise batch generalisation of leachate from PFBC and AFBC solid residues: characterisation and comparison with field and laboratory column leachates'. Paper to US Department of Energy, 7th International Conference, Fluidised Bed Combustion, Philadelphia, USA, October 1982. US Department of Energy, Proceedings, DOE/METC/83–48, 1983 (January), 544–558.

572 Bennett O.L., Reid R.L., Mays D.L., Whitsel T.J., Mitchell D.M., Stout W.L. and Hern J.L. (US Department of Agriculture). 'Animal feeding trials using feed and food produced with fluidised bed combustion residues'. Paper to US Department of Energy, 7th International Conference, Fluidised Bed Combustion, Philadelphia, USA, October 1982. US Department of Energy Proceedings, DOE/METC/83–48, 1983 (January), 559–566.

573 Grimshaw T.W., Garner W.F., Holland W.F. (Radian Corp.) and Kirchgessner D.A. (EPA). 'Laboratory and field studies of pressurised FBC waste leachate generation and attentuation'. Paper to Sixth International Conference on Fluidised Bed Combustion, Atlanta, USA, April 1980, US Dept. of Energy, DOE/CONF-800428, 925–938.

574 Boesmans B. (TNO). 'Applications to fluidised bed combustion ashes from coal'. Paper to CEGB Conference, Ashtech 84, London, September 1984. Paper No. 95.

575 Keairns D.L., Sun C.C., Peterson C.H. and Newby R.A. (Westinghouse). 'Fluid-bed combustion and gasification solids disposal'. *J. Energy Div., Proc. Am. Soc. Civil Eng.*, 1980, **106**, 213–228.

576 Cope D.R. and Dacey P.W. (IEA Coal Research). 'Solid residues from coal use – disposal and utilisation'. Report No. ICEAS/B3, IEA Coal Research, London, July 1984.

577 Chui S. (Argonne Nat. Lab.) 'An evaluation of the environmental and technical aspects of pressurised fluidised bed combustion'. Report No. AN1/EES-TM-257 to US Department of Energy, October 1983.

PAPERS ON FBC

Principles of FBC

578 Anon. 'Industrial power plants: natural gas makes inroads against oil, solid fuels'. *Power*, 1986 (November) 154–155.

Types of FBC

579 Sahagian J. and Weldman G.B. 'Pyroflow circulating bed boilers – a comprehensive overview'. Paper to Coal Technology '84, Houston, Texas, November 13–15, 1984.

580 Plass, O. and Anders, R. 'Fluid bed technology applied for the generation of steam and electrical power by burning cheap solid fuels in a CFB boiler plant'. Paper to Institute of Energy, 3rd International Fluidised Combustion Conference, London, October 1984. Proceedings, The Institute of Energy, London, Paper KN/II/1.

Features of Bubbling Bed Combustors

581 Battcock, W.V. and Pillai, K.K. 'Particle size in pressurised combustors'. Paper to US Department of Energy, 5th International Conference on Fluidised Bed Combustion, Washington, DEC, December 1977. Proceedings, The Mitre Corporation M78–68, 1978, **2**, 642–649.

582 Bass, J.W. and High, M.D. 'Operating and performance summary for TVA's 20-MW AFBC pilot plant'. Paper to Institute of Energy, 3rd International Fluidised Combustion Conference, London, October 1984. Proceedings, The Institute of Energy, London, Paper KN/IB/2.

583 Duqum, J.N., Tang, J.T., Morris, T.A., Esakov, J.L. and Howe, W.C. 'AFBC performance comparison for underbed and overbed feed systems'. Paper to US Department of Energy, 8th International Conference on Fluidised-Bed Combustion, Houston, Texas, March 1985. Proceedings, DOE/METC-85/6021, 1985 (July), **1**, 255–278.

584 Modrak, T.M., Tang, J.T. and Aulisio, C.J. 'Sulphur capture and nitrogen oxide reduction on the 6′ × 6′ atmospheric fluidised bed combustion test facility'. *ACS Div. Fuel Chem.*, 1982, **27** (1), 226–242.

Features of CFBC

585 Kullendorff, A. and Andersson, S. 'A general review on combustion in circulating fluidised beds'. Pages 83–96, 'Circulating fluidised bed technology' (Basu, P. Editor). Toronto: Pergamon Press, 1986.

586 Mooson Kwauk, Wang Nigde, Li Youchui, Chen Bingyu and Shen Zhiyuan (Inst. Chem. Metallurgy, Academia Sinica, Beijing). 'Fast fluidisation at ICM'. Pages 33–62, 'Circulating fluidised bed technology' (Basu, P. Editor). Toronto: Pergamon Press, 1986.

587 Grace, J.R. 'Heat Transfer in Circulating Fluidised beds'. Pages 63–81, 'Circulating Fluidised Bed Technology'. Toronto: Pergamon Press, 1986.

588 Thomas, J.F. 'The cost of coal-fired fluidised bed combustion boilers for industry'. Report by IEA Coal Research, London, Economic Assessment Service. (Publication pending.)

PFBC–Bubbling Beds

589 Rubow L.N., Borden M., Buchanan T.L. 'Cost and Performance Evaluation of Air and Steam Cooled PFBC Power Plants. Proceedings of the 7th International Conference on Fluidised Bed Combustion', Philadelphia, October 1982, Vol. 1, 198–210.

590 Hoy H.R., Roberts A.G., Phillips R.N., Carpenter L.K. 'Performance of a Small Combustor at Pressures up to 20 Atmospheres, Proceedings of the 7th International Conference on Fluidised Bed Combustion', Philadelphia, October 1982, Vol. 1, 473–481.

591 NCB (IEA Grimethorpe) Ltd. 'Grimethorpe Pressurised Fluidised Bed Combustion Project–Overall Project Review'.

Disposal of FBC Residues

592 Hubble, B.R. 'Fluidised-bed combustion: a review of environmental aspects'. Report No. ANL/ECT-12 to US Department of Energy, January 1982.

593 Sun, C.C., Peterson, C.H. and Keairns, D.L. 'Experimental/engineering support for EPA's FBC program: final report; Volume 3: solid residue study'. Report No. EPA-600/7-80-015c to US Environmental Protection Agency, January 1980.

594 Grimshaw, T.W., Garner, D.N., Holland, W.F. and Kirchgessner, O.A. 'Laboratory and field studies of pressurised FBC waste leachate generation and attenuation'. Paper to US Department of Energy, 6th International Conference on Fluidised-Bed Combustion, Atlanta, Georgia, April 1980. Proceedings, DOE/CONF-800428, 925–938.

595 Keairns, D.L., Sun, C.C., Peterson, C.H. and Newby, R.A. 'Fluid-bed combustion and gasification solids disposal'. *J. Energy Div., Proc. Am. Soc. Civil Eng.*, 1980, **106**, 213–228.

596 Schilling, H.D. 'Cost considerations for emission control in fluidised bed combustion'. Pages 236–240, 'Costs of coal pollution abatement' (Rubin, E.S. and Torrens, I.M. Editors). Paris: OECD, 1982.

597 Boesmans, B. 'Applications to fluidised bed combustion ashes from coal'. Paper to Central Electricity Generating Board Conference, Ashtec '84, London, September 1984. Proceedings, CEGB, London, Paper No. 95.

598 Minnick, L.J. 'Development of potential uses for the residue from FBC processes'. Report No. DOE/ET/10415-T6 to US Department of Energy, 1982.

Areas of Uncertainty in FBC

599 Vincent, R.Q. 'Mechanical/materials performance of the TVA 20-MW AFBC boiler'. Paper to US Department of Energy, 8th International Conference on Fluidised-Bed Combus-

tion, Houston, Texas, July 1985. Proceedings, DOE/METC-85/6021, **1**, 218–230.

600 Schwieger, R. 'Fluidised-bed boilers achieve commercial status worldwide'. *Power*, Special Report, 1985 (February), S1–S16.

601 Rogers, E.A., Page, A.J. and La Nauze, R.D. 'The corrosion performance of heat exchanger alloys in fluidised combustion systems'. Paper to Eurocor 77 (6th European Congress on Corrosion) London, 1977.

602 Stringer, J. and Minchener, J. 'Materials issues in the development of fluidised bed combustion'. Paper to Institute of Energy, 3rd International Fluidised Combustion Conference, London, October 1984. Proceedings, The Institute of Energy, London, Paper DISC/29.

603 Ehrlich, S. 'Fluidised combustion–is it achieving its promise?' Paper to Institute of Energy, 3rd International Fluidised Combustion Conference, London, October 1984. Proceedings, The Institute of Energy, London, Paper KA/I/1.

Future Developments in FBC

604 Energy and Environmental Engineering Inc. 'The staged cascading fluidised bed combustor'. Report No. DOE/MC/19327–1407, vols. 1 and 2 (DE 83010938 and 9) to US Department of Energy, March 1983.

605 Kono, H.O. 'Multistage fluidised beds with underflow downcomer–characteristics and application possibility'. Paper to US Department of Energy, 8th International Conference on Fluidised-Bed Combustion, Houston, Texas, March 1985. Proceedings, DOE/METC-85-6021, 1985, **3**, 1282–1299.

606 Fraley, L.D., Hsiao, L.H., Solbakken, A., Sadhukhan, P. and Yung-Yi Lin. 'Circulating multistage fluidised-bed combustor design concept'. Paper to US Department of Energy, 8th International Conference on Fluidised-Bed Combustion, Houston, Texas, March 1985. Proceedings, DOE/METC-85-6021, 1985, **3**, 1260–1269.

607 Fraley, L.D. and seven other authors. 'System design study to reduce capital and operating costs and bench-scale testing of a circulating-bed AFB advanced concept'. Report No. DOE/MC/21173–2062 (DE 86001637) to US Department of Energy, August 1985.

608 Westinghouse R & D Center, Pittsburgh. 'Advanced atmospheric fluidised bed combustion design–internally circulating AFBC'. Final Report No. DOE/MC/19329–1405 (DE 83010973) to US Department of Energy, January 1983.

609 Kalfadelis, C.D. 'Continuing development of regenerable sorbents for fluidised bed combustion'. Report Nos. DOE/ET-15166-1238 and DOE/ET-15166-1105 to US Department of Energy, 1981.

610 Yoo, H.J., McGauley, P.J. and Steinberg, M. 'Calcium silicate cements for desulphurisation of combustion gas'. Report No. METC/FT12/82–23 to US Department of Energy, 1982.

611 Gavalas, G.R., Weston, T.A. and Stephanopoulous, M.F. 'Alkali-alumina sorbents for regenerable SO_2 removal in fluidised coal combustion'. Paper to US Department of Energy, 8th International Conference on Fluidised-Bed Combustion, Houston, Texas, July 1985. Proceedings, DOE/METC-85/6021, 1985, **3**, 1178–1185.

612 Castleman, J.M. 'Process performance of the TVA 20-MW atmospheric fluidised-bed combustion (AFBC) pilot plant'. Paper to US Department of Energy, 8th International Conference on Fluidised-Bed Combustion, Houston, Texas, 1985 (July). Proceedings, DOE/METC-85/6021, **1**, 196–217.

613 Smith, G.W., Hajicek, D.R., Myles, K.M., Goblirsh, G.M., Mowry, R.W. and Teats, F.G. 'Demonstration of a hydration process for reactivating partially sulphated limestone sorbents'. Report No. ANL/CEN/FE-80-23 to US Department of Energy, October 1981.

614 Ulerich, N.H., O'Neill, E.P. and Keairns, D.L. 'The influence of limestone calcination on the utilisation of the sulphur-sorbent in atmospheric-pressure fluid-bed combustors'. Report No. EPRI/FP-426 to Electric Power Research Institute, August 1977.

615 Van Houte, C., Delmon, B., Maon, J.C. and Dumont, Ph. 'Desulphurisation of flue gases in a fluidised bed of modified limestone'. *J. Air. Poll. Control Ass.*, 1978, **28**, 1030–1033.

616 Desal, N.J. and Yang, R.T. 'Catalytic fluidised-bed combustion: enhancement of sulphation of calcium oxide by iron oxide'. *Ind. Eng. Chem. Process Des. Dev.*, 1983, **22**, 119–123.

Evaluation of Selected Types of FBC

617 Dunlop, W. and Van Slambrook, R.T. 'Assessment of alternative atmospheric fluidised bed combustion steam generator design'. Report No. RP1860-3 to Electric Power Research Institute, 1985.

618 Beacham, B. and Marshall, A.R. *J. Inst. Energy*, 1979, **52**, 59–64.

619 Shimizu, T., Tatebayashi, J., Terada, H., Furusawa, T. & Horio, M. 'The combustion characteristics of different types of coal in the 20 t/h fluidised-bed boiler'. Paper to US Department of Energy, 8th International Conference on Fluidised-Bed Combustion, Houston, Texas, March 1985. Proceedings, DOE/METC-85/6021, 1985, **1**, 231–240.

620 Yoo, H.J., McGauley, P.J. and Steinberg, M. 'Calcium silicate cements for desulphurisation of combustion gas'. Report No. METC/FT12/82-23 to US Department of Energy, 1982.

621 Tang, J.T., Duqum, J.N., Modrak, T.M. and Aulisio, C.J. 'An overall review of the EPRI/B & W 6′ × 6′ fluidised bed combustion test facility'. Paper to US Department of Energy, 7th International Conference on Fluidised-Bed Combustion, Philadelphia, USA, October 1982. Proceedings, DOE/METC/83-48, 1983, 373–380.

622 Vincent, R.Q. 'Mechanical/materials performance of the TVA 20-MW AFBC boiler'. Paper to US Department of Energy, 8th International Conference on Fluidised-Bed Combustion, Houston, Texas, July 1985. Proceedings, DOE/METC-85/6021, **1**, 218–230.

623 Virr, M.J. 'Operating experience with industrial package firetube FBC boilers'. Paper to Institute of Energy, 3rd International Fluidised Combustion Conference, London, October 1984. Proceedings, The Institute of Energy, London, Paper KN/IB/1.

624 Wormser, A. and Beckwith, W. 'The Wormser grate installation at Iowa Beef's plant in Amarillo, Texas'. Paper to US Department of Energy, 7th International Conference on Fluidised-Bed Combustion, Philadelphia, USA, October 1982. Proceedings, DOE/METC/83-48, 1983.

625 Sadowski, R.S. 'The first 10,000 hours of operation of the Wormser grate at Amarillo'. Paper to US Department of Energy, 8th International Conference on Fluidised-Bed combustion, Houston, Texas, March 1985. Proceedings, DOE/METC-85/6021, 1985, **2**, 634–643.

626 Peduto, E.F., Fennelly, P.F. and Milliken, J. 'Pollution control performance of a dual-bed fluidised-bed combustor'. Paper to US Department of Energy, 8th International Conference on Fluidised-Bed Combustion, Houston, Texas, March 1985. Proceedings, DOE/METC-85/6021, 1985, **3**, 1462.

627 Engstrom, F. and Yip, H.H. 'Operating experience of commercial scale Pyroflow circulating fluidised bed combustion boilers'. Paper to US Department of Energy, 7th International Conference on Fluidised-Bed Combustion, Philadelphia, USA, October 1982. Proceedings, DOE/METC/83-48, 1983, **2**, 1136–1143.

628 Oakes, E.J., Javen, J. and Engstrom, F. 'Start-up and operating experience at the Kauttua 22-MW(e) cogeneration circulating fluidised bed combustion plant'. Paper to American Power Conference, Chicago, Illinois, April 26–28, 1982.

629 Engstrom, F. and Sahagian, J. 'Operating experiences with circulating fluidised bed technology'. Pages 309–316, 'Circulating fluidised bed technology' (Basu, P. Editor). Toronto: Pergamon Press, 1986.

630 Taylor, E.S. 'Development, design and operational aspects of a 150 MW(e) circulating fluidised bed boiler plant for the Nova Scotia Power Corporation'. Pages 363–376, 'Circulating fluidised bed technology' (Basu, P. Editor). Toronto: Pergamon Press, 1986.

631 Anon. 'US's largest commercial CFB burns coal cleanly in California'. *Power*, 1986 (October), 43–44.

632 Kullendorff, A., Jansson, B. and Olofsson, J. 'Operating experience of circulating fluidised bed boilers'. Paper to US Department of Energy, 8th International Conference on Fluidised-Bed Combustion, Houston, Texas, March 1985. Proceedings, DOE/METC-85/6021, 1985, **2**, 574–583.

633 Stromberg, L., Kobro, H., Brereton, C., Morris, T.A. and Walker, D.J. 'The fast fluidised bed–a true multifuel boiler'. Paper to US Department of Energy, 8th International Conference on Fluidised-Bed Combustion, Houston, Texas, March 1985. Proceedings, DOE/METC-85/6021, 1985, **1**, 415–422.

634 D'Acierno, J.P., Garver, D.L. and Fisher, B. 'Design concepts for industrial coal-fired fluidised bed steam generators'. Paper to US Department of Energy, 8th International Conference on Fluidised-Bed Combustion, Houston, Texas,

March 1985. Proceedings, DOE/METC-85/6021, 1985, **1**, 406–414.

635 Basu, P. 'Design considerations for circulating fluidised bed combustors'. J. Inst. Energy, 1986 (December), **59** (441), 179–183.

636 Grace, J.R. 'Heat transfer in circulating fluidised beds'. Pages 63–81. 'Circulating fluidised bed technology' (Basu, P. Editor). Toronto: Pergamon Press, 1986.

637 Bolin, C. 'International separator for circulating fluidised bed combustion boiler'. Pages 385–396. 'Circulating fluidised bed technology' (Basu, P. Editor). Toronto: Pergamon Press, 1986.

638 Fusey, I., Lim, C.J. and Grace, J.R. 'Fast fluidisation in a concentric circulating bed'. Pages 409–416, 'Circulating fluidised bed technology'. (Basu, P. Editor). Toronto: Pergamon Press, 1986.

639 Jones, O., Litt, R.D. and Davis J.S. 'Performance of Conoco's prototype MS-FBC oil field steam generator'. Paper to US Department of Energy, 8th International Conference on Fluidised-Bed Combustion, Houston, Texas, March, 1985. Proceedings, DOE/METC-85-6021, 1985, **2**, 555–563.

640 Jones, O. and Seber, E.C. 'Initial operating experience at Conoco's South Texas multisolids FBC steam generator'. Paper to US Department of Energy, 7th International Conference on Fluidised-Bed Combustion, Philadelphia, USA, October 1982. Proceedings, DOE/METC/83–48, 1983.

641 Zielinski, E.A. and Bush, F. 'Conceptual design of a 500MW(e) circulating fluidised-bed plant'. Paper to US Department of Energy, 8th International Conference on Fluidised-Bed Combustion, Houston, Texas, March 1985. Proceedings, DOE/METC-85/6021, 1985, **1**, 385–394.

642 Turek, D.G., Spoko, S.J. and Janssen, K. 'A generic circulating fluidised-bed system for cogenerating steam, electricity and hot air'. Paper to US Department of Energy, 8th International Conference on Fluidised-Bed Combustion, Houston, Texas, March 1985. Proceedings, DOE/METC-85/6021, 1985, **1**, 395–405.

643 Beisswenger, H., Darling, S., Plass, L. and Wechsler, A. 'Burning multiple fuels and following load in the Lurgi/Combustion Engineering circulating fluid-bed boiler'. Paper to US Department of Energy, 8th International Conference on Fluidised-Bed Combustion, Houston, Texas, March 1985. Proceedings, DOE/METC-85/6021, **2**, 619–633.

644 Darling, S.L., Beisswenger, H. and Wechsler, A. 'The Lurgi/Combustion Engineering circulating fluidised bed boiler: design and operation'. Pages 297–308, 'Circulating fluidised bed technology' (Basu, P. Editor). Toronto: Pergamon Press, 1986.

645 Proceedings of the EPRI Second Biennial Pressurized-Bed Combustion Power Plants Utility Conference, Milwaukee, Wisconsin, June 1986.

646 Robertson A.S. 'Research and Development of a Second Generation PFB Combustion Power Plant Cycle', 14th Energy Technology Conference and Exposition, Washington DC, April 1987.

647 Bialkin E.P., Drenker S.G., Goiditch S.J., Wolowodiuk W. 'Design and Economic Evaluation of a Turbocharged PFBC Boiler for Utility Application'. Proceedings of the 8th International Conference on Fluidised-Bed Combustion, Houston, July 1985, Vol. 1, Pages 455–467.

648 Derdiger J.A., Saliga J.J., Koza H. 'Turbocharged PFBC Powerplants: A Technical and Economic Assessment'. Proceedings of the 8th International Conference on Fluidised-Bed Combustion, Houston, July 1985, Vol. 1, Pages 501–511.

649 Mukherjee D.K. 'Turbocharged Pressurized Fluidized Bed Boiler Design and Economics', Proceedings of the 8th International Conference on Fluidised-Bed Combustion, Houston, July 1985, Vol. 1, Pages 485–500.

BATTELLE BI-MONTHLY REPORTS

650 Battelle Bi-Monthly Report No. 37, 10 March 1980

651 Battelle Bi-Monthly Report No. 38, 10 May 1980

652 Battelle Bi-Monthly Report No. 39, 10 July 1980

653 Battelle Bi-Monthly Report No. 40, 10 September 1980

654 Battelle Bi-Monthly Report No. 41, 10 November 1980

655 Battelle Bi-Monthly Report No. 42, 10 January 1981

656 Battelle Bi-Monthly Report No. 43, 10 March 1981

657 Battelle Bi-Monthly Report No. 44, 10 May 1981

658 Battelle Bi-Monthly Report No. 45, 10 July 1981

659 Battelle Bi-Monthly Report No. 46, 10 September 1981

660 Battelle Bi-Monthly Report No. 47, 10 November 1981

661 Battelle Bi-Monthly Report No. 48, 10 January 1982

662 Battelle Bi-Monthly Report No. 49, 10 March 1982

663 Battelle Bi-Monthly Report No. 50, 10 May 1982

664 Battelle Bi-Monthly Report No. 51, 10 July 1982

665 Battelle Bi-Monthly Report No. 52, 10 September 1982

666 Battelle Bi-Monthly Report No. 53, 10 November 1982

667 Battelle Bi-Monthly Report No. 54, 10 January 1983

668 Battelle Bi-Monthly Report No. 55, 10 March 1983

669 Battelle Bi-Monthly Report No. 56, 10 May 1983

670 Battelle Bi-Monthly Report No. 57, 10 July 1983

671 Battelle Bi-Monthly Report No. 58, 10 September 1983

672 Battelle Bi-Monthly Report No. 59, 10 November 1983

673 Battelle Bi-Monthly Report No. 60, 10 January 1984

674 Battelle Bi-Monthly Report No. 61, 10 March 1984

675 Battelle Bi-Monthly Report No. 62, 10 May 1984

676 Battelle Bi-Monthly Report No. 63, 10 July 1984

677 Battelle Bi-Monthly Report No. 64, 10 September 1984

678 Battelle Bi-Monthly Report No. 65, 10 November 1984

679 Battelle Bi-Monthly Report No. 66, 10 January 1985

680 Battelle Bi-Monthly Report No. 67, 10 March 1985

681 Battelle Bi-Monthly Report No. 68, 10 May 1985

682 Battelle Bi-Monthly Report No. 69, 10 July 1985

683 Battelle Bi-Monthly Report No. 70, 20 September 1985

684 Battelle Bi-Monthly Report No. 71, 10 November 1985

685 Battelle Bi-Monthly Report No. 72, 10 January 1986

686 Battelle Bi-Monthly Report No. 73, 10 March 1986

Appendix 2

Vendor Information

(a) FGD AND SCR PROCESSES

AIR-FROHLICH AG
Romanshornerstrasse 100
CH-9320 Arbon
Switzerland
Tel: 010-41-71-465525
Tlx: 71400 AIRAG CH
Mr. R. Allemann

ANDERSEN 2000 INC.
306 Dividend Drive
Peachtree City
Georgia 30269
USA
Tel: 010-1-404-997-2000
Tlx: 542858 ANDERSEN PECH
Mr. J.D. Brady, President

BABCOCK-HITACHI K.K.
6, 2 Chome Ote Mashi
Chiyoda Ku
Tokyo
Japan
Tel:
Tlx: 3822467 BHKYW J

BABCOCK POWER LTD
165 Great Dover Street
London SE1 4YB
Tel: 01-407-8383
Tlx: 884151/2/3
Mr. R.G.J. Baker

BERGBAU-FORSCHUNG GmbH
Franz-Fischer-Weg 61
D-4300 Essen 13
West Germany
Tel: 010-49-201-1059456
Tlx:

BIONEER OY
PO Box 537
SF-13111 Hämeenlinna
Finland
Tel: 010-358 17 23371
Tlx: 2314 PERA SF
Mr. K. Harsunen
Managing Director

G. BISHOFF GmbH & CO. AG
Posf. 10 05 33,
Gartnerstrasse 44
D-4300 Essen
West Germany
Tel: 010-49-201-8112-0
Tlx: 857779 GASBI D
T Niess DV/AG

CHIYODA CORPORATION
PO Box 10
Tsurumi
Yokahama
Japan
Tel: 010-81-45-521-1231
Tlx: CHIYO J47726
Mr. K. Kohya
General Manager
Licensing Group

DAVY McKEE 15, Portland Place London W1A 4DD	Tel: 01-637 2821 Tlx: 22604 Mr. R.C. Akroyd Chief Executive Environmental Projects
DB GAS CLEANING CORP. 14 Orinda Way PO Box 944 Orinda CA 94563 USA	Tel: 010-1-415-254-4164 Tlx: 171256 Mr. P.B. Slakey, Vice President
DEUTSCHE BABCOCK ANLAGEN AG Postf. 4 & 6 Krefeld 11 West Germany	Tel: 010-49-2151-448571 Tlx: 853824 BSHK D D-4150 Dr. C. Hemmer, Sales Director
DOWA MINING CO. 8–2 Marounouchi 1-chome J26298 Chiyoda-Ku Tokyo Japan	Tel: Tlx: DOWAMICO
EBARA CORPORATION Haneda Asahicho EBARAC J Ota-Ku Tokyo 144 Japan	Tel: Tlx: 2466091 Mr. Keita Kawamura
ESTS BV PO Box 1000 1970 CA Ijmuiden The Netherlands	Tel: 010-31-2510-9922 Tlx: 35211 HOVS NL Mr. H Daalder, Sales Manager
FLAKT INDUSTRI AB S-35187 Vaxjo Sweden	Tel: 010-46-470-87000 Tlx: 52132 FLAKTV S Mr. Per Jacobson
GENERAL ELECTRIC ENVIRONMENTAL SERVICES 200 North 7th Street Lebanon–PA. 17042 USA	Tel: 010-1-717-274-7218 Tlx: 842332 GEESI A LEBA Mr. R. Snaddon, Manager International Marketing
HALDOR TOPSOE A/S Nymollevej 55 DK-2800 Lyngby Denmark	Tel: 010-45-2878100 Tlx: 37444 HTAS DK Mr. Frands E. Jensen Area Sales Manager
INSTITUT FRANCAIS DU PETROLE 1 et 4 Ave. de Bois-Preau 92506 Rueil Malmaison Cedex France	Tel: 010-33-47526000 Tlx: 203050 F Alphonse Hennico

ISHIKAWAJIMA-HARIMA HEAVY INDUSTRIES CO.
30–13, 5-Chome
Toyo,
Koto-Ku
Tokyo 135
Japan

Tel:
Tlx: IHIHET J22232
Mr. H. Ikeno, Air Pollution Control Design Department

KAWASAKI HEAVY INDUSTRIES
2-16-1 Nakamachidori
Ikeda-Ku
Kobe
Japan

Tel:
Tlx:

KRC UMWELTTECHNIK GmbH
Alfred Nobel Strasse 20
D-8700 Würzburg
West Germany

Tel: 010-49-931-90890
Tlx: 9318129 KRCWZB
Mr. W Zabel

LINDE AG
Werksgruppe TVT Muenchen
Dr. Carl-von-Linde-Strasse 6–14
D-8023 Hoellriegelskreuth
West Germany

Tel: 010-49-89-72731
Tlx: 5283270 LI D
Dr. H Becker

LODGE COTTRELL
Division of Dresser UK Ltd
George Street Parade
Birmingham
West Midlands B3 1QQ

Tel: 021-236-3388
Tlx: 338458
Dr. M J Ashley
Deputy Managing Director

LURGI GmbH
Postfach 111231
Gervinusstrasse 17/19
D-6000 Frankfurt
West Germany

Tel: 010-49-69-1571
Tlx: 412360 IG D
Mr. Wenzel Von Jordan, Hauptbevollmachtigter

MITSUBISHI HEAVY INDUSTRIES LTD
Bow Bells House
Bread Street
London EC4M 9BQ

Tel: 01-248-8821
Tlx: 888994 MHI LN G
Mr. T. Ono,
Senior Manager

MITSUI MIIKE MACHINERY CO. LTD
Mitsui Building
1-1 Nihonbashi Muromachi 2-chome
Chuo-Ku
Tokyo 103
Japan

Tel: 010-81-270-3481
Tlx: MMMCO J24529
Mr. K. Nagamatsu
Deputy General Manager

NEI INTERNATIONAL COMBUSTION LTD
Sinfin Lane
Derby

Tel: 0332 760223
Tlx:
Mr. V D Trimm

A/S NIRO ATOMIZER
Gladsaxevej 0305
DK-2860 Soeborg
Denmark

Tel: 010-45-169-1011
Tlx: 15603 ATOMN DK
Mr. R. Madhok

OTTO H. YORK CO. INC
Box 3100, 42 Intervale Road
Parsippany
New Jersey 07054-0918
USA

Tel: 010-1-201-299-9200
Tlx: 139134 OTTO YORK FFLD
Mr. K. Schifftner,
Scrubber Department

PEABODY PROCESS SYSTEMS
201 Merritt 7, Corporate Park
Box 6037
Norwalk CT 06852
USA

Tel: 010-1-203-846-1600
Tlx: 965870 PEABODYSYS STD
Dr. E. Baake,
Vice President

SAARBERG-HOLTER UMWELTTECHNIK GmbH
Hafenstrasse 6
D-6600 Saarbrücken 2
West Germany

Tel: 010-49-681-32104/5/6/7
Tlx: 4421124 SHU D
Mr. Michael Esche

L & C STEINMULLER GmbH
Posf. 100855/100865
Fabrikstrasse 1
D-5270 Gummersbach 1
West Germany

Tel: 010-49-2261-852920
Tlx: 884 5510 SG D
Dr. Ing H. Voos,
General Manager, Environmental Protection

SUMITOMO HEAVY INDUSTRIES LTD
2-1 Otemachi, 2-chome
Chiyoda-Ku
Tokyo 100
Japan

Tel: 010-81-3-245-4321
Tlx: SUMIJUKI J22264

TAMPELLA LTD
Boiler Division
PO Box 626
SF-33101 Tampere 10
Finland

Tel: 010-358-31-32400
Tlx: 22666 TABOI SF

TENNESSEE VALLEY AUTHORITY
3N 78A Missionary Ridge Place
Chattanooga
Tennessee 37401
USA

Tel:
Tlx: 9103333745 ZARSKI/CHATTAN
Dr. M. D. High
Director of Energy
Demonstrations & Technology

THYSSEN ENGINEERING GmbH
Postf. 10 38 54
Am Thyssenhaus 1
D-4300 Essen 1
West Germany

Tel: 010-49-201-1061
Tlx: 8579881-0 TI D
Mr. U. Gebhard
Sales Dept.

UHDE GmbH
Technical Dept. V1
Friedrich-Uhde-Strasse 15
D-4600 Dortmund 1
West Germany

Tel: 010-49-231-547 2373
Tlx: 822841–26 UD D
Dr. Ulrich Neumann,
Technical Div. V1

WAAGNER-BIRO AG
Waagner-Biro-Strasse 98
Postf. 1004
A-8021 Graz
Austria

Tel: 010-43-316-5010
Tlx: 31316 WABIG A
Mr. Weitzer

WALTHER & CIE AG
Postf. 85 05 61
Wattherstrasse 51
D-5000 Köln 80
West Germany

Tel: 010-49-221-67850
Tlx: 8873341 WAL D
Mr. S E Christeleit,
Export Sales Manager

(b) MANUFACTURERS OF FLUIDISED COMBUSTION EQUIPMENT

A. AHLSTROM CORPORATION
Boiler Works
PO Box 184
SF-78201, Varkaus
Finland

Tel: 010-358-72-211
Tlx: 4319 ALMEK SF

AHLSTROM PYROPOWER LTD
17 East Parade
Harrogate
North Yorkshire HG1 5LF

ALLIED BOILERS LTD
Unit 7, Belgrave Mills
Honeywell Lane
Oldham
Lancashire

Tel: 061-633-1131
Tlx: 668054

ANSALDO COMPONENTI
Breda Steam Generating,
Viale Sarca 336,
Milano,
Italy

Tel: 010-39-2-699-7208
Tlx: 331280

ASEA PFBC AB
S-61220, Finspong
Sweden

Tel: 010-46-12-281000
Tlx: 645045

ASEA LTD
48 Leicester Square
London WC2H 7NN

Tel: 01-930-5411
Tlx: 261243

BABCOCK HITACHI KK
6-2, 2-chome Ote-machi, Chiyoda-ku
Tokyo 100
Japan

Tel: 010-81-3-270-7351
Tlx: 222 3502

BABCOCK POWER LTD
165 Great Dover Street
London SE1 4YB

Tel: 01-407-8383
Tlx: 884151

BABCOCK POWER LTD
Shell Boiler Division
65 Livery Street
Birmingham B3 1HA

Tel: 021-236-7881
Tlx: 337069

BABCOCK & WILCOX CO.
20 S. Van Buren Avenue
Barberton
Ohio 44203
USA

Tel: 010-1-216-860-2721
Tlx: 98 6406

BABCOCK WORSLEY LTD
Worsley House
Liverpool Rd,
St. Helens
Lancashire WA10 1PQ
Tel: 0744-612111
Tlx: 627531

C.E. POWER SYSTEMS
1000 Prospect Hill Road
Windsor
Connecticut 06095
USA
Tel: 010-1-203-285-9973
Tlx: 99297

COMBUSTION POWER CO.
1346 Willow Road
Menlo Park
California 94025
USA
Tel: 010-1-415-324-4744
Tlx: 843471

DEDERT CORPORATION
Thermal Processes Div.
Governors Drive
Olympia Fields
Illinois 60461
USA
Tel: 010-1-312-747-7000
Tlx: 25-3889 20000

DELTAK CORPORATION
13330 12th Avenue North
Plymouth
Minnesota 54440
USA
Tel: 010-1-612-544-3371
Tlx: 29 0812

DEUTSCHE BABCOCK WERKE AG
Duisburger Strasse 375
D-4200 Oberhausen
West Germany
Tel: 010-49-208-8330
Tlx: 856951

EMS THERMPLANT LTD
Pheasey House
Farrier Rd, Great Barr
Birmingham B43 7JN
Tel: 021-360-8888
Tlx: 335176

ENERGIE U. VERFAHRENSTECHNIK GmbH
Johannesstrasse 37-45
7000 Stuttgart 1,
West Germany
Tel: 010-49-711-669-4276
Tlx: 723656A EVT D

ENERGY EQUIPMENT CO. LTD
Stephenson House
Brunel Centre
Bletchley
Milton Keynes MK2 2QX
Tel: 0908-74988
Tlx: 826387

ENERGY PRODUCTS OF IDAHO
4006 Industrial Avenue
Coeur d'Alene,
Idaho 83814
USA
Tel: 010-1-208-765
Tlx: 152319 EPI COEU

FIVES-CAIL BABCOCK
7 rue Montalivet, F-75383 Paris
France
Tel: 010-33-1-266-3525
Tlx: 650 328 Cedex 08

FLUIDISED COMBUSTION LTD
Broadcasting House, Newport Triangle
Newport Road
Middlesbrough
Cleveland TS1 5JA
Tel: 0642-226731
Tlx: 58622 FLUCOM G

FLUIDYNE ENGINEERING CORP
3900 Otson Memorial Highway
Minneapolis,
Minnesota 55422
USA
Tel: 010-1-612-544-2721
Tlx: 290518

FOSTER WHEELER CORPORATION
110 South Orange Avenue
Livingstone, New Jersey 07039
USA
Tel: 010-1-201-533-2310
Tlx: 138568

FOSTER WHEELER POWER PRODUCTS LTD
Greater London House, PO Box 160
Hampstead Road
London NW1 7QN
Tel: 01-388-1212
Tlx: 263984

GENERATOR INDUSTRIE AB
PO Box 95
S-43322 Partille
Sweden
Tel: 010-46-31-269000
Tlx: 21634 GENAB S

GIBSONS WELLS LTD
2 Town Gate
Calverley
Leeds LS28 5NF
Tel: 0532-550455
Tlx: 557584

GOTAVERKEN ENERGY SYSTEMS AB
PO Box 8734, S-40275
Goteborg
Sweden
Tel: 010-46-31-501000
Tlx: 2283 GOTAENY S

GWB INDUSTRIAL BOILERS LTD
PO Box 4, Burton Works
Dudley
West Midlands DY3 2AD
Tel: 0384-55455
Tlx: 337388

IHI (ISHIKAWAJIMA-HARIMA HEAVY INDUSTRIES) CO. LTD
2-16, 3-Chome, Toyosu
Tokyo 135
Japan
Tel: 010-81-3-649-1111
Tlx: J23507 HITOY J
Mr. H Ikeno
Air Pollution
Control Design Dept

INTEGRAL ENGINEERING INDUSTRIEBEDARF GmbH
Grosse Neugasse 8,
A-1041 Vienna
Austria
Tel: 010-43-222-579681
Tlx: 112026

INTERNATIONAL BOILER WORKS CO.
460 Birch Street
East Stroudsburg
Pennsylvania 18301
USA
Tel: 010-1-717-421-5100
Tlx: 510 671 4520 IBW ESTB

KAWASAKI HEAVY INDUSTRIES CO. LTD Simaya 4-chome, 1-31 Konohana-ku Osaka 554 Japan	Tel: 010-81-6-461-8001 Tlx: 5622355 KAWAJU J
KEELER/DORR OLIVER BOILER CO PO Box 548 West Street Williamsport Pennsylvania 17703 USA	Tel: 010-1-717-327-3166 Tlx: 841-435 238
LURGI GmbH Gervinusstrasse 17-19 D-6000 Frankfurt am Main 1, West Germany	Tel: 010-49-611-1571 Tlx: 41236 O LGD
ME BOILERS LTD ME House Fengate Peterborough PE1 5BE	Tel: 0733-68471 Tlx: 32138
MITSUBISHI HEAVY INDUSTRIES LTD 5-1 Marunouchi 2-chome Chiyoda-ku Tokyo 100 Japan	Tel: 010-81-3-212-3111 Tlx: J22443
MITSUI ENGINEERING & SHIPBUILDING CO. LTD 6-4 Tsukiji 5-chome Chuo-ku Tokyo 104 Japan	Tel: 010-81-3-544-3306 Tlx: J22821 MITZOSEN
NEI INTERNATIONAL COMBUSTION LTD The Cochran Unit, Newbie Works Annan Dumfries & Galloway DG12 5OU Scotland	Tel: 046-12-2111 Tlx: 778183
POWER RECOVERY SYSTEMS INC. 181 Bridge Avenue Extension Cambridge Massachusetts 02140 USA	Tel: 010-1-617-576-1900 Tlx:
RILEY STOKER CORPORATION 9 Neponset Street Worcester Massachusetts 01606 USA	Tel: 010-1-617-852-7100 Tlx: 920 426
SENIOR GREEN LTD Calder Vale Road Wakefield West Yorkshire WF1 5PF	Tel: 0924-378211 Tlx: 55368
STAL-LAVAL S-58101 Linkoping Sweden	Tel: 010-46-13-129400 Tlx: 50068 APPARAT S

STEINMULLER (L&C) GmbH Fabrikstrasse 1 D-5270 Gummersback 1 West Germany	Tel: 010-49-2261-852710 Tlx: 884551
STONE DANKS LTD Stone House, Tipton Road, Tividale, Warley West Midlands B69 3HR	Tel: 021-557-3977 Tlx: 338711
STONE JOHNSTON CORPORATION 300 Pine Street, Ferrysburg Michigan 49409 USA	Tel: 010-1-616-842-5050 Tlx: 228 406
STRUTHERS WELLS CORPORATION 1003 Pennsylvania Avenue West Warren, Pennsylvania 16365 USA	Tel: 010-1-814-726-1000 Tlx: 91 4455
STUDSVIK AB S-61182 Nykoping Sweden	Tel: 010-46-155-21000 Tlx: 64013 STUDS S
SULZER (UK) LTD Westmead Farnborough Hampshire GU14 7LP	Tel: 0252-544311 Tlx: Mr. S. Bowcutt Process Engineering Dept
TAMPELLA LTD Boiler Division PO Box 626 SF-33101 Tampere 10 Finland	Tel: 010-358-31-32400 Tlx: 22666 TABOI SF
VEREINIGTE KESSELWERKE AG Werdener Strasse 3, D-4000 Düsseldorf 1 West Germany	Tel: 010-49-211-781-4458 Tlx: 8582729 VKWD
WALLSEND SLIPWAY ENGINEERS LTD PO Box 8, Wallsend, Tyne and Wear NE28 6EN	Tel: 0632-628961 Tlx: 53611
WORMSER ENGINEERING INC 225 Merrimac Street Woburn Massachusetts 01801 USA	Tel: 010-1-617-938-9380 Tlx: 232 005 ASAS

(c) DENOX BURNER MANUFACTURERS

AIROIL BURNER CO. (GB) LTD Horton Road West Drayton Middlesex UB7 8BG	Tel: 08954-44031 Tlx: 23923 AIROIL G Mr. D.W. Harckham Sales/Marketing Director

BABCOCK POWER LTD
165 Great Dover Street
London SE1 4YB

Tel: 01-407-8383
Tlx: 884151/2/3
Mr. J.S. Goldsack
Chief Engineer,
Engineering Services

DEJONG COEN BV
PO Box 5
3100 AA Schiedam
The Netherlands

Tel: 010-376166
Tlx: 24372
Arie W. Spoormaker

DUNPHY OIL & GAS BURNERS LTD
Queensway
Rochdale
Lancashire OL11 2SL

Tel: 0706-49217
Tlx: 635071 A/B DUNPHY G
Mr. M.P. Dunphy
Managing Director

HAMWORTHY COMBUSTION LTD
Fleets Corner
Poole
Dorset BH17 7LA

Tel: 0202-675123
Tlx: 41226
Mr. J B Champion
Engineering & Development Manager

NU-WAY LTD
PO Box 1
Vines Lane
Droitwich
Worchester WR9 8NA

Tel: 0905-772331
Tlx: 338551 NUWAY G
Mr. J W Findlay

PEABODY HOLMES LTD
(Combustion Division)
Brenchley House
123–135 Week Street
Maidstone
Kent ME14 1RF

Tel: 0622-671381
Tlx: 965850 PHMAID G
Mr. J Lisowski
Chief Combustion Engineer

SAACKE LTD
Fitzherbert Road
Farlington
Portsmouth PO6 1RX

Tel: 0705-383111
Tlx: 86212 SAACKE G
Mr. J N Bartlam

STORDY COMBUSTION ENGINEERING LTD
Heath Mill Road
Wombourne
Wolverhampton WV5 8BD

Tel: 0902-89217
Tlx: 338528 STORCO G
Mr. R.A. Freeman
Managing Director

JOHN ZINK CO LTD
Alban Park
Hatfield Road
St Albans
Herts AL4 0JJ

Tel: 0727-61451
Tlx: 265930

Appendix 3

Cost Estimates and Procedures

1. PROCEDURE FOR ESTIMATING CAPITAL COSTS FOR NEW-BUILD FGD PLANT

(a) *Source*
Full details of the costing procedure adopted are available in EPRI Report Number CS-3342, Volumes 1–3, December 1983 [95].

(b) *Base Plant*

- 1000 MWe size (2×500 MWe)
- Location: Wisconsin, USA, 200 metres above sea-level
- Seismic Zone 1 (US), i.e. minor risk of damage
- Plant design life of 30 years
- Coal analysis: 4.0% Sulphur
  HHV of 23.5 MJ/kg

(c) *Base Capital Costs*
The total constructed cost of on-site FGD and related facilities, including direct and indirect construction costs. Items included are:

- Civils
- Process equipment
- Piping
- Electrics
- Instruments and control
- Insulation and painting
- Direct field costs
- Indirect field costs (taxes, insurance, construction supplies and equipment, temporary facilities, vendor fees)

The estimate is divided according to functional systems, some of which may not be applicable to certain processes:

- Reagent feed
- Sulphur dioxide removal
- Flue gas
- Regeneration
- By-product
- Waste handling
- General support
- Particulate removal

(d) *Process Adjustments*
Adjustment factors are applicable for the following process variations:

- Unit size (100 to 700 MWe)
- Flue gas flowrate (1.1 to 2.5 m^3/s per net MWe)
- Sulphur content of coal (1.0 to 6.0%)

(e) *Location*
Location adjustment factors are available for the following:

- Seismic zone
- Climate
- Soil conditions
- Material and labour cost index

(f) *Escalation Adjustments*
The EPRI cost estimation procedure is based on December 1982 costs. Although adjustments can be made for start-up dates up to 1993, they are based on EPRI's standard escalation rate of 8.5%.

(g) *Project Contingency*
The project contingency covers additional equipment or other costs which would result from a more detailed design. The contingency factors are based on EPRI Class II guidelines (e.g. 15–30%) and are applied on a system basis. Higher factors are applied to equipment items of special design, and lower factors to standard items.

(h) *Process Contingency*
The process contingency applies to new technology to quantify the design uncertainty and cost of the commercial-scale equipment. The contingency factors are based on EPRI guidelines (e.g. 10–50%) and are applied on a system basis.

(i) *General Facilities*
It is assumed that a major paved road will have to be built along with the necessary area drainage. A new laboratory, office building and a warehouse are to be constructed. This equates to 10 percent of the Escalated Total Process Capital in the EPRI procedure.

(j) *Engineering and Office Fees*
The engineering hours spent by the equipment supplier, architect-engineer and utility to place a

total system in operation. A base fee of 10 percent of the Total Process Capital is used, increasing to a maximum of 15 percent depending on the effects of location and retrofit factors.

(k) *Allowance for Funds During Construction*
The duration of the engineering, procurement and construction phases are expected to vary from 1 to 3 years.

(l) *Royalty Allowance*
The royalty allowance as established by EPRI is 0.5% of the process capital.

(m) *Pre-production Costs*
These costs are intended to cover operator training, equipment check-out, major changes in plant equipment, extra maintenance and inefficient use of materials during plant start-up. In addition, it covers fixed and variable operating costs for one month. This equates to 2 percent of the Total Retrofit Investment.

(n) *Inventory Capital*
The inventory capital includes the value of raw material and other consumables on a capitalised basis. A raw material supply of 60 days is assumed for a base-loaded unit at full capacity operation.

2. PROCEDURES FOR ESTIMATING CAPITAL COSTS FOR RETROFIT FGD PLANT

(a) *Source*
Full details of the costing procedure are available in EPRI Report Number CS-3696, October 1984 [96].

(b) *Base Plant*

- 1000 MWe size (2×500 MWe)
- Location: Wisconsin, USA, 200 metres above sea-level
- Seismic Zone 1 (US), i.e. minor risk of damage

(c) *Base Capital Costs*
The total constructed cost of on-site FGD and related facilities, including direct and indirect construction costs. Items included are:

- Civils
- Process equipment
- Piping
- Electrics
- Instruments and control
- Insulation and painting
- Direct field costs
- Indirect field costs (taxes, insurance, construction supplies and equipment, temporary facilities, vendor fees)

The estimate is divided according to functional systems, some of which may not be applicable to certain processes:

- Reagent feed
- Sulphur dioxide removal
- Flue gas
- Regeneration
- By-product
- Waste handling
- General support
- Particulate removal

(d) *Scope Adjustments*
Adjustment factors are available for the following:

- Chimney work
- Boiler reinforcement
- Draft controls
- Demolition and relocation (buildings, ductwork, piping and electrics)

(e) *Process Adjustments*
Adjustment factors are applicable for the following process variations:

- Unit size (100 to 700 MWe)
- Flue gas flowrate (1.1 to 2.5 m^3/s per net MWe)
- Sulphur content of coal (1.0 to 6.0%)

(f) *Location*
Location adjustment factors are available for the following:

- Seismic zone
- Climate
- Soil conditions
- Material and labour cost index

(g) *Retrofit Adjustments*
Difficulties associated with plant retrofit are site specific and the EPRI Report deals with a number of factors including:

- Accessibility and congestion
- Underground obstructions
- Ductwork length and distance from scrubber to tie-ins

(h) *Escalation Adjustments*
The EPRI cost estimation procedure is based on December 1982 costs. Although adjustments can be made for start-up dates up to 1993, they are based on EPRI's standard escalation rate of 8.5%.

(i) *Project Contingency*
The project contingency covers additional equipment or other costs which would result from a more detailed design. Higher factors are applied to equipment items of special design, and lower factors to standard items. The contingency factors are based

on EPRI Class II guidelines (e.g. 15–30%) and are applied on a system basis.

(j) *Process Contingency*
The process contingency applies to new technology to quantify the design uncertainty and cost of the commercial-scale equipment. The contingency factors are based on EPRI guidelines (e.g. 10–50%) and are applied on a system basis.

(k) *General Facilities*
It is assumed that a major paved road will have to be built along with the necessary area drainage. A new laboratory, office building and a warehouse are to be constructed. This equates to 10 percent of the Escalated Total Process Capital in the EPRI procedure.

(l) *Engineering and Office Fees*
The engineering hours spent by the equipment supplier, architect-engineer and utility to place a total system in operation. A base fee of 10 percent of the Total Process Capital is used, increasing to a maximum of 15 percent depending on the effects of location and retrofit factors.

(m) *Allowance for Funds During Construction*
The duration of the engineering, procurement and construction phases are expected to vary from 1 to 3 years.

(n) *Royalty Allowance*
The royalty allowance as established by EPRI is 0.5% of the process capital.

(o) *Pre-production Costs*
These costs are intended to cover operator training, equipment check-out, major changes in plant equipment, extra maintenance and inefficient use of materials during plant start-up. In addition, it covers fixed and variable operating costs for one month. This equates to 2 percent of the Total Retrofit Investment.

(p) *Inventory Capital*
The inventory capital includes the value of raw material and other consumables on a capitalised basis. A raw material supply of 60 days is assumed for a base-loaded unit at full capacity operation.

3. PROCEDURE FOR ESTIMATING GENERALISED FIRST-YEAR OPERATING COSTS FOR FGD PLANT

(a) *Fixed O & M Costs*
These costs include:

- Operating Labour
- Maintenance Labour
- Maintenance Material
- Administrative and Support Labour

It is assumed that these costs remain constant and do not vary with capacity factor.

(i) *Operating Labour Costs*
Operating Labour Cost
= (MW × LR × 40 × 51)/(Plant net kW)

Where MW = Number of man-weeks per week. It is based on the number of operators required per 40 hour week, assuming 4.2 shifts are needed per week for 24 hour coverage of a single job.

LR = Average labour rate at FGD system start-up. It is assumed to be $18.30 (£11.00) per hour (EPRI) for January 1983 start-up.

(ii) *Maintenance Labour Costs*
Maintenance Labour Costs = PC × MF × 0.40

Where PC = Process Capital Cost ($/kW)
The base fixed operating costs quoted by EPRI are based on maximum process capital costs for a specific FGD system.

MF = Maintenance Factor
Based on EPRI guidelines. An average of 3.75% of the total process capital is used, and is based on a weighted average of all the process area costs.

N.B. It is assumed that 40% of maintenance costs are attributable to labour.

(iii) *Maintenance Material Costs*
Maintenance Material Costs = PC × MF × 0.60

Where PC and MF are defined above, and it is assumed that 60% of maintenance costs are for material.

(iv) *Administrative and Support Labour Costs*
Administrative and support labour costs are assumed to be 30% of the total maintenance and operating labour costs.

(b) *Variable Costs*
These costs include (where applicable):

(i) Reagent consumption
(ii) Reheat steam
(iii) Power usage
(iv) Water usage

(v) Methane usage
(vi) Waste disposal costs

These costs vary directly with unit size, percentage SO_2 removal, percent sulphur in coal and capacity factor. Except for spray drying, all base costs assume 90% SO_2 removal, 4% S in coal, 1.15 reagent stoichiometry and 65% capacity factor. Adjustments are made for deviations from these base cases. For spray drying base costs assume 70% SO_2 removal and 0.5% S in coal.

The following costs (December 1982) have been assumed:

Limestone	$ 14	(£8.40)	per tonne
Lime	$ 60	(£36)	per tonne
Sodium Carbonate	$140	(£84)	per tonne
High pressure steam	$ 8.27	(£5)	per tonne
Low pressure steam	$ 6.66	(£4)	per tonne
Power	$ 0.045	(2.7p)	per kWh
Methane	$ 0.30	(18p)	per m^3

Sludge disposal costs are based on 10 years remaining unit life (in the case of Retrofit plant) and 10 miles to disposal site.

(c) *Escalation Factor*
Fixed and variable operating costs are escalated at EPRI's standard escalation rate (8.5% per annum) over the period of engineering and construction (1–3 years).

4. CAPITAL COST ESTIMATES

4.1a Limestone Gypsum Process–S21.1 (Newbuild)

	£/kWe
1) Base capital cost (see Table 2.67)	51.29
2) Scope adjustments–none	
Total process Newbuild capital	51.29
3) Process adjustments (a) Unit size × 1.00 (b) Gas flow/MWe × 0.81 (c) Percentage S in Coal × 0.87	
4) Location factors (a) Cost index × 1.043 (b) Soils × 1.05	
5) Retrofit factors–none	
6) Escalation adjustments × 1.158 Total escalation factor = 0.893 Escalated total process capital (excluding contingencies)	45.82
7) Project contingencies (10%)	4.58
8) Process contingency (5%)	2.29
Escalated total process capital (ETPC)	52.70
9) General facilities (5% of ETPC)	2.63
10) Engineering and office fees (12% of ETPC)	6.32
Total plant Newbuild costs (TPC)	61.65
11) Allowance for funds during construction (2.2% of TPC)	1.36
Total plant Newbuild investment (TPI)	63.01
12) Royalties (5% of TPI)	3.15
13) Pre-production costs (2% of TPI)	1.26
14) Inventory capital	0.71
	68.13

Total capital Newbuild cost for 2000 MWe
£136 million

4.1b Limestone Gypsum Process–S21.1 (Retrofit)

	£/kWe
1) Base capital cost (see Table 2.68)	51.29
2) Scope adjustments (4%)	2.05
Total process Retrofit capital	53.34
3) Process adjustments (a) Unit size × 1.00 (b) Gas flow/MWe × 0.81 (c) Percentage S in Coal × 0.87	
4) Location factors (a) Cost index × 1.043 (b) Soils × 1.05	
5) Retrofit factors (a) Site accessibility and congestion × 1.25 (b) Underground obstruction × 1.02 (c) Ductwork length and scrubber location × 1.12	
6) Escalation adjustments × 1.158 Total escalation factor = 1.276 Escalated total process capital (excluding contingencies)	68.05
7) Project contingencies (10%)	6.81

8) Process contingency (5%)	3.40
Escalated total process capital (ETPC)	78.26
9) General facilities (5% of ETPC)	3.91
10) Engineering office fees (12% of ETPC)	9.39
Total plant Retrofit costs (TPC)	91.56
11) Allowance for funds during construction (2.2% of TPC)	2.01
Total plant Retrofit investment (TPI)	93.58
12) Royalties (5%)	4.68
13) Pre-production costs (2% of TPI)	1.87
14) Inventory capital	0.71
	100.84

Total capital Retrofit cost for 2000 MWe £202 million

4.2a Chiyoda CT121 Process–S21.1 (Newbuild)

	£/kWe
1) Base capital cost (see Table 2.67)	47.35
2) Scope adjustments–none	
Total process Newbuild capital	47.35
3) Process adjustments	
(a) Unit size × 1.00	
(b) Gas flow/MWe × 0.81	
(c) Percentage S in Coal × 0.87	
4) Location factors	
(a) Cost index × 1.043	
(b) Soils × 1.05	
5) Retrofit factors–none	
6) Escalation adjustments × 1.158	
Total escalation factor = 0.893	
Escalated total process capital (excluding contingencies)	42.30
7) Project contingencies (10%)	4.23
8) Process contingency (5%)	2.12
Escalated total process capital	48.65
9) General facilities (5%)	2.43
10) Engineering and office fees (12%)	5.84
Total plant Newbuild costs	56.92
11) Allowance for funds during construction (2.2%)	1.25
Total plant Newbuild investment	58.17
12) Royalties (5%)	2.91
13) Pre-production costs (2%)	1.16
14) Inventory capital	0.63
	62.87

Total capital Newbuild cost for 2000 MWe £126 million

4.2b Chiyoda CT121 Process–S21.1 (Retrofit)

	£/kWe
1) Base capital cost (see Table 2.68)	47.35
2) Scope adjustments (4%)	1.89
Total process Retrofit capital	49.24
3) Process adjustments	
(a) Unit size × 1.00	
(b) Gas flow/MWe × 0.81	
(c) Percentage S in Coal × 0.87	
4) Location factors	
(a) Cost index × 1.043	
(b) Soils × 1.05	
5) Retrofit factors	
(a) Site accessibility and congestion × 1.25	
(b) Underground obstruction × 1.02	
(c) Ductwork length and scrubber location × 1.12	
6) Escalation adjustments × 1.158	
Total escalation factor = 1.276	
Escalated total process capital (excluding contingencies)	62.82
7) Project contingencies (10%)	6.28
8) Process contingency (5%)	3.14
Escalated total process capital	72.25
9) General facilities (5%)	3.61
10) Engineering office fees (12%)	8.67
Total plant Retrofit costs	84.53
11) Allowance for funds during construction (2.2%)	1.86
Total plant Retrofit investment	86.39

12) Royalties (5%)	4.32
13) Pre-production costs (2%)	1.73
14) Inventory capital	0.63
	93.07
Total capital Retrofit cost for 2000 MWe	£186 million

4.3a Saarberg-Holter Process–S21.1 (Newbuild)

	£/kWe
1) Base capital cost (see Table 2.67)	46.54
2) Scope adjustments–none	
Total process Newbuild capital	46.54
3) Process adjustments (a) Unit size × 1.00 (b) Gas flow/MWe × 0.81 (c) Percentage S in Coal × 0.87	
4) Location factors (a) Cost index × 1.043 (b) Soils × 1.05	
5) Retrofit factors–none	
6) Escalation adjustments × 1.158 Total escalation factor = 0.893 Escalated total process capital (excluding contingencies)	41.58
7) Project contingencies (10%)	4.16
8) Process contingency (5%)	2.08
Escalated total process capital	47.82
9) General facilities (5%)	2.39
10) Engineering and office fees (12%)	5.74
Total plant Newbuild costs	55.94
11) Allowance for funds during construction (2.2%)	1.23
Total plant Newbuild investment	57.18
12) Royalties (5%)	2.86
13) Pre-production costs (2%)	1.14
14) Inventory capital	0.71
	61.89
Total capital Newbuild cost for 2000 MWe	£124 million

4.3b Saarberg-Holter Process–S21.1 (Retrofit)

	£/kWe
1) Base capital cost (see Table 2.68)	46.54
2) Scope adjustments (4%)	1.86
Total process Retrofit capital	48.40
3) Process adjustments (a) Unit size × 1.00 (b) Gas flow/MWe × 0.81 (c) Percentage S in Coal × 0.87	
4) Location factors (a) Cost index × 1.043 (b) Soils × 1.05	
5) Retrofit factors (a) Site accessibility and congestion × 1.25 (b) Underground obstruction × 1.02 (c) Ductwork length and scrubber location × 1.12	
6) Escalation adjustments × 1.158 Total escalation factor = 1.276 Escalated total process capital (excluding contingencies)	61.75
7) Project contingencies (10%)	6.17
8) Process contingency (5%)	3.09
Escalated total process capital	71.01
9) General facilities (5%)	3.55
10) Engineering office fees (12%)	8.52
Total plant Retrofit costs	83.08
11) Allowance for funds during construction (2.2%)	1.83
Total plant Retrofit investment	84.91
12) Royalties (5%)	4.25
13) Pre-production costs (2%)	1.70
14) Inventory capital	0.71
	91.57
Total capital Retrofit cost for 2000 MWe	£183 million

4.4a Lime-Spray Drying Process–S22.1 (Newbuild)

	£/kWe
1) Base capital cost (see Table 2.67)	40.65
2) Scope adjustments–none	
Total process Newbuild capital	40.65
3) Process adjustments	
(a) Unit size × 1.00	
(b) Gas flow/MWe × 0.81	
(c) Percentage S in Coal × 1.08	
4) Location factors	
(a) Cost index × 1.043	
(b) Soils × 1.05	
5) Retrofit factors–none	
6) Escalation adjustments × 1.158	
Total escalation factor = 1.109	
Escalated total process capital (excluding contingencies)	45.08
7) Project contingencies (10%)	4.51
8) Process contingency (5%)	2.25
Escalated total process capital	51.85
9) General facilities (5%)	2.59
10) Engineering and office fees (12%)	6.22
Total plant Newbuild costs	60.66
11) Allowance for funds during construction (2.2%)	1.33
Total plant Newbuild investment	61.99
12) Royalties (5%)	3.10
13) Pre-production costs (2%)	1.24
14) Inventory capital	0.24
	66.57

Total capital Newbuild cost for 2000 MWe £133 million

4.4b Lime-Spray Drying Process–S22.1 (Retrofit)

	£/kWe
1) Base capital cost (see Table 2.68)	40.65
2) Scope adjustments (4%)	1.63
Total process Retrofit capital	42.28
3) Process adjustments	
(a) Unit size × 1.00	
(b) Gas flow/MWe × 0.81	
(c) Percentage S in Coal × 1.08	
4) Location factors	
(a) Cost index × 1.043	
(b) Soils × 1.05	
5) Retrofit factors	
(a) Site accessibility and congestion × 1.25	
(b) Underground obstruction × 1.02	
(c) Ductwork length and scrubber location × 1.12	
6) Escalation adjustments × 1.158	
Total escalation factor = 1.584	
Escalated total process capital (excluding contingencies)	66.95
7) Project contingencies (10%)	6.70
8) Process contingency (5%)	3.85
Escalated total process capital	77.00
9) General facilities (5%)	3.85
10) Engineering office fees (12%)	9.24
Total plant Retrofit costs	90.09
11) Allowance for funds during construction (2.2%)	1.98
Total plant Retrofit investment	92.07
12) Royalties (5%)	4.60
13) Pre-production costs (2%)	1.84
14) Inventory capital	0.24
	98.75

Total capital Retrofit cost for 2000 MWe £198 million

4.5a Wellman-Lord Process–S31.1 (Newbuild)

	£/kWe
1) Base capital cost (see Table 2.67)	76.54
2) Scope adjustments–none	
Total process Newbuild capital	76.54
3) Process adjustments	
(a) Unit size × 1.00	
(b) Gas flow/MWe × 0.81	
(c) Percentage S in Coal × 0.695	

4) Location factors
(a) Cost index × 1.043
(b) Soils × 1.05

5) Retrofit factors–none

6) Escalation adjustments × 1.158
Total escalation factor = 0.714
Escalated total process capital
(excluding contingencies) — 54.63

7) Project contingencies (10%) — 5.46

8) Process contingency (5%) — 2.73

Escalated total process capital — 62.82

9) General facilities (5%) — 3.14

10) Engineering and office fees (12%) — 7.54

Total plant Newbuild costs — 73.50

11) Allowance for funds during construction (2.2%) — 1.62
Total plant Newbuild investment — 75.12

12) Royalties (5%) — 3.76

13) Pre-production costs (2%) — 1.50

14) Inventory capital — 1.81

82.18

Total capital Newbuild cost for 2000 MWe £164 million

4.5b Wellman-Lord Process–S31.1 (Retrofit)

	£/kWe
1) Base capital cost (see Table 2.68)	76.54
2) Scope adjustments (4%)	3.06
Total process Retrofit capital	79.60

3) Process adjustments
(a) Unit size × 1.00
(b) Gas flow/MWe × 0.81
(c) Percentage S in Coal × 0.695

4) Location factors
(a) Cost index × 1.043
(b) Soils × 1.05

5) Retrofit factors
(a) Site accessibility and congestion × 1.25
(b) Underground obstruction × 1.02
(c) Ductwork length and scrubber location × 1.12

6) Escalation adjustments × 1.158
Total escalation factor = 1.019
Escalated total process capital
(excluding contingencies) — 81.13

7) Project contingencies (10%) — 8.11

8) Process contingency (5%) — 4.06

Escalated total process capital — 93.30

9) General facilities (5%) — 4.66

10) Engineering office fees (12%) — 11.20

Total plant Retrofit costs — 109.16

11) Allowance for funds during construction (2.2%) — 2.40
Total plant Retrofit investment — 111.56

12) Royalties (5%) — 5.58

13) Pre-production costs (2%) — 2.23

14) Inventory capital — 1.81

121.18

Total capital Retrofit cost for 2000 MWe £242 million

4.6a Magnesia (MgO) Scrubbing Process–S41.1 (Newbuild)

	£/kWe
1) Base capital cost (see Table 2.67)	87.48
2) Scope adjustments–none	
Total process Newbuild capital	87.48

3) Process adjustments
(a) Unit size × 1.00
(b) Gas flow/MWe × 0.81
(c) Percentage S in Coal × 0.72

4) Location factors
(a) Cost index × 1.043
(b) Soils × 1.05

5) Retrofit factors–none

6) Escalation adjustments × 1.158
Total escalation factor = 0.739
Escalated total process capital
(excluding contingencies) — 64.68

7) Project contingencies (10%)	6.47
8) Process contingency (5%)	3.23
Escalated total process capital	74.38
9) General facilities (5%)	3.72
10) Engineering and office fees (12%)	8.93
Total plant Newbuild costs	87.03
11) Allowance for funds during construction (2.2%)	1.91
Total plant Newbuild investment	88.94
12) Royalties (5%)	4.45
13) Pre-production costs (2%)	1.78
14) Inventory capital	1.81
	96.98

Total capital Newbuild cost for 2000 MWe £194 million

4.6b Magnesia (MgO) Scrubbing Process–S41.1 (Retrofit)

	£/kWe
1) Base capital cost (see Table 2.68)	87.48
2) Scope adjustments (4%)	3.50
Total process Retrofit capital	90.98

3) Process adjustments
 (a) Unit size × 1.00
 (b) Gas flow/MWe × 0.81
 (c) Percentage S in Coal × 0.72

4) Location factors
 (a) Cost index × 1.043
 (b) Soils × 1.05

5) Retrofit factors
 (a) Site accessibility and congestion × 1.25
 (b) Underground obstruction × 1.02
 (c) Ductwork length and scrubber location × 1.12

6) Escalation adjustments × 1.158
 Total escalation factor = 1.056

Escalated total process capital (excluding contingencies)	96.06
7) Project contingencies (10%)	9.61
8) Process contingency (5%)	4.80
Escalated total process capital	110.47
9) General facilities (5%)	5.52
10) Engineering office fees (12%)	13.26
Total plant Retrofit costs	129.24
11) Allowance for funds during construction (2.2%)	2.84
Total plant Retrofit investment	132.09
12) Royalties (5%)	6.60
13) Pre-production costs (2%)	2.64
14) Inventory capital	1.81
	143.14

Total capital Retrofit cost for 2000 MWe £286 million

5. OPERATING COST ESTIMATES

5.1 Limestone Gypsum Process–S21.1

	\$/kW-year
1) Total fixed cost	9.7
2) Limestone	2.16
3) High-pressure steam	4.8
4) Power	7.9
5) Waste disposal	1.84
Total Operating Costs (Dec '82)	26.4

Total (Dec '86) Costs = £20.75/kW-year

Escalated Total Operating Costs = £24.02/kW-year

5.2 Chiyoda CT-121 Process–S21.1

	\$/kW-year
1) Total fixed cost	9.6
2) Limestone	1.92
3) High-pressure steam	4.9
4) Power	7.4

5) By-product credit	(1.6)
6) Waste disposal	1.84
Total Operating Costs (Dec '82)	24.05

Total (Dec '86) Costs = £18.90/kW-year

Escalated Total Operating Costs = £21.88/kW-year

5.3 Saarberg-Holter Process–S21.1

	$/kW-year
1) Total fixed cost	6.9
2) Lime	4.8
3) Formic acid	0.12
4) High-pressure steam	4.9
5) Power	6.6
6) Waste disposal	1.84
Total Operating Costs (Dec '82)	25.16

Total (Dec '86) Costs = £19.78/kW-year

Escalated Total Operating Costs = £22.90/kW-year

5.4 Lime Spray Drying Process–S22.2

	$/kW-year
1) Total fixed cost	7.0
2) Lime	7.41
3) Power	3.9
4) Waste disposal	13.37
Total Operating Costs (Dec '82)	31.68

Total (Dec '86) Costs = £24.90/kW-year

Escalated Total Operating Costs = £28.82/kW-year

5.5 Wellman-Lord Process–S31.1

	$/kW-year
1) Total fixed cost	10.6
2) Sodium carbonate	0.52
3) Cooling water	1.9
4) High-pressure steam	5.1
5) Low-pressure steam	3.28
6) Power	9.8
7) Fuel	5.12
8) By-product credit	(2.8)
9) Waste disposal	1.95
Total Operating Costs (Dec '82)	35.47

Total (Dec '86) Costs = £27.88/kW-year

Escalated Total Operating Costs = £32.27/kW-year

5.6 Magnesia Scrubbing Process–S41.1

	$/kW-year
1) Total fixed cost	11.7
2) Magnesia	2.88
3) Fuel oil	4.76
4) Cooling water	2.7
5) High-pressure steam	4.7
6) Power	6.3
7) Waste disposal	1.84
8) By-product credit	(6.32)
Total Operating Costs (Dec '82)	28.56

Total (Dec '86) Costs = £22.34/kW-year

Escalated Total Operating Costs = £25.86/kW-year

6. EQUIPMENT COST ESTIMATE SHEETS

Appendix B of EPRI Report No. CS-3342 [95] includes detailed lists (and 1982 costs) of equipment for several FGD processes. These lists were used, with little modification, for estimating UK-supplied equipment costs for six selected processes. The equipment was budget-priced by potential vendors, wherever possible, based on the limited information available. A large percentage of the vendors' quotes were on a supply-and-erect basis, so an allowance of

25% of the total equipment costs was made for erection. The costs are based on December 1986 budget prices, and the accuracy cannot be considered to be anything less than 25% due to the limited information supplied.

Exclusions from the total equipment costs include:

- commissioning
- spares (running and commissioning)
- 'first fill' (initial catalysts and chemicals)

Only a summary of the estimated equipment costs for each process appears in this Appendix; a confidential annex to the Manual, with more detailed cost data, has been supplied to the Fellowship of Engineering.

6.1 Limestone with Forced Oxidation

	£
Total Equipment Erected Cost	38,377,000
plus 5% for Miscellaneous Process Equipment	1,918,850
Sub-Total	40,295,850
Excluding 25% allowance for erection:	
Total Equipment Cost	30,221,890

NB. No flue gas prescrubbing or gypsum de-watering facilities are included in the costs, although EPRI estimates indicate that their inclusion could increase capital costs by about 13%.

6.2 Chiyoda

	£
Total Equipment Erected Cost	43,122,230
Plus 5% for Miscellaneous Process Equipment	2,156,110
Sub-Total	45,278,340
Excluding 25% allowance for erection:	
Total Equipment Cost	33,958,760

6.3 Saarberg-Holter

	£
Total Equipment Erected Cost	41,470,130
Plus 5% for Miscellaneous Process Equipment	2,073,510
Sub-Total	43,543,646
Excluding 25% allowance for erection:	
Total Equipment Cost	32,657,740

6.4 Lime Spray Drying

	£
Total Equipment Erected Cost	34,993,730
Plus 5% for Miscellaneous Process Equipment	1,749,690
Sub-Total	36,743,420
Excluding 25% allowance for erection:	
Total Equipment Cost	27,557,570

6.5 Wellman-Lord

	£
Total Equipment Erected Cost	66,691,700
Plus 5% for Miscellaneous Process Equipment	3,334,590
Sub-Total	70,026,290
Excluding 25% allowance for erection:	
Total Equipment Cost	52,519,720

6.6 Magnesia (MgO) Scrubbing

	£
Total Equipment Erected Cost	98,972,020
Plus 5% for Miscellaneous Process Equipment	4,948,600
Sub-Total	103,920,620
Excluding 25% allowance for erection:	
Total Equipment Cost	77,940,470

Index

Appendix 4

Printed in the United Kingdom for HMSO.
Dd.291337, 6/91, C3, 3390/3, 5673, 150569.